VOLUME FOUR TWENTY FOUR

METHODS IN ENZYMOLOGY

RNA Editing

METHODS IN ENZYMOLOGY

Editors-in-Chief

JOHN N. ABELSON AND MELVIN I. SIMON

Division of Biology
California Institute of Technology
Pasadena, California

Founding Editors

SIDNEY P. COLOWICK AND NATHAN O. KAPLAN

VOLUME FOUR TWENTY FOUR

METHODS IN
ENZYMOLOGY

RNA Editing

EDITED BY

JONATHA M. GOTT
Center for RNA Molecular Biology
Case Western Reserve University
Cleveland, Ohio

ELSEVIER

AMSTERDAM • BOSTON • HEIDELBERG • LONDON
NEW YORK • OXFORD • PARIS • SAN DIEGO
SAN FRANCISCO • SINGAPORE • SYDNEY • TOKYO
Academic Press is an imprint of Elsevier

CONTENTS

Section VI. Editing in Plant Organelles   437

Contributors

Ruslan Aphasizhev
Department of Microbiology and Molecular Genetics, University of California, Irvine, California

Inna Aphasizheva
Department of Microbiology and Molecular Genetics, University of California, Irvine, California

Alejandro Araya
Laboratoire Microbiologie Cellulaire Moléculaire et Pathogenicité, UMR 5234, Centre National de la Recherche Scientifique and Université Victor Segalen, Bordeaux, France

Arpi Barsamian
Department of Molecular and Cell Biology, The University of Texas at Dallas, Richardson, Texas

Brenda L. Bass
Department of Biochemistry and Howard Hughes Medical Institute, University of Utah, Salt Lake City, Utah

Peter A. Beal
Department of Chemistry, University of Utah, Salt Lake City, Utah

Valerie Blanc
Division of Gastroenterology, Department of Medicine, Washington University School of Medicine, St. Louis, Missouri

Axel Brennicke
Molekulare Botanik, Universität Ulm, Ulm, Germany

Ralf Bundschuh
Department of Physics, The Ohio State University, Columbus, Ohio

Elaine M. Byrne
Trinity Centre for Engineering, Trinity College Dublin, Dublin, Ireland

Jason Carnes
Seattle Biomedical Research Institute, Seattle, Washington

Yu-Wei Cheng
Department of Microbiology and Immunology, Weill Medical College of Cornell University, New York, New York

Soo-Jin Cho
Division of Gastroenterology, Department of Medicine, Washington University School of Medicine, St. Louis, Missouri

David Choury
Laboratoire Microbiologie Cellulaire Moléculaire et Pathogenicité, UMR 5234, Centre National de la Recherche Scientifique and Université Victor Segalen, Bordeaux Cedex, France

Jorge Cruz-Reyes
Department of Biochemistry & Biophysics, Texas A&M University, College Station, Texas

Tao Cui
Department of Chemistry, University of Utah, Salt Lake City, Utah

Nicholas O. Davidson
Division of Gastroenterology, Department of Medicine, Washington University School of Medicine, St. Louis, Missouri

LaHoma M. Easterwood
Department of Chemistry, University of Utah, Salt Lake City, Utah

Ronald B. Emeson
Departments of Pharmacology and Molecular Physiology and Biophysics, Center for Molecular Neuroscience, Vanderbilt University School of Medicine, Nashville, Tennessee

Jean-Claude Farré
Section of Molecular Biology, Division of Biological Sciences, University of California, San Diego, La Jolla, California

Jonatha M. Gott
Center for RNA Molecular Biology, Case Western Reserve University, Cleveland, Ohio

Michael W. Gray
Department of Biochemistry and Molecular Biology, Dalhousie University, Halifax, Nova Scotia, Canada

Maureen R. Hanson
Department of Molecular Biology and Genetics, Cornell University, Ithaca, New York

Michael L. Hayes
Department of Molecular Biology and Genetics, Cornell University, Ithaca, New York

James E. C. Jepson
Department of Molecular Biology, Cell Biology, and Biochemistry, Brown University, Providence, Rhode Island

Liam P. Keegan
MRC Human Genetics Unit, Western General Hospital, Edinburgh, United Kingdom

Uma Krishnan
Department of Molecular and Cell Biology, The University of Texas at Dallas, Richardson, Texas

Amanda J. Lohan
Department of Biochemistry and Molecular Biology, Dalhousie University, Halifax, Nova Scotia, Canada

Kerry A. Lutz
Waksman Institute of Microbiology, Rutgers, The State University of New Jersey, Piscataway, New Jersey

Mark R. Macbeth
Department of Biochemistry and Howard Hughes Medical Institute, University of Utah, Salt Lake City, Utah

Pal Maliga
Waksman Institute of Microbiology, Rutgers, The State University of New Jersey, Piscataway, New Jersey

Dmitri A. Maslov
Department of Biology, University of California, Riverside, California

Olena Maydanovych
Department of Chemistry, University of Utah, Salt Lake City, Utah

Dennis L. Miller
Department of Molecular and Cell Biology, The University of Texas at Dallas, Richardson, Texas

Mary A. O'Connell
MRC Human Genetics Unit, Western General Hospital, Edinburgh, United Kingdom

Marie Öhman
Department of Molecular Biology and Functional Genomics, Stockholm University, Stockholm, Sweden

Johan Ohlson
Department of Molecular Biology and Functional Genomics, Stockholm University, Stockholm, Sweden

Aswini K. Panigrahi
Seattle Biomedical Research Institute, Seattle, Washington

Michel Pelletier
Department of Microbiology and Immunology, SUNY Buffalo School of Medicine, Buffalo, New York

Subhash Pokharel
Department of Chemistry, University of Utah, Salt Lake City, Utah

Laurie K. Read
Department of Microbiology and Immunology, SUNY Buffalo School of Medicine, Buffalo, New York

Robert A. Reenan
Department of Molecular Biology, Cell Biology, and Biochemistry, Brown University, Providence, Rhode Island

Amy C. Rhee
Center for RNA Molecular Biology, Case Western Reserve University, Cleveland, Ohio

Loretta M. Roberson
Department of Biology, University of Puerto Rico-Rio Piedras, San Juan, Puerto Rico

Joshua J. Rosenthal
Institute of Neurobiology, University of Puerto Rico-Medical Science Campus, San Juan, Puerto Rico

Elizabeth Y. Rula
Department of Pharmacology, Vanderbilt University School of Medicine, Nashville, Tennessee

Achim Schnaufer
Seattle Biomedical Research Institute, Seattle, Washington

Larry Simpson
Department of Microbiology, Immunology and Molecular Genetics, David Geffen School of Medicine at UCLA, University of California, Los Angeles, California

Jaimie Sixsmith
Department of Molecular Biology, Cell Biology, and Biochemistry, Brown University, Providence, Rhode Island

Harold C. Smith
Departments of Biochemistry and Biophysics, Pathology and Toxicology, University of Rochester School of Medicine and Dentistry, Rochester, New York

Kenneth D. Stuart
Seattle Biomedical Research Institute, Seattle, Washington

Mizuki Takenaka
Molekulare Botanik, Universität Ulm, Ulm, Germany

Eduardo A. Véliz
Department of Chemistry, University of Utah, Salt Lake City, Utah

Linda Visomirski-Robic
Department of Biological Sciences, Carnegie Mellon University, Pittsburgh, Pennsylvania

PREFACE

The phenomena encompassed by the term RNA editing range from the subtle to the bizarre. Initially, the term was coined by Benne and colleagues to describe alterations at the RNA level consisting of the insertion of "extra" nucleotides (or deletion of encoded nucleotides). This definition was subsequently expanded to include specific base changes, which often involve the deamination of cytidine to uridine or adenosine to inosine within mRNAs. The feature that these processes have in common is that they result in site-specific changes within an RNA sequence.

Not surprisingly, the various types of editing events differ mechanistically, and comprise both co-and posttranscriptional processes, sometimes within the same organism or organelle. There is also interplay between editing and other cellular processes, including transcription, splicing, and RNA silencing.

The diverse array of editing types has made it necessary to develop a range of different approaches for dissecting editing mechanisms. This volume covers most of the principal methods employed in the field, with the exception of RNA interference, which has been extensively covered elsewhere. Hopefully, the chapters in this volume will lead to a greater appreciation of the breadth of approaches taken by "RNA editors" as they meet the considerable challenges posed by these various experimental systems.

This volume and its companion (Volume 425: RNA Modification) are meant to complement one another. The fields of RNA modification and editing overlap extensively, and it is often impossible to make clear distinctions between changes that are classified as RNA modifications and those that are referred to as RNA editing events (e.g., A-to-I changes in tRNAs versus mRNAs). For the purpose of this series, I have designated chapters for the modification or editing volumes based on common conventions, which are largely historical rather than scientific in origin.

I wish to express my sincere appreciation to the authors for their thoughtful contributions and willingness to share their expertise. This volume is dedicated to William, Katherine, Eric, and the rest of my family in gratitude for their love and encouragement.

JONATHA M. GOTT

Methods in Enzymology

TRYPANOSOME U INSERTION/DELETION

CHAPTER ONE

ISOLATION AND COMPOSITIONAL ANALYSIS OF TRYPANOSOMATID EDITOSOMES

Aswini K. Panigrahi, Achim Schnaufer, *and* Kenneth D. Stuart

Contents

Abstract

Most mitochondrial (mt) mRNAs in trypanosomes undergo posttranscriptional RNA editing, which inserts and deletes uridines (Us) to produce the mature and functional mRNA. The editing process is catalyzed by multiple enzymatic steps and is carried out by an ~20S macromolecular complex, the editosome. Editosomes have been purified from *Trypanosoma brucei* using various techniques including combinations of column chromatography, gradient sedimentation, monoclonal antibody affinity, and TAP-tag affinity approaches. This article describes in detail the methods for editosome purification and identification of protein components by mass spectrometry analyses. It also describes the methods for isolation and analysis of TAP-tagged mutagenized complexes.

Seattle Biomedical Research Institute, Seattle, Washington

Methods in Enzymology, Volume 424
ISSN 0076-6879, DOI: 10.1016/S0076-6879(07)24001-7

1. INTRODUCTION

The trypanosomatid mitochondrial (mt) genome is composed of maxi-circle and minicircle DNAs commonly known as kinetoplast DNA (kDNA). The maxicircle DNA encodes mt rRNAs and several proteins of the oxidative phosphorylation system including subunits of complexes I (NADH dehydrogenase), III (cytochrome bc1), IV (cytochrome c oxidase), and V (ATP synthase). However, most of the mitochondrially encoded mRNAs undergo a posttranscriptional maturation process, termed RNA editing, by insertion and deletion of uridines (Us) at multiple sites to generate mature mRNAs (Madison-Antenucci *et al.*, 2002; Simpson *et al.*, 2004; Stuart *et al.*, 2000, 2005). The edited sequence is specified by small mt guide RNAs (gRNAs) that are transcribed from minicircle DNAs. The 5′ region of the gRNA (anchor) forms a duplex by base pairing with pre-mRNA upstream (3′) of the editing site, the central region of gRNA contains the informational sequence for editing and the 3′ poly(U) tail may provide stability for the mRNA + gRNA interaction. The edited mRNA sequence will become complementary to the guiding sequence by forming A:U and G:U base pairs. Editing is very precise and can be very extensive; in some cases, hundreds of Us are inserted and tens are deleted in a single mRNA. Overall, 3583 Us are inserted and 322 deleted in 12 mRNAs in *Trypanosoma brucei*. RNA editing is catalyzed by a multiprotein complex, the editosome. Like almost all the proteins in mitochondria, the editosome proteins are nuclearly encoded and imported into the mitochondria. This article describes the methods for the purification and characterization of editosomes and identification of their component proteins. It expands on an earlier methods chapter by us (Stuart *et al.*, 2004) and discusses new advances in studying the composition of editosomes.

2. ISOLATION OF ENZYMATICALLY ACTIVE EDITOSOMES

2.1. Overview

Multiple enzymatic steps are involved during the editing process. The mechanism of the editing process was determined using full round *in vitro* insertion and deletion RNA editing assays (Kable *et al.*, 1996; Seiwert *et al.*, 1996) that can achieve editing of only one site using mitochondrial extract. In the first step, the pre-mRNA is specifically cleaved by endonuclease at the editing site that is upstream of the base paired pre-mRNA and gRNA anchor region. Subsequently, Us are added by terminal uridylyltransferase (TUTase) in the case of insertion editing or Us are removed by exonuclease in the case of deletion editing as specified by the guiding sequences. Finally, the 5′ and 3′

fragments of the mRNA are ligated by RNA ligase. Several cycles of this U insertion and/or deletion editing occur until the gRNA is fully decoded. This results in full complementary between the gRNA and edited mRNA. In addition, multiple gRNAs are required to edit most mRNAs by a process in which one gRNA creates the sequence that can duplex with a subsequent gRNA. Several processes that occur during the complete decoding of the gRNA and the utilization of the multiple gRNAs to edit a single mRNA are as yet uncharacterized. RNA helicases may play a role in this process and/or in the processes by which editing by a single gRNA is accomplished (Stuart and Panigrahi, 2002).

Studies using glycerol gradient fractionated mitochondrial lysate showed that editosomes with full round *in vitro* editing activities sediment at $\sim$20S and another peak of editing activities is variably seen in the $\sim$40S region (Corell *et al.*, 1996; Pollard *et al.*, 1992). The $\sim$20S editosome has been well characterized, but its relationship to the $\sim$40S peak is unclear. In size exclusion chromatography, the 20S editosomes elute in the $\sim$1600 kDa region (Panigrahi *et al.*, 2001a). Different laboratories have used various biochemical approaches to isolate editosomes or editing-associated activities. These studies resulted in isolation of specific editing-associated activities or editing active complexes containing a variable number of component proteins (Aphasizhev *et al.*, 2003; Madison-Antenucci *et al.*, 1998; Panigrahi *et al.*, 2001a,b, 2003b; Rusche *et al.*, 1997). In our laboratory we used a full round *in vitro* deletion editing assay as the standard to follow the purification of editosomes, which is described below. A schematic of different methods used for purification of editosomes is shown in Fig. 1.1.

2.2. Conventional chromatography protocol

1. Grow *Trypanosoma brucei* procyclic form (PF) cells at 27° in SDM-79 media containing 10% fetal bovine serum (FBS) and hemin (7.5 mg/ml). We use strain IsTaR 1.7a for editosome isolation, but other laboratory-adapted strains may be used. See Cross lab webpage (http://tryps.rockefeller.edu/) for detailed protocols and media compositions for growing *T. brucei* cells. We have not attempted to isolate editosomes from blood form (BF) cells using column chromatography techniques as yet because these cells are smaller with fewer editosomes and grow to lower density in culture, making such studies expensive and likely to have low yield.

2. Harvest cells at mid-log phase of growth by centrifugation at $6000 \times g$ for 10 min at 4° and isolate the mitochondrial vesicles as described (Harris *et al.*, 1990). Routinely, we use 10–12 liters of cultured cells ($\sim$2 $\times$ 10^{11} cells). Wash the cell pellets once with 300 ml SBG buffer (20 mM phosphate buffer, pH 7.9, 150 mM NaCl, 6 mM glucose) and resuspend in a total of 200 ml of DTE buffer (1 mM Tris–HCl, pH 8.0, 1 mM EDTA). Very quickly disrupt the cells by five strokes in a Dounce

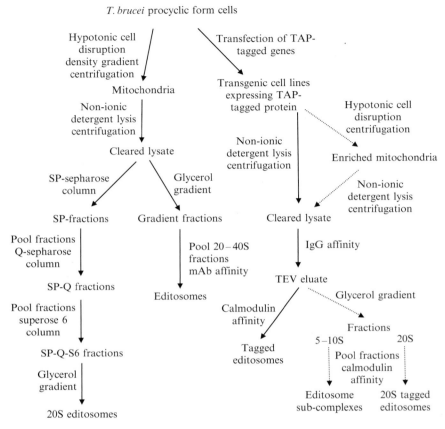

Figure 1.1 A schematic of the methods used for editosome purification. The optional steps are indicated by dotted arrows.

homogenizer (use 100 ml Dounce homogenizers with tight–fitting pestle B) and immediately add sucrose to a final concentration of 0.25 M (33.4 ml of 60% sucrose solution). Seal the top of the homogenizer with parafilm and mix by inverting. Centrifuge the cell lysate at 15,000×g for 10 min at 4°, pour off the supernatant, and resuspend the pellet by pipetting in 39 ml of STM buffer (20 mM Tris–HCl, pH 8.0, 250 mM sucrose, 2 mM MgCl$_2$). Add MgCl$_2$ to 3 mM, CaCl$_2$ to 0.3 mM, and DNase to 9 μg/ml final concentrations and mix by inverting. Incubate the tubes in ice for 60 min, add an equal volume of STE buffer (20 mM Tris–HCl, pH 8.0, 250 mM sucrose, 2 mM EDTA), and mix by inverting. Centrifuge as above and pour off the supernatant. The pellet contains the organelles and contaminating flagella.

3. Pour 32 ml of 20–35% linear Percoll gradients. Resuspend the pellets in 2–4 ml of 70% Percoll per original liter of culture and homogenize using a small Dounce homogenizer with tight-fitting pestle B for five strokes. Layer 5 ml of the suspension at the bottom of the gradient using a long needle and centrifuge these floatation gradients at $103,900 \times g$ for 60 min at $4°$. Using a syringe and 18-gauge needle, collect the mitochondria-enriched fraction, which appears as a broad smear in the density range of 1.052 to 1.069 g/ml between a prominent band at the top and another prominent band close to the bottom of the tube. Wash the mitochondrial fraction four times with STE buffer. In each wash fill the centrifuge tubes with STE buffer, mix well by inverting, and pellet mitochondrial vesicles by centrifugation at $32,530 \times g$ for 15 min. Carefully pipette off the supernatant without disturbing the pellet. The mitochondrial vesicles can be flash frozen in liquid nitrogen and stored at $-70°$ for future use. Depending on the experimental requirements, smaller-scale isolation of mitochondrial vesicles can be carried out from 1–2 liters of cells, and in such cases use correspondingly less volumes of reagents described above.

4. Lyse the mitochondrial vesicles with nonionic detergent Triton X-100 for subsequent fractionation. Resuspend mitochondrial vesicles isolated from above in 30 ml of SP-A buffer (10 mM Tris–HCl, pH 7.0, 10 mM MgCl$_2$, 50 mM KCl, 1 mM DTT). It is very important to add protease inhibitors before the lysis. Add a cocktail of protease inhibitors (10 μg/ml leupeptin, 5 μg/ml pepstatin, 1 μM Pefabloc), mix by inverting, and add Triton X-100 to a final concentration of 0.5% (from 10% stock solution). Carry out the lysis with bidirectional mixing for 15 min at $4°$. Centrifuge the tubes at $17,500 \times g$ for 30 min at $4°$ and collect the clear supernatant.

5. Fractionate the cleared mitochondrial lysate from above by sequential ion-exchange and gel filtration column chromatography for isolation of editosomes (Panigrahi et al., 2001a). We use an automated FPLC system, and all steps are carried out at $4°$. Filter the lysate through a 0.2 μm membrane using a syringe and disposable filters, and load the filtrate onto SP-A buffer equilibrated 5 ml SP Sepharose HR column (Pharmacia) at 1 ml/min flow rate. Wash the column with 25 ml of buffer SP-A at 1 ml/min flow rate, which washes away any unbound proteins and, as a result, the UV absorbance A_{280} of the flowthrough should drop to the background level. Elute the bound proteins with linear salt gradients using buffer SP-B (10 mM Tris–HCl, pH 7.0, 10 mM MgCl$_2$, 1 M KCl, 1 mM DTT). At 1 ml/min flow rate run a 40 ml linear gradient of 0–30% B followed by another 40 ml gradient of 30–100% B. Collect fractions of 2 ml size and assay for in vitro RNA editing activities (as described in Chapter 2, this volume).

6. Pool all positive fractions (typically fractions 9–19 contain in vitro deletion RNA editing activities) and further purify on a Q Sepharose column. Connect two HiTrap Q 1-ml columns (Pharmacia) in series and equilibrate

with buffer Q-A (10 mM Tris–HCl, pH 8.3, 10 mM MgCl$_2$, 50 mM KCl, 1 mM DTT). Dilute the pooled fractions from the SP Sepharose column and adjust the pH to the same conditions as buffer Q-A (a 3-fold dilution using 10 mM Tris–HCl, pH 8.8, 10 mM MgCl$_2$, 1 mM DTT makes the sample compatible for binding to the Q column). Load the sample, preferably using a super-loop, onto the Q-Sepharose column at 1 ml/min flow rate. Wash the column with 10 ml of buffer Q-A and elute the proteins using linear salt gradients in buffer Q-B (10 mM Tris–HCl, pH 8.3, 10 mM MgCl$_2$, 1 M KCl, 1 mM DTT) as above; run a 16 ml linear gradient of 0–30% B, followed by a 14 ml linear gradient of 30–100% B at 0.5 ml/min flow rate. Collect fractions of 1 ml size and assay for RNA editing activities as above. Pool the positive fractions (fractions between 11 and 20 contain deletion RNA editing activities) and purify further as described below for structural analysis or production of monoclonal antibodies. For routine *in vitro* assays, samples from this step may be used without any further purification. In this case, add glycerol to a final concentration of 10% (v/v), make aliquots, flash freeze in liquid nitrogen, and store at −70°.

7. Purify the editing complex further by size exclusion chromatography. Add Triton X-100 to the pooled sample from above to a final concentration of 0.1%, mix, and concentrate to 1/10–1/20 volume using a Centricon-YM50 membrane (Amicon) at 3000×g at 4°. Transfer the concentrated sample into a 1.5-ml tube, resuspend any protein aggregates by pipetting, and clarify the sample by centrifugation at 15,000×g for 30 min at 4°. Equilibrate a Superose 6 HR 10/30 column (Pharmacia) with 75 ml of S6 buffer (10 mM Tris–HCl, pH 7.0, 10 mM MgCl$_2$, 200 mM KCl, 1 mM DTT) at a 0.2 ml/min flow rate. In each run, load a 250 μl sample onto the column and elute at 0.2 ml/min flow rate with a total of 25 ml S6 buffer. Collect 500 μl fractions and assay for RNA editing activities. Most of the *in vitro* deletion RNA editing activities elute in fraction 19–22.

8. To determine the approximate size of the purified complex, run high molecular weight protein standards (Pharmacia) on the same Superose 6 column, and estimate the size of the complex based on the retention time in the column compared to that of the standards using the criteria provided by the manufacturer. The editosome peak corresponds to approximately 1600 kDa in size.

9. Pool the peak positive fractions from the Superose 6 column, concentrate as above, and further purify by sedimentation on a 10–30% linear glycerol gradient. Pour an 11-ml 10–30% linear glycerol gradient using a gradient maker in 10 mM Tris, pH 7.0, 10 mM MgCl$_2$, 100 mM KCl buffer. Layer 500 μl of the sample on top of the gradient. Balance the tube by weight and centrifuge at 38,000 rpm for 5 h at 4° (SW40 rotor, Beckman). Carefully collect 500 μl of fractions either from the top using a pipette or from the bottom using a needle and a valve connected to a fraction collector. Assay individual fractions for RNA editing activities

and store the positive fractions at $-70°$ as above. Determine the sedimentation value of the complex by comparing to parallel gradients containing standards. The functional editosomes sediment at $\sim20S$ in the gradient.

2.3. Monoclonal antibody affinity purification of editosomes

Monoclonal antibodies (MAbs) provide very useful tools for purification of native complexes from cellular fractions by affinity chromatography and for studying their physical and functional properties. MAbs can be generated against native or recombinant proteins. We purified editosomes as above in multiple batches and concentrated the 20S complexes using a Centricon-YM50 membrane and used as immunogen to generate a panel of MAbs (Panigrahi et al., 2001a). This approach was taken since none of the proteins of the 20S editosome was identified at that stage and we hoped to obtain MAbs against different component proteins from one batch of hybridoma generation. The study resulted in four well-characterized MAbs against four different editosome proteins KREPA1 (TbMP83), KREPA2 (TbMP63), KREL1 (TbMP52) and KREPA3 (TbMP42) (Panigrahi et al., 2001a,b). We also obtained some MAbs that are not directed against editosome proteins, but perhaps against some contaminating proteins in the editosome preparations. The sedimentation profiles of their target proteins were different from editosome MAbs whose target proteins cosediment together in the $\sim 20S$ region along with the peak editing activities (Panigrahi et al., 2001b, 2003a). The complexes pulled down by one of the editosome MAbs contain the three other proteins as determined by Western analysis and in vitro RNA editing activities. Three of these MAbs (except anti-KREPA1) immunoprecipitated functional editing complexes from mitochondrial lysate, and the lower efficacy of KREPA1 MAb may be due to its epitope being buried in the complex. Pull-down experiments using mitochondrial 20S fractions resulted in highly pure complexes (Panigrahi et al., 2001b, 2003b).

2.4. Protocol

1. Isolate mitochondrial vesicles from PF cells and lyse with 1% Triton X-100, as described above. Clear by centrifugation and fractionate on a 10–30% glycerol gradient. Probe alternate fractions by Western blot with MAbs of interest, i.e., those to be used for immunoprecipitation experiments. Pool positive fractions from the 20–40S region for subsequent purification.
2. Use anti-mouse IgG coated magnetic beads (Dynabeads M-450) for immunoprecipitation experiments. Incubate 4×10^7 beads with 1–2 ml of anti-KREPA2 MAb tissue culture supernatant and 1% bovine serum albumin (BSA) at $4°$ with bidirectional mixing for 1 h (a pool of different MAbs specific to the complex can also be used). Wash the beads three

times (each wash of 5 min duration with bidirectional mixing) with IP buffer (10 mM Tris, pH 7.2, 10 mM MgCl$_2$, 200 mM KCl, 0.1% Triton X-100). Add 50–100 μl of the pooled mitochondrial 20S fraction (about 50 μg proteins) and 1 × IP buffer and 1% BSA to a final volume of 500 μl and incubate for 1 h at 4° with bidirectional mixing. Wash the beads four times with IP buffer. It is important to completely resuspend the beads in IP buffer during the washes without making bubbles.

3. Following the final wash, pipette off the IP buffer. The complex bound beads can be added directly to the reaction tubes for assaying editing activities, presence of editosome proteins by adenylation, and Western analyses. Carry out the purification in a 5-fold concentration for protein identification by mass spectrometry.

3. IDENTIFICATION OF PROTEIN COMPONENTS

3.1. Overview

With progress in the development of mass spectrometric proteomics technologies, proteins can be analyzed in a high throughput, automated manner similar to how the genes are studied (Aebersold and Goodlett, 2001; Griffin *et al.*, 2001). Proteins separated by 1D or 2D gels or protein mixtures in solution can be analyzed by mass spectrometry for protein identification. Mass spectrometric proteomic analyses require the genome sequence of the source organism and are now routinely used to identify the molecular components of organelles, subcellular structures, and biological macromolecular complexes, to compare the level of protein expression between two different cell states and to study various posttranslational modifications that control regulatory pathways. A protein can be identified by determining the masses of peptides derived from it by a specific proteolytic cleavage (which represents a fingerprint for the protein), or by MS/MS spectra generated by collision induced dissociation (CID) of a peptide followed by correlating these with a sequence database. Masses of intact peptides for peptide mass mapping are more frequently determined by matrix-assisted laser desorption/ionization–time of flight mass spectrometry (MALDI-TOF), and CID spectra (MS/MS) analysis of peptides are carried out by electrospray ionization (ESI) ion-trap or TOF/TOF instruments. The MS/MS spectra contain complete or partial sequence information of individual peptides and so are more accurate in identifying the parent protein than peptide mass mapping. Application of ESI coupled with high-performance liquid chromatography systems (LC-MS/MS) has improved the sensitivity and the speed to analyze the peptides at femtomole levels. Powerful search algorithms like SEQUEST (Yates *et al.*, 1995) and Mascot (Perkins *et al.*, 1999) allow correlating the CID data with *in silico* CIDs generated from sequence databases to identify a correct peptide match from all other sequences in the database. Automated analysis of MS/MS data has been made possible by software developed from different laboratories. We use SEQUEST for analyzing the MS/MS data against databases and subsequently PeptideProphet (Keller *et al.*, 2002) and ProteinProphet (Nesvizhskii *et al.*, 2003) for downstream analysis and filtration of MS/MS database search results. To identify the proteins in purified complexes, the samples can be prepared from proteins separated on SDS–PAGE gel or from gel-free samples in solution. The immunoprecipitated samples are analyzed by SDS–PAGE followed by LC-MS/MS analysis. They will contain immunoglobulin heavy and light chains, which, however, do not interfere with the identification of complex proteins. Cross-linking the immunoglobulins to magnetic beads will decrease the abundance of immuno-noglobulin peptides in eluted complex samples, but, in our experience, this is not a critical step.

3.2. Protocol

3.2.1. In-gel digestion

1. Denature the proteins in $1\times$ SDS–PAGE sample buffer and separate the proteins on 10% SDS–PAGE gel. If the interest is in identifying each protein separately in individual protein bands, then run a full-length gel, otherwise, stop once the bromophenol blue tracking dye has migrated 1 cm into the resolving gel. The later method is a convenient and economic way of identifying all the proteins present in a sample.
2. Preferably stain the proteins in the gel with Sypro-Ruby Red and, after the staining/destaining reaction is complete, wash with nanopure water. If staining with Coomassie blue, destain the gel thoroughly and wash with nanopure water. Always wear gloves and wash them under running DI water before handling the samples/gels.
3. Transfer the gel onto a clean glass plate. Cut out the protein bands of interest from the full length gel (avoid cutting extra gel outside the sample) or, in case of the short length gels, divide the gel area containing the proteins horizontally into three equal parts. Cut each of the gel slices further into small pieces of approximately 1 mm square and transfer into 0.5-ml tubes.
4. Add 100–200 μl of acetonitrile (CH_3CN) to each tube to dehydrate the gel pieces. Incubate for 10 min and pipette off the liquid using gel-loading tips. The gel pieces will appear opaque white; if not, repeat the procedure once. Dry the samples in Speedvac for 20 min.
5. Prepare fresh 50 mM NH_4-HCO_3 buffer containing 12.5 ng/μl of sequencing-grade modified trypsin. Add just enough trypsin solution to cover the gel pieces (about 20–30 μl and never use more than 1 μg of

trypsin) and incubate the tubes in ice for 45 min to reswell the gel pieces. Check if there is liquid over gel pieces and if required cover the gel pieces with 50 mM NH$_4$-HCO$_3$ solution. Incubate at 37° overnight to digest the proteins.

6. Centrifuge the gel pieces for 5 min and collect the supernatant that contains peptides into fresh 0.5-ml tubes. Add 10–20 μl of 20 mM NH$_4$-HCO$_3$ to just cover the gel, vortex every 3–4 min, and incubate 20 min at RT to extract the peptides from gel pieces. Centrifuge and collect the supernatant into peptide-containing tubes. Repeat the extraction procedure twice with 5% formic acid in 50% acetonitrile and collect the supernatant to their respective tubes. Dry the sample in speedvac to about 5 μl or less and store at −20°.

The above method is suitable for quick identification of protein samples. However, it will not allow for identification of Cys-containing peptides that need to be modified by alkylation using the following protocol:

1. Excise the protein bands of interest and dehydrate as described above (steps 1–4).
2. Add 10 mM DTT in 100 mM NH$_4$-HCO$_3$ buffer so as to cover the gel pieces, and incubate at 56° for 1 h to reduce the proteins. Cool the tubes to RT, pipette off DTT, and add an equal volume of 55 mM iodoacetamide in 100 mM NH$_4$-HCO$_3$ buffer. Incubate in the dark at RT for 45 min.
3. Wash the gel pieces with 100 μl of 100 mM NH$_4$-HCO$_3$ for 10 min. Dehydrate with acetonitrile and reswell in 100 mM NH$_4$-HCO$_3$ as described above. Dehydrate again with acetonitrile, digest with Trypsin, and extract the peptides as described above.

3.2.2. In-solution digestion

Protein mixtures in solution (e.g., purified complex or subcellular fractions) can be digested with trypsin for subsequent mass spectrometry analysis without any prior separation of proteins on SDS–PAGE gel.

1. Determine the protein concentration in the sample using the Bradford assay or estimate the concentration based on the stained gel. Add 600 μl of prechilled (−20°) acetone to 100 μl of sample (6:1 v/v), mix by inverting the tube, and incubate on ice for 5 min. Usually, 100 μl of purified complex (~5 μg of proteins) isolated as described above is enough, but increase the sample volume if the protein amount is very low.
2. Centrifuge the samples in a microcentrifuge at full speed at 4° for 30 min. Carefully drain off the liquid and air-dry the pellet (protein pellet may not be visible) or alternatively dry in speedvac for about 2–3 min.

3. Add 10 μl of 8 M urea, 1 mM DTT solution, and dissolve the pellet by pipetting. Incubate the sample at 50° for 1 h. Add 30 μl of 50 mM ammonium bicarbonate buffer (the final concentration of urea should be 2 M or less) and mix the sample. Add 5 μl trypsin solution (20 ng/μl $= 100$ ng) to the sample (the ratio of trypsin to protein should be 1:50 to 1:100), mix, and incubate at 37° for 5 h to O/N.

4. Purify the peptides using C_{18} beads, preferably Magnetic Dynabeads RPC$_{18}$ (Invitrogen) or C_{18} Zip-tip (Millipore), as per the manufacturer's protocol. Briefly adsorb the peptides to C_{18} matrix in buffer containing 2–5% acetonitrile and 0.1–0.4% acetic acid. Wash out salt and urea with 5% acetonitrile, 0.4% acetic acid. Elute the peptides with 60% acetonitrile, 0.4% acetic acid, and dry in speedvac.

3.2.3. LC-MS/MS analysis

Analyze the peptides prepared from protein samples by LC-MS/MS. We use an electrospray ionization technique using LCQ DECA XP or LTQ mass spectrometers that are connected to Surveyor HPLC systems (Thermo Finnigan). The peptides prepared from gels and purified complexes containing an estimate of less than 100 proteins can be fractionated by C_{18} reverse-phase chromatography and analyzed directly by mass spectrometry. However, additional chromatography steps are required to fractionate peptides prepared from more complex samples, e.g., subcellular fractions containing hundreds to a few thousand proteins.

1. Dissolve isolated peptides from gel pieces or purified complexes in 5% acetonitrile, 0.4% acetic acid buffer, and inject 2 μl of the sample into a 10-cm long × 75-μm ID C_{18} capillary column at a flow rate of 200 nl/min. The nanoflow rate is achieved by using a split valve in the buffer flow line. Elute the peptides from the column using a linear acetonitrile gradient and analyze in-line by mass spectrometer, by switching between the MS mode for peptide detection and the MS/MS mode for peptide fragmentation and sequence analysis. Collect the data using a Dynamic exclusion method where a specific ion is sequenced only twice and is excluded from the list for 1 min. We typically collect three MS/MS scans of ions following each scan in DECA-XP or five MS/MS scans in LTQ mass spectrometer. The recommended gradient for analysis of complexes is as follows: 5 min wash with 5% acetonitrile and 0.4% acetic acid followed by elution of peptides with a 45 min linear gradient of 5–40% acetonitrile in 0.4% acetic acid, and a 5 min linear gradient of 40–80% acetonitrile in 0.4% acetic acid. For peptides isolated from gel bands, a 30 min linear gradient of 10–60% acetonitrile in 0.4% acetic acid buffer can be used.

2. An optional step is recommended for analysis of highly complex protein mixtures, e.g., fractions collected during chromatographic steps or subcellular fractions. Digest 100–200 μg of proteins with trypsin as above.

Prior to RP-LC-MS/MS analysis, reduce the complexity of peptides using strong cation-exchange (SCX) column chromatography. We carry out this step using the HPLC system and collect multiple fractions for subsequent analysis. Load the peptide sample to a 10-cm long × 2.1-mm ID polysulfoethyl column (PolyLC Inc.) at a flow rate of 200 μl/min. Wash the column for 10–20 min with 5% acetonitrile in 0.4% acetic acid at a 200 μl/min flow rate until the A_{280} of the flowthrough comes to base line. Elute the peptides at 200 μl/min flow rate with a 20 min linear gradient of 0–200 mM of ammonium acetate in 5% acetonitrile and 0.4% acetic acid, followed by a 5 min linear gradient of 200–500 mM of ammonium acetate in 5% acetonitrile and 0.4% acetic acid. The gradient length can be increased based on the complexity of the sample. Collect 200 μl fractions and dry in speedvac. Analyze the peptides in each fraction sequentially by C_{18} RP-LC-MS/MS analysis as above.

3. Analyze the acquired mass spectrometry data (MS/MS spectra) with the source organism (*T. brucei*) protein database using SEQUEST. Additionally, a nucleotide sequence database can also be used for the search and in such a case translate the genome sequence in all six open reading frames (ORFs of 50 or more amino acids stop to stop) to build a predicted peptide sequence database. The SEQUEST program compares the theoretical MS/MS spectra from the database with the acquired MS/MS spectra from the samples.

4. The SEQUEST program assigns a score to each of the peptides. A protein is identified based on multiple peptide matches (recommended two or more peptide hits) that pass the cut-off threshold corresponding to the specific ORF. Protein identification with a single peptide hit needs expert interpretation.

5. Results from SEQUEST analysis can be compiled and filtered more readily using PeptideProphet and ProteinProphet software (www.systemsbiology.org).

It is expected that all or most of the proteins identified in MAb immuno-precipitated complexes would be part of the complex. Their specific association can be assessed by increasing the stringency, e.g., salt and nonionic detergent, during the pull down experiments. A combination of genetic and biochemical approaches [see below for details on the tandem affinity purification tag (TAP-tag) approach] can be taken for further validation of their association and characterization of the proteins and complexes. Stable association of specific proteins with editosomes is validated by their presence in complexes isolated by at least two independent purification methods. Using these criteria we have identified 20 proteins in *T. brucei* editosomes using different approaches (Table 1.1). Most editosome proteins can be grouped into different pairs or sets based on their sequence similarities and/or presence of specific motifs and domains (Stuart *et al.*, 2005).

Table 1.1 Protein components of editosomes isolated by various methods

Protein	KREN1 editosome	KREPB2 editosome	KREN2 editosome	Tagged KREPB5	KREPA2 mAb-IP	SP-Q-S6 column
KREPA1	√	√	√	√	√	√
KREPA2	√	√	√	√	√	√
KREPA3	√	√	√	√	√	√
KREPA4	√	√	√	√	√	√
KREPA5	√	ND[a]	√	√	√	√
KREPA6	√	√	√	√	√	√
KREN1	√	ND	ND	√	√	√
KREPB2	ND	√	ND	√	√	√
KREN2	ND	ND	√	√	√	√
KREPB4	√	√	√	√	√	√
KREPB5	√	√	√	√	√	√
KREPB6	ND	√[b]	ND	√	ND	ND
KREPB7	ND	ND	√	√	ND	√
KREPB8	√	ND	ND	√	ND	√
KREX1	√	ND	ND	√	√	√
KREX2	√	√	√	√	√	√
KREL1	√	√	√	√	√	√
KREL2	√	√	√	√	√	√
KRET2	√	√	√	√	√	√
KREH1	ND	ND	ND	ND	√	√

[a] ND, not detected by mass spectrometry.
[b] Preliminary data.

Thus a newly identified protein in isolated editosomes having sequence similarities to other editosome proteins may suggest that it is an editosome component.

4. Tandem Affinity Purification of Editosomes

4.1. Overview

Originally developed for yeast (Rigaut *et al.*, 1999), the tandem affinity purification (TAP) procedure is now one of the most widely used methods for the isolation of protein complexes under native conditions (Burckstummer *et al.*, 2006). Several laboratories, including ours, have adapted the method for the purification of multiprotein complexes from trypanosomatids with slightly different strategies for tagging and purification (Aphasizhev *et al.*, 2003; Estevez *et al.*, 2001; Panigrahi *et al.*, 2003b; Schimanski *et al.*, 2005; Schnaufer *et al.*, 2003). In the following, we will focus on the inducible system that we have developed, but other systems will be discussed where appropriate. In the original system, which is the one we have adapted, the protein of interest is fused to a tandem affinity tag consisting of a calmodulin-binding peptide (CBP) and two repeats of the protein A domain, in that order (Rigaut *et al.*, 1999). These two tags are separated by a recognition site for the TEV protease. The sequential purification over IgG Sepharose and calmodulin resin under native conditions results in the isolation of active editosome complexes suitable for functional studies and identification of components by mass spectrometry (Panigrahi *et al.*, 2003b; Schnaufer *et al.*, 2003). We frequently combine the TAP procedure with fractionation based on S-values by loading the TEV eluate onto a glycerol gradient (see above for a description of the procedure) and pooling selected fractions for further purification over a calmodulin column. Importantly, the ability to tag different proteins in the complex has resulted in the isolation of specific subcomplexes (Schnaufer *et al.*, 2003) and the discovery that 20S editosomes purified by conventional chromatography (see above) represent a mixture of complexes with distinct composition and catalytic characteristics (Panigrahi *et al.*, 2006).

Before tagged editosomes can be purified, the protein of interest needs to be selected, its coding sequence cloned into the TAP expression plasmid, pLew79-TAP (Panigrahi *et al.*, 2003b), and the resulting plasmid introduced into a *T. brucei* cell line set up for tetracycline (tet)-regulated expression. We routinely use the 29.13 and "single marker" cell lines as procyclic and bloodstream form hosts for expression, respectively, which express suitable levels of T7 RNA polymerase and tet repressor protein (Wirtz *et al.*, 1999). Plasmid pLew79-TAP is based on a plasmid developed by George Cross' laboratory (Wirtz *et al.*, 1999) and provides an *Hind*III/*Bam*HI cloning cassette. Expression is driven by an inducible EP promoter, resulting in high and moderate expression rates in procyclic and bloodstream stages, respectively. Advantages of the inducible system are that it allows expression of mutated versions of a protein with potentially dominant negative effects and the fine tuning of expression levels according to the experimental question. Recently, Marilyn Parsons' laboratory has developed a modified version of this plasmid, pLew79-MH-TAP, which features additional cloning sites and myc and his tags between the sequence of interest and the calmodulin-binding peptide (Jensen *et al.*, 2007). Depending on the conformation of the tagged protein, the additional tags may also result in better exposure of the CBP tag and improved recovery (B. Jensen, personal communication). An alternative system for *T. brucei* results in tagging of an endogenous allele *in situ* and replaces the CBP with the protein C epitope, which reportedly resulted in increased recovery in a number of cases (Schimanski *et al.*, 2005). *Leishmania* editosomes were purified using a constitutive, episomal expression system (Aphasizhev *et al.*, 2003).

Expression and purification of tagged proteins and complexes can be monitored with antibodies specific for editosome components, such as the ones described above, or for the tag itself. We routinely use the PAP reagent (peroxidase–antiperoxidase complex, Sigma-Aldrich, Inc.), which binds to the protein A part of the tag and allows detection of tagged proteins and complexes before cleavage with TEV protease, or an anti-CBP antiserum (Upstate USA, Inc.), which allows detection throughout the purification.

4.2. Purification protocol

1. Induce transgenic *T. brucei* culture, grown as described above, with tet. Induction must be timed in such a way that by the time of harvesting, 2 liters of cell culture has reached mid-log phase ($\sim$2 $\times$ 10^7 cells per ml). For instance, for standard purification of 20S editosomes using the *Tb*REL1-TAP cell line (Schnaufer *et al.*, 2003), we use 10 ng/ml tetracycline, induce 2 liters of cell culture at a density of 2.5 $\times$ 10^6 cells/ml, and harvest the cells 48 h later. Depending on the particular protein and experimental question, use a different tet concentration and induce for shorter or longer periods of time. Tet concentrations >200 ng/ml will result in full induction of the promoter (Wirtz *et al.*, 1999) and, in procyclic cells, overexpression of the tagged protein. This can be useful, for instance, for the production of subcomplexes (Aphasizhev *et al.*, 2003; Schnaufer *et al.*, 2003). Conditions can be optimized by performing small-scale one-step purifications (e.g., 10 ml over IgG Sepharose) with tet concentrations ranging from 0.01 to 200 ng/ml and induction times up to 72 h. For these pilot studies, follow the protocol below and adjust volumes accordingly.
2. Harvest cells by centrifugation at 1300$\times$*g* for 10 min at room temperature and discard the supernatant. Wash the cells once with 200 ml PBS-G (150 m*M* NaCl, 0.2 *M* sodium phosphate buffer, pH 7.4, 6 m*M* glucose), pellet by centrifugation, and discard the supernatant.
3. Resuspend cells in 18 ml ice-cold IPP150 (10 m*M* Tris–HCl, pH 8.0, 150 m*M* NaCl, 0.1% NP40), supplemented with two dissolved tablets of "complete mini EDTA-free" protease inhibitors (Roche Applied Science). *Note*: all subsequent steps (with the exception of the TEV digest) should be carried out at 4°. Precool all buffers. Save aliquots of all lysates, washes, and eluates for subsequent analysis.
4. Lyse cells by adding Triton X-100 to a final concentration of 1% and inverting the tube several times. Incubate on ice for 20 min or until cells are completely lysed (monitor lysis under a microscope).
5. Centrifuge the lysate at 15,000$\times$*g* for 15 min to separate soluble proteins and complexes from debris. Collect the supernatant carefully.
6. In a 20-ml disposable column (e.g., Econo-pac, Bio-Rad), mix 10 ml IPP150 and 200 μl IgG Sepharose beads ("fast flow," Pharmacia) and drain by gravity flow. Plug the bottom of the column, add the cleared lysate, and close the top of the column. Slowly rotate for 2 h.

7. Remove the top plug first, then the bottom plug to drain the liquid by gravity flow, and wash the beads three times with 20 ml of IPP150 and once with 10 ml of TEV cleavage buffer (IPP150, 0.5 mM EDTA, 1 mM DTT).

8. Plug the bottom of the column and add 1 ml TEV cleavage buffer and 10 μl AcTEV protease (Invitrogen, 10 units/μl). Plug the top and carefully mix beads and protease by swirling the column. Incubate for 2 h at 16°, carefully mixing the contents every 30 min.

9. Remove top and bottom plugs and collect the eluate by gravity flow into a 1.5-ml tube. Carefully layer 200 μl TEV cleavage buffer on the beads and collect the flowthrough (dead volume) and combine with the first eluate. The eluate contains the tagged proteins and associated complexes.

10. *Optional step*: The tagged complexes can be purified by glycerol gradient sedimentation prior to the second affinity step. In such a case, load the TEV eluate on an 11-ml 10–30% glycerol gradient and fractionate as described above in the chromatography section. The duration of centrifugation can be varied based on the experimental requirements (see the next section). Monitor the sedimentation profile of tagged protein/complexes by Western analysis.

11. Measure the total volume of the eluate or (pooled) glycerol gradient fractions. Add 3 volumes calmodulin-binding buffer (IPP150, 10 mM fresh 2-mercaptoethanol, 1 mM Mg acetate, 1 mM imidazole, 2 mM CaCl$_2$) and 0.003 volumes 1 M CaCl$_2$.

12. Mix 5 ml calmodulin-binding buffer and 200 μl calmodulin resin (Stratagene) in a disposable column (e.g., 5-ml Econo-column or 20-ml Econo-pac column, Bio-Rad) and drain the liquid by gravity flow. Plug the bottom of the column, add the sample, close the top of the column, and slowly rotate for 1 h at 4°.

13. Drain the liquid by gravity flow and wash the beads with 30 ml calmodulin-binding buffer.

14. Elute four fractions of 250 μl with calmodulin elution buffer (contains 2 mM EGTA instead of CaCl$_2$ in the binding buffer).

15. The eluate and the various fractions obtained during the purification can be analyzed by Western blot (Fig. 1.2). Identify the proteins in the purified fraction by LC-MS/MS analysis as described above. The functional characteristics of the complexes can be monitored by *in vitro* RNA editing assays as described in Chapter 2 (this volume).

5. CHARACTERIZATION OF COMPLEXES

In our laboratory we have isolated editosomes by tagging various component proteins. The tag is placed at the C-terminus since the mt import signal is N-terminal. In most cases, the tagged protein incorporates into the 20S editosome. Also, in many instances, partial tagged complexes

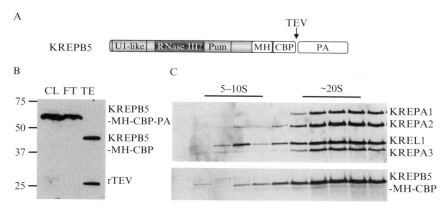

Figure 1.2   Isolation of editosomes via KREPB5-MH-TAP. (A) Schematic representation of the motifs and domains in KREPB5 and the TAP tag that contains the myc epitope (M), His6 (H), calmodulin-binding protein (CBP), and protein A (PA) regions. (B) Western blot analysis with anti-His antibody of the tagged complexes: cleared lysate (CL), the unbound portion of tagged protein in the IgG column flowthrough (FT), and TEV eluate (TE). The size of the tagged protein is reduced following treatment with TEV protease that cleaves off the protein A (PA) tag. The his-tagged TEV protease (rTEV) is also detected. (C) Western blot analysis showing the sedimentation profile of the TEV eluate from (B) on the glycerol gradient. Editosome proteins KREPA1, KREPA2, KREL1, and KREPA3 were detected with monoclonal antibodies (top panel), and the tagged protein KREPB5 was detected with an antiserum against the calmodulin-binding peptide (bottom panel). Most of the editosomes sediment in the ~20S region, and what appears to be subcomplexes sediment in the 5–10S region.

are observed in the 5–10S region (Fig. 1.2), most likely due to the overexpression of the tagged protein. Purification and analysis of these smaller subcomplexes provide tools to study the structural and functional associations of a specific tagged protein. Analysis of tagged RNA ligase proteins showed that KREL1 is associated with KREPA2 and KREX2, and in this subcomplex KREX2 can catalyze U removal and KREL1 the RNA ligation steps of deletion RNA editing. Reciprocally KREL2 is associated with KREPA1 and KRET2 where KRET2 can catalyze U addition and KREL2 the RNA ligation steps of insertion RNA editing (Schnaufer et al., 2003). To characterize the smaller subcomplexes/partial complexes, fractionate the TEV eluate from the IgG affinity column on 10–30% glycerol gradient for 9–14 h. Pool the fractions from the 5–10S region and the 20S region separately and further purify by calmodulin affinity column. Determine the composition of the purified 5–10S and 20S complexes by mass spectrometry and correlate their protein composition with the presence of specific RNA editing activities by in vitro assays.

Mass spectrometry analysis of editosomes isolated by various methods described above have identified 20 proteins that constitute most, but perhaps not all, protein components of the ~20S T. brucei editosomes (Panigrahi et al., 2006). The orthologs of all of these proteins have been identified in *L. major* and *T. cruzi* sequence databases using PSI-BLAST algorithm and syntenic alignments of gene sequences (Panigrahi *et al.*, 2006; Worthey *et al.*, 2003), and 14 of these proteins have been identified in REL1-tagged *Leishmania tarentolae* complexes (Simpson *et al.*, 2004). Eighteen of the 20 editosome proteins occur in pairs or sets based on sequence similarities and/or motifs (Panigrahi *et al.*, 2003b; Stuart *et al.*, 2005). Tagging and analysis of each specific set of proteins may provide new insights into editosome compositions and functions. Such analysis of tagged RNase III complexes (KREN1, KREPB2, and KREN2) showed the presence of three compositionally and functionally distinct 20S editosomes, each unique to one of the endonucleases (Panigrahi *et al.*, 2006).

Mutation(s) of amino acids can be incorporated into tagged proteins of interest and their effect on the function(s) of the tagged protein and editosome structure can be analyzed. Incorporation of point mutations in the putative catalytic domain of tagged endonucleases (KREN1 deletion endonuclease, KREN2 insertion endonuclease) abolished the specific endonuclease catalytic activities of tagged complexes (Panigrahi *et al.*, 2006). Point mutations in zinc finger domains of tagged KREPA2 in *Leishmania* affected the structure of editosomes (Kang *et al.*, 2004). A brief description is given below on how to isolate and analyze mutagenized complexes using KREN2 as an example:

1. Incorporate point mutations into TAP-KREN2 plasmid (Panigrahi *et al.*, 2006) by site-directed mutagenesis, using the QuikChange XL site-directed mutagenesis kit (Stratagene). Use primer sets 4862 (CGTTCGTGGGAATTC*GTA*CG TCTCGAATGGATTG) and 4863 (CAATCCATTCGAGACGTACGAATT CCCACGAACG) to mutagenize E_{227} to V (the mutagenized codon is in italic).

2. To incorporate two point mutations E_{227} to V and G_{233} to V, use primer sets 4864 (CGTGGGAATTC*GTA*CGTCTCGAATGGATT G*T*GGATA-ATGTTGTG) and 4865 (CACAACATTATCCACAATC CATTC-GAGACGTACGAATTCCCACG).

3. Verify the sequences of mutagenized constructs by sequencing and transfect the plasmid into *T. brucei* cells.

4. Monitor the expression of tagged protein by Western blot analysis following tet induction as described above, and determine the cell division rate by counting cells at regular intervals.

5. Isolate the TAP-tagged proteins and complexes as described above by sequential IgG Sepharose affinity chromatography, glycerol gradient sedimentation, and calmodulin affinity chromatography. In this case, we fractionated the TEV eluate on a glycerol gradient for 14 h. Monitor the fractionation and sedimentation profile by Western blot analysis using MAbs against editosome proteins KREPA1, KREPA2, KREL1, and KREPA3.

Western blot analysis of glycerol gradient fractionated tagged complexes showed that single mutant KREN2 (mut-$E_{227}V$) incorporates into 20S editosomes. However, incorporation of double mutations in KREN2 (mut-$E_{227}V$-$G_{233}V$) resulted in generation of only $\sim$10S mutant complexes, indicating it affects the structure of the protein and/or its association with 20S editosomes (Fig. 1.3). Similar mutations in KREN1 endonuclease did not affect the structure of the editosome (Panigrahi *et al.*, 2006). We have studied the effect of double mutations on the catalytic function, and in both cases the specific nuclease function was abolished in mutant

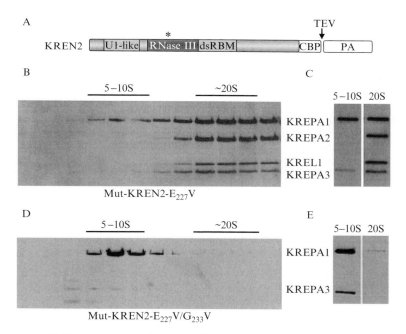

Figure 1.3 TAP-tag mutational analysis of KREN2 complexes. (A) Schematic representation of the motifs and domains in KREN2 and the position of the tag. The mutations were incorporated into the RNase III motif as indicated by the asterisk. (B and D) Western blot analyses of glycerol gradient fractions of IgG affinity-purified mutant complexes using MAbs specific to editosome proteins KREPA1, KREPA2, KREL1, and KREPA3. Most of the mutant KREN2 editosomes that contain a single point mutation (B) sediment are in the $\sim$20S region, but some subcomplexes are in the 5–10S region. Protein bands corresponding to KREPA1, KREPA2, KREL1, and KREPA3 are indicated. KREN2 complexes containing two point mutations (D) primarily sediment at $\sim$10S and editosome proteins KREPA2 and KREL1 are not detected. Pooled fractions from the 5 to 10S region and the $\sim$20S region as indicated by the line were further purified by calmodulin affinity column. (C and E) Western analysis of calmodulin affinity-purified samples from (B) and (D), respectively, using MAbs against editosome proteins KREPA1, KREPA2, KREL1, and KREPA3.

complexes. The KREN1 editosomes catalyze deletion RNA editing, while KREN2 editosomes catalyze insertion RNA editing. This specificity may be provided by the structure of the catalytic proteins and/or their structural organization and interactions with specific proteins in the complex. The diversity of editosome composition helps explain some of the complexity of the editing process.

ACKNOWLEDGMENTS

We thank Michelle Fleck and Rosemary Proff for technical assistance, and Yuko Ogata for help with mass spectrometry. This work was supported by NIH Grants AI 14102 and GM 42188.

REFERENCES

Aebersold, R., and Goodlett, D. R. (2001). Mass spectrometry in proteomics. *Chem. Rev.* **101,** 269–295.

Aphasizhev, R., Aphasizheva, I., Nelson, R. E., Gao, G., Simpson, A. M., Kang, X., Falick, A. M., Sbicego, S., and Simpson, L. (2003). Isolation of a U–insertion/deletion editing complex from *Leishmania tarentolae* mitochondria. *EMBO J.* **22,** 913–924.

Burckstummer, T., Bennett, K. L., Preradovic, A., Schutze, G., Hantschel, O., Superti-Furga, G., and Bauch, A. (2006). An efficient tandem affinity purification procedure for interaction proteomics in mammalian cells. *Nat. Methods* **3,** 1013–1019.

Corell, R. A., Read, L. K., Riley, G. R., Nellissery, J. K., Allen, T. E., Kable, M. L., Wachal, M. D., Seiwert, S. D., Myler, P. J., and Stuart, K. D. (1996). Complexes from *Trypanosoma brucei* that exhibit deletion editing and other editing-associated properties. *Mol. Cell. Biol.* **16,** 1410–1418.

Estevez, A. M., Kempf, T., and Clayton, C. (2001). The exosome of *Trypanosoma brucei*. *EMBO J.* **20,** 3831–3839.

Griffin, T. J., Goodlett, D. R., and Aebersold, R. (2001). Advances in proteome analysis by mass spectrometry. *Curr. Opin. Biotechnol.* **12,** 607–612.

Harris, M. E., Moore, D. R., and Hajduk, S. L. (1990). Addition of uridines to edited RNAs in trypanosome mitochondria occurs independently of transcription. *J. Biol. Chem.* **265,** 11368–11376.

Jensen, B. C., Kifer, C. T., Brekken, D. L., Randall, A. C., Wang, Q., Drees, B. L., and Parsons, M. (2007). Characterization of protein kinase CK2 from *Trypanosoma brucei*. *Mol. Biochem. Parasitol.* **151,** 28–40.

Kable, M. L., Seiwert, S. D., Heidmann, S., and Stuart, K. (1996). RNA editing: A mechanism for gRNA-specified uridylate insertion into precursor mRNA. *Science* **273,** 1189–1195.

Kang, X., Falick, A. M., Nelson, R. E., Gao, G., Rogers, K., Aphasizhev, R., and Simpson, L. (2004). Disruption of the zinc finger motifs in the *Leishmania tarentolae* LC-4 (=TbMP63) L-complex editing protein affects the stability of the L-complex. *J. Biol. Chem.* **279,** 3893–3899.

Keller, A., Nesvizhskii, A. I., Kolker, E., and Aebersold, R. (2002). Empirical statistical model to estimate the accuracy of peptide identifications made by MS/MS and database search. *Anal. Chem.* **74,** 5383–5392.

Madison-Antenucci, S., Sabatini, R. S., Pollard, V. W., and Hajduk, S. L. (1998). Kineto-plastid RNA-editing-associated protein 1 (REAP-1): A novel editing complex protein with repetitive domains. *EMBO J.* **17,** 6368–6376.

Madison-Antenucci, S., Grams, J., and Hajduk, S. L. (2002). Editing machines: The complexities of trypanosome RNA editing. *Cell* **108,** 435–438.

Nesvizhskii, A. I., Keller, A., Kolker, E., and Aebersold, R. (2003). A statistical model for identifying proteins by tandem mass spectrometry. *Anal. Chem.* **75,** 4646–4658.

Panigrahi, A. K., Gygi, S., Ernst, N., Igo, R. P., Jr., Palazzo, S. S., Schnaufer, A., Weston, D., Carmean, N., Salavati, R., Aebersold, R., and Stuart, K. D. (2001a). Association of two novel proteins, *Tb*MP52 and *Tb*MP48, with the *Trypanosoma brucei* RNA editing complex. *Mol. Cell. Biol.* **21,** 380–389.

Panigrahi, A. K., Schnaufer, A., Carmean, N., Igo, R. P., Jr., Gygi, S. P., Ernst, N. L., Palazzo, S. S., Weston, D. S., Aebersold, R., Salavati, R., and Stuart, K. D. (2001b). Four related proteins of the *Trypanosoma brucei* RNA editing complex. *Mol. Cell. Biol.* **21,** 6833–6840.

Panigrahi, A. K., Allen, T. E., Haynes, P. A., Gygi, S. P., and Stuart, K. (2003a). Mass spectrometric analysis of the editosome and other multiprotein complexes in *Trypanosoma brucei. J. Am. Soc. Mass. Spectrom.* **14,** 728–735.

Panigrahi, A. K., Schnaufer, A., Ernst, N. L., Wang, B., Carmean, N., Salavati, R., and Stuart, K. (2003b). Identification of novel components of *Trypanosoma brucei* editosomes. *RNA* **9,** 484–492.

Panigrahi, A. K., Ernst, N. L., Domingo, G. J., Fleck, M., Salavati, R., and Stuart, K. D. (2006). Compositionally and functionally distinct editosomes in *Trypanosoma brucei. RNA* **12,** 1038–1049.

Perkins, D. N., Pappin, D. J., Creasy, D. M., and Cottrell, J. S. (1999). Probability-based protein identification by searching sequence databases using mass spectrometry data. *Electrophoresis* **20,** 3551–3567.

Pollard, V. W., Harris, M. E., and Hajduk, S. L. (1992). Native mRNA editing complexes from *Trypanosoma brucei* mitochondria. *EMBO J.* **11,** 4429–4438.

Rigaut, G., Shevchenko, A., Rutz, B., Wilm, M., Mann, M., and Seraphin, B. (1999). A generic protein purification method for protein complex characterization and proteome exploration. *Nat. Biotechnol.* **17,** 1030–1032.

Rusche, L. N., Cruz-Reyes, J., Piller, K. J., and Sollner-Webb, B. (1997). Purification of a functional enzymatic editing complex from *Trypanosoma brucei* mitochondria. *EMBO J.* **16,** 4069–4081.

Schimanski, B., Nguyen, T. N., and Günzl, A. (2005). Highly efficient tandem affinity purification of trypanosome protein complexes based on a novel epitope combination. *Eukaryot. Cell* **4,** 1942–1950.

Schnaufer, A., Ernst, N., O'Rear, J., Salavati, R., and Stuart, K. (2003). Separate insertion and deletion sub-complexes of the *Trypanosoma brucei* RNA editing complex. *Mol. Cell* **12,** 307–319.

Seiwert, S. D., Heidmann, S., and Stuart, K. (1996). Direct visualization of uridylate deletion *in vitro* suggests a mechanism for kinetoplastid RNA editing. *Cell* **84,** 831–841.

Simpson, L., Aphasizhev, R., Gao, G., and Kang, X. (2004). Mitochondrial proteins and complexes in *Leishmania* and *Trypanosoma* involved in U-insertion/deletion RNA editing. *RNA* **10,** 159–170.

Stuart, K., and Panigrahi, A. K. (2002). RNA editing: Complexity and complications. *Mol. Microbiol.* **45,** 591–596.

Stuart, K., Panigrahi, A. K., and Salavati, R. (2000). RNA editing in kinetoplastid mitochondria. *In* "RNA Editing: Frontiers in Molecular Biology" (B.L. Bass, ed.), pp. 1–19. Oxford University Press, Oxford.

Stuart, K., Panigrahi, A. K., and Schnaufer, A. (2004). Identification and characterization of trypanosome RNA editing complex components. *In* "Methods in Molecular Biology" (J. M. Gott, ed.), Vol. 265, RNA Interference, Editing, and Modification: Methods and Protocols, pp. 273–291. Humana Press, Inc., Totowa, NJ.

Stuart, K. D., Schnaufer, A., Ernst, N. L., and Panigrahi, A. K. (2005). Complex management: RNA editing in trypanosomes. *Trends Biochem. Sci.* **30,** 97–105.

Wirtz, E., Leal, S., Ochatt, C., and Cross, G. A. M. (1999). A tightly regulated inducible expression system for conditional gene knock-outs and dominant-negative genetics in *Trypanosoma brucei. Mol. Biochem. Parasitol.* **99,** 89–101.

Worthey, E. A., Schnaufer, A., Mian, I. S., Stuart, K., and Salavati, R. (2003). Comparative analysis of editosome proteins in trypanosomatids. *Nucleic Acids Res.* **31,** 6392–6408.

Yates, J. R., Eng, J. K., and McCormack, A. L. (1995). Mining genomes: Correlating tandem mass spectra of modified and unmodified peptides to sequences in nucleotide databases. *Anal. Chem.* **67,** 3202–3210.

CHAPTER TWO

Uridine Insertion/Deletion Editing Activities

Jason Carnes *and* Kenneth D. Stuart

Contents

Abstract

The uridine nucleotide insertion and deletion editing of trypanosomatid mitochondrial mRNAs is catalyzed by a macromolecular complex, the editosome. Many investigations of RNA editing involve some assessment of editosome activity either *in vitro* or *in vivo*. Assays to detect insertion or deletion editing activity on RNAs *in vitro* have been particularly useful, and can include the initial endonucleolytic step (full-round) or bypass it (precleaved). Additional assays to examine individual catalytic steps have also proved useful to dissect particular steps in editing. Detection of RNA editing activity *in vivo* has been significantly advanced

Seattle Biomedical Research Institute, Seattle, Washington

Methods in Enzymology, Volume 424
ISSN 0076-6879, DOI: 10.1016/S0076-6879(07)24002-9

by the application of real-time PCR technology, which can simultaneously assay several edited and pre-edited targets. Here we describe these assays to assess editing both *in vitro* (full-round insertion and deletion; precleaved insertion and deletion; individual TUTase, ligase, or helicase activity) and *in vivo* (real-time PCR).

1. INTRODUCTION

RNA editing in trypanosomatids occurs by the insertion and deletion of uridine nucleotides (Us), and this editing transforms nonfunctional RNA transcripts into mature mRNAs that can be translated into proteins (Stuart *et al.*, 2005). Editing occurs by a coordinated series of catalytic steps: cleavage of the mRNA editing site, U addition or removal, and ligation. A multiprotein complex called the editosome catalyzes RNA editing, and uses partially complementary guide RNAs (gRNAs) to direct the process. The sequences of gRNAs specify the sites and numbers of Us that are inserted or deleted. Each gRNA specifies editing at multiple sites, and a single gRNA can specify both U insertion and deletion at different sites. The editosome sediments at ~20S in glycerol gradients, and contains at least 20 different proteins. Its catalytic activities include endoribonuclease, U–specific exoribonuclease (exoUase), $3'$ terminal uridine transferase (TUTase), RNA ligase, and RNA helicase. Known editosome substrates are transcribed from the mitochondrial maxi-circle DNA, and primarily encode proteins of the oxidative phosphoryla-tion pathway. Twelve of the 20 identified *Trypanosoma brucei* maxicircle gene transcripts undergo posttranscriptional RNA editing. Editing can be extensive, as in the case of COIII mRNA, where 547 uridine nucleotides are inserted and 41 are deleted to generate translatable mRNA (Feagin *et al.*, 1988). Such extensively edited mRNAs require numerous gRNAs to complete editing.

Since the discovery of RNA editing in trypanosomes roughly two decades ago, several standard experimental protocols have been developed to assay editing activity both *in vitro* and *in vivo*. Initial *in vitro* RNA editing assays tracked a single cycle of "full-round" editing activity on a single editing site, where gRNA specified either insertion or deletion editing (Kable *et al.*, 1996; Seiwert *et al.*, 1996). In these full-round editing assays, mitochondrial extracts or biochemically purified editosomes are combined with a radiolabeled substrate RNA and its cognate gRNA. These assays simultaneously assess endonucleolytic cleavage of the substrate RNA, U insertion or deletion as specified by gRNA, and ligation to generate RNA edited at a single site. Currently, *in vitro* editing assays are limited to a single catalytic cycle, and editing at multiple sites or with more than one gRNA remains elusive. The endonucleolytic step of full–round assays is relatively inefficient, and can be bypassed using "precleaved" editing assays, in which the RNA is provided as two fragments to mimic substrate after the endonucleolytic cleavage step (Igo *et al.*, 2000, 2002). Precleaved assays are particularly helpful in assessing editing activities when endonuclease activity is either impaired or absent. Simple modifications to precleaved assays can be used to detect individual catalytic activities of RNA editing, making them a common alternative to specific assays for individual catalysts. More recently, an assay that uses real-time polymerase chain reaction (PCR) was developed to assess the editing of several RNAs *in vivo* (Carnes *et al.*, 2005). This assay essentially replaces the more cumbersome poison primer extension and Northern analysis methods. Real-time PCR permits assessment of multiple RNAs in a single experiment, requires less time and RNA than prior methods, and does not use radioactivity. Together, these assays represent a standard toolkit for the investigation of trypanosomatid RNA editing.

The ability to genetically manipulate *T. brucei* has provided a means to dissect the functions of the proteins responsible for the editing activities as well as the RNA editing process *in vivo*. Two approaches have been particularly informative: knock-in of tagged alleles to facilitate biochemical purification of editing complexes and repression of editing genes by either conditional inactivation of expression or RNAi using tetracycline (tet) regulatable promoters. In the first approach, tandem affinity purification (TAP) of editosome proteins was used to reveal three types of editosomes that differ in composition (see Chapter 1, this volume; Panigrahi *et al.*, 2006). *In vitro* RNA editing assays were used to show that the editosomes differ in their enzymatic capabilities. In the second approach, the repression of editing genes *in vivo* has revealed that almost all editosome genes are essential to survival of the trypanosomes (Carnes *et al.*, 2005; Schnaufer *et al.*, 2001; Trotter *et al.*, 2005; Wang *et al.*, 2003). Real-time PCR has been used to assess the consequences to the editing of specific RNAs *in vivo* following repression of editosome protein genes. The *in vitro* RNA editing assays are also useful to assess the effects of repressing a particular editosome gene to elucidate its function.

The following sections describe the full-round and precleaved assays for RNA editing activity *in vitro*, assays for individual TUTase, ligase, and helicase activities, as well as a real-time PCR assay for assessing *in vivo* RNA editing.

 ## 2. FULL-ROUND INSERTION AND DELETION

2.1. Overview

Full-round insertion and deletion assays recapitulate a single cycle of RNA editing *in vitro*, which entails three enzymatic activities (Fig. 2.1). A cycle of full-round insertion begins with endonuclease activity that cleaves substrate RNA, followed by TUTase activity that adds uridine nucleotides to the 3' end

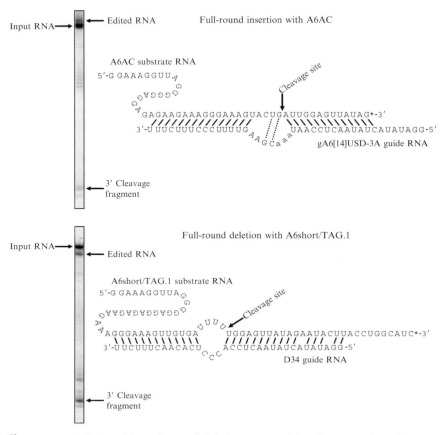

Figure 2.1 Full-round insertion and deletion assays with substrate and guide RNA schematics. Asterisk in sequences denotes ^{32}P radiolabel at the 3′ end of the RNA. In both assays, RNAs are incubated with mitochondrial extracts and then resolved by electrophoresis on 11% polyacrylamide 7 M urea 1 × TBE sequencing gel. The top panel shows full-round insertion editing of A6AC substrate RNA. Input A6AC RNA shifts three nucleotides higher after a single full cycle of insertion editing, including endonucleolytic cleavage, U addition as specified by guide RNA, and ligation. A faint band corresponding to the 3′ cleavage fragment is also observed in this assay. The bottom panel shows full-round editing of A6short/TAG.1 substrate RNA. The band 4 nucleotides below the input band is A6short/TAG.1 RNA after a single full cycle of deletion editing, including endonucleolytic cleavage, U removal as specified by guide RNA, and ligation. A band corresponding to the 3′ cleavage fragment is also observed in this assay.

of the 5′ cleavage fragment, and then ligase activity that rejoins the edited substrate RNA. A cycle of full–round deletion similarly entails endonuclease and ligase activities, but an exoUase activity removes uridine nucleotides from the 3′ end of the 5′ cleavage fragment. Several RNA substrates have been used along with their cognate gRNAs (Cifuentes–Rojas et al., 2005; Cruz–Reyes *et al.*, 2001; Kable *et al.*, 1996; Piller *et al.*, 1995, 1997; Seiwert *et al.*, 1996; Stuart *et al.*, 2004). This illustrates that the protocol that follows can be used with various substrates and their gRNAs. We will focus here on two substrates based on ATPase subunit 6 mRNA.

2.1.1. Purification of substrate RNA

The substrate RNAs for these assays are transcribed *in vitro* using T7 polymerase. This polymerase frequently adds a random extra base to the 3′ end of transcripts, thus generating a mixture of the intended RNA and variants with one additional 3′ nucleotide (nt). Both the intended RNA and the +1 nt RNA perform equally well in full-round assays. However, use of the transcribed and radiolabeled RNA directly results in "shadow bands" at +1 nt upon analytical electrophoresis, which are especially evident next to the cleavage product. This makes interpretation more complicated, and should be avoided by separating these two substrate RNAs using preparative electrophoresis with single nt resolution.

2.1.2. Quantification and optimization

Full-round assays are reasonably sensitive but they have a very narrow dynamic range, and thus quantification must be approached with caution. Serial dilutions of extracts being tested for RNA editing activity are often required in order to find a level of activity within the dynamic range of the assay. These assays involve multiple catalytic activities; therefore, the optimal conditions *in vitro* for one of these activities may have detrimental effects on one of the others. Differences in salt concentrations can significantly alter the activity observed in these assays.

2.1.3. Radiolabel considerations

RNA is typically radiolabeled at the 3′ end by ligation with pCp. The terminal 3′ phosphate also prevents ligation (circularization) that can result from the robust ligase activity often found in mitochondrial extracts. The 3′ labeling does not permit visualization of the addition or removal of Us at the 3′ end of the 5′ cleavage fragment. Alternatively, a 5′ labeled mRNA substrate can be made by capping with [^{32}P]GTP and guanylyltransferase, although the sensitivity is reduced due to lower efficiency of capping compared to pCp labeling.

2.1.4. Examination of endonuclease activity

Endonuclease activity appears to be inefficient when assessing full-round editing, but the standard full-round protocol outlined below can be modified to enhance cleavage product abundance. The master mix for both insertion and deletion assays includes ATP, which is a cofactor for RNA ligase activity. However, adenosine nts also affect RNA editing endonuclease activities. Adenosine nts stimulate deletion cleavage activity, but conversely inhibit insertion cleavage activity (Cruz-Reyes et al., 1998). Thus, if observing robust ligase activity is not desired, the adenosine nucleotide added to each master mix can be changed to enhance cleavage. We substitute 30 mM ADP for 30 mM ATP to enhance cleavage in deletion assays, and remove the 200 μM ATP entirely to enhance cleavage in insertion assays. Some residual ligase activity is often observed in these assays, presumably due to ligase that is already adenylated.

2.1.5. RNA editing controls

The response of RNA editing nsertion and deletion cleavage activities to adenosine nucleotide can also be utilized as a control in many experiments. Contaminating nucleases can potentially generate bands similar to those expected for RNA editing endonucleolytic cleavage. However, unlike RNA editing insertion and deletion cleavage activities, these potential artifacts are unlikely to be stimulated or inhibited by adenosine nucleotide. Another control that is commonly included is the omission of gRNA to assess gRNA dependence. The specificity of the gRNA can also be changed to determine if the number of Us inserted or deleted reflects real RNA editing. Finally, omission of cofactors such as UTP for insertion or ATP for ligation can be used as controls for these editing assays.

2.1.6. Isolation of RNA editing samples

The details of the various methods developed to extract RNA editing activity from cells are not discussed since they are covered in depth elsewhere in this volume (see Chapter 1). The following protocol was originally developed to test extracts from purified mitochondria and fractions from 10–30% glycerol gradients that contain the RNA editing complexes. However, they are also effective in assaying immunoprecipitated or TAP-tag purified editosomes using modifications to compensate for differences in buffer conditions. The TAP-tagged eluates are assayed by changing the 10× buffer for insertion to 250 mM HEPES, pH 7.9, and 100 mM Mg(OAc)$_2$; the 10× buffer for deletion is changed to 250 mM Tris–HCl, pH 7.0, and 100 mM Mg(OAc)$_2$. The final reaction volumes remain 30 μl with water used to adjust the volume.

2.1.6.1. Protocol

1. Make DNA templates for T7 transcription using PCR (see Table 2.1 for the sequences). For A6AC insertion substrate RNA, use primers *Eco*RI T7 and A6ACdTAG with an A6-eES1 DNA template. For A6short/TAG.1 deletion substrate RNA, use primers *Eco*RI T7 and 3′ TAG A6 with an A6short/TAG.1 DNA template. For gA6[14]USD-3A insertion gRNA, use primers *Eco*RI T7 and gUSDi + 3 with a gA6[14]wt DNA template. For D34 deletion gRNA, use primers *Eco*RI T7 and gA6USDdccc with a D34 DNA template. The thermocycler conditions are 94° for 4 min,

Table 2.1 Oligonucleotide sequences for full-round, precleaved, and helicase assays

Name	Purpose	Sequence
*Eco*RI T7	Forward primer	5'-CGGCGGGAATTCTCTGTAATACGACTCAC-3'
Full-round insertion		
A6ACdTAG	Reverse primer	5'-CTATAACTCCAATCAGTACTTTC-3'
A6–eES1 DNA	PCR template	5'-TAATACGACTCACTATAGGAAAGGTTAGGGGGAGGAGAGAAA-GGGAAAGTTGTGATTGGAGTTATAGAATACTTACCTGGCATC-3'
A6AC RNA	Substrate RNA	5'-GGAAAGGUUAGGGGGAGGAGAGAAAGGGAAAGUACUGAUU-GGAGUUAUAG-3'
gUSDi+3	Reverse primer	5'-AAAGAAAGGGAAAACTTCGTTTATTGGAGTTATAG-3
gA6[14]wt DNA	PCR template	5'-TAATACGACTCACTATAGGATATACTATAACTCCGATAACGAAT-CAGATTTTGACAGTGATATGATAATTATTTTTTTTTTTT-3'
gA6[14]USD-3A	Guide RNA	5'-GGAUAUACUAUAACUCCAAUaaaCGAAGUUUUCCCUUCUUU-3'
Full-round deletion		
3' TAG A6	Reverse primer	5'-GATGCCAGGTAAGTATTC-3'
A6short/ TAG.1 DNA	PCR template	5'-TAATACGACTCACTATAGGAAAGGTTAGGGGGAGGAGAGAAGA-AAGGGAAAGTTGTGATTTTTGGAGTTATAGAATACTTACCTGGC-ATC-3'
A6short/ TAG.1 RNA	Substrate RNA	5'-GGAAAGGUUAGGGGGAGGAGAGAAGAAAGGGAAAGUUGUGAUU-UUUGGAGUUAUAUAGAAUACUUACCUGGCAUC-3'
gA6USDdccc D34 template	Reverse primer	5'-AAGAAAGTTGTGAGGGTGGAGTTATAG-3'
	PCR template	5'-TAATACGACTCACTATAGGATATACTATAACTCCACCCTCACAAC-TTTCTT-3'

(continued)

Table 2.1 (*continued*)

Name	Purpose	Sequence
D34	Guide RNA	5'-GGAUAUACUAUAAACUCCACCCUCACAACUUUCUU-3'
Precleaved insertion		
5'CL18	PCi 5' RNA	5'-GGAAGUAUGAGACGUAGG_{OH}-3'
3'CL13pp	PCi 3' RNA	5'-_pAUUGGAGUUAUAG_p-3'
ArtgA6 2A	Reverse primer	5'-GGAAGTATGAGACGTAGGTTATCGGAG-3'
gPCA6-2A-Tmpl	PCR template	5'-CGGCGGAATTCTGTAATACGACTCACTATAGGATATACTATAAC-TCCGATAACCTACGTCTCATACTTCC-3'
gPCA6-2A	Guide RNA	5'-GGAUAUACUAUAAACUCCGAUAACCUACGUCUCAUACUUCC-3'
Precleaved deletion		
U5–5'CL	PCd 5' RNA	5'-GGAAAGGGAAAGUUGUGAUUUU_{OH}-3'
U5–3'CL	PCd 3' RNA	5'-_pGCGAGUUAUAGAAUA_p-3'
A6comp1	Reverse primer	5'-GGAAAGGGAAAGTTGTGAGCGAGTTATAGAACCTATAGAAC-CTATAGTGAGTCGTATTAC-3'
gA6[14]PC-del	Guide RNA	5'-GGUUCUAUAACUCGCUCACAACUUUCCCUUUCC-3'
Helicase assay		
3' A6 edited	Substrate RNA	5'-GGUUUAGUUUUGUAUUUGAUUUUUGAUAGUUAUUAUUG-UUGUUGAAAUUUGGGUUUGUUAUUGGAGUUAUAGAAUAAGAU-3'
gA6[14]	guide RNA	5'-GGAUAUACUAUAAACUCCGAUAACGAAUCAGAUUUGACAGU-GAUAUGAUAAUUAUUUUUUUUUUUUUUUUU-3'

followed by 30 cycles of 94° for 30 sec, 42° for 1 min, and 72° for 30 sec, followed by a final 7 min at 72°.

2. Generate the substrate and guide RNAs by T7 transcription. Allow the transcription reagents to equilibrate to room temperature, then combine the following in order to create a 100 μl transcription reaction: 55 μl of PCR product (total of 2–5 μg DNA), 30 μl 25 mM (each) rNTPs, 10 μl 10× T7 transcription buffer (400 mM Tris–HCl, pH 7.6; 240 mM MgCl$_2$; 20 mM spermidine; 0.1% Triton X-100), 1 μl 1M DTT, 1 μl 40 U/μl RNasin, and 3 μl 80U/μl T7 RNA polymerase. Incubate the reaction overnight at 37°. Precipitation of inorganic phosphate is frequently observed following transcription. After transcription, add 5 μl of 1 U/μl RQ1 DNase (Promega) and incubate at 37° for 15 min. Add 300 μl of water, 40 μl of 3 M NaOAc, 1 μl of 20 mg/ml glycogen, and 1 ml of 95% ethanol; invert to mix. Centrifuge at 17,000×g for 15 min at 4°. Wash the pellet once with 75% ethanol and allow the RNA pellet to air dry. Resuspend RNA in 40 μl of 7 M urea loading dye (7 M urea, 1× TBE, 0.05% bromophenol blue, 0.05% xylene cyanol) and heat for 1 min at 65°.

3. Purify transcribed RNAs by electrophoresis in 9% polyacrylamide, 7 M urea, 1× TBE gel 40 cm in length. On 9% gel, bromophenol blue dye runs ~14 nt, while xylene cyanol runs ~65 nt. A6AC is 53 nt, gA6[14] USD-3A is 42 nt, A6short/TAG.1 is 73 nt, and D34 is 34 nt.

4. Place the gel between two sheets of plastic wrap and place it on a reflection screen. Visualize RNAs in gel by UV shadowing with a short-wave UV lamp, marking the bands to be cut out with a felt-tip pen. Cut out the bands with a razor blade or scalpel. Place the excised bands in a tube with 400 μl of elution buffer [0.3 M NaOAc, pH 5.2; 0.1% sodium dodecyl sulfate (SDS); 1 mM EDTA] and elute overnight, rotating at room temperature.

5. Transfer the RNA to a fresh tube, add 1 μl of 20 mg/ml glycogen, and precipitate with 1 ml 95% ethanol. Centrifuge at 17,000×g for 15 min at 4° to pellet and resuspend in 20 μl water.

6. Label the substrate RNA by ligation to ^{32}P pCp [cytidine 3′,5′-bis(phosphate), [5′-^{32}P], 3000 Ci/mmol; Perkin Elmer]. Combine 40 pmol of RNA in 7 μl with 2 μl of 10× pCp ligation buffer [250 mM HEPES, pH 8.3; 50 mM MgCl$_2$; 16 mM dithiothreitol (DTT); 0.25 mM ATP], 2 μl of dimethyl sulfoxide (DMSO), 3 μl of 100% glycerol, 3 μl of T4 RNA ligase, and 5 μl (50 μCi) of ^{32}P pCp. Incubate overnight at 4°.

7. Most unincorporated radiolabel can be removed either by precipitation or by spin column. For precipitation, add 280 μl of water, 30 μl of 3 M NaOAc, and 825 μl of 95% ethanol to a 20 μl reaction. Air dry the RNA pellet and resuspend it in 20 μl of 7 M urea loading dye. Alternatively, add 10 μl of water to 20 μl ligation reaction and spin it on a G-25 spin column (Amersham Biosciences). Add 20 μl of 7 M urea loading dye.

8. Load RNA on a 60-cm-long 9% polyacrylamide 7 M urea 1× TBE gel
   to purify ^{32}P-labeled RNA to single nucleotide resolution. After electrophoresis, cover one side of the gel with plastic wrap and expose a
   portion of the gel with radiolabeled RNA to X-ray film for ~1 min.
   Develop the film to determine the location of bands to cut out. Usually,
   two primary bands that are 1 nt different in length are present. Precisely
   cut each band out, taking care to separate the bands from each other.

9. Place the excised bands in a tube and elute and precipitate as in step 5.
   Resuspend the RNA in 32 μl water. Assuming 80% of input RNA was
   labeled and recovered, the concentration is now 1 pmol/μl. Use a Geiger
   counter to estimate cpm in 1 μl. Use 10,000–40,000 cpm per editing assay.

10. For each editing assay plus two extra, combine 1 μl 2.5 pmol/μl gRNA
    and 1 μl 0.25 pmol/μl ^{32}P-labeled substrate RNA and heat to 65° for
    2 min. Allow it to cool at room temperature for 5–10 min.

11. Assemble the master mixes for editing assays, making enough for all
    reactions plus two extra. Each final editing reaction will be 30 μl:20 μl of
    master mix plus 10 μl of extract to be tested. The final concentration of
    KCl in each 30 μl reaction should be between 30 and 50 mM; for mito-
    chondrial extracts that contain 100 mM KCl, no additional KCl needs to
    be added. Each 20 μl of insertion master mix contains 2 μl 10× HHE
    buffer [250 mM HEPES, pH 7.9; 100 mM Mg(OAc)$_2$, 5 mM DTT,
    10 mM EDTA], 1.5 μl 2 mM UTP, 1.5 μl of 200 μM ATP, 0.8 μl
    100 mM CaCl$_2$, 1 μl 500 ng/μl $Torula$ type VI RNA, 0.2 μl 40 U/μl
    RNasin, 11 μl of water, and 2 μl of combined ^{32}P substrate RNA/
    gRNA. Each 20 μl of deletion master mix contains 2 μl of 10× Tris buffer
    [500 mM Tris–HCl, pH 7.0; 100 mM Mg(OAc)$_2$, 5 mM DTT, 10 mM
    EDTA], 1 μl 30 mM ATP, 0.8 μl 100 mM CaCl$_2$, 1 μl 500 ng/μl $Torula$
    type VI RNA, 0.2 μl 40 U/μl RNasin, 13 μl of water, and 2 μl of
    combined ^{32}P substrate RNA/gRNA. The volume of water can be
    altered in each master mix to accommodate either larger or smaller samples
    of extract to be added in the following step.

12. Add 10 μl of mitochondrial extract (or 7 μl TAP-tag-purified complex;
    see overview for modifications).

13. Incubate in a heat block at 27° for 3 h.

14. Add 75 μl of a master mix containing 2 μl stop buffer (130 mM EDTA,
    2.5% SDS), 1 μl 20 mg/ml glycogen, 10 μl 3 M NaOAc, and 62 μl
    water per editing reaction.

15. Add 100 μl of phenol:chloroform:isoamyl alcohol to the stopped editing
    reaction, mix by vortexing, centrifuge briefly to separate phases, and
    transfer the aqueous phase (top) to a fresh tube.

16. Precipitate RNA by addition of 250 μl 95% ethanol. Mix well by
    inversion and incubate at −70° for 20 min or at −20° for at least 1 h .

17. Pellet the RNA by centrifugation at 17,000×g for 20 min at 4°. Care-
    fully remove the supernatant using a Geiger counter to confirm the

RNA is not accidentally removed. *Optional*: wash the RNA pellet with 150 μl 75% ethanol, recentrifuge, and remove the supernatant. Allow the RNA to air dry completely. Residual ethanol will distort subsequent gel electrophoresis.

18. Resuspend the RNA in 10 μl of 7 *M* urea loading dye. Incubate at 100° for 2 min and cool to room temperature.

19. Create an RNase T_1 ladder of substrate RNA to act as a marker in gel analysis. Make a 1:300 dilution of RNase T_1 (Roche) in water. Combine 1 μl of diluted RNase T_1, 3 μl of RNase T_1 buffer (33 m*M* sodium citrate, pH 5.5; 7 *M* urea; 1.7 m*M* EDTA; 1 mg/ml yeast tRNA; 0.05% bromophenol blue; 0.05% xylene cyanol), and 1 μl of ^{32}P-labeled substrate RNA (equivalent to the amount in editing reactions). Incubate the T_1 digest at 55° for 15 min, place it on ice, and add 5 μl of 7 *M* urea loading dye. Keep the T_1 ladder cold to prevent further digestion.

20. Create a partial alkaline hydrolysis ladder of substrate RNA (optional). Combine 0.5 μl of 10× alkaline hydrolysis buffer (0.5 *M* NaHCO$_3$, pH 9.3; 10 m*M* EDTA), 1 μl of 1 mg/ml yeast tRNA, 2.5 μl of water, and 1 μl of ^{32}P-labeled substrate RNA (equivalent to the amount in the editing reactions). Incubate alkaline hydrolysis at 95° for 5 min, place it on ice, and add 5 μl of 7 *M* urea loading dye. Some optimization of alkaline hydrolysis may be required, which can be accomplished by varying the incubation time or the amount of tRNA added.

21. Load 5 μl of each editing reaction and ladders on 0.35-mm-thick 9% (or alternatively 11%) polyacrylamide, 7 *M* urea, 1× TBE gel for S2 sequencing electrophoresis apparatus (GibcoBRL). We use 32-well combs for best resolution.

22. Resolve the editing reaction products by electrophoresis so that the expected cleavage product (13 nt for A6AC; 27 nt for A6short/TAG.1) is 5–10 cm from the end of the gel.

23. Transfer the gel onto Whatman 3MM paper, cover with plastic wrap, and dry the gel on a gel-drying apparatus. Drying time depends on thickness and percent polyacrylamide. Dry a 0.35-mm-thick 9% gel at 80° for at least 30 min, or an 11% gel at 80° for at least 1 h .

24. Expose the dried gel to a PhosphorImager (GE Healthcare) screen overnight and scan the exposed screen to produce data.

3. PRECLEAVED INSERTION AND DELETION

3.1. Overview

Precleaved insertion and deletion assays tend to be more robust and easier to perform in comparison to full-round assays. Each precleaved assay examines two enzymatic activities: editing site U addition and ligation for insertion or

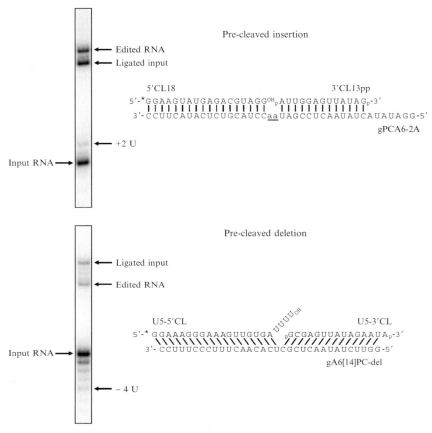

Figure 2.2 Precleaved insertion and deletion assays with substrate and guide RNA schematics. Asterisk in sequences denotes ^{32}P radiolabel at the 5′ end of the RNA. In both assays, RNAs are incubated with mitochondrial extracts and then resolved by electrophoresis on 11% polyacrylamide 7 M urea 1× TBE sequencing gel. The top panel shows precleaved insertion editing of 5′CL18/3′CL13pp substrate RNA. Four bands are highlighted: radiolabeled input 5′CL18 RNA, input RNA 2 nucleotides higher after U addition, ligated input of 5′CL18 to 3′CL13pp without U addition, and edited RNA of 5′CL18 with 2 Us added before ligation to 3′CL13pp. The bottom panel shows precleaved deletion editing of U5-5′CL/U5-3′CL substrate RNA. Four bands are highlighted: radiolabeled input U5-5′CL RNA, input RNA 4 nucleotides lower after U removal, ligated input of U5-5′CL to U5-3′CL without U removal, and edited RNA of U5-5′CL with four Us removed before ligation to U5-3′CL.

U removal and ligation for deletion (Fig. 2.2). Because endonuclease activity is necessarily required in full-round assays in order to observe the subsequent steps of RNA editing, precleaved assays are appropriate when only the U addition, U removal, or RNA ligation catalytic steps are under investigation. As with full-round substrates, the precleaved RNAs were derived from A6 pre-mRNA. To more precisely mimic an RNA that has been cleaved by endonuclease, the 3' fragment RNA (3'CL13pp or U5-3'CL) has a 5' monophosphate. Because a 3' OH at the end of the 3' fragment could serve as a substrate for ligation, it has been replaced with a 3' monophosphate.

Ligase activity is often very robust in mitochondrial extracts, and can lead to additional bands in both precleaved assays. The radiolabeled 5' RNA can be circularized or its 3' OH can sometimes ligate to the 5' end of the guide RNA. Fortunately, these additional ligation products do not interfere with the results of the editing assay, and are relatively minor in comparison to the editing bands.

3.1.1. Modifications to examine individual activities

Simple modifications to the precleaved assay protocol can be used to specifically examine individual enzymatic activities of the editosome. To examine U addition (TUTase) activity, the ATP is omitted from the precleaved insertion protocol to reduce ligase activity. Addition of Us to a single-strand RNA (ssRNA) template can also be assessed by using only 5' CL18 RNA in a reaction. To specifically examine U removal (exoUase) activity, the ATP is omitted from the precleaved deletion protocol. Some residual ligation remains, probably due to existing adenylated ligase. To specifically examine ligation, the gPCA6-2A gRNA is replaced with gPCA6-0A in the precleaved insertion protocol and UTP is omitted (Igo et al., 2000). The gA6PC-0A gRNA is a more efficient, nicked substrate for ligase.

3.1.2. Isolation of RNA editing samples

Similar to the protocol for full-round assays, the precleaved assays described below were originally developed to test mitochondrial extracts and fractions from 10–30% glycerol gradients. The details of multiple methods to extract RNA editing activity from cells are covered in depth elsewhere in this volume (see Chapter 1). The most common alternative to glycerol gradient fractions is editosomes isolated via TAP-tag. To assay TAP-tag isolated samples, use 10× HHE buffer without 5 mM DTT or 10 mM EDTA, and adjust the final volume of reaction to 30 μl using water.

3.2. Protocol

1. Both 5' and 3' RNA fragments used in both insertion and deletion are synthesized by Dharmacon (see Table 2.1 for the sequences). For precleaved insertion, the 5' RNA is 5'CL18 and the 3' RNA is 3'CL13pp'. For precleaved deletion, the 5' RNA is U5-5'CL and the 3' RNA is U5-3'CL. We order 2'-ACE stabilized RNA oligos, deprotected using the manufacturer's protocol, and gel purified prior to use.

2. The gRNAs for both precleaved insertion and deletion are T7 transcripts (see Table 2.1 for the sequences). Make DNA templates for T7 transcription using PCR or primer extension. For gPCA6-2A insertion gRNA, use primers *Eco*RI T7 and gA6PCiREV to PCR amplify the gPCA6-2A-Tmpl DNA template to make a T7 transcription template. Thermocycler conditions are 94° for 4 min, followed by three cycles of 94° for 30 sec, 30° for 1 min, and 72° for 30 sec, followed by 27 cycles of 94° for 30 sec, 40° for 1 min, and 72° for 30 sec, followed by a final 7 min at 72°. For gA6[14]PC-del deletion gRNA, anneal primers *Eco*RI T7 and A6comp1 and extend with *Taq* polymerase or Klenow fragment to make the T7 transcription template.

3. Generate gRNAs by T7 transcription. Allow transcription reagents to equilibrate to room temperature, then combine in order to create a 100 μl transcription reaction: 55 μl of PCR product (total of 2–5 μg RNA), 30 μl 25 mM (each) rNTPs, 10 μl 10× T7 transcription buffer (400 mM Tris–HCl, pH 7.6; 240 mM MgCl$_2$; 20 mM spermidine; 0.1% Triton X-100), 1 μl 1 M DTT, 1 μl 40 U/μl RNasin, and 3 μl of 80U/μl T7 RNA polymerase. Incubate the reaction overnight at 37°. Precipitation of inorganic phosphate is frequently observed following transcription. After transcription, add 5 μl of 1 U/μl RQ1 DNase (Promega) and incubate at 37° for 15 min. Add 300 μl of water, 40 μl of 3 M NaOAc, 1 μl of 20 mg/ml glycogen, and 1 ml of 95% ethanol; invert to mix. Centrifuge at 17,000×g for 15 min at 4°. Wash the pellet once with 75% ethanol and allow the RNA pellet to air dry. Resuspend the RNA in 40 μl of 7 M urea loading dye (7 M urea, 1× TBE, 0.05% bromophenol blue, 0.05% xylene cyanol) and heat for 1 min at 65°.

4. Purify the transcribed RNAs by electrophoresis on 9% polyacrylamide 7 M urea 1× TBE gel 40 cm in length. On 9% gel, bromophenol blue dye runs ~14 nt, while xylene cyanol runs ~65 nt. gPCA6 is 40 nt and gA6[14]PC-del is 33 nt. Purify synthetic RNAs by electrophoresis on 15% polyacrylamide 7 M urea 1× TBE gel 40 cm in length. On 15% gel, bromophenol blue dye runs ~10 nt, while xylene cyanol runs ~41 nt. U5-5′ is 22 nt, 5′CL18 is 18 nt, U5-3′pp is 15 nt, and 3′Cle13pp is 13 nt.

5. Place the gel between two sheets of plastic wrap and place it on a reflection screen. Visualize RNAs in gel by UV shadowing with a short-wave UV lamp, marking the bands to be cut out with a felt-tip pen. Cut out the bands with a razor blade or scalpel. Place the excised bands in a tube with 400 μl of elution buffer (0.3 M NaOAc, pH 5.2; 0.1% SDS; 1 mM EDTA) and elute overnight, rotating at room temperature.

6. Transfer the RNA to a fresh tube, add 1 μl of 20 mg/ml glycogen, and precipitate with 1 ml 95% ethanol. Centrifuge at 17,000×g for 15 min at 4° to pellet and resuspend in 20 μl water.

7. Label the 5′ RNA fragment using T4 kinase and $[\gamma\text{-}^{32}P]ATP$. Combine 1 μl of 20 pmol/μl 5′ RNA fragment with 6 μl of 10 μCi/μl $[\gamma\text{-}^{32}P]ATP$, 2 μl of 5× forward reaction buffer, and 1 μl of 10 U/μl T4 kinase (Invitrogen). Incubate for 1 h at 37°. Add 20 μl of 7 M urea loading dye.

8. Load the RNA on a 40-cm-long 9% polyacrylamide 7 M urea 1× TBE gel to purify ^{32}P-labeled RNA. After electrophoresis, cover one side of the gel with plastic wrap and expose the portion of the gel with radiolabeled RNA to X-ray film for ∼1 min. Develop the film to determine the location of the band and carefully cut out with a razor blade.

9. Place the excised bands in a tube and elute and precipitate as in step 6. Resuspend the RNA in 16 μl water. Assuming 80% of input RNA was labeled and recovered, the concentration is now 1 pmol/μl. Use a Geiger counter to estimate cpm in 1 μl. Use 10,000–40,000 cpm per editing assay.

10. For each editing assay plus two extra, combine 1 μl 2.5 pmol/μl guide RNA and 1 μl of 0.1–0.25 pmol/μl ^{32}P-labeled 5′ RNA fragment and 1 μl 1 pmol/μl 3′ RNA fragment and heat to 65° or 2 min. Allow to cool at room temperature for 5–10 min.

11. Assemble master mixes for editing assays, making enough for all reactions plus two extra. Each final editing reaction will be 30 μl:20 μl of master mix plus 10 μl of extract to be tested. The final concentration of KCl in each 30 μl reaction should be between 30 and 50 mM; for mitochondrial extracts that contain 100 mM KCl, no additional KCl needs to be added. Each 20 μl of insertion master mix contains 2 μl 10× HHE buffer [250 mM HEPES, pH 7.9; 100 mM Mg(OAc)$_2$, 5 mM DTT, 10 mM EDTA], 1.5 μl 2 mM UTP, 1.5 μl 200 μM ATP, 1 μl 100 mM CaCl$_2$, 1 μl 500 ng/μl *Torula* type VI RNA, 0.2 μl 40 U/μl RNasin, 9.8 μl of water, and 3 μl of combined ^{32}P substrate RNA/gRNA. Each 20 μl of deletion master mix contains 2 μl of 10× HHE buffer, 1.5 μl 200 μM ATP, 0.8 μl 100 mM CaCl$_2$, 1 μl 500 ng/μl *Torula* type VI RNA, 0.2 μl 40 U/μl RNasin, 11.5 μl of water, and 3 μl of combined ^{32}P substrate RNA/gRNA. The volume of water can be altered in each master mix to accommodate either larger or smaller samples of extract to be added in the following step.

12. Add 10 μl of mitochondrial extract (or 7 μl TAP-tag-purified complex; see overview for modifications).

13. Incubate in a heat block at 27° for 3 h.

14. Add 75 μl of a master mix containing 2 μl stop buffer (130 mM EDTA, 2.5% SDS), 1 μl 20 mg/ml glycogen, 10 μl 3 M NaOAc, 2 μl of 5 pmol/μl competitor RNA (5′ CL18 for insertion; U5-5′ for deletion), and 60 μl water per editing reaction. Competitor RNA does not need to be gel purified.

15. Add 100 μl of phenol:chloroform:isoamyl alcohol to the stopped editing reaction, mix by vortexing, centrifuge briefly to separate phases, and transfer the aqueous phase (top) to a fresh tube.

16. Precipitate RNA by addition of 250 μl 95% ethanol. Mix well by inversion, and incubate at −70° for 20 min or at −20° for at least 1 h.

17. Pellet RNA by centrifugation at 17,000×g for 20 min at 4°. Carefully remove the supernatant using a Geiger counter to confirm that the RNA is not accidentally removed. *Optional*: wash the RNA pellet with 150 μl 75% ethanol, recentrifuge, and remove the supernatant. Allow the RNA to air dry completely. Residual ethanol will distort subsequent gel electrophoresis.

18. Resuspend RNA in 10 μl of 7 M urea loading dye. Incubate at 100° for 2 min and cool to room temperature.

19. Load 5 μl of each editing reaction on 0.35-mm-thick 11% polyacrylamide 7 M urea 1× TBE gel for S2 sequencing electrophoresis apparatus (GibcoBRL). We use 32-well combs for best resolution.

20. Resolve editing reaction products by electrophoresis so that the bromophenol blue dye is 5–10 cm from the end of the gel.

21. Transfer the gel onto Whatman 3MM paper, cover with plastic wrap, and dry the gel on a gel-drying apparatus. Drying time depends on thickness and percent polyacrylamide. Dry a 0.3-mm-thick 11% gel at 80° for at least 1 h.

22. Expose the dried gel to a PhosphorImager (GE Healthcare) screen overnight, and scan the exposed screen to produce data.

4. TUTase Activity

4.1. Overview

Although the precleaved insertion assays previously discussed are frequently used to assess RNA editing TUTase activity, alternative TUTase assays are sometimes used. Multiple TUTases have been identified, and which TUTase is under examination plays a significant role in which assay is used to assess TUTase activity. The best characterized TUTases are RET1 and RET2: RET1 adds U tails to gRNAs, while RET2 is the editosome TUTase that adds Us during RNA editing (Aphasizhev *et al.*, 2002, 2003; Ernst *et al.*, 2003). RET2 primarily adds a single U to single-stranded RNA (ssRNA) substrates, or the number of Us specified by gRNA in double-strand RNA (dsRNA) editing substrates. In contrast, RET1 processively adds many Us to ssRNA, and works less well with dsRNA editing templates. Because RET1 is relatively inactive on dsRNA templates, an alternative TUTase assay in which the addition of radiolabeled UTP onto either yeast tRNA or an ssRNA primer can be used. While this assay can be used to assess RET2, the precleaved insertion assay is preferred, since the addition of a single U to ssRNA produces a much weaker signal than the incorporation of multiple Us by RET1.

In the TUTase assay described below, radiolabeled UTP is added to yeast tRNA substrate. The 10× HHE buffer used here assumes that the mitochondrial extract contains 100 mM KCl; the final concentration of KCl in the reaction should be 25–50 mM.

4.2. Protocol

1. Assemble a master mix at room temperature for enough reactions plus two extra. Each reaction requires 2 μl of 10× HHE buffer [250 mM HEPES, pH 7.9; 100 mM Mg(OAc)$_2$, 5 mM DTT, 10 mM EDTA], 1 μl of 1 mg/ml yeast tRNA (Sigma), 0.5 μl of [α-^{32}P]UTP (800 Ci/mmol), and 11.5 μl of nuclease-free water.
2. Add 5 μl of mitochondrial extract to 15 μl master mix and mix by pipetting.
3. Incubate reaction at 27° for 30–60 min.
4. Place 5 μl of reaction on a glass fiber disc (Whatman GF/C).
5. Wash each disc in 10 ml of 4° 10% trichloroacetic acid (TCA) containing 50 mM disodium pyrophosphate for 10 min. Repeat the wash two additional times.
6. Rinse each disc once with 4° 95% ethanol.
7. Let the disc dry and count using a scintillation counter.

5. LIGASE ACTIVITY

5.1. Overview

Direct assessment of ligase activity can be achieved using an adenylation assay, which is an efficient alternative to assessing ligase activity by modified precleaved assay. In this assay, mitochondrial extracts are incubated with radiolabeled ATP, and the radiolabel is covalently transferred onto either REL1 or REL2 RNA ligase by adenylation (Rusché et al., 2001; Sabatini and Hajduk, 1995; Schnaufer et al., 2001). The adenylated ligases are subsequently resolved by electrophoresis, and radiolabeled bands are visualized by exposure to a PhosphorImager screen. Although this assay assesses only the first step in catalysis for ligation, it is convenient and simple to perform.

Two RNA editing ligases are readily observed in this assay: REL1 at 52 kDa and REL2 at 48 kDa. Accurate quantification of adenylated REL1 or REL2 is not straightforward, since the amount of preadenylated ligase can vary, and ligase that is preadenylated will not be observed in this assay. To circumvent preadenylated ligase, the mitochondrial extract can be deadenylated prior to use in an adenylation assay (Cruz-Reyes *et al.*, 2002). However, the efficiency of deadenylation can also be quite variable. Therefore, quantification of adenylated REL1 and REL2 should be approached with caution. The 10× buffer used here assumes that the mitochondrial extract contains 100 mM KCl; the final concentration of KCl in the reaction should be 25–50 mM (Palazzo *et al.*, 2003).

5.2. Protocol

1. Assemble a master mix at room temperature for enough reactions plus two extra. Each reaction requires 2 μl of 10× buffer [125 mM HEPES, pH 7.9; 50 mM Mg(OAc)$_2$, 2.5 mM DTT], 0.5 μl of [α-^{32}P]ATP (3000 Ci/mmol), and 12.5 μl of nuclease-free water.
2. Add 5 μl of mitochondrial extract to 15 μl master mix and mix by pipetting.
3. Incubate the reaction at 27° for 15 min.
4. Add 7 μl of 3× SDS–PAGE dye [100 mM Tris–HCl, pH 6.8; 4% (w/v) SDS; 0.2% (w/v) bromophenol blue; 20% (v/v) glycerol, 200 mM DTT].
5. Place lid locks on the tubes, and boil the reactions for 5 min.
6. Centrifuge the tubes briefly at room temperature to collect liquid condensed on the lids after boiling.
7. Load 10 μl on 10% SDS–PAGE gel and run until the bromophenol blue dye reaches the end of the gel.
8. Transfer the gel onto Whatman 3MM paper, cover with plastic wrap, and dry the gel on a gel-drying apparatus at 80° for at least 1 h.
9. Expose the dried gel to a PhosphorImager (GE Healthcare) screen overnight and scan the exposed screen to produce data.

6. HELICASE ACTIVITY

6.1. Overview

Helicase activity is not directly observed in either full-round or precleaved editing assays, but it has been implicated in RNA editing *in vivo* (Panigrahi *et al.*, 2003). Helicase KREH1 is a component of the editosome, and other helicases may also play a role in RNA editing. Because RNA helicases function by unwinding dsRNA, their activity can be assessed by measuring the amount of dsRNA converted into ssRNA (Corell *et al.*, 1996; Missel and Göringer, 1994).

The assay for helicase activity involves the separation of A6–derived dsRNA that is equivalent to fully edited substrate RNA in duplex with cognate gRNA. The substrate RNA is radiolabeled, and separation of the radiolabeled strand from the gRNA alters migration on native PAGE to reveal helicase activity. The 10× buffer used here assumes that the mitochondrial extract contains 100 mM KCl; the final concentration of KCl in the reaction should be 25–100 mM.

6.2. Protocol

1. Generate RNAs for gA6[14] and the radiolabeled complementary fully-edited 3′ domain of A6 (3′A6 edited) as previously described (see Table 2.1 for the sequences). For this assay, 3′A6 edited RNA can be internally radiolabeled during transcription if desired.
2. For each reaction plus two extra, combine 1 μl of 0.5 pmol/μl gA6[14] RNA and 1 μl of 0.25 pmol/μl ^{32}P-labeled 3′A6 edited RNA. Heat to 65° for 2 min and then cool at room temperature for 5–10 min.
3. For each reaction, combine 2 μl of 10× buffer [60 mM HEPES, pH 7.9; 100 mM Mg(OAc)$_2$, 10 mM DTT], 2 μl of gRNA/substrate RNA mix, 2 μl of 10 mM ATP, 0.2 μl 40 U/μl RNasin, 8.8 ml of nuclease-free water, and 5 μl of mitochondrial extract. Incubate 30–60 min at 27°.
4. Put the tube on ice and stop the reaction by addition of 2 μl of 4% SDS and 2 μl 100 mM EDTA.
5. Add 6 μl of 6× native load dye [40% sucrose (w/v), 1× TBE, 0.25% bromophenol blue, 0.25% xylene cyanol].
6. Load 10 μl on nondenaturing 9% polyacrylamide (acrylamide-bisacrylamide, 19:1) 1× TBE 0.1% SDS gel and resolve by electrophoresis.
7. Transfer the gel onto Whatman 3MM paper, cover with plastic wrap, and dry the gel on a gel-drying apparatus at 80° for at least 1 h.
8. Expose the dried gel to a PhosphorImager (GE Healthcare) screen overnight and scan the exposed screen to produce data.

7. REAL-TIME PCR OF EDITING *IN VIVO*

7.1. Overview

Analysis of *in vivo* RNA editing by real-time PCR (Carnes *et al.*, 2005) has several distinct advantages over previous methods, such as poison primer extension (Lambert *et al.*, 1999; Wang *et al.*, 2003) and Northern blotting (Feagin *et al.*, 1987, 1988). It is very sensitive, can assay multiple targets simultaneously, requires much less RNA, and uses no radioactivity (Wong and Medrano, 2005). The protocol described here determines specific transcript abundance in an experimental sample (treated) relative to a control sample (untreated). Relative comparisons are simpler to perform, because absolute quantitation by real-time PCR requires generation of standard curves using known amounts of template for each target RNA, which is time-consuming when multiple targets are investigated.

7.1.1. Relative quantitation by real-time PCR

In real-time PCR, isolated cellular RNA is converted into cDNA by reverse transcription, and a 50- to 150-base-pair target sequence is amplified by PCR using specific primers. The progress of the amplification is measured in real time by fluorescence. RNA integrity is important and thus should be examined prior to cDNA synthesis (Fig. 2.3). During the logarithmic phase of PCR amplification, the amount of fluorescence signal is quantitatively proportional to the amount of cDNA template in the original sample. Amplification of an internal reference, such as β-tubulin or 18S rRNA, is used to normalize the data to control for unequal amounts of total RNA in the samples being compared.

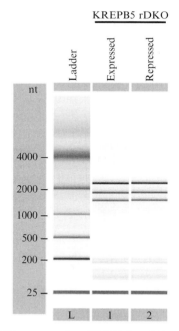

Figure 2.3 Analysis of DNase-treated total RNA from *T. brucei*. Total RNA was isolated from KREPB5 regulatable knockout bloodstream from *T. brucei* cells, in which KREPB5 was either expressed or repressed. These RNAs were treated with DNase I, and subsequently analyzed on an Agilent Technologies Bioanalyzer LabChip. For both RNA samples, ribosomal RNA bands are intact, and no degradation is apparent.

Real-time detection of the logarithmic phase of target cDNA amplification during PCR is used to extrapolate the relative amount of original template in the samples. The amount of amplified target DNA can be tracked either with fluorogenic oligonucleotides (e.g., Taqman probes) or with dyes that fluoresce only when they bind double-stranded DNA (SYBR green). Fluorogenic probes have the advantage of requiring annealing to the target sequence to generate fluorescence, which decreases potential false-positive artifacts. Fluorogenic probes can also be used to monitor multiple targets in the same reaction tube. However, specific fluorogenic probes are required for each intended target, which increases the expense and thus reduces the experimental flexibility. SYBR green dye, on the other hand, can be used to monitor the amplification of any target amplicon (amplified PCR product). This significantly reduces experimental costs, but has the disadvantage of potential false-positive signals generated from nonspecific amplification because any double-stranded DNA will fluoresce in the presence of SYBR green dye. Therefore, confirming amplicon identity is critical when performing real-time PCR using SYBR green.

7.1.2. Identifying PCR amplicons

The identity of the amplified DNA from a real-time PCR reaction can be assessed by various means. A thermal dissociation curve can be generated using the real-time PCR machine. Single or multiple peaks will reveal if the amplified DNA contains one sequence or mixed sequences, respectively, but does not indicate whether a single sequence is the intended one. Gel electrophoretic analysis can indicate if the PCR amplicons are the expected size, and if restriction endonuclease digestion results in fragments with the expected size. However, the DNA amplified with each primer pair must be sequenced to be confident that it represents amplification of the intended target. Amplicons generated by the primer pairs in Table 2.2 have been so analyzed and their specificity confirmed. Special care must be taken when designing primers to ensure specificity to the target, and to adhere to the requirements of the uniform cycling conditions of real-time PCR.

7.1.3. Real-time PCR primer design

Primer Express 2.0 software (Applied Biosystems) is useful for designing primers for real-time PCR analysis of both edited and pre-edited targets, but some primers must be designed manually. For example, this software frequently cannot find suitable primer sequences for the unusually U-rich sequences in some edited mRNAs. Primer pairs for several edited, pre-edited, and never-edited targets, as well as internal control primers, are shown in Table 2.2. Since RNA editing of mRNAs progresses 3′ to 5′, the primers for edited targets anneal at the 5′ end of the edited portion of the RNA, to restrict amplification to completely edited RNAs. By the same logic, primers for pre-edited RNAs anneal at the 3′ end of the region to be

Table 2.2 Primers for real-time PCR with sizes of amplicons in base pairs (bp)

Target amplicon	Forward primer	Reverse primer	Amplicon size (bp)
ND7pre	GCGGGCGGAGCATTATT	GATCTACGGTCCCCTCTTTCCT	68
ND7edit	GCATCCCGCAGCACATG	CTGTACCACGATGCAAATAACCTATAAT	101
COIIIpre	GAAACCAGATGAGATTGTTTGCA	TTCATTCCAACTAAACCCTTTCC	153
COIIIedit	TTGTGTTTTATTACGTTGTATCCAGTATTG	CGAAAGCAAACTCACACACAAA	128
CYbpre	ATATAAAAGCGGAGAAAAAAGAAAG	CCCATATATTCTATATAAACAACCTGACA	64
CYbedit	AAATATGTTTCGTTGTAGATTTTTATTATTT	CCCATATATTCTATATAAACAACCTGACA	94
A6pre	TTGCCTTTGCCAAACTTTTAGAAG	ATTCTATAACTCCAAATCACAACTTTCC	90
A6edit	GATTTATTTTGGTTGCGTTTGTTATTATG	CAAACCAACAAACAAATACAAATCAAAC	144
COIIpre	ATTACAGTGTAACCATGTATTGACATT	TTCATTACACCTACCAGGTTCTCT	135
COIIedit	ATTACAGTGTAACCATGTATTGACATT	ATTTCATTTACACCTACCAGGTATACAA	141
Murf2pre	GATTTTAAGATTGGCTTTGATTGA	AATATAAAATCTAGATCAAACCATCACA	97
Murf2edit	GATTTTAATGTTTGGTTGTTTAATTTAG	AATATAAAATCTAGATCAAACCATCACA	126
CR4pre	GGGTTTAGAAGAGGACGAAATTGA	CCACACTTTCTCCCCAACTAAAA	67
CR4edit	GTGTTTGTGTTTTATGTACAGTTTATGG	AACATACCCACACAAACAAACAACA	84
ND3pre	GAATGGGAGATGGGTTTTGG	AACAAATCTCTTTACCCCCTTCAG	72
ND3edit	TGTTTTCGTTGTTGTTGTGGTT	CAATGTATAAAACACCAAACGTGAATT	72
RPS12pre	CGACGGAGAGCTTCTTTTGAATA	CCCCCACCCAAATCTTT	68
RPS12edit	CGTATGTGATTTTTGTATGGTGTTG	ACACGTCGGTTACCGGAACT	98
COI	CCCGATATGGTATTTCCTCGTATAAA	CCCCATACCCTCTTCAGTCA	103
ND4	CAATCTGACCATTCCATGTGTGA	TTTCAGCACAATACTTGCTAATAAAAACA	89
18S rRNA	CGGAATGGCACCACAAGAC	TGGTAAAGTTCCCGTGTTGA	64
β-tubulin	TTCCGCACCCTGAAACTGA	TGACGCCGGACAACAACAG	76

edited, to restrict amplification to pre-edited RNAs. The primers shown in Table 2.2 are not designed to detect the significant population of *in vivo* RNA that is partially .edited namely, edited in the 3' region but not completely edited in the 5' region. Primers can be designed to detect partially edited RNAs if desired.

7.1.4. The importance of internal references

A critical issue for relative comparisons in real-time PCR is the internal reference RNA. Ideally, an internal reference RNA would be expressed equally in all cells, and would not be altered by the experimental treatment under investigation. In addition, the amount of this internal reference RNA should be similar to that of the target RNAs to be measured. Because no such ideal RNA exists, at least two independent internal references are used to decrease the possibility of artifacts arising from unforeseen changes in internal reference RNA levels due to the experimental treatment. Obtaining similar results with both internal references strongly supports the validity of the data. Various "housekeeping" RNAs are typically selected as internal references since their expression is presumed to be unlikely to change. Such RNAs are often expressed at much higher levels than the RNAs under investigation, as is the case with the β-tubulin and 18S rRNA internal reference RNAs that we routinely use.

7.1.5. Essential PCR controls

Elimination of genomic DNA from RNA samples prior to cDNA synthesis is critical to ensure that the subsequent real-time PCR amplification signal reflects cellular RNA levels. Thus, *T. brucei* RNA samples are subjected to rigorous DNase I treatment to eliminate any contaminating genomic DNA. Inclusion of a control to assess for potential contamination in each RNA sample is essential. This control requires a mock cDNA synthesis reaction that lacks reverse transcriptase, which is then used as a template in one well of the assay plate for each primer pair. This control well can also detect other artifacts, such as contaminating plasmid or PCR product, that can compromise the integrity of the assay. Contamination by a previously generated PCR template is a major concern. Thus, strict separation of pre-PCR setup and post-PCR analyses is strongly suggested.

7.1.6. Performing real-time PCR calculations

The real-time PCR data require analysis to determine the relative RNA levels between the treated and untreated samples. Only the general concepts of data analysis are summarized here since many real-time PCR systems contain specific embedded software for data analysis. The raw PCR amplification data are typically plotted on a graph with the PCR cycle number on the *x*-axis and the amount fluorescence signal representing the PCR product on the *y*-axis (Fig. 2.4). An arbitrary threshold line is placed parallel to

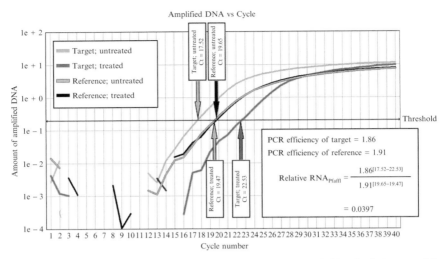

Figure 2.4 Sample real-time PCR amplification plot and Pfaffl calculation. In this plot, the amount of amplified DNA (double-stranded DNA bound by SYBR green dye) detected by a fluorescence signal increases as the number of PCR cycles increases. In this example, two primer pairs (target and reference) are used to amplify either cDNA from a treated source (experimental) or an untreated source (control). Amplification by target primers (e.g., ND7 edited) on the untreated sample (e.g., KREPB5 expressed) is in light gray, while amplification on the treated sample (e.g., KREPB5 repressed) is in dark gray. Amplification by reference primers (e.g., 18S rRNA) on the untreated sample (e.g., KREPB5 expressed) is in gray with black outline, while amplification on the treated sample (e.g., KREPB5 repressed) is in black. The threshold is set so that all amplification plots cross during the exponential phase of the PCR reaction. The point where the amount of amplified DNA crosses the threshold (arrows) is the Ct value. Linear regression of the exponential phase of the PCR reaction is used to calculate PCR efficiency, which is subsequently used to calculate the relative amount of RNA in the treated sample versus the untreated sample. The calculation here reveals that the treated sample has only ~4% of the target RNA found in the untreated sample.

the x-axis such that it crosses the logarithmic phase of each amplification plot. The value at the point at which each amplification curve intersects the threshold line is referred to as its cycle threshold (Ct) or crossing point (CP). The simplest method for analyzing real-time PCR data is the $\Delta\Delta$Ct approximation method [Eq. (2.1)]. Only four Ct values are needed to perform $\Delta\Delta$Ct calculations. This method assumes that the PCR product is perfectly doubled after each cycle.

$$\text{Relative RNA}_{\Delta\Delta\text{Ct}} = 2^{([A-B]-[C-D])} \qquad \begin{aligned} A &= Ct_{\text{untreated}} \text{ of target} \\ B &= Ct_{\text{untreated}} \text{ of reference} \\ C &= Ct_{\text{treated}} \text{ of target} \\ D &= Ct_{\text{treated}} \text{ of reference} \end{aligned} \qquad (1)$$

$$\text{Relative RNA}_{\text{Pfaffl}} = \frac{X^{[E-F]}}{Y^{[G-H]}} \quad \begin{aligned} E &= Ct_{\text{untreated}} \text{ of target} \\ F &= Ct_{\text{treated}} \text{ of target} \\ G &= Ct_{\text{untreated}} \text{ of reference} \\ H &= Ct_{\text{treated}} \text{ of reference} \\ X &= \text{PCR efficiency of target} \\ Y &= \text{PCR efficiency of reference} \end{aligned} \quad (2)$$

Since most reactions fail to achieve perfect doubling at each cycle, many researchers prefer to analyze real-time PCR data using the method of Pfaffl (2001). Unlike the $\Delta\Delta Ct$ method, this method uses a calculated PCR efficiency for each primer pair [Eq. (2)]. The PCR efficiency is experimentally determined by creating a standard curve for each primer pair with a dilution series of known amounts of template. This can be burdensome, particularly if multiple primer pairs are to be tested. Alternatively, PCR efficiency can be determined by performing linear regression analysis on the amplification data collected from each real-time PCR reaction. The LinRegPCR application can calculate PCR efficiencies on real-time PCR data for use in Pfaffl calculations (Ramakers et al., 2003). Accurate calculation of the PCR efficiencies requires that the linear regression is fit to the logarithmic phase of amplification. The PCR efficiencies are then used in the calculations of the amounts of RNA in the experimental (treated) relative to the control (untreated) samples. For both the $\Delta\Delta Ct$ and Pfaffl methods, results are expressed such that greater than 1 indicates more RNA in the experimental sample compared to the control, less than 1 indicates less, and 1 indicates no difference. Log-scale bar graphs provide an intuitive way to plot real-time PCR data, since bars representing relatively more RNA in the experimental sample go up, and those representing relatively less go down.

7.1.7. Example of RNA editing analysis by real-time PCR

The effects of in vivo repression of KREPB5 expression on edited, pre-edited, and never edited RNAs are shown as an example of a real-time PCR experiment on RNA editing. In this experiment, RNA was isolated form the bloodstream from T. brucei cells that have had both endogenous alleles of the editosome gene KREPB5 eliminated by homologous recombination and a tetracycline–regulatable ectopic KREPB5 allele integrated into the rDNA locus (Wang et al., 2003). Our laboratory has previously shown that repression of KREPB5 expression for 72 h results in disruption and loss of editosomes, and a dramatic reduction in edited RNA in vivo as shown by poison primer extension (Wang et al., 2003). For real-time PCR, the experimental (treated) RNA sample comes from cells in which KREPB5 expression was repressed for 72 h, and the control (untreated) RNA comes from cells in which KREPB5 was expressed. Isolated cellular RNA was treated with DNase I to eliminate any contaminating genomic DNA.

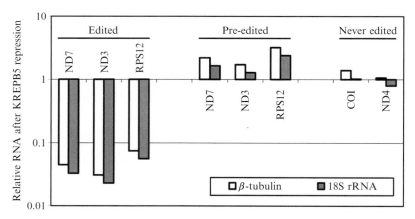

Figure 2.5 Real-time PCR analysis of editing *in vivo* after repression of KREPB5. Repression of KREPB5 in the bloodstream from *T. brucei* cells results in a significant loss of edited RNAs and an increase of pre-edited RNAs relative to internal controls, while never edited RNAs are relatively unchanged. Two internal controls are used, β-tubulin (left bars, in white) and 18S rRNA (right bars, in gray).

Figure 2.3 illustrates the integrity of the DNase I-treated RNAs as revealed by the intact rRNAs upon analysis with a Bioanalyzer (Agilent Technologies) using an RNA 6000 Nano LabChip. Such an assessment of RNA integrity is required for real-time PCR prior to cDNA synthesis due to the lability of RNA *in vitro*. The real-time PCR amplification curves in Fig. 2.4 reveal less edited ND7 mRNA (Ct 22.53) in cells in which KREPB5 was repressed than in those in which KREPB5 was expressed (Ct 17.52), while the amounts of the reference are similar in cells in which KREPB5 was repressed or expressed (Ct 19.47 and 19.65, respectively). Analysis of the data for ND7 and other edited and unedited RNAs by the Pfaffl method and normalization of the data to the 18S rRNA and β-tubulin RNAs allow representation of the relative abundance of several RNAs (Fig. 2.5). This shows that repression of KREPB5 expression resulted in a dramatic reduction in the amounts of edited ND7, ND3, and RPS12 RNAs, a modest increase in the corresponding pre-edited RNAs, and essentially no change in the never edited COI and ND4 RNAs.

The protocol that follows is designed for use with the ABI Prism 7000 system; however, the general concepts and primer pairs should be readily adaptable to other real-time PCR systems.

7.2. Protocol

1. Isolate total RNA from ∼1 × 10^8 *T. brucei* cells. We use 5 ml of Trizol (Invitrogen) for 1 to 3 × 10^8 *T. brucei* cells, which typically yields 45–135 μg of total RNA. Care should be taken to avoid accidental

collection of material at the interface of the aqueous and organic phases, as this may contribute to genomic DNA contamination of the RNA. Precipitated RNA from Trizol purification is resuspended in 100 μl of nuclease-free water and stored at $-80°$ when not in use.

2. Determine the total RNA concentration with a spectrophotometer.

3. Eliminate contaminating genomic DNA from 10 μg of total RNA by DNase I treatment. We use DNAfree kit (Ambion). We use the rigorous DNase treatment protocol, with 1 μl of DNase I in a 50 μl volume for 30 min at 37°, followed by addition of another 1 μl of DNase I and 30 more minutes at 37°. After addition of 10 μl of inactivation reagent, centrifuge for 2 min at 10,000×g, and transfer the supernatant to a new tube and precipitate the RNA with 1 μl of 20 mg/ml glycogen, 0.1 volumes of 3 M NaOAc, and 2.5 volumes of ethanol. Resuspend the RNA in 10 μl of water.

4. RNA quality should be assessed to make sure that RNA has not been degraded. We use a Bioanalyzer (Agilent Technologies). Add 0.3 μl of DNase-treated RNA to 2.7 μl water, and load 1 μl on an RNA 6000 Nano LabChip. Analysis of high-quality *T. brucei* total RNA should show three distinct rRNA bands at ~2.2 kb (18S small subunit), ~1.8 kb (28S large subunit α fragment), and ~1.5 kb (28S large subunit β fragment) (Fig. 2.3).

5. Make cDNA by reverse transcription of total RNA using random hexamer priming. We use Taqman Reverse Transcription Reagents (Applied Biosystems). Our 30 μl cDNA synthesis reaction contains 4.5 μl of 1 μg/μl DNase-treated RNA, 6 μl of 25 mM MgCl$_2$, 3 μl of 10× reaction buffer, 12 μl of 10 mM dNTPs, 1.5 μl of 50 μM random hexamers, 1.5 μl of RNasin (200U/μl), and 1.5 μl of reverse transcriptase (500 U/μl). Less RNA can also be used; however, lower abundance RNAs may become difficult to detect. A duplicate reaction in which the reverse transcriptase is replaced by water should be made to serve as a negative control in every experiment. Reactions are incubated at 25° for 10 min, then 42° for 20 min, and finally at 98° for 5 min.

6. Add 170 μl H$_2$O to 30 μl cDNA synthesis reaction to make a starting template dilution. Take 2 μl from this starting template dilution and add to 98 μl water (1:50 dilution) to be used as a template cDNA for internal controls.

7. Make a 1.5 μM primer mix consisting of both forward and reverse primers, with each at a final concentration of 0.75 μM.

8. Each 25 μl real-time PCR reaction consists of 10 μl 1.5 μM primer mix, 2.5 μl cDNA, and 12.5 μl 2× SYBR Green PCR Master Mix (Applied Biosystems). Add 10 μl primer mix to each well of a 96-well plate. Then add 15 μl of master mix containing both cDNA and SYBR Green PCR mix to the appropriate wells. Make the master mix for at least two more wells than will be needed. We typically test each pair of primers and

cDNA in triplicate, with a single well for the minus reverse transcriptase negative control. For internal reference real-time PCR reactions with β-tubulin or 18S rRNA primers, we use a 1:50 dilution of cDNA so that the Ct values for these reactions are closer to the Ct values for less abundant editing transcripts.

9. Place the 96-well plate in the ABI Prism 7000 PCR machine. Change the reaction volume from 50 μl to 25 μl, and check the box for the dissociation protocol. Use the default thermal cycler protocol conditions of 50° for 2 min, then 95° for 10 min, followed by 40 cycles of 95° for 15 sec and 60° for 1 min.

10. Analyze the real-time PCR data by either the $\Delta\Delta$Ct or Pfaffl method.

ACKNOWLEDGMENTS

This work was supported by NIH Grants AI 14102 and GM 42188. The authors would like to thank Nancy Lewis Ernst for helpful discussions concerning the manuscript.

REFERENCES

Aphasizhev, R., Sbicego, S., Peris, M., Jang, S. H., Aphasizheva, I., Simpson, A. M., Rivlin, A., and Simpson, L. (2002). Trypanosome mitochondrial 3′ terminal uridylyl transferase (TUTase): The key enzyme in U-insertion/deletion RNA editing. *Cell* **108**, 637–648.

Aphasizhev, R., Aphasizheva, I., and Simpson, L. (2003). A tale of two TUTases. *Proc. Natl. Acad. Sci. USA* **100**, 10617–10622.

Carnes, J., Trotter, J. R., Ernst, N. L., Steinberg, A. G., and Stuart, K. (2005). An essential RNase III insertion editing endonuclease in *Trypanosoma brucei*. *Proc. Natl. Acad. Sci. USA* **102**, 16614–16619.

Cifuentes-Rojas, C., Halbig, K., Sacharidou, A., De Nova-Ocampo, M., and Cruz-Reyes, J. (2005). Minimal pre-mRNA substrates with natural and converted sites for full-round U insertion and U deletion RNA editing in trypanosomes. *Nucleic Acids Res.* **33**, 6610–6620.

Corell, R. A., Read, L. K., Riley, G. R., Nellissery, J. K., Allen, T. E., Kable, M. L., Wachal, M. D., Seiwert, S. D., Myler, P. J., and Stuart, K. D. (1996). Complexes from *Trypanosoma brucei* that exhibit deletion editing and other editing-associated properties. *Mol. Cell. Biol.* **16**, 1410–1418.

Cruz-Reyes, J., Rusché, L. N., Piller, K. J., and Sollner-Webb, B. (1998). *T. brucei* RNA editing: Adenosine nucleotides inversely affect U-deletion and U-insertion reactions at mRNA cleavage. *Mol. Cell* **1**, 401–409.

Cruz-Reyes, J., Zhelonkina, A., Rusché, L., and Sollner-Webb, B. (2001). Trypanosome RNA editing: Simple guide RNA features enhance U deletion 100-fold. *Mol. Cell. Biol.* **21**, 884–892.

Cruz-Reyes, J., Zhelonkina, A. G., Huang, C. E., and Sollner-Webb, B. (2002). Distinct functions of two RNA ligases in active *Trypanosoma brucei* RNA editing complexes. *Mol. Cell. Biol.* **22**, 4652–4660.

Ernst, N. L., Panicucci, B., Igo, R. P., Jr., Panigrahi, A. K., Salavati, R., and Stuart, K. (2003). TbMP57 is a 3' terminal uridylyl transferase (TUTase) of the *Trypanosoma brucei* editosome. *Mol. Cell* **11**, 1525–1536.

Feagin, J. E., Jasmer, D. P., and Stuart, K. (1987). Developmentally regulated addition of nucleotides within apocytochrome b transcripts in *Trypanosoma brucei*. *Cell* **49**, 337–345.

Feagin, J. E., Abraham, J. M., and Stuart, K. (1988). Extensive editing of the cytochrome c oxidase III transcript in *Trypanosoma brucei*. *Cell* **53**, 413–422.

Igo, R. P., Jr., Palazzo, S. S., Burgess, M. L. K., Panigrahi, A. K., and Stuart, K. (2000). Uridylate addition and RNA ligation contribute to the specificity of kinteoplastid insertion RNA editing. *Mol. Cell. Biol.* **20**, 8447–8457.

Igo, R. P., Jr., Weston, D. S., Ernst, N. L., Panigrahi, A. K., Salavati, R., and Stuart, K. (2002). Role of uridylate-specific exoribonuclease activity in *Trypanosoma brucei* RNA editing. *Eukaryot. Cell* **1**, 112–118.

Kable, M. L., Seiwert, S. D., Heidmann, S., and Stuart, K. (1996). RNA editing: A mechanism for gRNA-specified uridylate insertion into precursor mRNA. *Science* **273**, 1189–1195.

Lambert, L., Muller, U. F., Souza, A. E., and Goringer, H. U. (1999). The involvement of gRNA-binding protein gBP21 in RNA editing—an *in vitro* and *in vivo* analysis. *Nucleic Acids Res.* **27**, 1429–1436.

Missel, A., and Göringer, H. U. (1994). *Trypanosoma brucei* mitochondria contain RNA helicase activity. *Nucleic Acids Res.* **22**, 4050–4056.

Palazzo, S. S., Panigrahi, A. K., Igo, R. P., Jr., Salavati, R., and Stuart, K. (2003). Kinetoplastid RNA editing ligases: Complex association, characterization, and substrate requirements. *Mol. Biochem. Parasitol.* **127**, 161–167.

Panigrahi, A. K., Schnaufer, A., Ernst, N. L., Wang, B., Carmean, N., Salavati, R., and Stuart, K. (2003). Identification of novel components of *Trypanosoma brucei* editosomes. *RNA* **9**, 484–492.

Panigrahi, A. K., Ernst, N. L., Domingo, G. J., Fleck, M., Salavati, R., and Stuart, K. D. (2006). Compositionally and functionally distinct editosomes in *Trypanosoma brucei*. *RNA* **12**, 1038–1049.

Pfaffl, M. W. (2001). A new mathematical model for relative quantification in real-time RT-PCR. *Nucleic Acids Res.* **29**, e45.

Piller, K. J., Decker, C. J., Rusché, L. N., and Sollner-Webb, B. (1995). *Trypanosoma brucei* mitochondrial guide RNA-mRNA chimera-forming activity cofractionates with an editing-domain-specific endonuclease and RNA ligase and is mimicked by heterologous nuclease and RNA ligase. *Mol. Cell. Biol.* **15**, 2925–2932.

Piller, K. J., Rusché, L. N., Cruz-Reyes, J., and Sollner-Webb, B. (1997). Resolution of the RNA editing gRNA-directed endonuclease from two other endonucleases of *Trypanosoma brucei* mitochondria. *RNA* **3**, 279–290.

Ramakers, C., Ruijter, J. M., Deprez, R. H. L., and Moorman, A. F. M. (2003). Assumption-free analysis of quantitative real-time polymerase chain reaction (PCR) data. *Neurosci. Lett.* **339**, 62–66.

Rusché, L. N., Huang, C. E., Piller, K. J., Hemann, M., Wirtz, E., and Sollner-Webb, B. (2001). The two RNA ligases of the *Trypanosoma brucei* RNA editing complex: Cloning the essential Band IV gene and identifying the Band V gene. *Mol. Cell. Biol.* **21**, 979–989.

Sabatini, R., and Hajduk, S. L. (1995). RNA ligase and its involvement in guide RNA/mRNA chimera formation. Evidence for a cleavage-ligation mechanism of *Trypanosoma brucei* mRNA editing. *J. Biol. Chem.* **270**, 7233–7240.

Schnaufer, A., Panigrahi, A. K., Panicucci, B., Igo, R. P., Jr., Salavati, R., and Stuart, K. (2001). An RNA ligase essential for RNA editing and survival of the bloodstream form of *Trypanosoma brucei*. *Science* **291**, 2159–2162.

Seiwert, S. D., Heidmann, S., and Stuart, K. (1996). Direct visualization of uridylate deletion *in vitro* suggests a mechanism for kinetoplastid RNA editing. *Cell* **84,** 831–841.

Stuart, K., Salavati, R., Igo, R. P., Jr., Ernst, N. L., Palazzo, S. S., and Wang, B. (2004). *In vitro* assays for kinetoplastid U insertion-deletion editing and associated activities. *In* "Methods in Molecular Biology" (J. M. Gott, ed.), Vol. 265, pp. 251–272. RNA Interference, Editing, and Modification: Methods and Protocols. Humana Press, Inc., Totowa, NJ.

Stuart, K. D., Schnaufer, A., Ernst, N. L., and Panigrahi, A. K. (2005). Complex management: RNA editing in trypanosomes. *Trends Biochem. Sci.* **30,** 97–105.

Trotter, J. R., Ernst, N. L., Carnes, J., Panicucci, B., and Stuart, K. (2005). A deletion site editing endonuclease in *Trypanosoma brucei. Mol. Cell* **20,** 403–412.

Wang, B., Ernst, N. L., Palazzo, S. S., Panigrahi, A. K., Salavati, R., and Stuart, K. (2003). TbMP44 is essential for RNA editing and structural integrity of the editosome in *Trypanosoma brucei. Eukaryot. Cell* **2,** 578–587.

Wong, M. L., and Medrano, J. F. (2005). Real-time PCR for mRNA quantitation. *Biotechniques* **39,** 75–85.

CHAPTER THREE

RNA EDITING URIDYLYLTRANSFERASES OF TRYPANOSOMATIDS

Ruslan Aphasizhev *and* Inna Aphasizheva

Contents

Abstract

Terminal RNA uridylyltransferases (TUTases) catalyze the transfer of UMP residues to the 3′ hydroxyl group of RNA. These enzymes belong to the DNA polymerase β superfamily, which also includes poly(A) polymerases, CCA-adding enzymes, and other nucleotidyltransferases. Studies of uridylyl insertion/deletion RNA editing in mitochondria of trypanosomatids provided the first examples of

Department of Microbiology and Molecular Genetics, University of California, Irvine, California

Methods in Enzymology, Volume 424
ISSN 0076-6879, DOI: 10.1016/S0076-6879(07)24003-0

biological functions for TUTases: posttranscriptional uridylylation of guide RNAs by *RNA editing TUTase 1* (RET1) and U-insertion mRNA editing by *RNA editing TUTase 2* (RET2). The editing TUTases are unified by the presence of conserved catalytic and nucleotide base recognition domains, yet differ substantially in auxiliary function-specific domains, quaternary structure, RNA substrate specificity, and processivity. This chapter describes isolation of TUTases and their complexes from trypanosomatids, methods used for analysis of interactions involving RET1 and RET2, purification of recombinant proteins, and enzyme kinetic assays.

1. INTRODUCTION

Terminal uridylyltransferases (TUTases) add UMP residues to the $3'$ hydroxyl group of RNA in a template-independent polymerization reaction. These activities have been described in mammalian cells, plants, and trypanosomatids (Aphasizhev, 2005). Previous work on protein complexes involved in uridine insertion/deletion RNA editing in mitochondria of trypanosomes identified two TUTases (Simpson *et al.*, 2003; Stuart *et al.*, 2005). The RNA Editing TUTase 1, or RET1 (Aphasizhev *et al.*, 2002), was implicated in the posttranscriptional addition of the nonencoded $3'$ oligo (U) tail to guide RNAs (Aphasizhev *et al.*, 2003c). In addition, a possible role for RET1 in mRNA turnover has been suggested (Ryan and Read, 2005). The RNA Editing TUTase 2, or RET2 (Aphasizhev *et al.*, 2003a; Ernst *et al.*, 2003), was shown to be responsible for the major U-insertion mRNA editing activity in mitochondria of *Trypanosoma brucei* (Aphasizhev *et al.*, 2003c). Editing TUTases vary substantially in polypeptide size, RNA substrate specificity, processivity, and quaternary structure. RET1 from *Leishmania tarentolae* is a homotetramer of 121-kDa subunits whereas RET2 is a 57-kDa integral component of the core editing complex, the 20S editosome (Aphasizhev *et al.*, 2002, 2003c; Ernst *et al.*, 2003). Immunochemical and fractionation analysis of RET1 interactions revealed its participation in several high-molecular-weight complexes ranging from 10 to 30S, including RNA-mediated interactions with MRP1/2 complex of RNA binding proteins (Aphasizhev *et al.*, 2003b) and the 20S editosome (Aphasizhev *et al.*, 2002). Proteomic analysis of RET1 interactions remains the main challenge in assessing the plurality of this enzyme's functions in mitochondria of trypanosomatids. Conversely, RET2 has been found exclusively in the 20S editosome. Within the editosome, RET2 interacts with a defined partner, an MP81 structural protein that possesses two zinc-finger motifs (Schnaufer *et al.*, 2003). Furthermore, depletion of the TbRET2 by RNAi led to a loss of the U-insertion subcomplex, which indicates a requirement for assembly of the U-insertion subcomplex. Remarkably, the rest of the core editing complex, including U–deletion activity, remained unaffected (Aphasizhev *et al.*, 2003c).

The presence of a conserved N-terminal domain, which contains three invariant carboxylate amino acid residues, classifies TUTases as members of the nucleotidyltransferase (NT) superfamily typified by DNA polymerase β (Pol β) (Holm and Sander, 1995). The triad of aspartate coordinates phosphates and divalent metals, which are essential for catalysis. The UTP specificity *in vitro* is nearly absolute in RET2-catalyzed reactions, whereas RET1 has been found to add nucleosides other than UMP, albeit with low efficiency. The polymerization efficiency of editing enzymes also differs dramatically. RET1 acts preferentially on a single-stranded RNA substrate and is highly processive, while RET2 is equally active on a double-stranded and single-stranded RNA and is distributive. These distinctions may reflect the presumed RNA specificities of editing TUTases *in vivo*: poorly structured ones guide RNAs for RET1 and a double-stranded region formed by annealing of the mRNA cleavage fragment to the antisense guides RNA for RET2.

Recent crystallographic studies of RET2 (Deng *et al.*, 2005) and nonediting trypanosomal TUTase 4 (Stagno *et al.*, 2006) with bound UTP revealed a distinct domain organization previously unseen in NTs. The N-terminal catalytic domain (NTD) forms extensive surface contact with the C-terminal nucleotide base recognition module (CTD), essentially creating a compact globular bidomain. The presence of a CTD may be used to define a subfamily of enzymes within the Pol β superfamily, which includes animal Gld-2 type cytoplasmic poly(A) polymerases, yeast Trf4/5 nuclear poly (A) polymerase, archaeal CCA-adding enzymes, and 2′-5′-oligoadenylate synthases. Additional domains may be inserted between NTD and CTD, as reported for RET1 (Aphasizhev, 2005) and RET2 (Aphasizhev *et al.*, 2003a; Deng *et al.*, 2005). The RET2 structure reveals that such an insertion may fold into a compact domain (Deng *et al.*, 2005). Other functional motifs, such as zinc fingers and oligomerization regions, have been located at the N- and C-termini of RET1, respectively (Aphasizheva *et al.*, 2004). Human U6 snRNA-specific TUTase (Trippe *et al.*, 2006) shares the overall organization of the core bidomain with trypanosomal TUTases, but also possesses an insertion in a different location within NTD (Stagno *et al.*, 2006), as well as the RNA binding domain.

A deep cleft formed at the interface of the NTD and CTD bears the UTP binding site. The phosphate, sugar, and uracil base moieties of the bound UTP are coordinated via water-mediated and direct hydrogen bonds, and stacking interactions with amino acid side chains. The UTP specificity of RET2 and TbTUT4 is determined primarily by the two carboxylic residues, which coordinate a water molecule to accept a proton from the endocyclic nitrogen N(3) of the uracil base (Deng *et al.*, 2005; Stagno *et al.*, 2006). In addition, a critical π-stacking interaction of the aromatic tyrosine residue and the pyrimidine ring contributes to UTP binding. Remarkably, the amino acids involved in UTP binding are highly conserved among four currently known trypanosomal TUTases, which provides a strong rationale for exploring the UTP binding site as a target for trypanocide development. Conversely, the available crystallographic data give little insight on the structural basis of RNA specificity displayed by TUTases, in particular RET2's ability to discriminate a mismatch of the mRNA cleavage fragment with the guide RNA, which is probably one of the major factors facilitating the overall fidelity of U–insertion editing reaction (Igo *et al.*, 2002).

Apparently, some editing steps may be reconstituted by mixing recombinant enzymes with RNA substrates (Kang *et al.*, 2006), which establishes individual enzymes as the main carriers of specificity, but brings into question the functional role of complex formation. Unlike multiple endonucleases, exonucleases, and RNA ligases, RET2 is the only enzyme that does not have homologs in the 20S editosome. Therefore, the availability of the editosome-embedded and individual RET2 proteins provides an attractive system for comparative kinetic studies.

2. PURIFICATION OF THE RNA EDITING TUTASE 1 FROM MITOCHONDRIA OF *LEISHMANIA TARENTOLAE*

2.1. Overview

Purification of a homogeneous RET1 from an enriched mitochondrial fraction is accomplished by a series of chromatographic steps that remove most of the interacting proteins and RNA molecules. Depending on the experimental objective, it may not be necessary to purify a homogeneous polypeptide. For example, the fraction obtained by size exclusion chromatography typically does not contain nucleases or other editing activities, but retains approximately 3-fold more of RET1 than the phenyl Sepharose-purified material. RET1 is a low abundance protein, but its robust enzymatic activity and relative stability make purification feasible. Because several polypeptides of similar molecular mass copurify with RET1 until the last step (hydrophobic interactions chromatography), all fractions should be screened by activity assays rather than SDS–polyacrylamide gel. The RET2 activity has not been found to interfere with RET1 assays. *Leishmania tarentolae* remains the organism of choice for this purpose due to moderate costs of cultivation and the availability of a relatively straightforward method for mitochondria purification (this volume, Chapter 4). In addition to isolation of the RET1 homotetramer, the branching purification scheme may be used to isolate a stable ∼700-kDa complex of RET1 with several unidentified components termed TUTII (Aphasizhev *et al.*, 2002). For anticipated purification yield, recovery, and protein profile at each step; refer to Aphasizhev *et al.* (2002). All chromatography runs should be performed on a programmable protein purification work-station capable of generating precise gradients, e.g., Akta platform (GE Healthcare). The uninterrupted workflow is recommended for the poly [U] Sepharose and anion-exchange chromatography steps.

2.2. Methods

2.2.1. RET1 uridylyltransferase activity assays

2.2.1.1. *Filter assay* The RET1 enzymatic activity during purification is measured via incorporation of an [α-^{32}P]UMP into synthetic RNA. This substrate should be at least 15 nucleotides long with six or more uridylyl residues at the 3' end. Stable secondary structures in RNA may inhibit RET1 activity. Typically, 1 μl of protein fraction, or 10–20 fmol of the purified protein, is added to a 20 μl reaction mixture containing 50 mM Tris–HCl (pH 8.0), 10 mM MgoAc, 1 mM dithiothreitol (DTT), 0.1 mM [α-^{32}P]UTP (1000–2000 cpm/pmol), and 0.5 μM RNA primer. The reaction is incubated at 27° for 10–20 min and is terminated with 15 μl of 0.5 M sodium phosphate, 0.5% sodium dodecyl sulfate (SDS). The entire mix is spotted on DEAE filters (DE81, Whatman) and washed three times for 15 min with 0.5 M sodium phosphate (100 ml for 10 filters). Filters are rinsed with 95% ethanol, dried, and quantified by liquid scintillation counting. Note that exceeding 30–50 mM monovalent salt in the reaction will inhibit uridylyltransferase reaction.

2.2.1.2. *Acrylamide/urea gel assay* This method is highly sensitive but more time-consuming. The synthetic RNA, e.g., 6[U] RNA (GCUAUGUCUGUCA ACUUGUUUUUU), must be gel purified and 5' labeled with T4 polynucleotide kinase in the presence of [γ-^{32}P]ATP. The purified protein (10–20 fmol), or 0.5 μl of the fraction, is added to a 10 μl reaction containing 50 nM of the radioactive RNA and 100 μM of UTP. Buffer conditions are the same as for the filter assay. Samples are incubated at 27° for 30 min and the reaction is stopped by 20 μl of 95% formamide, 0.05% bromophenol blue, 0.05% xylene cyanol FF, and 10 mM EDTA solution. Reaction products are analyzed on a 15% acrylamide/urea gel. Typical NTP incorporation patterns are shown on Fig. 3.1C.

2.2.2. Extract preparation

The enriched mitochondrial fraction (40 g wet weight, ~2.2 g of protein) is defrosted on ice and resuspended in 120 ml (final volume) of SA buffer [50 mM HEPES (pH 7.5), 5 mM MgCl$_2$, 50 mM KCl, 1 mM DTT, and 2 mM CHAPS (3-[(3-cholamidopropyl) dimethylammonio]-1-propanesul-fonate, Roche)]. Three tablets of Complete Protease Inhibitors (Roche) are

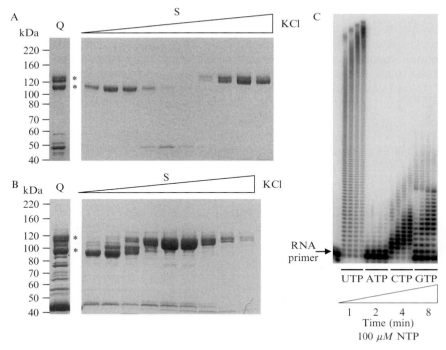

Figure 3.1 Purification of the recombinant RET1. Expression of *L. tarentolae* (A) and *T. brucei* (B) RET1 in *E. coli* produces the full-length and truncated forms (shown by asterisks), which become apparent after the anion-exchange chromatography step (Q). These forms can be separated by cation-exchange chromatography (S). The origin of the truncation, which occurs at the C-terminus, is unknown. Fractions were separated on 8–16% polyacrylamide–SDS gel and stained with Coomassie blue R250. (C) The radiolabeled 6[U] RNA was incubated with purified LtRET1 for the indicated time periods in the presence of ribonucleoside triphosphates and the products were separated on 15% acrylamide urea gel. (See color insert.)

added and the extract is sonicated in a plastic beaker (large tip, three bursts for 30 sec at 36 W). The temperature during sonication should not exceed 10–15°. The extract is passed through a French pressure cell at 12,000 psi and clarified by centrifugation for 1 h at 200,000×g. The pellet is reextracted in 100 ml (final volume) of SA buffer with additional sonication (six bursts for 30 sec at 36 W) and centrifugation is repeated. Both supernatants are pooled and filtered through a 0.45-μm low-protein-binding filter. Expect to recover ~200 ml of clarified extract.

2.2.3. Affinity chromatography on poly[U] sepharose

The first chromatography step should be performed immediately after extract preparation; for time considerations, it may be programmed to run overnight. The clarified mitochondrial extract is loaded at 2 ml/min on a freshly prepared poly[U] Sepharose 4B (GE Healthcare). The 20 ml (bed volume) of resin may be packed into a 26/20 XK column (GE Healthcare) or comparable. The column is washed with 100 ml of SA buffer, or until a stable UV reading is achieved. Five-milliliter fractions are collected from this point on. The concentration of KCl in SA buffer is linearly increased from 50 to 250 mM over three column volumes, and the column is washed with 250 mM KCl. It is essential to continue the 250 mM wash until OD_{280} stabilizes, which may take up to 150–200 ml. The flow rate is reduced to 1 ml/min and the RET1 is eluted with a 250–500 mM gradient of KCl in five column volumes. Active fractions ($\sim$20 ml, eluted at 300–400 mM KCl) may be supplemented with glycerol to 10% and flash-frozen in liquid nitrogen, although immediately proceeding with anion-exchange chromatography is recommended.

2.2.4. Anion-exchange chromatography

Fractions recovered from poly[U] Sepharose are desalted at 5 ml/min with the HiPrep 26/20 desalting column (GE Healthcare) into a QA buffer [20 mM Tris–HCl (pH 8.0), 50 mM KCl, 10% glycerol, and 1 mM DTT] and immediately loaded onto a Poros 20 HQ 46/100 column (Roche) at 2 ml/min. Other anion-exchange resins, such as Sepharose Q HP (GE Healthcare), may be used, but we advise against ubiquitous MonoQ columns (GE Healthcare) for this particular application. The Poros column is developed with a 50–500 mM linear gradient of KCl in 25 column volumes of QA buffer and 0.5-ml fractions are collected. Two peaks of activity are observed. The sharp peak of RET1 eluting at $\sim$200 mM KCl (TUTI) typically contains over 80% of the total UTP-incorporating activity. These fractions may be flash-frozen without loss of activity and subjected to further purification to obtain the homogeneous RET1 fraction. The second broad peak of activity eluting between 300 and 400 mM KCl contains the RET1 complexes (TUTII) (Aphasizhev et al., 2002). Notably, this peak also includes a significant amount of the 20S editosome, which may be separated from the TUTII by sedimentation on a glycerol gradient (below).

2.2.5. Size fractionation

The TUTI fractions ($\sim$3 ml) are concentrated with Microcon YM30 (Millipore) or another high-recovery concentrator device to 200 μl and loaded at 0.1 ml/min on Superose 12 and Superose 6 columns (GE Healthcare) joined sequentially and equilibrated with 50 mM HEPES (pH 7.5), 300 mM KCl, 5% glycerol, and 1 mM DTT. Fractions of 300 μl are collected and analyzed for activity and protein composition; RET1 is typically eluted in the first peak of optical density. Silver or Sypro Ruby (Invitrogen) staining may be required to visualize protein bands. The RET1 migrates on SDS–polyacrylamide gel at $\sim$140 kDa, often appearing as a doublet.

2.2.6. Hydrophobic interactions chromatography

A 4 M ammonium sulfate solution is added to the combined fractions from the size fractionation step (~3 ml) to a final concentration of 0.5 M and loaded onto a 5/5 Phenyl Superose column (GE Healthcare). The Phenyl Superose should be preequilibrated with 50 mM HEPES, pH (7.5), 0.7 M ammonium sulfate, 0.1 mM EDTA. The column is developed with 15 ml of a reverse gradient from 0.7 to 0 M of ammonium sulfate at 0.2 ml/min and 0.3-ml fractions are collected. RET1-containing fractions are detected by SDS–gel electrophoresis. Fractions are pooled and dialyzed in Slide-a-Lyzer cassettes (Pierce) against 50 mM HEPES (pH 7.5), 0.1 mM EDTA, 1 mM DTT, 50% glycerol for 2 h. Protein is stable at $-20°$ for several months. Use of filter-based concentration devices may result in significant protein loss and is not recommended.

2.2.7. Isolation of the TUTII complex

The ~700-kDa TUTII particle represents the RET1 complex, which can withstand the high salt conditions of affinity and anion-exchange chromatography. It is separated from the unassociated RET1 protein by anion-exchange chromatography but requires further purification for proteomics or kinetic analysis. Size fractionation by gel filtration chromatography has been found to result in low protein yields, which leaves sedimentation on the glycerol gradient as the preferred alternative. The second broad peak of RET1 activity (TUTII) eluting from an anion-exchange column at 300–400 mM is concentrated by reverse dialysis against the Slide-A-Lyzer Concentrating Solution in the Slide-a-Lyzer cassettes (Pierce) to ~300 μl and loaded on a 10–30% glycerol gradient as described elsewhere in this volume (see Chapter 4). Changes to the protocol include maintaining potassium chloride concentration at 100 mM and collecting 0.5-ml fractions.

3. EXPRESSION AND PURIFICATION OF RECOMBINANT RET1

3.1. Overview

Enzymatically active recombinant RET1 protein may be isolated from *E. coli*, but it is important to note that both expression and purification yields for the full-size RET1 are much lower than typically observed with bacterial systems. Significantly higher expression levels may be obtained with truncated RET1 polypeptides (Aphasizheva *et al.*, 2004). Six-histidine affinity tag at the C-terminus is only marginally effective while larger tags or those placed at the N-terminus are virtually dysfunctional. The pET series of vectors from Novagen, e.g., pET29a, are a reasonable starting point to construct a cassette for expression of the C-terminally 6 His-tagged RET1 fusion protein. The purification may be monitored by filter activity assay or Western blotting analysis with Penta-His monoclonal antibodies (Qiagen). The use of *E. coli* strain BL 21(DE3) Codon Plus RIL (Stratagene), or other strains bearing additional tRNAs for expression of genes with G-rich codons, is essential for RET1 expression. Purification protocols differ between RET1 from *L. tarentolae* (LtRET1) and *T. brucei* (TbRET1), therefore, both methods are described.

3.2. Purification of the recombinant LtRET1

Four liters of bacterial culture are grown in 2YT media in the presence of appropriate antibiotic at $37°$ to 0.3 OD_{600} and the temperature is lowered to $20°$ over 30–60 min. Protein expression is induced with 1 mM IPTG and cultivation is continued for 3 h. The *E. coli* cell pellet ($\sim$15 g) is resuspended in 100 ml of 50 mM HEPES buffer (pH 7.5), 50 mM NaCl, 1 mM DTT, and lysed in the presence of 0.1 mg/ml of lysozyme and 50 U/ml of DNase I (Sigma) for 30 min on ice. The cell extract is passed through a French pressure cell at 12,000 psi and centrifuged at 200,000×g for 30 min. The supernatant is filtered through a low-protein-binding 0.45-μm filter and loaded onto a 15-ml column prepacked with Sepharose S Fast Flow cation-exchange resin (GE Healthcare). DTT should be avoided in chromatography buffers for compatibility with metal affinity chromatography. The column is washed with 100 mM NaCl until the UV reading is stable and the RET1 is eluted with 50 mM HEPES (pH 8.0), 300 mM NaCl. The Sepharose S fraction is immediately loaded onto a 3-ml Talon metal affinity column (Clontech), washed with 50 ml of 50 mM HEPES (pH 8.0), 300 mM NaCl, and eluted with the same buffer containing 50 mM imidazole. A single protein peak is recovered in 4–5 ml. The fraction may be kept at $-20°$ for several days after addition of glycerol to 40%.

The anion-exchange chromatography is performed on a Mono Q 5/5 column (GE Healthcare) as follows: the sample in 40% glycerol is diluted 3-fold with 50 mM Tris (pH 8.0), 1 mM DTT, 0.1 mM EDTA and loaded onto Mono Q 5/5 column at 0.5 ml/min. The column is developed with a 20 ml gradient from 50 to 500 mM KCl. Fractions containing RET1 are desalted into 50 mM HEPES (pH 7.5), 50 mM KCl, 0.1 mM EDTA, 1 mM DTT using gravity-flow PD10 column (GE Healthcare), and loaded onto a Mono S 5/5 column (GE Healthcare). A linear gradient of potassium chloride (20 column volumes, 50–400 mM) separates the full length and truncated forms of RET1 (Fig. 3.1A). The obtained protein is typically $\sim$95% pure as estimated by Sypro Ruby staining of the SDS–acrylamide gel. The final fraction is supplemented with glycerol to 10%, flash frozen in liquid nitrogen, and stored at $-80°$. The expected yield is 0.2–0.5 mg/liter of bacterial culture.

3.3. Purification of the recombinant TbRET1

The cell growth and lysis are performed as for LtRET1. After passage through the French pressure cell step, the potassium chloride is adjusted to 300 mM and the extract is centrifuged at $200,000 \times g$ for 30 min. The pH of the extract should be adjusted to 7.5–8.0 with 0.1 M NaOH. It is important to maintain the concentration of potassium chloride above 80 mM during purification, as protein tends to precipitate under low-salt conditions. The clarified extract is loaded at 1 ml/min onto a 5-ml column prepacked with Talon metal affinity resin (Clontech) in 50 mM HEPES, pH (7.8), 300 mM KCl. The column is washed with 75 ml of the loading buffer and TbRET1 is eluted with 20 mM Tris (pH 7.5), 300 mM NaCl, 150 mM imidazole. The eluted material may be used for further purification or stored at −20° for up to 1 week after addition of glycerol to 40%. For anion-exchange chromatography, the 40% glycerol fraction is diluted 2-fold with 50 mM Tris–HCl (pH 8.0) and loaded at 1 ml/min onto a 5-ml HiTrap Q column (GE Healthcare), which has been preequilibrated with 50 mM Tris–HCl (pH 8), 100 mM KCl, 1 mM DTT, 0.1 mM EDTA. A 75 ml gradient from 100 to 500 mM KCl is applied to the column and 1-ml fractions are collected. Fractions containing TbRET1 elute at ~300 mM salt. For the final purification step, the combined fraction is diluted 3-fold or desalted against 50 mM HEPES (pH 7.5), 10% glycerol, 100 mM KCl, 0.1 mM EDTA, 1 mM DTT using a PD 10 column, and loaded onto a 1-ml MonoS 5/5 column (GE Healthcare). The 50 to 400 mM gradient of KCl in 50 mM HEPES (pH 7.5) 10% glycerol, 0.1 mM EDTA, 1 mM DTT is applied over 20 column volumes. The full-length TbRET1-containing fractions eluting at ~300 mM KCl (Fig. 3.1B) may be flash-frozen, or supplemented with glycerol to 50% and stored at −20°. The expected yield is approximately 0.5 mg/liter of bacterial culture.

4. ISOLATION OF THE EDITOSOME-EMBEDDED AND RECOMBINANT RNA EDITING TUTASE 2

4.1. Overview

The RNA Editing TUTase 2 has been identified via proteomic analysis of the purified core RNA editing complex, the 20S editosome (Aphasizhev *et al.*, 2003a; Panigrahi *et al.*, 2003). The uridylyltransferase activity of the individual 57-kDa polypeptide has been demonstrated upon transcription/translation of the corresponding gene in reticulocyte lysate (Ernst *et al.*, 2003) and by expression and purification of RET2 lacking a mitochondria importation signal in *L. tarentolae* (Aphasizhev *et al.*, 2003c). Expression and purification of active and folded protein from bacteria allowed determination of the high-resolution structure of the TbRET2–UTP complex (Deng *et al.*, 2005).

Conventional (Ernst *et al.*, 2003) and affinity (Aphasizhev *et al.*, 2003c) purification studies established that in the mitochondria RET2 exists exclusively in the editosome-bound form. The RNA editing TUTase 2 is a nonprocessive enzyme and transfers UMP residue to both single-stranded and double-stranded RNAs, although the presumed structure of the editing intermediate argues for the dsRNA to be a more authentic substrate. The approaches described below are intended to provide the researcher with sources of individual recombinant RET2 and the RET2-containing core editing complex, the 20S editosome.

5. Affinity Isolation of RET2-Containing Complex from *T. brucei*

The trypanosomal vectors designed for inducible, high-level protein expression in procyclic *T. brucei*, such as pLew79 in the 29–13 cell line (Wirtz *et al.*, 1999), appear to be the current system of choice for expression of the editosome components. Significant overexpression of the TbRET2 fused with the affinity TAP tag (Puig *et al.*, 2001) at the C-terminus evidently does not present a problem of mistargeting or aberrant complex formation (Aphasizhev *et al.*, unpublished observations). The unincorporated polypeptide is probably degraded, which controls the RET2 protein level by the amount of the available 20S editosome. In practice, this translates into isolation of the 20S editosome bearing the tagged version of the RET2 in near stoichiometric amounts to other complex components (Fig. 3.2A). Trace amounts of the endogenous RET2 are also detectable in the affinity-purified complex. Clonal and nonclonal cell lines have been used with equal success. If copurifying RNA is important for further analysis, the concentration of monovalent salt may be lowered to 60 mM and EDTA may be replaced with 2 mM MgCl$_2$ during all purification steps. Any additional treatments intended to assess complex stability, e.g., exposure to high salt, detergent, or RNase, may be performed while the complex is bound to IgG beads. Finally, aliquots should be collected at each step for Western blotting analysis in order to evaluate binding efficiency and recovery at each step. Limited amounts of antigen-purified polyclonal antibodies against TbRET2 are available from the authors' laboratory.

5.1. Cell growth

Cell density should be kept between 2 and 20 × 10^6/ml during the culture build up. *Day 1*: Cells are grown in 50 ml of SDM79 (this volume, Chapter 4), 10% serum, 50 μg/ml of Geneticin (G418), 50 μg/ml of hygromycin, 2.5 μg/ml of phleomycin to 10^7 cells/ml and diluted to 200 ml (~2 × 10^6 cells/ml). *Day 2*: Cells are diluted to 400 ml (~2 × 10^6 cells/ml). *Day 3*: Protein expression is induced with 1 μg/ml

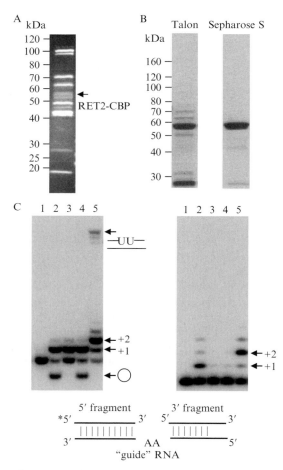

Figure 3.2 Purification of the 20S editosome-associated and recombinant RET2. (A) Protein profile of the affinity isolated RET2 complex. The fraction obtained from the calmodulin column (10 μl) was separated on a 10–20% gradient acrylamide gel and stained with Sypro Ruby (Invitrogen). The position of the RET2 fused with a calmodulin binding peptide (CBP) at the C-terminus is indicated by an arrow. (B) Purification of the recombinant RET2. Fractions after the metal affinity step (Talon) and cation-exchange chromatography (Sepharose S) were separated on 8–16% polyacrylamide–SDS gel and stained with a Coomassie blue R250. (C) Precleaved insertion activity of the affinity isolated RET2 complex (left panel) and recombinant RET2 enzyme (right panel). Schematic representation of the RNA substrate is shown underneath. Migration positions of the extended and ligated, extended by +1 and +2 uridylyl residues, and the-circularized 5' cleavage fragment are indicated by arrows. (1) Control RNA with no protein added; (2) 5' fragment; (3) 5' fragment plus "guide" RNA; (4) 5' fragment plus 3' fragment only; (5) fully assembled substrate. Note +2 additions occurring with fully assembled substrate and 5' fragment circularization in the absence of the antisense "guide" RNA. (See color insert.)

tetracycline at $\sim$4 $\times$ 10^6 cells/ml. *Day 4*: Cell suspension, which typically reaches $\sim$15–20 $\times$ 10^6 cells/ml, is diluted to 4 liters total ($\sim$2 $\times$ 10^6 cells/ml) while maintaining tetracycline at 1 μg/ml. At this point, the phleomycin concentration can be decreased to 1 μg/ml, and Geneticin (G418) and hygromycin can be omitted in order to minimize costs. *Day 6*: Cells should be at $\sim$15 $\times$ 10^6 cells/ml. *Day 7*: Cells are collected at $\sim$25–30 $\times$ 10^6 cells/ml in a 1 liter fixed-angle rotor at 5000$\times$g for 10 min. The cell pellet is gently washed with 100 ml of cold phosphate-buffered saline (pH 7.6) and centrifuged at 3000$\times$g for 10 min, flash-frozen in liquid nitrogen, and stored at $-80°$.

5.2. Solutions for TAP purification

Extraction Buffer. 50 mM Tris–HCl (pH 8.0), 150 mM KCl, 2 mM EDTA. *IgG binding buffer (IBB):* 25 mM Tris–HCl (pH 7.6), 150 mM KCl, 1 mM EDTA, 0.1% NP-40. *TEV cleavage buffer:* IgG binding plus 1 mM DTT. *Calmodulin binding buffer (CBB):* 25 mM Tris–HCl (pH 7.6), 150 mM KCl, 0.1% NP-40, 10 mM 2-mercaptoethanol, 1 mM magnesium acetate, 1 mM imidazole, 2 mM CaCl$_2$. *Calmodulin elution buffer 1* (CEB1, for activity assays): 25 mM Tris–HCl (pH 7.6), 150 mM KCl, 3 mM EGTA, 1 mM EDTA, 0.1% NP40. The NP40 is replaced with 3 mM CHAPS if planning to concentrate the final fraction by reverse dialysis. *Calmodulin elution buffer 2* (CEB2, for concentration with StrataClean Beads prior to preparative SDS gel): 25 mM Tris–HCl (pH 7.6), 100 mM KCl, 3 mM EGTA, 1 mM EDTA.

5.3. Purification

Day 1. All steps are performed at $4°$. Cell paste ($\sim$5 g, wet weight) is resuspended in 30 ml of extraction buffer (final volume); NP40 is added to 0.3% plus one tablet of Complete protease inhibitor (Roche). The extract is prepared by passing the cell suspension through a French pressure cell at 12,000 psi. Approximately 10 ml of the extraction buffer lacking any detergent may be used to rinse the pressure cell. Sonication is an alternative lysis method if a pressure cell is not available, but partial degradation of the complex may occur. The final concentration of NP40 in the extract should not exceed 0.2%. The pH should be $\sim$7.5 and may be adjusted with 1 M Tris–HCL (PH8.0). The extract is centrifuged at 200,000$\times$g for 45 min and filtered through a syringe-driven 0.45-μm low protein binding filter. The expected recovery is $\sim$35 ml. The IgG Sepharose resin (GE Healthcare) is prepared during centrifugation as follows: transfer 0.3 ml of suspension (packed volume $\sim$0.2 ml) into a 50-ml Falcon tube and wash twice with 50 ml of IBB. Centrifugation at 1000$\times$g for 2 min with brakes off may be used to collect the resin. The extract is incubated with the resin for 1 h with gentle mixing and transferred into a 1-ml disposable column with an extender (Pierce). The flow-through is reloaded twice. The column with extender is washed once with IBB (full volume). The extender is then removed and the column is washed sequentially with

30 ml of IBB and 10 ml of TEV cleavage buffer, drained, and plugged. Full resuspension of the resin with each wash is recommended. Protease cleavage solution is prepared by adding 150 U of TEV protease (Invitrogen) and 15 μl of fresh Complete protease inhibitor (Roche) stock solution (one tablet dissolved in 1 ml of water) to 1.5 ml of TEV cleavage buffer. The protease solution is added to the IgG resin. The upper end of the column is sealed with Parafilm M and the column is incubated overnight at $+4°$ with gentle agitation.

Day 2. To improve protein recovery at the calmodulin chromatography step, pretreatment of the 1-ml disposable column (Pierce) and 15-ml conical tube with 2% Tween solution is recommended. Traces of the detergent are removed by extensive rinsing in water. The calmodulin resin suspension (0.25 ml, Stratagene) is transferred into the pretreated 15-ml plastic tube and washed twice with CBB. The IgG column is drained directly into the 15-ml plastic tube with calmodulin resin. The IgG column is further washed with 4.5 ml of CBB and the wash is collected in the tube with calmodulin. Expect to recover ~6 ml. Calcium chloride is added to supplement the EDTA present in solution (4 μl of 1 M stock). The calmodulin suspension is incubated for 1 h on a Nutator mixer and transferred to the 1-ml pretreated disposable column (Pierce). The flow-through is collected into the same 15-ml tube and reloaded three times. The column is washed with 30 ml of CBB and closed with a bottom cap. The elution is performed by sequential application of 0.5 ml of the CEB. Allow a 10-min incubation at $4°$ for each elution; periodic gentle mixing is optional. Use of high-recovery 1.5-ml microcentrifuge tubes is recommended. Typically, fractions 1–3, which contain most of the material, are supplemented with glycerol to 10% and frozen in 100-μl aliquots in liquid nitrogen and stored at $-80°$. The protein profile of the calmodulin fraction is shown in Fig. 3.2A.

Protein separation by SDS–polyacrylamide gel is often required for subsequent mass spectrometry analysis. For this application, the CEB2 buffer lacking any detergent is used for elution from the calmodulin column. Protein-containing fractions are reduced to 1 ml on a vacuum concentrator and incubated with 10 μl of Strataclean beads (Stratagene) for 30 min at room temperature. Beads are pelleted by centrifugation at $3000 \times g$ for 5 min and incubated with 15 μl of 1.5$\times$ SDS loading buffer for 3 min at $95°$. The entire mixture is transferred into 8–16% gradient polyacrylamide–SDS gel and separated by standard methods.

6. EXPRESSION AND PURIFICATION OF THE RECOMBINANT TbRET2

The RET2 gene lacking 21 codons at the 5' end, which correspond to the predicted mitochondria importation signal, was cloned into pET15b+ vector (Novagen) to generate N-terminal 6 His-RET2 fusion protein. The transformed *E. coli* strain [BL 21(DE3) Codon Plus RIL (Stratagene)] is grown in 5 liters of 2xYT media with 0.1 mg/liter of ampicillin as described for LtRET1, except that the IPTG induction and subsequent cultivation for 3 h are performed at 10°. The bacterial cell paste (~20 g wet weight) is resuspended in 120 ml of lysis buffer [50 mM HEPES (pH 8.0), 50 mM NaCl] and incubated in the presence of 0.2 mg/ml of lysozyme and 50 U/ml of DNase I (EMD) for 30 min on ice. The cell extract is passed through a French pressure cell at 12,000 psi. Sodium chloride and Triton X-100 are then added to 300 mM and 0.1%, respectively. The extract is centrifuged at 200,000×g for 30 min, filtered through a 0.45-μm low protein binding filter (Nalgene), and loaded onto a 3-ml Talon affinity column preequilibrated with Talon buffer [50 mM HEPES (pH 8.0), 300 mM NaCl]. The column is washed sequentially with 60 ml of the Talon buffer and 30 ml of Talon buffer plus 10 mM imidazole. The protein is eluted with 200 mM imidazole in Talon buffer. The RET2 fraction obtained by metal affinity chromatography typically requires further purification (Fig. 3.2B). The material eluted from the Talon column (~4 ml) is diluted 6-fold with 25 mM HEPES (pH 7.5), 1 mM DTT, 2 mM CHAPS, and 0.1 mM EDTA buffer to achieve a 50 mM concentration of KCl. The sample is loaded immediately onto a 1-ml HiTrap Sepharose S column (GE Healthcare) preequilibrated with 25 mM HEPES, pH (7.5), 1 mM DTT, 0.1 mM EDTA, 50 mM KCl. The column is developed with a 15 ml gradient from 50 to 400 mM KCl at 1 ml/min. The protein-containing fractions (Fig. 3.2B) are supplemented with glycerol to 10% or 50% for storage at −80° or −20°, respectively. The expected yield is approximately 1 mg of protein per liter of bacterial culture. The presence of the N-terminal 6 His affinity tag has not been found to affect enzyme activity. The recombinant RET2 has a solubility limit of ~2 mg/ml and remains fully active at −20° in 50% glycerol up to 3 months. The RET2 mutations that increased solubility and allowed protein crystallization are described in Deng *et al.* (2005).

7. Uridylyltransferase Activity Assays for the 20S Editosome and Recombinant RET2

The exact structure of the RET2's RNA substrate *in vivo* has yet to be determined, but most likely represents the mRNA fragments scaffolded by antisense guide RNA. It has been demonstrated that the partially purified 20S editosome prefers a stable double-stranded RNA as substrate and inserts uridines according to the number of guiding purines. Figure 3.2C depicts RNA substrates for this type of editing assay, which was termed by the authors as precleaved insertion editing (Igo *et al.*, 2000). The recombinant RET2 is active on both single-stranded and double-stranded RNA substrates (Aphasizhev *et al.*, 2003c; Deng *et al.*, 2005; Ernst *et al.*, 2003). Remarkably, uridylyltransferase reactions catalyzed by the purified 20S editosome and the recombinant RET2 deliver virtually identical products (Fig. 3.2C). In the presence of a single-stranded RNA, predominantly one uridylyl residue is added (Fig. 3.2C, lane 2), whereas two nucleosides are added to fill in the gap between cleavage fragments as specified by the number of guiding nucleotides (Fig. 3.2C, lane 5). The RNA ligase activities in the 20S editosome also catalyze ligation of the cleavage fragments and circularization of the single-stranded RNA.

The RNA substrates listed below represent arbitrary sequences that, upon annealing, will form a precleaved substrate depicted on Fig. 3.2C. Chemical synthesis of all RNAs is recommended while maintaining a dideoxynucleotide at the 3′ end of the 3′ fragment and +2 guide RNA is optional. This modification prevents uridylyl incorporation outside of the editing site. Any RNA containing A or G at the 3′ end may be used as a single-stranded RNA substrate. Reduced levels of activity are observed with RNAs bearing pyrimidines at the 3′ end.

5′ fragment: CGACUACACGAUAAAUAUAAAAAG
3′ fragment: AACAUUAUGCUUCUUCddG
+2 gRNA: AAGAAGCAUAAUGUU**AA**CUUUUUAUAUUUAUC-
GUGUA- GUCddG

Guiding nucleotides are in bold. Alternative RNA substrates for the precleaved insertion assay are described in Igo *et al.* (2000, 2002). RNAs are purified to a single-nucleotide resolution by electrophoresis in 15% polyacrylamide/8 M urea gel and excised under UV shadow. A minimal gel length of 40 cm is recommended. RNA is eluted into 0.1 M sodium acetate, pH 5.0, 0.1% SDS, 5% phenol, 1 mM EDTA for 10–15 h at +4°, extracted with phenol/chloroform in Phase-Lock tubes (Eppendorf), and ethanol precipitated. The 5′ fragment is labeled with T4 polynucleotide kinase in the presence of [γ-^{32}P]ATP assuming 0.5 pmol of RNA per reaction. Because RNA annealing proceeds more efficiently at higher RNA concentrations, it is advisable to hybridize RNAs for multiple reactions in the smallest possible volume. For example, 5 pmol of the radiolabeled 5′ fragment, 10 pmol of gRNA, and 20 pmol of 3′-fragment are mixed in 10 μl of 10 mM HEPES (pH 8.0), 50 mM KCl (final concentrations) and heated to 80° for 2 min. Gradual cooling to room temperature over 30 min using a PCR cycler with heated block cover and adjustable ramping is recommended.

Precleaved reactions are performed in 20 μl of 50 mM HEPES (pH 8.0), 10 mM MgCl$_2$, 50 mM KCl, 1 mM DTT, 100 μM UTP buffer. The assembled RNA substrate (1 μl) is added to the reaction buffer and incubated at 27° for 10 min. The reaction is started by addition of protein and incubated at 27° for 1 h. For the affinity-purified 20S editosome, the optimal volume of the calmodulin fraction should be determined empirically. The recommended concentration of the recombinant RET2 is 25–50 nM. At any rate, the final concentration of the KCl in the reaction should not exceed 60–70 mM. If visualization of the ligation activity is necessary, ATP may be added to a final concentration of 100 μM after the uridylyltransferase step and incubation continued for 30 min more. The reaction is terminated by adding 50 μl of 0.5 M sodium acetate, pH 5.0, 10 mM EDTA, 0.1% SDS, and 2 μg of glycogen (Ambion). RNA is extracted with phenol-chloroform, precipitated with ethanol, and resuspended in 5 μl of 95% formamide, 0.05% bromophenol blue, 0.05% xylene cyanol FF, and 10 mM EDTA. The sample is denatured by heating at 80° for 2 min and 1–2 μl analyzed on 15% acrylamide/urea gel. For a single-stranded RNA substrate, no pretreatment is necessary. The 5′ labeled RNA is added directly to the reaction mix to 100 nM. The 5′ cleavage fragment used in the precleaved assay and the 6[A] substrate 5′-GCUAUGUCUG-CUAACUUGAAAAAA-3′ (Aphasizhev et al., 2002) have been found to work equally well.

Kinetic analysis of the RET2-catalyzed reaction cannot be performed by filter-based assay due to the enzyme's low catalytic efficiency and processivity. Instead, application of the high-resolution denaturing acrylamide urea gel is required. A 40 cm/0.4 mm 15% gel is recommended. The specific activity of the RNA substrate is calculated based on the known amount of RNA in the lane and signal intensity units generated by image acquisition software. The molar amount of extended RNA directly translates into the amount of incorporated UMP, which is then used to calculate initial velocities at various UTP concentrations. In order to measure steady-state catalytic parameters for UTP incorporation, not more than 10% of the input RNA should be converted into extension products.

8. *In Vitro* Analysis of Protein–Protein Interactions Involving TbRET2

The protein of interest is typically cloned into the bacterial expression vector under control of the T7 RNA polymerase promoter. Depending on the available antibodies, immunoaffinity tags may be added. Such a construct is then used to synthesize a radiolabeled polypeptide in the TNT Quick Coupled Transcription/Translation System (Promega). Typically, 40 μl of TNT Quick Master Mix is combined with 1 μg of supercoiled plasmid DNA and 2 μl of [^{35}S]methionine (1000 Ci/mmol) in the 50 μl reaction. Upon incubation for 90 min at 30°, 1 pmol of the recombinant RET2 purified from *E. coli* is added to the reaction and the incubation is continued for 15 min. Alternatively, RET2 may be synthesized in a separate

TNT setup and two reactions combined for 15 min. Synthesis of two proteins in the same reaction mix is not recommended. The addition of purified RET2 to the TNT-synthesized probable interacting partner gives an option of using higher concentrations of RET2 in order to detect low-affinity interactions. Monoclonal, antigen-purified monospecific polyclonal or antigen-purified polyclonal antibodies are recommended for coimmunoprecipitation. For one reaction, the magnetic Dynabeads Protein G (Invitrogen) are coupled with purified antibodies for 1 h at 4° with mixing (5 μl of suspension/1 μg of total immunoglobulin) in 50 μl of IP buffer [20 mM Tris–HCl (pH 7.8), 150 mM NaCl, 0.1% Triton X-100] plus 10 mg/ml of bovine serum albumin. Beads are collected and washed twice with 0.5 ml of the IP buffer. Beads are resuspended in 100 μl of IP buffer plus 10 mg/ml of BSA and incubated with the TNT reaction (50–100 μl, depending on whether one or both proteins were synthesized) for 1 h at 4° with constant mixing. Beads are washed three times for 15 min in 1 ml of IP buffer, resuspended in 20 μl of 1× SDS loading buffer, and separated on SDS–polyacrylamide gel, which can be dried and exposed to a phosphor storage screen or transferred to a nitrocellulose membrane.

ACKNOWLEDGMENTS

The work on trypanosomal uridylyltransferases in the authors' laboratory is supported by NIH Grant AI064653. We thank Gene-Errol Ringpis for valuable suggestions on the manuscript.

REFERENCES

Aphasizhev, R. (2005). RNA uridylyltransferases. *Cell. Mol. Life Sci.* **62,** 2194–2203.

Aphasizhev, R., Sbicego, S., Peris, M., Jang, S. H., Aphasizheva, I., Simpson, A. M., Rivlin, A., and Simpson, L. (2002). Trypanosome mitochondrial 3′ terminal uridylyl transferase (TUTase): The key enzyme in U-insertion/deletion RNA editing. *Cell* **108,** 637–648.

Aphasizhev, R., Aphasizheva, I., Nelson, R. E., Gao, G., Simpson, A. M., Kang, X., Falick, A. M., Sbicego, S., and Simpson, L. (2003a). Isolation of a U-insertion/deletion editing complex from *Leishmania tarentolae* mitochondria. *EMBO J.* **22,** 913–924.

Aphasizhev, R., Aphasizheva, I., Nelson, R. E., and Simpson, L. (2003b). A 100-kD complex of two RNA-binding proteins from mitochondria of *Leishmania tarentolae* catalyzes RNA annealing and interacts with several RNA editing components. *RNA* **9,** 62–76.

Aphasizhev, R., Aphasizheva, I., and Simpson, L. (2003c). A tale of two TUTases. *Proc. Natl. Acad. Sci. USA* **100,** 10617–10622.

Aphasizheva, I., Aphasizhev, R., and Simpson, L. (2004). RNA-editing terminal uridylyl transferase 1: Identification of functional domains by mutational analysis. *J. Biol. Chem.* **279,** 24123–24130.

Deng, J., Ernst, N. L., Turley, S., Stuart, K. D., and Hol, W. G. (2005). Structural basis for UTP specificity of RNA editing TUTases from *Trypanosoma brucei*. *EMBO J.* **24,** 4007–4017.

Ernst, N. L., Panicucci, B., Igo, R. P., Jr., Panigrahi, A. K., Salavati, R., and Stuart, K. (2003). TbMP57 is a 3' terminal uridylyl transferase (TUTase) of the *Trypanosoma brucei* editosome. *Mol. Cell* **11,** 1525–1536.

Holm, L., and Sander, C. (1995). DNA polymerase beta belongs to an ancient nucleotidyl-transferase superfamily. *Trends Biochem. Sci.* **20,** 345–347.

Igo, R. P., Jr., Palazzo, S. S., Burgess, M. L., Panigrahi, A. K., and Stuart, K. (2000). Uridylate addition and RNA ligation contribute to the specificity of kinetoplastid insertion RNA editing. *Mol. Cell. Biol.* **20,** 8447–8457.

Igo, R. P., Jr., Lawson, S. D., and Stuart, K. (2002). RNA sequence and base pairing effects on insertion editing in *Trypanosoma brucei*. *Mol. Cell. Biol.* **22,** 1567–1576.

Kang, X., Gao, G., Rogers, K., Falick, A. M., Zhou, S., and Simpson, L. (2006). Reconstitution of full-round uridine-deletion RNA editing with three recombinant proteins 1. *Proc. Natl. Acad. Sci. USA* **103,** 13944–13949.

Panigrahi, A. K., Schnaufer, A., Ernst, N. L., Wang, B., Carmean, N., Salavati, R., and Stuart, K. (2003). Identification of novel components of *Trypanosoma brucei* editosomes. *RNA* **9,** 484–492.

Puig, O., Caspary, F., Rigaut, G., Rutz, B., Bouveret, E., Bragado-Nilsson, E., Wilm, M., and Seraphin, B. (2001). The tandem affinity purification (TAP) method: A general procedure of protein complex purification. *Methods* **24,** 218–229.

Ryan, C. M., and Read, L. K. (2005). UTP-dependent turnover of *Trypanosoma brucei* mitochondrial mRNA requires UTP polymerization and involves the RET1 TUTase. *RNA* **11,** 763–773.

Schnaufer, A., Ernst, N. L., Palazzo, S. S., O'Rear, J., Salavati, R., and Stuart, K. (2003). Separate insertion and deletion subcomplexes of the *Trypanosoma brucei* RNA editing complex. *Mol. Cell* **12,** 307–319.

Simpson, L., Sbicego, S., and Aphasizhev, R. (2003). Uridine insertion/deletion RNA editing in trypanosome mitochondria: A complex business. *RNA* **9,** 265–276.

Stagno, J., Aphasizheva, I., Rosengarth, A., Luecke, H., and Aphasizhev, R. (2006). UTP-bound and apo structures of a minimal RNA uridylyltransferase. *J. Mol. Biol.* **366,** 882–899.

Stuart, K. D., Schnaufer, A., Ernst, N. L., and Panigrahi, A. K. (2005). Complex management: RNA editing in trypanosomes. *Trends Biochem. Sci.* **30,** 97–105.

Trippe, R., Guschina, E., Hossbach, M., Urlaub, H., Luhrmann, R., and Benecke, B. J. (2006). Identification, cloning, and functional analysis of the human U6 snRNA-specific terminal uridylyl transferase 1. *RNA* **12,** 1494–1504.

Wirtz, E., Leal, S., Ochatt, C., and Cross, G. A. (1999). A tightly regulated inducible expression system for conditional gene knock-outs and dominant–negative genetics in *Trypanosoma brucei*. *Mol. Biochem. Parasitol.* **99,** 89–101.

CHAPTER FOUR

Isolation of RNA Binding Proteins Involved in Insertion/Deletion Editing

Michel Pelletier,[†] Laurie K. Read,[†] *and* Ruslan Aphasizhev*

Contents

* Department of Microbiology and Molecular Genetics, University of California, Irvine, California
† Department of Microbiology and Immunology, SUNY Buffalo School of Medicine, Buffalo, New York

Methods in Enzymology, Volume 424   © 2007 Elsevier Inc.
ISSN 0076-6879, DOI: 10.1016/S0076-6879(07)24004-2   All rights reserved.

Abstract

RNA editing is a collective term referring to a plethora of reactions that ultimately lead to changes in RNA nucleotide sequences apart from splicing, 5′ capping, or 3′ end processing. In the mitochondria of trypanosomatids, insertion and deletion of uridines must occur, often on a massive scale, in order to generate functional messenger RNAs. The current state of knowledge perceives the editing machinery as a dynamic system, in which heterogeneous protein complexes undergo multiple transient RNA–protein interactions in the course of gRNA processing, gRNA–mRNA recognition, and the cascade of nucleolytic and phosphoryl transfer reactions that ultimately change the mRNA sequence. Identification of RNA binding proteins that interact with the mitochondrial RNAs, core editing complex, or contribute to mRNA stability is of critical importance to our understanding of the editing process. This chapter describes purification and characterization of three RNA binding proteins from kinetoplastid mitochondria that have been genetically demonstrated to affect RNA editing.

1. INTRODUCTION

Uridine insertion/deletion editing in kinetoplastid protozoa entails massive processing of pre-mRNAs. Many pre-mRNAs undergo hundreds of uridine additions and less frequent deletions, and require the sequence information from dozens of *trans*-acting guide RNA (gRNAs) molecules for the creation of a translatable mRNAs. Precise editing requires that cognate mRNA–gRNA pairs form in the presence of hundreds of potential substrates. The editing reaction then necessitates a vast number of RNA–protein and RNA–RNA interactions and extensive rearrangements as editing progresses along an mRNA. In addition, editing is regulated during the trypanosome life cycle in the face of relatively constant mRNA and gRNA levels, implying that gRNA and/or mRNA usage is regulated in a specific manner (Stuart *et al.*, 2005). All of these scenarios suggest that RNA binding proteins will have a major impact on insertion/deletion editing, as is the case for other complex RNA processing events such as RNA splicing (Sanford *et al.*, 2005). The editing machinery is emerging as a dynamic system, in which heterogeneous protein complexes undergo multiple transient RNA–protein interactions. Thus, the identification of RNA binding proteins that interact with the editosome and/or impact the editing of specific RNAs is of critical importance to our understanding of the editing process. To date, three RNA binding proteins from kinetoplastid mitochondria have been genetically demonstrated to affect RNA editing. These are the related MRP1 and MRP2 proteins (Vondruskova et al., 2005) and RBP16 (Pelletier and Read, 2003).

1.1. MRP1 and MRP2

MRP1, originally designated as gBP21, is an arginine-rich mitochondrial protein that was identified by UV cross-linking to a synthetic radiolabeled gRNA and purified from *Trypanosoma brucei* (Koller et al., 1997). Purification of the poly(U) binding proteins from *Crithidia fasciculata* (Blom et al., 2001) and proteins cross-linking to a model gRNA–mRNA hybrid in the mitochondrial extracts of *Leishmania tarentolae* (Aphasizhev et al., 2003) independently led to identification of gBP21 orthologs in these organisms, gBP29 and p28. Studies in both *Crithidia* and *Leishmania* also identified a second RNA binding protein, which copurified with gBP21 orthologues. The copurifying polypeptides were termed gBP27 and p26, respectively. Under the current nomenclature, the gBP21 orthologs have been renamed as MRP1 and gBP27(p26) protein, designated as MRP2 (Simpson et al., 2003). The two highly positively charged proteins are homologous, although the sequence similarity is relatively low (18% overall identity) (Blom et al., 2001). MRP1 and MRP2 homologs can be identified in several kinetoplastids, but are apparently absent from other species. *In vivo*, MRP1 and MRP2 associate to form an $\alpha_2\beta_2$ heterotetramer (Aphasizhev et al., 2003; Schumacher et al., 2006). The MRP1/2 complex interacts with mitochondrial gRNAs, preedited mRNAs, and edited mRNAs, as shown by coimmunoprecipitation analysis (Allen et al., 1998; Aphasizhev et al., 2003; Blom et al., 2001). Additional indirect evidence of MRP1/2 involvement in editing came from coimmunoprecipitation and affinity purification experiments demonstrating that some fraction of MRP1 is associated with an *in vitro* editing activity, RNA editing TUTase 1 (RET1), and RNA editing ligase (REL) activities (Allen et al., 1998; Aphasizhev et al., 2003; Lambert et al., 1999). Importantly, all interactions between MRP1/2 and the aforementioned activities appeared to be sensitive to RNase treatment, suggesting an RNA-mediated nature of these contacts.

MRP1 null mutants in the bloodstream from *T. brucei* are viable, but incapable of transforming into procyclic forms (Lambert et al., 1999). Editing is essentially unaffected in the bloodstream from MRP1 knockout cells, although the stabilities of some never edited, preedited, and edited RNAs are substantially decreased. However, a more dramatic phenotype is observed in the procyclic form of trypanosomes in which RNAi-mediated knockdown of MRP1, MRP2, or both leads to a slow growth phenotype that is especially pronounced in cells with reduced MRP2 or MRP1/2 (Vondruskova *et al.*, 2005). Depletion of MRP2 mRNA causes a reduction in both the MRP2 and MRP1 proteins, presumably because the complex formation is required for MRP1 stabilization. In MRP1/2-depleted cells, editing of the apocyto-chrome *b* (CYb) RNA is dramatically reduced (96% reduction compared to uninduced levels) and RPS12 RNA editing is moderately decreased (51% reduction). The editing of several other RNAs is unaffected. The transcript-specific effects on editing due to MRP1/2 depletion by RNAi remains a mystery, as the complex binds RNA in a nonspecific fashion *in vitro* (Aphasizhev *et al.*, 2003). Editing complexes enriched from MRP1/2 knockdown cells demonstrate normal *in vitro* editing activity, indicating that MRP1/2 depletion has no effect on core editosome activity. In addition to causing decreased editing of specific RNAs, MRP1/2 knockdown also leads to destabilization of several never-edited RNAs. Therefore, the possibility of the MRP1/2 complex being involved in selective stabilization of some fully edited mRNAs may not be excluded.

The mechanism by which MRP1/2 participates in editing is presumed to be through facilitation of gRNA–mRNA annealing. The recombinant MRP1 stimulates annealing of cognate mRNA–gRNA pairs within the anchor duplex region up to 30-fold compared to spontaneous hybridization (Muller *et al.*, 2001). MRP1 can stimulate annealing of both RNA/RNA and RNA/DNA duplexes, but not DNA/DNA duplexes, with an apparent lack of sequence specificity. The recombinant MRP2, native and recon-stituted MRP1/2 complexes were also shown to facilitate annealing of complementary RNAs (Aphasizhev *et al.*, 2003). Biochemical studies on MRP1 suggest a matchmaker type of annealing function in which the protein binds to one of the RNA reactants and converts it to an annealing-active conformation (Muller and Goringer, 2002). Reduction of electrostatic repulsion between the two RNAs is also thought to con-tribute to formation of the RNA–RNA hybrid. The matchmaker model of MRP1/2 annealing was recently validated by the crystal structure of the MRP1/2 bound to gRNA (Schumacher *et al.*, 2006). The structure reveals that within the heterotetramer, MRP1 and MRP2 both adopt the "whirly" conformation previously seen in plant transcription factor p24, which binds to a single-stranded DNA in a sequence-specific manner (Desveaux *et al.*, 2002). The MRP1/2 heterotetramer contacts an extended region of the gRNA, in which MRP2 binds stem loop II of the gRNA, the basic region between MRP1 and MPR2 interacts with nucleotides between the anchor sequence and stem loop II, while MRP1 contacts the anchor sequence that is required for interaction with mRNA. In the absence of MRP1/2, the gRNA anchor sequence forms stem loop I of the gRNA (Hermann *et al.*, 1997). Remarkably, MRP1/2 captures the gRNA in a form in which stem loop I is unfolded, thereby maintaining the anchor sequence in a structure competent for pre-mRNA annealing.

1.2. RBP16

RBP16 was initially identified as a gRNA binding protein from *T. brucei* mitochondria and purified based on its affinity for poly(U) (Hayman and Read, 1999). The protein is highly conserved in both *Leishmania* and *T. cruzi*, although its function has not been analyzed in these species. RBP16 is composed entirely of two RNA binding domains. At its N-terminus, the protein contains a cold shock domain (CSD), the most evolutionarily conserved nucleic acid binding domain (Kohno *et al.*, 2003). The CSD is homologous to the cold shock proteins of prokaryotes and comprises a feature of eukaryotic Y-box family proteins. At its C-terminus, RBP16 contains an RG-rich RNA binding domain (Miller and Read, 2003). MALDI-TOF analysis shows that three arginine residues within the RBP16 C-terminal domain can undergo posttranslational methylation, and suggests that differentially methylated forms of the protein exist *in vivo*, which may contribute to the regulation of RBP16 function (Goulah *et al.*, 2006; Pelletier *et al.*, 2001). Native RBP16 purified from *T. brucei* mitochondria binds gRNAs of different sequences *in vitro*, and oligo(U) tail binding is critical for this interaction (Hayman and Read, 1999). Detailed RNA binding studies are consistent with a model in which CSD–oligo(U) tail interactions provide the primary affinity and specificity for binding, while the RG-rich domain helps stabilize the interaction (Miller and Read, 2003; Pelletier *et al.*, 2001). *In vivo* cross-linking studies demonstrate that RBP16 is associated with gRNA within *T. brucei* mitochondria, and coimmunoprecipitation experiments consistently show that 30–40% of the total gRNA pool is associated with RBP16 *in vivo* (Hayman and Read, 1999; Militello *et al.*, 2000; Pelletier and Read, 2003). In addition, substantial amounts of 9S and 12S rRNAs and some mRNAs are also coimmunoprecipitated with RBP16 from mitochondrial extracts (Goulah *et al.*, 2006; Hayman and Read, 1999). Recent studies have shown that the majority of RBP16 is present within mitochondria in two multicomponent complexes of 5S and 11S (Goulah *et al.*, 2006). The 11S complex is a ribonucleoprotein complex containing RBP16, gRNA, and additional proteins whose identities are not currently known. The 5S complex is an entirely proteinaceous subcomponent of the 11S complex, and may represent a primary functional form of RBP16 that interacts with other cellular components to regulate its activity and specificity. Methylation of RBP16 arginine residues 78 and 85 by the TbPRMT1 protein arginine methyltransferase is required for assembly and/or stability of the 5S and 11S RBP16-containing complexes. RBP16 does not appear to be a stable component of the 20S editosome or the MRP1/2 complex, although transient interactions with these complexes cannot be ruled out.

The demonstration that RBP16 associates with gRNAs suggested that this protein might function in RNA editing *in vivo*. Indeed, targeted depletion of RBP16 by RNAi in procyclic form *T. brucei* leads to cessation of growth as well as a very dramatic and specific effect on the editing of CYb mRNA

(Pelletier and Read, 2003). In the absence of RBP16, CYb editing is decreased by 98%, while editing of other RNAs examined is essentially unaffected. Levels of CYb gRNAs do not change under these conditions, suggesting that RBP16 affects gRNA utilization. In addition, RBP16 knockdown alters the abundance of two never-edited RNAs, COI and ND4. The RBP16 RNAi phenotype and pattern of affected mRNAs in procyclic *T. brucei* are strikingly similar to that of the MRP1/2 knockdown, suggesting that these proteins cooperate or participate in the same pathway to facilitate CYb mRNA editing and never edited RNA stabilization. Finally, in support of *in vivo* studies demonstrating a role for RBP16 in RNA editing, recombinant RBP16 significantly stimulates RNA editing *in vitro* (Miller *et al.*, 2006). Editing stimulation is exerted primarily at, or prior to, the step of pre-mRNA cleavage as evidenced by increased accumulation of 3′ cleavage product intermediates in the presence of increasing RBP16. The mechanism by which RBP16 stimulates editing is not currently known. Interestingly, other CSD-containing proteins have been reported to modulate RNA structure and facilitate RNA annealing (Jiang *et al.*, 1997; Skabkin *et al.*, 2001). Therefore, a possible scenario is that RBP16 acts as an RNA chaperone that affects RNA structure and/or annealing and leads to an increase in the functional association of RNA with the editing machinery and, hence, increased pre-mRNA cleavage.

Here we present methods for conventional and affinity purification of the MRP1/2 complex and assessment of RNA–protein interactions in mitochondrial extracts. Expression and purification of the recombinant MRP1 and MRP2 proteins, reconstitution of the MRP1/2 complex, as well as RNA binding and RNA annealing assays are provided. We also describe methods for purification of RBP16 from procyclic and bloodstream form trypanosomes followed by methods for the isolation of the recombinant protein from *Escherichia coli* and an RNA binding assay.

2. PURIFICATION AND CHARACTERIZATION OF MITOCHONDRIAL RNA BINDING PROTEINS 1 AND 2 FROM *LEISHMANIA TARENTOLAE*

Purification of RNA binding proteins from mitochondria of trypanosomes historically has been aimed at the identification of factors involved in uridine insertion/deletion RNA editing. The UV-induced cross-linking of mitochondrial extracts with synthetic RNA molecules, which resemble gRNAs or gRNA–mRNA hybrids, was the primary assay used throughout biochemical fractionations. In addition, a hypothetical affinity of editing complexes for gRNA's universal features, such as the 3′ oligo(U) tail, warranted application of affinity chromatography on poly(U) resins. Notwithstanding the obstacles posed by a presence of a large number of abundant, nonspecific RNA binding proteins in the mitochondrial extracts (Bringaud *et al.*, 1995, 1997), efforts undertaken by the laboratories of Göringer (Koller *et al.*, 1997), Benne (Blom *et al.*, 2001), and Simpson (Aphasizhev *et al.*, 2003) ultimately led to identification of the stable 100-kDa $(\alpha + \beta)_2$ tetramer complex composed of two homologous proteins, MRP1 and 2. This chapter describes the purification of the MRP1/2 complex from *L. tarentolae* by conventional and affinity techniques, methods for analysis of MRP1/2 interactions in the mitochondrial extracts, reconstitution of the 100-kDa complex from recombinant proteins, and *in vitro* assays that may be relevant to the MRPs' function in mitochondrial RNA editing. As an experimental system, *L. tarentolae* remains an attractive option in kinetoplastid research due to its low cost and robust cultivation combined with the availability of protein expression tools.

3. PURIFICATION OF THE MRP1/2 COMPLEX FROM *LEISHMANIA TARENTOLAE*

3.1. Overview

The choice between conventional and affinity purification procedures depends on the specific experimental objective and available resources. The conventional protocol does not require plasmid construction, *Leishmania* transfections, and clonal selection; it is highly efficient and produces an essentially pure MRP complex. Exposure to high salt conditions effectively removes bound RNAs and disrupts association with RNA editing components, such as RNA Editing TUTase 1 (Chapter 2, this volume) and the 20S editosome (Chapter 1, this volume). Tandem affinity purification (TAP) can be performed under low ionic strength conditions and is the method of choice for analyzing MRP's transient interactions, including RNA-mediated contacts. This affinity method is particularly suitable for protein mass spectrometry and RNA detection. Finally, a mitochondrial fraction enriched by centrifugation in isopycnic gradients is recommended as a starting material for both methods. If growing sufficient volumes of *Leishmania* cultures is not feasible, the affinity techniques described for purification of the RNA editing TUTase 2 (RET2) from whole-cell lysate of *T. brucei* are applicable (Chapter 2, this volume). Contamination with cytoplasmic proteins, mostly tubulins and ribosomal proteins, may be expected.

3.2. Methods

3.2.1. Isolation of the mitochondrial fraction

L. tarentolae cells (UC strain) are grown in 10-liter batches of brain–heart infusion (BHI) medium (Gibco) supplemented with hemin (0.2 mg/liter) to a late logarithmic growth phase ($\sim$120 $\times$ 10^6 cells/ml). Roller bottles revolving at 6–8 rpm or a fermentor vessel can be used. The latter allows for yields of up to $200–250 \times 10^6$ cells/ml and requires proportional scaling of the protocol. The mitochondrial fraction is enriched by the modified procedure originally described in Braly *et al.* (1974). All steps are performed at 4°. Cells are collected by centrifugation at $5000 \times g$ for 10 min, washed in 1 liter of ice-cold phosphate-buffered saline (pH 7.6), and resuspended in DTE buffer [1 mM Tris–HCl (pH 8), 1 mM EDTA] at 1.4×10^9/ml to homogeneity. Carryover of PBS and extended incubation in DTE should be avoided. Addition of one tablet of the Complete protease inhibitor cocktail (Roche) is optional. The cell suspension is pressed out of a stainless-steel, or polycarbonate pressure device (Millipore) through a 26-gauge needle at 90 lb/in². In order to stabilize the mitochondria, a sucrose solution (60% w/v) is immediately added to the cell lysate at the ratio of 6 volumes of sucrose to 50 volumes of cell suspension. Upon lysis and mitochondria stabilization, the extract is supplemented with 20 mM Tris–HCl (pH 7.6) and 3 mM MgCl$_2$. DNase I (Sigma) is added to 20 U/ml and the lysate is incubated for 30 min on ice with occasional mixing. The membrane fraction is collected by centrifugation at $15,000 \times g$ for 20 min, resuspended in 35 ml of 76% Renografin [0.25 M sucrose, 0.5 mM EDTA dissolved in RenoCal 76 (Bracco Diagnostics)]. The membrane fraction (6–7 ml) is layered underneath density gradients using a syringe fitted with an 18-gauge needle and polyethylene tubing. Six density gradients should be prepared in advance in SW28 tubes (Beckman) as follows. For a single isolation, light and heavy solutions are obtained by dissolving 8.55 g of sucrose in 26.3 or 46.2 ml of RenoCal 76, respectively. The concentration of Tris–HCl (pH 7.6) is adjusted to 20 mM and EDTA to 0.1 mM in a final volume of 100 ml. The heavy solution is portioned into six tubes, 16 ml/tube, and frozen at $-20°$. The tubes are then filled with 16 ml of light solution and refrozen. Approximately 36–40 h prior to isolation, frozen tubes are placed at 4° . Extended light exposure should be avoided. Gradients are centrifuged for 2 h at 24,000 rpm ($76,221 \times g$) in an SW28 rotor (Beckman). The mitochondria-containing band typically sediments in the middle of the tube and should be clearly visible when placed in front of a light source. The band is extracted by puncturing the tube on the side and under the band with an 18-gauge needle fitted to a 10-ml syringe. The fraction is diluted 4- to 5-fold with isotonic STE buffer [20 mM Tris–HCl (pH 7.6), 0.25 M sucrose, 0.1 mM EDTA] and centrifuged for 20 min at $15,000 \times g$. The pellet is gently resuspended in 50 ml of STE, centrifuged again, and frozen at $-80°$. The expected yield is approximately 5–6 g of pellet (wet weight), or 0.3–0.5 g of total protein.

3.2.2. Conventional chromatography purification of the MRP1/2 complex

For successful conventional purification, an expeditious handling is essential due to the high sensitivity of the MRP proteins to proteolysis. Ideally, all chromatographic steps should be performed on a protein purification workstation capable of generating precise gradients. The recommended scheme takes 2 days to accomplish. The amount of starting material may vary between 5 and 8 g of wet weight pellet. On day 1, the enriched mitochondria are extracted in 45 ml of 50 mM HEPES (pH 7.6), 300 mM KCl, 1 mM dithiothreitol (DTT), 1 mM EDTA, 1 tablet of Complete protease inhibitors (Roche), and 2 mM CHAPS with sonication (five pulses, for 15 sec each, 36 W). The temperature of the extract during sonication should not exceed 10°. The efficiency of lysis may be monitored by fluorescent microscopy after staining with DAPI (4',6-diamidino-2-phenylindole), and the kinetoplastid DNA, which normally stains as a compact and bright cup-like structure, should be completely dispersed. The extract is centrifuged at 150,000×g in Ti 60 rotor (Beckman) for 30 min. Ammonium sulfate (AmS) is gradually added to 40% saturation with constant mixing over an approximately 20-min period, and the centrifugation is repeated in the same mode. The supernatant is recovered and the AmS is added to 60% saturation. After 1 h of gentle mixing, the precipitate is collected by centrifugation as before and dissolved in 25 ml of 50 mM HEPES (pH 7.6), 10 mM MgCl$_2$, 1 mM DTT, 5% glycerol. Conductivity is adjusted to ~15 mSi/cm^2 and the fraction is loaded onto a prepacked 5-ml HiTrap heparin column (GE Healthcare) pre-equilibrated with same buffer plus 150 mM KCl. Other heparin resins, e.g., Poros Heparin 20 and self-made columns, may be used with equal success. The column is developed with a 100-ml linear gradient from 150 to 1000 mM KCl at 1 ml/min and 1-ml fractions are collected. The MRP1/2 complex (~70% pure) typically elutes at ~600 mM as the last discrete peak of optical density at 280 nm. The proteins of interest are detected by SDS–gel electrophoresis as two protein bands of similar intensity migrating at 28 and 26 kDa. The sample can be flash-frozen in liquid nitrogen or left on ice overnight.

Preparations exceeding 90% purity may be obtained by hydrophobic interaction chromatography. On the second day, ammonium sulfate powder is gradually added to the combined fractions from the heparin column (~5–8 ml, 200–300 μg of protein) to 1.4 M and centrifuged 10 min at 100,000×g. The supernatant is loaded at 0.3 ml/min onto a 1-ml Phenyl Resource column (GE Healthcare), which was preequilibrated with 1.5 M AmS in 50 mM HEPES (pH 7.6), 0.1 mM EDTA. The column is developed with a 20-ml gradient from 1.5 M AmS to no salt. The complex elutes at ~1 M AmS as a single sharp peak and is readily detectable by SDS gel electrophoresis. Some of the later eluting fractions may contain three or four closely migrating bands, which are proteolysis products. Fractions containing MRP1 and 2 in equal amounts (Fig. 4.1A, I) are dialyzed against 50 mM HEPES (pH 7.6), 0.1 mM EDTA, 200 mM KCl, 10% glycerol, concentrated to 1–2 mg/ml with a 10-kDa cut-off Amicon centrifugal device (Millipore) and flash-frozen in liquid nitrogen in small aliquots. Multiple cycles of freezing and defrosting are not recommended. The expected yield is 50–100 μg of protein.

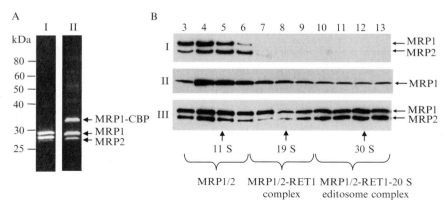

Figure 4.1 Purification and sedimentation analysis of the MRP1/2 complex. (A) The MRP1/2 complex obtained by conventional (I) or affinity (II) purification techniques was separated on 10–20% SDS–polyacrylamide urea gel and stained with Sypro Ruby (Invitrogen). The positions of the MRP proteins and MRP1–calmodulin binding peptide fusion (CBP) are indicated by arrows. (B) Fractions (I) and (II) from (A) and the mitochondrial extract (III) were separated on a 10–30% rate zonal glycerol gradient and analyzed by Western blotting. Polyclonal antibodies against MRP1/2 were used in (I) and (III). (II) was treated with antibodies raised against MRP1. Sedimentation zones of MRP1/2, MRP1/2–RET1, and MRP1/2–RET1–20S editosome particles, as determined by immunoprecipitation analysis of each fraction, are shown by brackets.

3.2.3. Affinity purification of the MRP complex

A construct for expression of the MRP1-TAP C-terminal fusion in *Leishmania* has been described (Aphasizhev *et al.*, 2003) and is available from the author's laboratory. The transfection protocol is adopted from Robinson and Beverley (2003). Briefly, *L. tarentolae* cells (UC strain) are grown to $5–8 \times 10^7$ cells/ml in BHI media and harvested by centrifugation at $2000 \times g$ for 10 min. The cell pellet is gently resuspendend in 1/2 of the culture's original volume in the Cytomix electroporation buffer [25 mM HEPES (pH 7.6), 120 mM KCl, 10 mM K_2HPO_4, 5 mM $MgCl_2$, 2 mM EDTA, 0.15 mM $CaCl_2$]. Cells are pelleted again and resuspendend in Cytomix at 1.5×10^8 cells/ml. Typically, 10 μl of plasmid DNA (1 mg/ml) is added into a 4-mm gap electroporation cuvette followed by 0.5 ml of cell suspension. After a 10-min incubation on ice, two pulses (1500 V, 25 μF) are applied in 10-sec intervals with a Gene Pulser apparatus (Bio-Rad). Cells are immediately transferred into 10 ml of BHI media prewarmed to 27°and incubated overnight without antibiotic. Geneticine (G418, Sigma) is added to 50 mg/liter on the next day and to 100 mg/liter the day after. After the culture has been growing for a total of 3 days, 3 ml is centrifuged at $2000 \times g$ for 10 min, resuspendend in 100 μl of residual media, and plated on BHI-agar. Plates are sealed with Parafilm and incubated at 27°

until 2- to 3-mm colonies appear (7–10 days). To prepare the plates, 37 g of BHI, 0.8 g of folic acid, and 8 g of agar are autoclaved in 1 liter of water for 15 min. Heat-inactivated serum (100 ml), hemin (5 g/liter), and 200 mg of G418 are added at ~50° prior to preparing plates. Plates are left to dry for 15–20 h at room temperature with covers on.

Individual colonies appear in 7–10 days. Colonies of 2–3 mm in diameter are transferred into 10-ml cultures and grown in the presence of 100 mg/liter of G418 to late-log phase (~100 × 10^6 cells/ml). One milliliter of culture is collected, washed with 1 ml of PBS, boiled in 100 μl of 1× SDS loading buffer, and centrifuged for 30 min at 15,000×g at room temperature. The supernatant equivalent to 10^7 cells is loaded on SDS–polyacrylamide gel. Blotting analysis with peroxide–antiperoxidase (PAP) reagent (Sigma), which detects the protein A moiety of the TAP cassette, can be used to test protein expression. For isolation, a 10-liter culture is grown in the presence of 100 mg/liter of G418 and the mitochondrial fraction is isolated as described above. The TAP protocol is adapted from Puig $et\ al.$ (2001) with modifications. The mitochondrial fraction is adjusted to 12 ml with TMK buffer (20 mM Tris–HCl, pH 7.6, 60 mM KCl, 10 mM MgCl$_2$) plus 0.4% of NP40. One tablet of Complete protease inhibitors (Roche) is added, the extract is resuspended, gently sonicated (three pulses for 10 sec, 12 W), and incubated on ice for 15 min. After centrifugation at 200,000×g for 15 min, the pellet is reextracted with sonication (three pulses for 30 sec, 12 W) in 12 ml of the TMK buffer without detergent and centrifugation is repeated. Supernatants from both extractions are pooled, filtered through 0.45-μm low-protein binding filter, and incubated with 0.3 ml (packed volume) of IgG Sepharose FF (GE Healthcare) for 1 h in a 50-ml conical tube. Suspension is then transferred to a disposable 1-ml column (Pierce) and the extract is reloaded three times. The column is washed sequentially with 50 ml of TMK buffer plus 0.1% of NP40, and 10 ml of TMK, 0.1% of NP40, 1 mM DTT. An attempt should be made to resuspend the resin with each wash. TMK buffer plus 0.1% of NP40 and 1 mM DTT (1.5 ml) with 200 U of TEV proteinase (Invitrogen) is added to the column, which is sealed and incubated overnight with constant mixing. The IgG resin is drained into a 15-ml conical tube containing 0.3 ml (packed volume) of calmodulin agarose (Stratagene) prewashed with CB buffer (20 mM Tris–HCl, pH 7.6, 60 mM KCl, 10 mM MgCl$_2$, 1 mM imidazole, 10 mM 2-mercaptoethanol, 2 mM CaCl$_2$, 0.1% NP40). To improve recovery, the IgG column is rinsed with 4.5 ml of CB buffer and the rinse is collected into the tube with calmodulin resin. Calcium chloride is added (5 μl of 1 M stock) and the eluted material (~6 ml) is incubated with calmodulin resin for 1 h, transferred to a disposable 1 ml column (Pierce), reloaded three times, and washed with 50 ml of CB buffer. Elution is performed with 5 ml of 20 mM Tris–HCl (pH 7.6), 60 mM KCl, 2 mM MgCl$_2$, 5 mM EGTA, 2 mM CHAPS. Fractions of 0.5 ml are collected and, if necessary, are concentrated with Slide-A-Lyzer solution (Pierce) to 300 μl. A typical protein profile is shown in Fig. 4.1A, II. The expected yield is ~50 μg of protein.

4. ANALYSIS OF MRP1/2 INTERACTIONS IN MITOCHONDRIAL EXTRACTS

4.1. Overview

Affinity isolation of the MRP1/2 complex provides an excellent starting point for assessing its apparently numerous interactions in the mitochondrial extract (Aphasizhev *et al.*, 2003) that may be occurring, in part, due to MRP's high affinity for RNA and lack of sequence specificity (Schumacher *et al.*, 2006). The high degree of purity and mild isolation conditions allow for immunochemical detection of interacting partners that are present in substoichiometric amounts and bind to MRP via an RNA component. Rapid isolation also preserves bound gRNAs that can be isolated from the affinity-purified material and selectively radiolabeled. In addition, expression of the TAP-tagged proteins followed by pull-down with IgG-Sepharose provides a highly specific generic alternative to immunoprecipitation analysis. For example, the mitochondrial extract may be subjected to gradient fractionation and IgG-Sepharose used for pull-downs in each fraction. RNA ligase adenylation and *in vitro* editing activity are readily detected in complexes bound to the beads, which may be further analyzed by SDS–gel electrophoresis and Western blotting.

4.2. Methods

4.2.1. Two-dimensional size fractionation of the affinity purified MRP1/2-containing particles

The concentrated calmodulin fraction after affinity purification is loaded on a 10–30% glycerol gradient, which allows for effective separation in the 5–35S range. The 10% and 30% (v/v) glycerol stocks are prepared in 25 mM HEPES (pH 7.6), 10 mM MgCl$_2$, 60 mM KCl, 1 mM CHAPS. Rate zonal sedimentation is performed in an SW41 rotor (Beckmann) for 20 h at 30,000 rpm (111,000×g). Fractions of 700 μl are collected. Each fraction may be assigned an S value by calibrating the gradient with catalase (11S), thyroglobulin (19S), and small ribosomal subunit from *Escherichia coli* (30S). Excellent resolution and reproducible results can be obtained with gradient maker/fractionator from Biocomp (Fredericton, NB, Canada) following the manufacturer's instructions. Whereas TAP affinity-purified MRP1/2 sediments over a broad range (10–35S), the MRP1/2 complex purified by combination of heparin and hydrophobic interaction chromatography sediments as a discrete peak at 10–15S (Fig. 4.1B). The fraction of MRP1/2 associated with the 20S

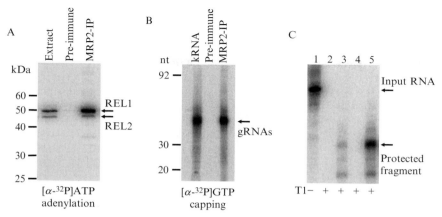

Figure 4.2 MRP1/2 interactions and RNA annealing activity. (A) Self-adenylation of RNA editing ligases, REL1 and REL2, may be used as a sensitive method to detect the 20S editosome in MRP1/2 complexes. Adenylation was performed in mitochondrial extract and material coimmunoprecipitated with preimmune serum and anti-MRP2 polyclonal antibody. (B) Guide RNAs are selectively labeled with vaccinia virus capping enzyme in the presence of [α-^{32}P]GTP. (C) RNA annealing assay. (1) Input RNA; (2) noncomplementary RNA added; (3) self-annealing in the absence of protein; (4) same polarity strand added; (5) annealing in the presence of the MRP1/2 complex.

editosome sediments in a 25–35S region. Two components of the 20S editosome, RNA Editing Ligases 1 and 2, may be radioactively labeled by self-adenylation with [α-^{32}P]ATP in the absence of RNA substrate, which provides a sensitive method for editosome detection (Fig. 4.2A).

Adenylation is performed for each fraction by mixing 0.5 μl of [α-^{32}P] ATP (6000 Ci/mmol) with 10 μl of the gradient fraction and incubating the reaction for 30 min at 27°. The adenylated complexes may be further analyzed under denaturing or native conditions using the 8–16% gradient polyacrylamide precast Tris–glycine gels (Invitrogen). For SDS gel fractionation, 10 μl of the loading buffer (0.25 M Tris–HCl, pH 6.8, 2% SDS, 100 mM DTT, 0.05% bromophenol blue, no glycerol) is added to the adenylation reaction followed by standard SDS electrophoresis. The proteins are transferred to a nitrocellulose membrane by semidry or tank electroblotting methods. Upon exposure to a phosphor storage screen, RNA ligases are visualized as two closely migrating bands of ~50 and 48 kDa (Fig. 4.2A). For separation under native conditions, 10 μl of 20 mM Tris–HCL (pH 7.6), with 0.01% of bromophenol blue is added to the adenylation reaction. Native Tris–glycine buffer (pH 8.2) (Invitrogen) is used for the electrophoresis at 4° (~16 h at 45 V). Native high-molecular-mass protein standards (GE Healthcare) may be run alongside to determine the apparent molecular mass of the adenylated complexes. A refrigerated chamber and a power supply capable of maintaining low current (<2 mA) are required. The gel is soaked in transfer buffer (10 mM Tris, 100 mM glycine, 10% methanol) for 30 min in the cold room and subjected to tank electrotransfer at 40 V for 8–10 h at 4°. In order to visualize molecular mass standards, the corresponding lane is cut off from membrane and stained with Ponceau S (Pierce). The membrane may be exposed to a phosphor storage screen to visualize the 20S editosome typically migrating at 1.2–1.6 MDa, or treated with antibodies. Because of MPRs' high isoelectric point, under native condition these proteins enter the gel only if associated with RNA or negatively charged partners.

4.2.2. Pull-down of TAP-tagged MRP1/2 complexes from gradient fractions

In this approach, a mitochondrial fraction is isolated from *L. tarentolae* cells that express TAP-tagged proteins. Rather than carrying out the full-scale TAP purification, the mitochondrial extract is fractionated on a 10–30% glycerol gradient followed by IgG pull down in each fraction. This protocol takes advantage of gradient fractionation as a tool to assess the sedimentation values for complexes that had not been affected by TAP purification, which usually removes weakly associated components. Another consideration that may contribute to the choice of running the gradient fractionation prior to TAP purification is the amount of protein that is being handled. If the TAP procedure produces less than 10 μg of protein, significant losses may be expected during subsequent gradient fractionation due to nonspecific binding.

The mitochondrial pellet ($\sim$200 mg wet weight, $\sim$15 mg of total protein) is resuspendend in 0.5 ml of 50 mM HEPES (pH 7.6), 60 mM KCl, 10 mM MgCl$_2$, 1/50 part of Complete protease inhibitor tablet, and 0.4% of NP40. All concentrations are final, adjusted for the mitochondrial pellet's volume. The extract is gently sonicated using a micro tip (three pulses, 5 sec, 6 W) and incubated on ice for 15 min. The membrane fraction is removed by centrifugation for 10 min at 200,000×g in a Beckman TL-100 ultracentrifuge and the soluble fraction is recovered. Gradient fractionation is performed using 200 μl of cleared extract. Each fraction may be analyzed in the second dimension by SDS–gel electrophoresis in order to provide references to the sedimentation position of the 20S core editosome, or other complexes of interest. For IgG pull-downs, 5 μl of bead suspension (settled volume) and 20 μl of 50 mg/ml of acetylated bovine serum albumin are added to each fraction and binding is performed for 1 h with constant rocking. The beads are collected by centrifugation at 1000×g for 1 min and washed three times for 15 min with 1 ml of buffer. Depending on the experimental objective, buffer composition may vary in KCl concentration from 50 to 500 mM of salt, the presence of Mg ions or EDTA, and from 0.1 to 0.5% of nonionic detergent without affecting IgG–protein A binding. Prior to the adenylation reaction, an extra wash with low salt buffer [50 mM

HEPES (pH 7.6), 50 mM KCl, 10 mM MgCl$_2$] is required. The reaction is initiated by adding 1 μCi of [α-^{32}P]ATP to pelleted beads and incubating at 27° for 30 min. The final wash is performed with low salt buffer and 10 μl of 2× SDS loading solution is added to the beads. Bound proteins are eluted from beads by heating to 95° for 2 min and analyzed on SDS–polyacrylamide gel followed by blotting onto a nitrocellulose membrane. If proteomic identification of the coprecipitating components becomes a necessity, the selected gradient fractions may be subjected to a full-scale TAP procedure following the above-described protocol, but downscaled 2-fold in terms of the amount of resin and volumes of washes. No prior treatment is required for gradient fractions.

4.2.3. Analysis of MRP1/2-associated proteins and RNAs

In order to identify proteins copurifying with the MRP1/2 complex through the TAP procedure, the polypeptides are recovered by binding to Strataclean resin (Stratagene) prior to SDS–gel electrophoresis. Typically, 10 μl of bead suspension is added to 1 ml of gradient fraction and incubated at room temperature for 30 min with constant mixing. The suspension is centrifuged at 3000×g for 5 min, the supernatant is removed, and 10 μl of 2× SDS loading buffer is added. The sample is boiled for 3 min and the entire mixture is transferred onto a precast gradient SDS–PAGE. Depending on the available mass spectrometry resources, the gel is stained with colloidal Coomassie Blue or Sypro Ruby for further analysis.

Guide RNAs are detected in gradient fractions after phenol/chloroform extraction and ethanol precipitation in the presence of glycogen (20 μg/ml) and 0.5 M sodium acetate, pH 5.0. Because gRNAs are primary transcripts bearing diphosphates or triphosphates at the 5′ end (Blum *et al.*, 1990), they are readily detected by incubation with vaccinia virus capping enzyme, a guanylyltransferase (Ambion), in the presence of [α-^{32}P]GTP. Upon separation on a 15% acrylamide/8 M urea gel, the radioactively labeled gRNAs migrate as a group of bands corresponding to ~60–65 nucleotides (Fig. 4.2B).

5. RECONSTITUTION OF THE MRP1/2 COMPLEX FROM RECOMBINANT PROTEINS

5.1. Overview

The MRP1 and 2 proteins form a stable complex in mitochondria, which prevents purification of individual proteins for biochemical studies from the native source. Furthermore, it has been shown that the RNA interference-mediated depletion of the MRP2 subunit in *T. brucei* leads to a decrease in abundance of MRP1 without affecting the respective mRNA

(Vondruskova *et al.*, 2005). Therefore, formation of the complex is likely to be essential for stability of these proteins *in vivo*. At the functional level, a certain degree of redundancy may be anticipated, as MRP2 or dual MRP1/2 RNAi knockouts, but not the MRP1 depletion, have been found to inhibit the parasite's growth. A possible explanation for this phenomenon may come from MRP2's ability to form dimers (Aphasizhev *et al.*, 2003), which are apparently the functional units in MRP1/2 complexes (Schumacher *et al.*, 2006). The methods described below allow purification of individual MRP proteins for biochemical studies and reconstitution of the complex. An alternative strategy for reconstitution of the MPP1/2 complex via coexpression of both subunits in *E. coli* has been reported by others (Schumacher *et al.*, 2006).

5.2. Methods

5.2.1. Purification of recombinant MRP1 and 2 from *E. coli*

The experimentally determined mitochondrial localization signals for *L. tarentolae* proteins, position -46 for MRP1 and position -30 for MRP2, were omitted during construction of pET15b-based (Novagen) *E. coli* expression vectors (Aphasizhev *et al.*, 2003). MRP1 forms inclusion bodies upon expression in bacteria and should be purified under denaturing conditions and refolded. The 2 liters of *E. coli* culture, strain BL21(DE3) CodonPlus RIL (Stratagene), are grown in 2YT media in the presence of 100 mg/liter of ampicillin at 37°. Expression is induced at $\sim$0.6 OD_{600} by adding isopropyl-β-D-thiogalactopyranoside (IPTG) to 1 mM and cultivation is continued for 3 h. The bacterial pellet is resuspendend in 40 ml of 6 M guanidine hydrochloride, 50 mM sodium phosphate (pH 7.5), and subjected to sonication (five pulses, 36 W for 30 sec). The extract is centrifuged for 30 min at 200,000$\times g$, filtered through a 0.45-μm low-protein binding filter, and loaded onto a 2-ml column with Talon metal affinity resin (Clontech). Followed by washing of the column with 50 ml of 8 M urea in 50 mM sodium phosphate (pH 7.5), the protein is refolded while still attached to the Talon resin. The 40 ml reverse gradient of urea from 8 to 0 M in 20 mM sodium phosphate (pH 7.5) 300 mM NaCl is run at 20° for 20 h at 2 ml/h. The protein is eluted with 200 mM imidazole, and further purified on a 1-ml HiTrap heparin column using the procedure described above for the MRP1/2 complex, but downscaled 5-fold. The expected yield is 1–2 mg of protein.

  MRP2 is partially soluble in *E. coli* extracts if the bacterial culture is induced and grown at 20°. The protein is purified at 4° under native conditions by Talon metal affinity and heparin chromatography. The cell pellet is resuspended in 50 ml of 50 mM HEPES (pH 8.0), 50 mM NaCl and passed through a French pressure cell at 12,000 psi. The sodium chloride is adjusted to 300 mM and the extract is centrifuged at 200,000$\times g$ for 30 min.

The supernatant is loaded into a 2-ml Talon column, which is washed sequentially with 50 ml of 50 mM HEPES (pH 8.0), 300 mM NaCl, and 20 ml of the same buffer with 10 mM imidazole. The MRP2 is step-eluted with 50 mM HEPES (pH 8.0), 300 mM NaCl, 200 mM imidazole, and is typically more than 95% pure. Additional purification by heparin chromatography is required for reconstitution, but not RNA activity assays (below). The fraction obtained from the Talon column is diluted 4-fold with 50 mM HEPES (pH 7.6), 10 mM MgCl$_2$, 1 mM DTT, 5% glycerol, and loaded on a 1-ml HiTrap (GE Healthcare) heparin column. A 40 ml linear gradient (0.1– 1 M) of sodium chloride is applied and 1-ml fractions are collected. A single protein peak containing 3–5 mg of protein is eluted at ~400 mM NaCl.

5.2.2. Reconstitution of the MRP1/2 complex

Purified recombinant MRP1 and 2 are mixed in 200 μl at a 1:2 molar ratio (final protein concentration of 2–4 mg/ml) and dialyzed against 20 mM sodium phosphate, 200 mM NaCl for 1 h at 25°. The excess of the MRP2 protein is necessary for *in vitro* complex formation, likely due to its ability to form a homodimer. After a 10-min centrifugation at 15,000×g, the sample is loaded on a Superose 12 column and eluted in the same buffer at 0.1 ml/min. Fractions of 0.2 ml are collected, separated on 12% acrylamide–SDS gel, and stained with Sypro Ruby (Invitrogen). Fractions eluting at ~11.5 ml that contain protein bands of equal intensity are pooled and concentrated, if necessary, with Slide-A-Lyzer solution (Pierce). The expected yield is 200–300 μg. The protein sample is supplemented with glycerol to 10%, flash-frozen in liquid nitrogen, and stored at –80°.

6. RNA Binding and RNA Annealing Activities of the MRP1/2 Complex

6.1. Overview

Several studies have addressed the specificity and affinity of RNA binding by the MRP1 (gBP21) protein (Koller *et al.*, 1997; Lambert *et al.*, 1999; Muller *et al.*, 2001) and MRP1/2 complex from *T. brucei* (Schumacher *et al.*, 2006) and the MRP1/2 complex from *L. tarentolae* (Aphasizhev *et al.*, 2003). Excellent agreement between results obtained for the complexes (Aphasizhev *et al.*, 2003; Schumacher *et al.*, 2006), combined with structural data (Schumacher *et al.*, 2006), makes it possible to conclude that MRP1/2– RNA interactions are sequence independent and driven by electrostatic interactions of positively charged residues with the phosphodiester backbone. Importantly, single-stranded and double-stranded RNAs are bound by the MRP1/2 complex with high affinity: apparent dissociation constants for both types of molecules fall into the 2–10 nM range. An *in vitro* RNA annealing activity, which was reported for the MRP1 (Muller *et al.*, 2001), MRP2, and MRP1/2 complex (Aphasizhev *et al.*, 2003), remains the most significant, albeit circumstantial, indication of MRP's function in RNA editing. The capacity of RNA binding proteins to promote the formation of a double-stranded RNA from complimentary single-stranded molecules is well documented, but may be occurring by a variety of mechanisms (Rajkowitsch *et al.*, 2005). Although folding of all guide RNAs into a similar secondary and tertiary structure remains to be established, the structure of the MRP1/2 complex with a fragment of the gND7–506 (Schumacher *et al.*, 2006) suggests that the interaction of MRP1/2–gRNA–mRNA presents the gRNA's "anchor" region in the unfolded state with bases exposed to the solvent and suitable for hybridization with preedited mRNA.

6.2. Methods

6.2.1. RNA binding

The affinity of RNA binding is measured as an equilibrium dissociation constant, K_d, by fitting nitrocellulose filter binding data from three independent experiments into an equilibrium binding model. The radioactively labeled guide RNA of interest is synthesized by run-off transcription from a linear DNA template using T7 RNA polymerase (Invitrogen). Typically, 1 μg of linearized plasmid DNA is incubated in the supplied buffer with 50 U of RNA polymerase in the presence of 1 mM GTP, CTP, and UTP and 50 μM ATP plus 50 μCi of [α-^{32}P]ATP. After a 2-h reaction, 2 U of RNase-free DNase is added and incubation is continued for 30 min. RNA is purified by 15% acrylamide/urea gel electrophoresis, eluted into 0.1 M sodium acetate, pH 5.0, 0.1% SDS, 1 mM EDTA, and ethanol precipitated. The RNA concentration is determined spectrophotometrically. Double-stranded RNA is assembled from complementary molecules, which are independently synthesized and purified. Prior to binding assays, RNAs are mixed at 10 μM each in a buffer with 10 mM HEPES (pH 7.6), 50 mM KCl, and 0.1 mM EDTA, heated to 90° for 2 min, and cooled to room temperature over a 30-min time period. Individual RNAs are subjected to the same folding procedure. RNA–protein binding is carried out in a 10-μl reaction mixture containing 10 mM HEPES (pH 7.6), 50 mM KCl, 2 mM MgCl$_2$, 0.5 mM DTT, 6% glycerol, and 0.05 mg/ml of bovine serum albumin (BSA) at 27° for 30 min. The entire reaction is spotted on a 33-mm (0.22-μm) nitrocellulose filter, which is held in a vacuum-driven manifold (Millipore). The filter must be prewetted and washed with binding buffer lacking BSA. Washing volumes required for a low background need to be determined empirically, but typically ~15 ml/filter is sufficient. Filters are dried and exposed to a phosphor storage screen followed by scanning and quantitation. Alternatively, scintillation counting may be used.

With either detection method, the amount of bound RNA is determined by calculating the RNA's specific activity. To obtain this value, the reaction mix is spotted on a separate filter and dried without washing. For apparent K_d determinations, the RNA concentration is kept constant at 5 nM and increasing amounts of protein are used. Recommended ranges are 0.1–1000 nM for the MRP complex purified from *L. tarentolae*, 0.1–2000 nM for the recombinant MRP1, and 1–4000 nM for the recombinant MRP2. The percentage of active protein in each preparation must be determined in reciprocal experiments in order to adjust the concentration values used for K_d calculations. This is done with 2 nM protein and increasing concentrations of RNA from 0.01 to 2 μM. The MRP1/2 complex purified from *L. tarentolae* by conventional chromatography is typically 60% active, and the recombinant LtMRP1 and LtMRP2 are 25–35% active, depending on the preparation. Prizm3 (GraphPad) or SigmaPlot 8 software packages are suitable for K_d calculation by nonlinear regression analysis.

6.2.2. RNA annealing

Stimulation of annealing of two complementary RNAs by RNA binding proteins may be measured quantitatively by monitoring an increase in the resistance of guanylyl residues to digestion by RNase T1, which cleaves a single-stranded RNA 3' of these nucleosides. Upon strand hybridization, Gs in the double-stranded region become inaccessible while single-stranded RNA is rapidly digested. Because the RNA annealing activity is nucleotide sequence–independent, substrates of arbitrary sequence or derived from trypanosomal gRNA–mRNA pairs may be utilized. The choice of a particular strand to be synthesized as radiolabeled RNA does not appear to affect the results. Importantly, the radioactively labeled strand should have non-complimentary termini containing at least one guanosine residue so that the protected fragment can be electrophoretically separated from the input RNA. Fragments of 30–60 bases are synthesized by run-off transcription as for RNA binding assays, except that for the synthesis of nonlabeled RNA concentrations all NTPs are maintained at 2 mM. To generate the fully annealed control, 1 pmol of radioactively labeled strand is incubated with 2 pmol of nonlabeled RNAs in a 10-μl reaction at 90° for 2 min in 20 mM HEPES (pH 7.5) 50 mM KCl, 0.1 mM EDTA and cooled to 20° over a 30-min time interval.

The RNA annealing reactions contain 0.5 nM labeled mRNA and 0.5–50 nM nonradioactive RNA, and the protein concentrations vary from 0 to 5 μM. Routinely, the protein is preincubated with radiolabeled RNA in 10 μl of annealing buffer [20 mM HEPES (pH 7.6), 50 mM KCl, 1 mM MgCl$_2$, 1 mM DTT] at 27° for 10 min. The nonlabeled RNA is treated identically, and the reaction is initiated by combining the two mixtures. After incubation for 0.5, 1, 2, 4, and 10 min, the reaction is stopped by the addition of 1 μl T1 RNase (18 U/μl) and incubated for

10 min at 27°, followed by phenol/chloroform extraction and ethanol precipitation. Products are analyzed on 15% polyacrylamide/urea gels and exposed to a phosphor storage plate (Fig. 4.2C). The time dependence of the signal intensity of the protected fragment at a given complementary RNA concentration is fitted into a single-phase exponential association model to obtain the first-order rate constant.

7. Purification of RBP16 from Procyclic Form *Trypansoma brucei*

Native RBP16 is posttranslationally methylated on arginine residues, and different methylated derivatives of the protein are apparently present in procyclic forms (Pelletier *et al.*, 2001). Methylation modulates some aspects of the RBP16 function *in vivo* (Goulah *et al.*, 2006). Since recombinant RBP16 is unmodified, comparison of native and recombinant proteins will be useful for analyzing the direct effects of methylation on RBP16 interaction with specific RNAs and proteins. It is not yet known whether the methylation state of RBP16 in the bloodstream forms differs from that in procyclic forms, but comparative analysis of RBP16 isolated from both life-cycle stages may reveal different forms of the proteins with different properties.

7.1. Cells and growth medium

Procyclic form *T. brucei* clone IsTAR1 stock EATRO 164 is routinely used in our laboratory. Cells are grown in SDM-79 media supplemented with fetal calf serum (10% final) (Brun and Schonenberger, 1979). Indicated quantities are per liter of media: 7 g of MEM F-14 (Invitrogen), 2 g Medium 199 TC 45 (Sigma-Aldrich), 1 g glucose, 8 g HEPES, 5 g 3-morpholinopropanesulfonic acid (MOPS), 2 g sodium bicarbonate, 200 mg L-alanine, 100 mg L-arginine, 300 mg L-glutamine, 70 mg L-methionine, 80 mg L-phenylalanine, 600 mg L-proline, 60 mg L-serine, 160 mg L-taurine, 350 mg L-threonine, 100 mg L-tyrosine, 10 mg adenosine, 10 mg guanosine, 50 mg glucosamine–HCl, 8 ml MEM amino acids (50×) without glutamine (Invitrogen), 6 ml MEM nonessential amino acids (100×) (Invitrogen), 9.08 ml sodium pyruvate (100×) (Invitrogen), 10 ml vitamins (100×) (Invitrogen), and 5 ml glycerol. Prior to sterile filtration on 0.22-μm filters, the pH of the media is adjusted to 7.3, and the volume is completed to 900 ml with double deionized water. Following filtration, 3 ml of 2.5 mg/ml hemin (in 0.05 M NaOH), 100 ml of fetal calf serum (denatured by heating at 55° for 30 min), and 10 ml of penicillin–streptomycin (10,000 units each/ml) (Invitrogen) are sterilely added. Cells are grown in this media at 27° with gentle agitation (100 rpm).

7.2. Purification of mitochondrial vesicles

The following buffers are used during the mitochondria purification procedure (Harris *et al.*, 1990): (1) DTE: 1 mM Tris (pH 8.0), 1 mM EDTA (pH 8.0); (2) SBG: 20 mM glucose, 150 mM NaCl, 20 mM sodium phosphate buffer (pH 7.9); (3) STM: 250 mM sucrose, 20 mM Tris (pH 8.0), 2 mM MgCl₂; and (4) STE: 250 mM sucrose, 20 mM Tris (pH 8.0), 2 mM EDTA (pH 8.0). For the lysis of the purified mitochondrial vesicles, we use a buffer composed of 25 mM Tris (pH 8.0), 15 mM MgOAc, and 50 mM KCl, which we designate Buffer A. All buffers are filter-sterilized.

Routinely, 12 liters of cells are grown up in SDM-79 media to a density of 1–3 × 10⁷ cells/ml. We also have isolated mitochondria from as little as 2 liters and as much as 20 liters of cells at a time (Ryan and Read, 2005). Cells are counted using a hemacytometer and pelleted at 6000×*g* for 10 min at 4°. The pellets are then resuspended in 200–300 ml of SBG and combined in one centrifuge bottle. Following a 10-min centrifugation at 6000×*g* at 4°, the SBG is carefully removed and the pellet is resuspended in DTE at 1.2 × 10⁹ cells/ml. The suspension is then dounced five times using a Kontes tissue grinder pestle SC40 (catalog no. 885302–0040) and a Wheaton Potter-Elvehjem tissue grinder (catalog no. 358049). The cells are then lysed by passage through a 26-gauge needle into a 60% sucrose solution (filter sterilized). We use a volume of 60% sucrose corresponding to one-seventh of the volume of DTE from the previous step. The membrane fraction is then pelleted by centrifugation at 15,800×*g* for 10 min at 4° and resuspended in a volume of STM buffer equal to one-sixth of the original cell lysate plus sucrose. We then add 3 mM MgCl₂ (final), 0.3 mM CaCl₂ (final), and 117 units of RNase-free DNase I (Invitrogen) per ml. Following a 30–60 min incubation on ice, an equal volume of STE is added and the crude mitochondrial fraction is pelleted at 15,800×*g* for 10 min at 4°. The pellet is resuspended in 75% sterile Percoll (prepared in STE buffer), dispersed as much as possible, and loaded at the bottom of a linear 32 ml 20–35% Percoll gradient (prepared in STE). For every 2 liters of cells, we use 4 ml of 75% Percoll, and this suspension is layered under one gradient. For loading, we use a 10-ml syringe and a 14-gauge, 15-cm-long needle. The gradients are formed in Beckman ultraclear centrifuge tubes (25 × 89 mm, catalog no. 344058) using a Hoefer SG50 gradient maker. The tubes are balanced with 20% Percoll (in STE) and centrifuged at 103,900×*g* for 50 min at 4° in a Beckman SW-28 rotor. The mitochondrial vesicles are present in a broad smear, which is located between two more distinct bands, one near the top of the gradient and a slightly less prominent one near the bottom of the gradient. An 18-gauge, 1.5-in. needle connected to a 30-ml syringe is used to puncture the centrifuge tube just below the smear containing mitochondrial vesicles. The vesicles are collected in the syringe and transferred to a 40-ml Oakridge tube. The tube is then filled with STE and centrifuged at 32,530×*g* for 15 min at 4° using a slow deceleration setting. About half of the supernatant is discarded, typically using a 10-ml pipette. This step must be performed carefully, so as not to disrupt the initial pellets, which are very soft. The STE washes are repeated three additional times. As the pellets become more solid with every wash, more of the supernatant can be discarding every time until after the final wash, when all the supernatant can be discarded. Finally, the mitochondria are resuspended in 50% glycerol in STE (1 part 100% glycerol, 1 part 2× STE) at a protein concentration of about 1 mg/ml. Typically, 1×10^{10} cells equals 5 mg of mitochondrial protein. Immediately snap freeze this in liquid nitrogen and store at −80°. This procedure usually results in 40–50 1-ml aliquots from 12 liters of culture at 1–3×10^7 cells/ml.

7.3. Mitochondrial vesicle lysis

Purified mitochondrial vesicles are thawed and centrifuged at 13,000 rpm in a Biofuge centrifuge (Heraeus Instruments) for 15 min at 4°. The pellet is resuspended in Buffer A [containing 1 m*M* phenylmethylsulfonyl fluoride (PMSF), 1 *μ*g/ml leupeptin, and 1 *μ*g/ml pestatin A to minimize proteolysis] at a concentration of 1×10^{11} cells/ml. Vesicles are lysed by the addition of NP-40 to 0.2% (final) and incubation on ice for 5 min. The insoluble material is then cleared from mitochondrial extracts by centrifugation at 13,000×*g* for 10 min at 4°. After addition of $CaCl_2$ to 1 m*M*, the cleared extract is incubated with micrococcal nuclease (Sigma–Aldrich) (100 units/1×10^{11} cells) at 27° for 15 min. Micrococcal nuclease is then inhibited by the addition of ethylene glycol bis(2-aminoethyl ether)-*N*,*N*, *N′N′*-tetraacetic acid (EGTA) to 5 m*M*.

7.4. Poly(U) sepharose chromatography

Tris-based buffers containing magnesium chloride, EGTA, and potassium chloride are used. A buffer containing 25 m*M* Tris (pH 8.0), 1.5 m*M* magnesium acetate, 5 m*M* EGTA, 50 m*M* KCl, and 10% glycerol is used as column equilibration and wash buffer ["Poly(U)50 Buffer"]. For the elution of RBP16, the same buffer, except that it contains either 300 m*M* or 800 m*M* KCl, is used to create a linear gradient ranging from 300 to 800 m*M* KCl. To prevent proteolysis, PMSF (1 m*M* final), leupeptin (1 *μ*g/ml final), and pestatin A (1 *μ*g/ml final) are added to all buffers.

Typically, 1-ml of poly(U) Sepharose (GE Healthcare) is sufficient to purify RBP16 from a mitochondrial fraction equivalent to 1–4×10^{11} cells. A 1-ml poly(U) Sepharose column is equilibrated with 20 column volumes of wash buffer. The cleared, micrococcal nuclease-treated mitochondrial extract is then loaded onto the column and allowed to flow by gravity at a rate of approximately 1 ml/min. The flow rate is usually controlled manually using a two-way stopcock. The column is washed with 10 column volumes of wash buffer containing 300 mM KCl at a flow rate of approximately 1 ml/min. Bound proteins are eluted with 10 column volumes linear gradient from 300 to 800 mM of KCl. To generate the salt gradient, we use a Hoefer SG15 gradient maker connected to a peristaltic pump. Five milliliters of wash buffer containing 800 mM KCl is poured into the chamber the farthest from the outlet connector. Five milliliters of wash buffer containing 300 mM KCl is then poured into the chamber closest to the outlet connector (the mixing chamber). A small magnetic stir bar is placed in the mixing chamber. The magnetic stirrer is started and the delivery stopcock is then opened. Next, the connector stopcock is opened and the pump is started simultaneously. Twenty 0.5-ml fractions are collected in 1.5-ml Eppendorf tubes. Fractions are analyzed by SDS–PAGE on a 15% acrylamide gel Coommassie staining and and by UV cross-linking to synthetic gRNA gA6[14] internally labeled with [α-^{32}P]UTP. Typically, the bulk of RBP16 is eluted in fractions 10–15 (Fig. 4.3A and B).

7.5. Poly(A) sepharose chromatography

RBP16 can be further purified by chromatography on a poly(A) Sepharose column. The protein is recovered in the flowthrough of the poly(A) Sepharose column, as RBP16 does not bind poly(A). Fractions from the poly(U) Sepharose column that contain RBP16 are pooled and dialyzed twice against 1 liter of wash buffer containing 150 mM KCl at 4° for at least 5 h. The dialyzed sample is then applied to a 0.5-ml (bed volume) poly(A) Sepharose column and RBP16 is allowed to flow through at a rate of ~1 ml/min. The presence of RBP16 in the flowthrough is confirmed by SDS–PAGE on a 15% acrylamide gel followed by silver staining (Ansorge, 1985) and by Western hybridization using polyclonal anti-RBP16 antibodies as described (Hayman and Read, 1999). RBP16 is the predominant protein in the flowthrough fraction (Fig. 4.3C). Anti-RBP16 antibodies for detection of the protein in column fractions are available from the author. Using this procedure, mitochondrial vesicles from 12 liters (1–3 × 10^{11} cells) yield approximately 40 μg of RBP16.

8. Purification of RBP16 from the Bloodstream Forms *Trypansoma brucei*

8.1. Cells and growth medium

T. brucei strain 221 cells are grown in medium HMI-9, which contains (per liter): 714 ml Iscove's modified Dulbecco's medium (IMDM) (contains L-glutamine, 25 mM HEPES buffer, 3024 mg/liter sodium bicarbonate,

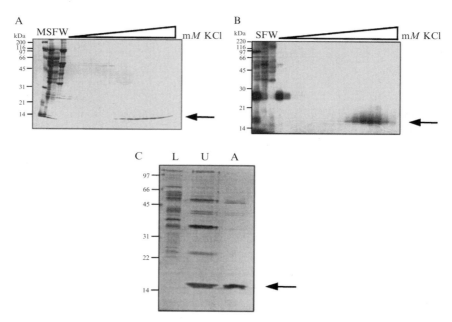

Figure 4.3 Purification of native RBP16 from procyclic form *T. brucei* by affinity chromatography on poly(U) and poly(A) Sepharose. *T. brucei* mitochondrial extract was loaded onto a poly(U) Sepharose column, and the column was washed with loading buffer containing 300 mM KCl. The column was then eluted with a linear 300–800 mM KCl gradient in loading buffer. (A) 0.5% of the starting material (S), 0.5% of the flow-through (F), 10% of the 300 mM wash (W), and 10% of eluted fractions were separated by SDS–PAGE on a 12.5% gel and stained with Coomassie Brilliant Blue. Molecular mass standards (M) are shown on the left. An arrow marks the position of an abundant 16-kDa protein (RBP16). (B) Fractions shown in (A) were assayed for gRNA-binding activities by UV cross-linking to radiolabeled gA6[14]. Ten femtomoles of gA6[14] was incubated with 4 μl each of total mitochondrial extract (S), poly(U) Sepharose flow-through (F), 300 mM KCl wash (W), and eluted poly(U) Sepharose fractions. Proteins were resolved by SDS–PAGE on a 15% gel and UV cross-linking proteins were detected by autoradiography. The positions of molecular mass markers are shown on the left. The position of gRNA UV cross-linking activity that corresponds in size and elution pattern to RBP16 is indicated with an arrow. (C) One microgram each of mitochondrial lysate (L), poly(U)-purified RBP16 (U), and the subsequent poly(A) Sepharose flowthrough (A) were separated by SDS–PAGE on a 15% gel. Proteins were detected by staining with silver. (Reprinted with permission from Hayman and Read, 1999.)

but no α-thioglycerol or 2-mercaptoethanol) (Invitrogen), 12.5 ml 100 mM sodium pyruvate (Invitrogen), 136 mg hypoxanthine (dissolved in 20 ml 0.1 N NaOH), 28 mg bathocuproine disulfonic acid (in 10 ml water), 182 mg cysteine (in 20 ml water), and 39 mg thymidine (in 10 ml water). The pH is adjusted to 7.6, and the medium is then sterilized by filtration on 0.22-μm filters. Following filtration, 100 ml of fetal calf serum (denatured at 55° for 30 min), 100 ml of Serum Plus (JRH Bioscience), and 10 ml of penicillin–streptomycin (10,000 units each/ml) (Invitrogen) are sterilely added. Prior to use, 2-mercaptoethanol is added to a final concentration of 0.2 mM. We typically prepare a 1000× stock solution (140 μl of 2-mercaptoethanol in 10 ml water). Cells are grown in this media at 37° in the presence of 5% CO_2 (Hirumi and Hirumi, 1989).

8.2. Isolation of mitochondria

Twelve liters of cells in HMI-9 are grown up to a density of 1×10^6 cells/ml. Cells are processed as described for the procyclic cells except that once the cells are resuspended in DTE, sucrose is added to a final concentration of 0.5%.

8.3. Purification of RBP16 from mitochondrial vesicles

RBP16 is purified from the bloodstream form *T. brucei* as described previously for the procyclic form. The typical yield is approximately 1–2 μg of RBP16 from 12 liters ($\sim 1 \times 10^{10}$ cells).

9. EXPRESSION AND PURIFICATION OF RECOMBINANT RBP16

Recombinant RBP16 is routinely expressed as a fusion with Maltose Binding Protein at the N-terminus (MBP-RBP16), or as a fusion protein with a C-terminal 6xHis–Tag sequence (His-RBP16) (Hayman and Read, 1999; Hayman et al., 2001). Expression as an MBP fusion protein yields nearly 10-fold more protein than that of the 6xHis-tagged version. However, since MBP is a large tag (42 kDa), production of RBP16 with the smaller 6xHis tag is often preferable. Expression of both fusion proteins is carried out in *Escherichia coli* BL21 pLyS cells (Novagen). Purification of MBP-RBP16 is achieved by chromatography over amylose resin followed by poly(U) Sepharose. Purification of His-RBP16 is carried out by chromatography over nickel–resin followed by poly(U) Sepharose.

9.1. Expression and purification of MBP-RBP16

The buffers used in the purification are (1) amylose column buffer: 20 mM Tris (pH 7.5), 200 mM NaCl, 1 mM EDTA, 10% glycerol and (2) amylose elution buffer: 20 mM Tris (pH 7.5), 200 mM NaCl, 1 mM EDTA, 10% glycerol, 20 mM maltose. For poly(U) Sepharose chromatography, we use the same buffers previously described for the purification of native RBP16. PMSF (1 mM final), leupeptin (1 μg/ml final), and pestatin A (1 μg/ml final) are added to all buffers.

The full-length RBP16 open reading frame was amplified by PCR from total procyclic cDNA and cloned into the *Bam*HI/*Sal*I sites of pMal-C2 (New England Biolabs) as described (Hayman and Read, 1999). Induction of MBP-RBP16 is performed as follows: *E. coli* BL21 pLyS cells containing pMal-C2-RBP16 are grown overnight in 20 ml of LB broth with 100 μg/ml ampicillin at 37°. The next day, 10 ml of this overnight culture is transferred to 1 liter of LB broth containing 100 μg/ml ampicilin and the cells are allowed to grow at 37° (225 rpm) up to an OD_{600} of 0.5–0.6. IPTG is then added to a final concentration of 0.3 mM, and the cells are grown for 2 more hours at 37° (225 rpm). Following induction, cells are harvested by centrifugation at 5000$\times g$ for 10 min at 4°, and the pellet is resuspended in 1/10 volume (100 ml) of amylose column buffer. The cells are then sonicated for four periods of 30 sec at intensity 5, 50% pulse (settings are for the Sonifier cell disruptor 300 from Branson Ultrasonics), and centrifuged at 14,000$\times g$ for 20 min at 4°. The supernatant is diluted 1:4 in amylose column buffer, NP-40 is added to 0.1%, and the supernatant is incubated with 2 ml of amylose resin (GE Healthcare) (preequilibrated with amylose column buffer) for 2 h at 4° with gentle rocking. The mixture is then applied to a 1.5-cm-diameter column and allowed to flow through at a rate of ~1 ml/min. The column is washed with 20 column volumes (40 ml) of amylose wash buffer at a flow rate of ~1 ml/min. Bound MBP-RBP16 is eluted with 5 column volumes (10 ml) of elution buffer. Ten 1-ml fractions are collected and electrophoresed on SDS–PAGE, and the presence of MBP-RBP16 determined by Coomassie staining (MBP-RBP16 has an electrophoretic mobility of 58 kDa) (Fig. 4.4). Fractions that contain MBP-RBP16 are pooled, diluted 1:2 in 25 mM Tris (pH 8.0), 1.5 mM magnesium acetate, 5 mM EGTA, 50 mM KCl, and 10% glycerol buffer and further purified by poly(U) Sepharose chromatography, as described above, using a column with a 3-ml bed volume. The resulting MBP-RBP16 is essentially pure (Fig. 4.4). The expected yield following these two chromatographic steps is approximately 10–30 mg of purified MBP-RBP16 per liter of *E. coli* cells.

9.2. Expression and purification of His-RBP16

The following buffers are used for the purification of His-RBP16. (1) Lysis buffer: 10 mM Tris (pH 6.8), 300 mM NaCl, 10 mM imidazole, 10% glycerol, 0.1% sodium deoxycholate, 0.01% NP40, 10 mM MgCl$_2$, 0.06 mg/ml lysozyme, and 0.06 mg/ml DNase I. (2) Wash buffer: 10 mM Tris (pH 6.8), 300 mM NaCl, 30 mM imidazole. (3) Elution buffer: 10 mM (pH 6.8), 300 mM NaCl, 250 mM imidazole. All buffers contain PMSF (1 mM final), leupeptin (1 μg/ml final), and pestatin A (1 μg/ml final).

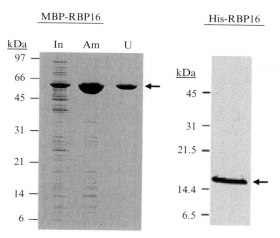

Figure 4.4   Expression and purification of recombinant RBP16 from *E. coli*. Left panel: *E. coli* cells harboring the pMal-C2 plasmid with the RBP16 gene in frame with MBP were grown in the presence of 0.3 mM isopropyl-1-thio-β-D-galactopyranoside. Input (In), amylose column eluate (Am), and subsequent poly(U)-purified protein (U) were separated by SDS–PAGE on a 12.5% gel and stained with Coomassie Blue. The 59-kDa fusion protein is indicated with an arrow. (Reprinted with permission from Hayman and Read, 1999.) Right panel: *E. coli* cells harboring the pET-21a plasmid with the RBP16 gene in frame with a C-terminal 6xHis-Tag sequence were grown in the presence of 0.1 mM isopropyl-1-thio-β-D-galactopyranoside. Purified His-RBP16 was electrophoresed on 15% SDS–PAGE and stained with Coomassie Blue. The position of the 20-kDa fusion protein is indicated with an arrow. (Reprinted with permission from Miller and Read, 2003.)

The full-length RBP16 open reading frame was amplified by PCR and cloned into the *NdeI/XhoI* sites of pET-21a (Novagen) as described (Miller and Read, 2003). Induction of His-RBP16 is achieved as follows: *E. coli* BL21 pLyS cells containing pET-21a-RBP16 are grown at 37° in 20 ml of LB broth containing 100 μg/ml ampicillin overnight at 37° with vigorous shaking (225 rpm). The next day, 10 ml of this overnight culture is transferred to 1 liter of LB broth containing 100 μg/ml ampicillin and the cells are allowed to grow at 37° (225 rpm) up to an OD$_{600}$ of 0.5–0.6. IPTG is then added to a final concentration of 0.1 mM, and the induction of His-RBP16 is allowed to proceed for 2.5 h at 30° (225 rpm). Cells are collected by centrifugation for 10 min at 5000×g at 4°. Cells are weighed and resuspended in 3 ml lysis buffer per gram of cells and sonicated for four periods of 30 sec at intensity 5, 50% pulse (settings are for the Sonifier cell disruptor 300 from Branson Ultrasonics). The supernatant fraction is collected after centrifugation at 12,000×g for 20 min at 4°. Polyethylenimine is added to the supernatant to a final concentration of 0.05%. The resulting suspension is rocked for 20 min at 4°, and then centrifuged at 12,000×g for

10 min at 4° and the supernatant is recovered. To this supernatant is added one–twentieth volume of ProBond resin (Invitrogen), which has been preequilibrated with wash buffer. This mixture is rocked for 1 h at 4° and then poured into a 1.5-cm-diameter column. The flow rate is set at ~1 ml/min, and the solution containing the unbound proteins is allowed to flow through. The column is washed with 20 bed volumes of wash buffer, and the recombinant His-RBP16 eluted from the column in 5 volumes of elution buffer. Fractions containing His-RBP16 are identified by SDS–PAGE and Coomassie staining. These fractions are pooled and dialyzed against 100 volumes Poly(U)50 buffer [see Poly(U) Sepharose chromatography]. His-RBP16 is then subjected to poly(U) Sepharose chromatography essentially as described above, using a column with a 3-ml bed volume, except that the column is first washed with 10 volumes of poly(U) Sepharose wash buffer containing 200 mM KCl. Bound His-RBP16 is eluted with 5 volumes of wash buffer containing 800 mM KCl. The presence and purity of His-RBP16 are determined by SDS–PAGE Coomassie staining (His-RBP16 has an electrophoretic mobility of ~20 kDa). His-RBP16 is the sole protein visible in the elution (Fig. 4.4). Starting with 1 liter of *E. coli* cells, the typical yield after Probond resin and poly(U) Sepharose chromatography is between 1.5 and 2.0 mg of purified His-RBP16.

9.3. Assay of RBP16 RNA binding activity by UV X-linking to synthetic gRNA

To determine that the purified protein retains RNA binding activity, the ability of RBP16 to bind to gRNA is assayed by UV X-linking to the synthetic gRNA gA6[14] (Fig. 4.3B). For synthesis of body-labeled gA6 [14], we use a construct encoding the gRNA gA6[14] with a 17 nucleotide oligo(U) tail as previously described (Read *et al.*, 1994). The plasmid is digested with *Dra*I (1 unit/μg plasmid) at 37° for at least 1 h. Following extraction with 1 volume phenol/chloroform, the digested plasmid is precipitated with one-tenth volume of ammonium acetate and 2.5 volumes of cold 95% ethanol for 30 min at −20°. The sample is then centrifuged for 30 min at 13,000 rpm and the pellet is washed with 500 μl of cold 70% ethanol, centrifuged again at 13,000 rpm for 5 min, and finally air-dried.

gA6[14] internally labeled with [α-^{32}P]UTP is generated by *in vitro* transcription using the MAXIscript T7/T3 kit (Ambion). A typical 20-μl reaction consists of the following: 1 μg *Dra*I-digested gA6[14] plasmid, 2 μl 10× transcription buffer, 2 μl 0.1 M DTT, 1 μl 10 mM ATP, 1 μl 10 mM CTP, 1 μl 10 mM GTP, 1 μl 0.05 mM UTP, 5 μl [α-^{32}P]UTP (10 μCi/ μl; 3000 Ci/mmol), and 1 μl RNase inhibitor (2 units/μl). The reaction is initiated by the addition of 2 μl T7 polymerase (15 units/μl) and allowed to proceed for 1 h at room temperature. The reaction is stopped by the addition of 20 μl of 90% formamide (in TBE) and boiled for 3 min.

The ^{32}P-labeled gRNA is gel-purified on a 19-cm tall, 6% acrylamide/7 M urea gel. Following electrophoresis in TBE for 1 h at 420 V, the gel is exposed to a film for 30–60 sec. The film is then developed and super-imposed on the gel. The band corresponding to the ^{32}P-labeled gA6[14] is excised from the gel using a clean razor blade and incubated in 750 μl of TE buffer (pH 7.5) containing ammonium acetate (0.75 M final) overnight at 4° with gentle rocking. The following day, the gRNA is precipitated with an equal volume of cold isopropanol at –20° for 1 h, then centrifuged for 60 min at 13,000 rpm, the pellet washed with 500 μl of cold 70% ethanol, centrifuged again at 13,000 rpm for 5 min, and finally air-dried and resuspended in 20 μl of DEPC-treated water. The concentration and specific activity of the labeled gRNA are determined by scintillation counting 10 μl of a 1:10 dilution.

To assess the RNA binding activity of RBP16 following chromatography, 10 μl of each fraction is incubated with 10 fmol of ^{32}P-labeled gA6[14] for 10 min at room temperature in a 6 mM HEPES buffer (pH 7.5) containing 2.1 mM MgCl$_2$, 0.5 mM DTT, 1.5 mM ATP, 5 mM creatine phosphate, 0.1 mM EDTA, 10 μg/μl torula yeast RNA, and 6% glycerol, in a total volume of 15 μl. Following incubation, samples are transferred to a 96-well plate on ice, positioned 4–5 cm to the light source, and irradiated for 10 min at 254 nm using a UV Stratalinker 2400 (Stratagene). One microgram of RNase A per microliter is then added, and samples are incubated at 37° for 15 min. SDS–PAGE loading buffer is added and the samples are boiled for 5 min before loading on a 15% polyacrylamide gel. The gel is dried and the ^{32}P-labeled proteins are detected by autoradiography or phosphorimager analysis.

ACKNOWLEDGMENTS

This work was supported by grants from the National Institute of Allergy and Infectious Diseases to Laurie K. Read (AI060260 and AI061580) and to Ruslan Aphasizhev (AI064653).

REFERENCES

Allen, T. E., Heidmann, S., Reed, R., Myler, P. J., Goringer, H. U., and Stuart, K. D. (1998). Association of guide RNA binding protein gBP21 with active RNA editing complexes in *Trypanosoma brucei*. *Mol. Cell. Biol.* **18,** 6014–6022.
Ansorge, W. (1985). Fast and sensitive detection of protein and DNA bands by treatment with potassium permanganate 1. *J. Biochem. Biophys. Methods* **11,** 13–20.
Aphasizhev, R., Aphasizheva, I., Nelson, R. E., and Simpson, L. (2003). A 100-kD complex of two RNA-binding proteins from mitochondria of *Leishmania tarentolae* catalyzes RNA annealing and interacts with several RNA editing components. *RNA* **9,** 62–76.

Blom, D., Burg, J., Breek, C. K., Speijer, D., Muijsers, A. O., and Benne, R. (2001). Cloning and characterization of two guide RNA-binding proteins from mitochondria of Crithidia fasciculata: gBP27, a novel protein, and gBP29, the orthologue of *Trypanosoma brucei* gBP21. *Nucleic Acids Res.* **29,** 2950–2962.

Blum, B., Bakalara, N., and Simpson, L. (1990). A model for RNA editing in kinetoplastid mitochondria: "Guide" RNA molecules transcribed from maxicircle DNA provide the edited information. *Cell* **60,** 189–198.

Braly, P., Simpson, L., and Kretzer, F. (1974). Isolation of kinetoplast-mitochondrial complexes from *Leishmania tarentolae. J. Protozool.* **21,** 782–790.

Bringaud, F., Peris, M., Zen, K. H., and Simpson, L. (1995). Characterization of two nuclear-encoded protein components of mitochondrial ribonucleoprotein complexes from *Leishmania tarentolae. Mol. Biochem. Parasitol.* **71,** 65–79.

Bringaud, F., Stripecke, R., Frech, G. C., Freedland, S., Turck, C., Byrne, E. M., and Simpson, L. (1997). Mitochondrial glutamate dehydrogenase from *Leishmania tarentolae* is a guide RNA-binding protein. *Mol. Cell. Biol.* **17,** 3915–3923.

Brun, R., and Schonenberger, M. (1979). Cultivation and *in vitro* cloning or procyclic culture forms of *Trypanosoma brucei* in a semi-defined medium. Short communication. *Acta Trop.* **36,** 289–292.

Desveaux, D., Allard, J., Brisson, N., and Sygusch, J. (2002). A new family of plant transcription factors displays a novel ssDNA-binding surface. *Nat. Struct. Biol.* **9,** 512–517.

Goulah, C. C., Pelletier, M., and Read, L. K. (2006). Arginine methylation regulates mitochondrial gene expression in *Trypanosoma brucei* through multiple effector proteins. *RNA* **12,** 1545–1555.

Harris, M. E., Moore, D. R., and Hajduk, S. L. (1990). Addition of uridines to edited RNAs in trypanosome mitochondria occurs independently of transcription. *J. Biol. Chem.* **265,** 11368–11376.

Hayman, M. L., and Read, L. K. (1999). *Trypanosoma brucei* RBP16 is a mitochondrial Y-box family protein with guide RNA binding activity. *J. Biol. Chem.* **274,** 12067–12074.

Hayman, M. L., Miller, M. M., Chandler, D. M., Goulah, C. C., and Read, L. K. (2001). The trypanosome homolog of human p32 interacts with RBP16 and stimulates its gRNA binding activity 1. *Nucleic Acids Res.* **29,** 5216–5225.

Hermann, T., Schmid, B., Heumann, H., and Goringer, H. U. (1997). A three-dimensional working model for a guide RNA from *Trypanosoma brucei. Nucleic Acids Res.* **25,** 2311–2318.

Hirumi, H., and Hirumi, K. (1989). Continuous cultivation of *Trypanosoma brucei* blood stream forms in a medium containing a low concentration of serum protein without feeder cell layers. *J. Parasitol.* **75,** 985–989.

Jiang, W., Hou, Y., and Inouye, M. (1997). CspA, the major cold-shock protein of *Escherichia coli,* is an RNA chaperone 4. *J. Biol. Chem.* **272,** 196–202.

Kohno, K., Izumi, H., Uchiumi, T., Ashizuka, M., and Kuwano, M. (2003). The pleiotropic functions of the Y-box-binding protein, YB-1 21. *BioEssays* **25,** 691–698.

Koller, J., Muller, U. F., Schmid, B., Missel, A., Kruft, V., Stuart, K., and Goringer, H. U. (1997). *Trypanosoma brucei* gBP21. An arginine-rich mitochondrial protein that binds to guide RNA with high affinity. *J. Biol. Chem.* **272,** 3749–3757.

Lambert, L., Muller, U. F., Souza, A. E., and Goringer, H. U. (1999). The involvement of gRNA-binding protein gBP21 in RNA editing–an *in vitro* and *in vivo* analysis. *Nucleic Acids Res.* **27,** 1429–1436.

Militello, K. T., Hayman, M. L., and Read, L. K. (2000). Transcriptional and post-transcriptional in organello labelling of *Trypanosoma brucei* mitochondrial RNA. *Int. J. Parasitol.* **30,** 643–647.

Miller, M. M., and Read, L. K. (2003). *Trypanosoma brucei*: Functions of RBP16 cold shock and RGG domains in macromolecular interactions 2. *Exp. Parasitol.* **105,** 140–148.

Miller, M. M., Halbig, K., Cruz-Reyes, J., and Read, L. K. (2006). RBP16 stimulates trypanosome RNA editing *in vitro* at an early step in the editing reaction 4. *RNA* **12,** 1292–1303.

Muller, U. F., and Goringer, H. U. (2002). Mechanism of the gBP21-mediated RNA/RNA annealing reaction: Matchmaking and charge reduction 2. *Nucleic Acids Res.* **30,** 447–455.

Muller, U. F., Lambert, L., and Goringer, H. U. (2001). Annealing of RNA editing substrates facilitated by guide RNA-binding protein gBP21. *EMBO J.* **20,** 1394–1404.

Pelletier, M., and Read, L. K. (2003). RBP16 is a multifunctional gene regulatory protein involved in editing and stabilization of specific mitochondrial mRNAs in *Trypanosoma brucei. RNA* **9,** 457–468.

Pelletier, M., Xu, Y., Wang, X., Zahariev, S., Pongor, S., Aletta, J. M., and Read, L. K. (2001). Arginine methylation of a mitochondrial guide RNA binding protein from *Trypanosoma brucei. Mol. Biochem. Parasitol.* **118,** 49–59.

Puig, O., Caspary, F., Rigaut, G., Rutz, B., Bouveret, E., Bragado-Nilsson, E., Wilm, M., and Seraphin, B. (2001). The tandem affinity purification (TAP) method: A general procedure of protein complex purification. *Methods* **24,** 218–229.

Rajkowitsch, L., Semrad, K., Mayer, O., and Schroeder, R. (2005). Assays for the RNA chaperone activity of proteins 15. *Biochem. Soc. Trans.* **33,** 450–456.

Read, L. K., Goringer, H. U., and Stuart, K. (1994). Assembly of mitochondrial ribonucleoprotein complexes involves specific guide RNA (gRNA)-binding proteins and gRNA domains but does not require preedited mRNA. *Mol. Cell. Biol.* **14,** 2629–2639.

Robinson, K. A., and Beverley, S. M. (2003). Improvements in transfection efficiency and tests of RNA interference (RNAi) approaches in the protozoan parasite *Leishmania. Mol. Biochem. Parasitol.* **128,** 217–228.

Ryan, C. M., and Read, L. K. (2005). UTP-dependent turnover of *Trypanosoma brucei* mitochondrial mRNA requires UTP polymerization and involves the RET1 TUTase. *RNA* **11,** 763–773.

Sanford, J. R., Ellis, J., and Caceres, J. F. (2005). Multiple roles of arginine/serine-rich splicing factors in RNA processing 3. *Biochem. Soc. Trans.* **33,** 443–446.

Schumacher, M. A., Karamooz, E., Zikova, A., Trantirek, L., and Lukes, J. (2006). Crystal structures of *T. brucei* MRP1/MRP2 guide-RNA binding complex reveal RNA matchmaking mechanism 24. *Cell* **126,** 701–711.

Simpson, L., Sbicego, S., and Aphasizhev, R. (2003). Uridine insertion/deletion RNA editing in trypanosome mitochondria: A complex business. *RNA* **9,** 265–276.

Skabkin, M. A., Evdokimova, V., Thomas, A. A., and Ovchinnikov, L. P. (2001). The major messenger ribonucleoprotein particle protein p50 (YB-1) promotes nucleic acid strand annealing 1. *J. Biol. Chem.* **276,** 44841–44847.

Stuart, K. D., Schnaufer, A., Ernst, N. L., and Panigrahi, A. K. (2005). Complex management: RNA editing in trypanosomes. *Trends Biochem. Sci.* **30,** 97–105.

Vondruskova, E., Van den, B. J., Zikova, A., Ernst, N. L., Stuart, K., Benne, R., and Lukes, J. (2005). RNA interference analyses suggest a transcript-specific regulatory role for mitochondrial RNA-binding proteins MRP1 and MRP2 in RNA editing and other RNA processing in *Trypanosoma brucei* 1. *J. Biol. Chem.* **280,** 2429–2438.

CHAPTER FIVE

RNA–PROTEIN INTERACTIONS IN ASSEMBLED EDITING COMPLEXES IN TRYPANOSOMES

Jorge Cruz-Reyes

Contents

Department of Biochemistry & Biophysics, Texas A&M University, College Station, Texas

Methods in Enzymology, Volume 424
ISSN 0076-6879, DOI: 10.1016/S0076-6879(07)24005-4

Abstract

Multisubunit RNA editing complexes recognize thousands of pre-mRNA sites in the single mitochondrion of trypanosomes. Specific determinants at each editing site must trigger the complexes to catalyze a complete cycle of either uridylate insertion or deletion. While a wealth of information on the protein composition and catalytic activities of these complexes is currently available, the precise mechanisms that govern substrate recognition and editing site specificity remain unknown. This chapter describes basic assays to visualize direct photocrosslinking interactions between purified editing complexes and targeted deletion and insertion sites in model substrates for full-round editing. It also illustrates how variations of these assays can be applied to examine the specificity of the editing enzyme/substrate association, and to dissect structural or biochemical requirements of both the substrates and enzyme complex.

1. INTRODUCTION

Trypanosome U insertion and U deletion RNA editing of mitochondrial pre-mRNAs is catalyzed by multisubunit editing complexes also termed ~20S editosomes or L-complexes (Simpson *et al.*, 2004; Stuart *et al.*, 2005). These reactions are directed by trans–acting guide RNAs (gRNAs) that form canonical and G–U base pairs with mature mRNA sequence. The basic enzymatic mechanisms and protein composition of these high-molecular-mass complexes have been under intense study, but the specific interactions with functional pre-mRNA/gRNA hybrid substrates are unknown. Major efforts in the coming years may unravel the molecular mechanisms of editing complex docking on bona fide substrates, discrimination of mRNA and gRNA strands in a hybrid, and identification of deletion and insertion editing sites (ESs). In such studies, the identification and characterization of specific associations between assembled editing complexes and functional substrates will be critical.

The editing process is believed to be initiated with the formation of an "anchor duplex" with pre-mRNA and gRNA. Catalysis of a complete editing cycle involves three basic activities: namely, mRNA endonuclease, $3'$ terminal uridylyltransferase (TUTase, in insertion) or $3'$ to $5'$ U–specific exoribonuclease (in deletion), and RNA ligase. Significant information has been accumulated on the functional and structural composition of editing complexes, including the identity of the catalytic subunits (Simpson *et al.*, 2004; Stuart *et al.*, 2005). The complexes are heterogeneous in protein composition but share most of the approximately 20 subunits identified (Panigrahi *et al.*, 2006). Several factors known or proposed to play auxiliary roles in editing (Aphasizhev *et al.*, 2003b; Blom *et al.*, 2001; Halbig *et al.*, 2006; Madison-Antenucci and Hajduk, 2001; Miller *et al.*, 2006; Missel *et al.*, 1997; Muller *et al.*, 2001; Panigrahi *et al.*, 2003; Pelletier and Read, 2003; Vanhamme *et al.*, 1998; Vondruskova *et al.*, 2005) do not copurify with tightly bound editing complexes and are dispensable *in vitro* (Allen *et al.*, 1998; Aphasizhev *et al.*, 2003a; Halbig *et al.*, 2006; Panigrahi *et al.*, 2003; Rusche *et al.*, 1997).

There is an increasing interest in different laboratories in identifying RNA-binding subunits of editing complexes and substrate determinants of editing function. Several subunits of editing complexes contain conserved motifs for potential nucleic acid binding; however, only isolated recombinant KREPA3 (MP42) and KREPA4 (MP24) have been shown to bind RNA (Brecht *et al.*, 2005; Salavati *et al.*, 2006). The latter exhibited specificity for poly(U), which is characteristic of the $3'$ terminus of natural gRNAs. Other studies have defined minimal substrates for full-round editing *in vitro* including a 43-nt pre-mRNA for insertion (Fig. 5.1A) (Cifuentes-Rojas *et al.*, 2005). Systematic mutagenesis of this transcript defined a completely artificial model substrate (Cifuentes-Rojas *et al.*, 2007) showing that secondary structure, rather than specific nucleotide sequence, is recognized for the assembly of ribonucleoprotein complexes during full-round editing. Such structural features include 11-bp helices flanking the internal loop region that contain the targeted ES (Fig. 5.1A). Furthermore, the structure of ESs for full-round editing *in vitro* conforms to a relatively simple geometry that can be manipulated to convert natural sites from deletion to insertion and vice versa (Fig. 5.1B) (Cifuentes-Rojas *et al.*, 2005). Thus, the basic determinants for the type of editing reside within the ES but sequence-independent flanking helices are also critical for efficient editing. All of these substrate features are likely to make important contacts with editing subunits. Finally, the first observations of direct editing complex interactions with substrates for full-round editing were based on the introduction of a highly photoreactive 4-thioU at targeted ESs (Fig. 5.2) (Cifuentes-Rojas *et al.*, 2007; Sacharidou *et al.*, 2006). Such an approach detected at least four specific photocrosslinking interactions by editing complex subunits, which are both conserved and relatively stable at deletion and insertion sites in pre-mRNA/gRNA hybrids. This chapter describes the methodology used in photocrosslinking studies and variations of the basic assays to examine the specificity of the editing enzyme/substrate association, as well as structural and biochemical requirements for efficient RNA–protein interactions in the context of assembled editing complexes. The methods first established in *Trypanosoma brucei* should be also suitable to study the very similar editing complexes in *Leishmania* species.

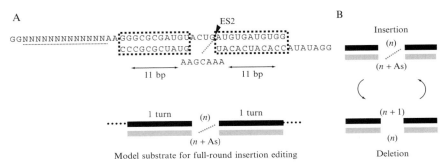

Figure 5.1 Editing complexes recognize the secondary structure of enhanced substrates for full-round editing. (A, top) A minimal 43-nt-long pre-mRNA substrate for U insertion forming completely artificial (11-bp) duplexes with gRNA in the lower strand; "Ns" indicate purines or pyrimidines at protruding positions (Cifuentes-Rojas *et al.*, 2005, 2007). One or two 5' most purines are needed for efficient *in vitro* T7 transcription, and potential pairing by the two residues indicated by a dotted line enhances insertion (Igo *et al.*, 2002). (A, bottom) An artificial model substrate for full-round U insertion with single-turn (11-bp) helices flanking the ES region. The internal loop with a targeted editing site in enhanced substrates exhibits a simple geometry in which the number (*n*) of residues, rather than their specific sequence, is critical. Importantly, potential pairing (conventional and noncanonical G-U) within the internal loop should be minimized, except for the relevant pair indicated by the dotted line. (B) Artificial interconversion of editing sites in substrates for full-round insertion and deletion. ESs in enhanced substrates for editing can be converted from deletion to insertion or vice versa (Cifuentes-Rojas *et al.*, 2005). "As" indicates the number of guiding purines, As or Gs, in insertion sites (typically two or three, but one or four are also efficient in model substrates), and *n* + 1 includes the chosen number of Us to be removed in deletion sites (one to five have been used in previous reports).

2. SITE-SPECIFIC PHOTOCROSSLINKING METHODOLOGY TO DETECT DIRECT INTERACTIONS OF PURIFIED EDITING COMPLEXES WITH SUBSTRATES FOR FULL-ROUND EDITING

2.1. Overview

Critical RNA–protein recognition events are expected to mediate several steps of the editing process, including docking of complexes onto potential substrates, sorting of productive and nonproductive mismatches in hybrids, distinguishing between pre-mRNA and gRNA strands, specifying steps of insertion or deletion editing, proofreading, and potential remodeling for dissociation or translocation between sites. All of these processes remain to be understood at the molecular level, and the responsible RNA–protein interactions in assembled editing complexes are unknown. The most efficient substrate known for full-round editing *in vitro* (Cruz-Reyes *et al.*, 2001) was recently found to support robust photocrosslinking with subunits

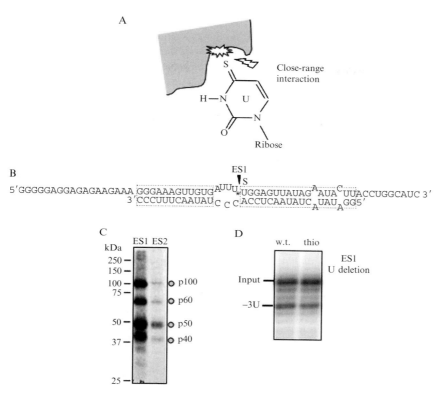

Figure 5.2 A common substrate for both site-specific photocrosslinking and full-round editing activities of purified native editing complexes. (A) A close-range interaction between a photoreactive thiolated uridine and a protein subunit of editing complexes. UV irradiation may produce a covalent linkage that can be used to radiolabel the RNA-binding protein. (B) Diagram of an enhanced 62-nt pre-mRNA/gRNA A6 substrate for full-round deletion and photocrosslinking at its first editing site (ES1). The ES scissile bond with a single ^{32}P (asterisk) is flanked by photoreactive 4-thioU ("s") and Us to be deleted. (C) RNA–protein photocrosslinking patterns and relative efficiencies at targeted sites for deletion (ES1) and insertion (ES2) in A6 substrates. Four prominent proteinase K-sensitive crosslinks are apparently common in these substrates, and relatively stable as shown in competition assays (Cifuentes-Rojas et al., 2007; Sacharidou et al., 2006). Based on their approximate molecular size, they are termed p40, p50, p60, and p100. (D) Full-round editing assays for deletion using ES1-singly labeled substrates either without additional modifications (w.t.) or including a 4-thioU at ES1 (thio). The input pre-mRNA and major product of 3U deletion (-3U) are indicated.

of purified editing complexes (Fig. 5.2A–C) (Sacharidou et al., 2006). Other substrates tested so far support less efficient photocrosslinking with editing complexes (Cifuentes-Rojas et al., 2007). In these photocrosslinking studies, the pre-mRNA strand was modified with a sensitive photoreactive 4-thioU residue at the targeted site. Either introduction of base pairing within the editing region or positioning of the 4-thioU in the flanking duplexes is strongly inhibitory. This suggests that the observed crosslinking interactions may exhibit structural selectivity for the single strandedness of the editing site. Importantly, 4-thioU at an ES does not interfere with editing *in vitro* (Fig. 5.2D). Thus, the same substrate can be used to directly compare catalytic and photocrosslinking activities of editing complexes.

2.2. Preparation of RNA substrates to analyze both catalytic and photocrosslinking activities of editing complexes

2.2.1. Full-round deletion substrate

The most efficient substrate for *in vitro* RNA editing so far reported in *T. brucei* consists of a model A6 pre-mRNA (Seiwert *et al.*, 1996) paired with the enhanced gRNA D33 (Cruz-Reyes *et al.*, 2001). This hybrid substrate supports full-round deletion at the first editing site (ES1) in which three uridylates are removed. The typical A6 pre-mRNA substrate for full-round deletion at ES1 is 73 nt long (Seiwert *et al.*, 1996). However, we have found a shorter 62-nt version that is equivalent in both editing and crosslinking assays (Fig. 5.2B) (Sacharidou *et al.*, 2006).

Relatively short transcripts such as those described in this chapter are simple to prepare using the Uhlenbeck single-stranded T7 transcription method (Milligan *et al.*, 1987). All RNA transcripts are gel purified and ethanol precipitated. Longer transcripts can be synthesized using linearized DNA plasmids or PCR templates (Seiwert *et al.*, 1996). Although plasmid templates generally yield higher transcription efficiencies, PCR templates expedite the preparation of multiple transcripts and facilitate sequence mutagenesis (Cruz-Reyes *et al.*, 1998a). Typically, pre-mRNA substrates for full-round deletion are 3′ end labeled by ligation of $[5'-^{32}P]pCp$ or 5′ end labeled by phosphorylation with $[\gamma-^{32}P]ATP$ (Seiwert *et al.*, 1996). However, pre-mRNA substrates containing a single internal ^{32}P are also easy to prepare and are similarly effective in deletion assays (Halbig *et al.*, 2006). In standard full-round deletion assay, the pre-mRNA 3′-terminus requires no special chemical treatments as is the case in full-round insertion *in vitro* (see below).

For efficient site-specific photocrosslinking with editing complexes, the pre-mRNA strand must bear a 4-thioU at the targeted ES with a ^{32}P at its 5′ position. Notably, this modification does not interfere with the deletion activity of editing complexes (Fig. 5.2D) (Sacharidou *et al.*, 2006). Such pre-mRNA is prepared by splinted ligation of two fragments (Moore and Sharp, 1992). To specifically target ES1 in the 62-nt A6 substrate, the following fragments are synthesized: acceptor 5′GGGGGAGGAGAGAAGAAA-GGGAAAGUUGUGAUUU3′ and donor 5′UGGAGUUAUAGAAU-ACUUACCUGGCAUC3′, respectively. The donor piece contains a

5′-terminal 4-thioU. The splinted ligation of RNA pieces is performed using the following DNA bridge oligonucleotide: 5′TATTCTA-TAACTCCAAAATCACAACTTTCC3′. A 3:1:2 molar ratio of acceptor/donor/bridge molecules works well for ligation at ES1, and routinely allows at least 50% incorporation of the radiolabeled donor piece into the ligation product. If necessary, other proportions of acceptor (or donor) can be tested to improve the ligation efficiency. This strategy can be easily adapted to target other sites in the same or different substrates. In general, we use a 30-nt bridge, although longer bridges may be required if the junction sequence is AU rich. The detailed protocol is as follows:

2.2.1.1. *The acceptor RNA piece*

Transcribe this piece using the single-stranded T7 transcription method (Milligan *et al.*, 1987). Typically, we use 0.4 μg of the DNA oligonucleotide template and 0.1 μg of the complementary T7 promoter oligonucleotide in a reaction with a high–yield transcription kit. After transcription, treat the mixture with 1 unit of DNase I for 2 h, perform a phenol extraction, concentrate the RNA by ethanol precipitation, and gel purify it using UV shadowing. All transcripts described here are well resolved in 9% denaturing polyacrylamide gels (19:1, 7 M urea in TBE), excised in discrete gel pieces, and recovered using standard elution and ethanol precipitation protocols.

2.2.1.2. *The donor RNA piece*

Prepare a freshly 5′-end radiolabeled donor fragment by treating 15 pmol of the RNA oligonucleotide with polynucleotide kinase and [γ-^{32}P]ATP 3000 μCi/mmol (using a 1:2 molar ratio of 5′ ends:ATP). Gel purify the relevant fragment as indicated above. Thiolated RNA fragments can be chemically synthesized by Dharmacon Inc. We store thiolated RNA as other transcripts at −20° and away from light, but do not use amber tubes or reduced light during most manipulations.

2.2.1.3. *Ligation of acceptor and donor RNA pieces*

We routinely use a rapid DNA ligation kit (Roche). Mix approximately 30 pmol of acceptor piece with 10 pmol of isolated donor piece and 20 pmol of bridge DNA oligonucleotide in ~3 μl of buffer TE. Incubate this mixture at 95° for 1 min and then leave it at 37° for 5 min to allow annealing. Add DNA dilution buffer, T4 DNA ligation buffer, and DNA ligase provided in the kit to complete a 10 μl reaction mixture as indicated by the manufacturer, and incubate at 26° for 2 h. The ligated RNA runs near the Xylene cyanol dye in a 9% denaturing polyacrylamide gel, and should be readily detected within 5–10 sec of autoradiography exposure. Note that shorter artifacts often form due to ligation of broken pieces. The full-length transcript can be recognized by including in the gel a homologous end-labeled pre-mRNA transcript as a size marker.

2.2.2. Full-round insertion substrate

A popular A6 hybrid substrate for full-round insertion at ES2 uses a 51-nt pre-mRNA (Igo et al., 2002). We found a minimal 45-nt pre-mRNA derivative that is similarly efficient for insertion and can be modified to perform photo-crosslinking studies (Cifuentes-Rojas et al., 2005, 2007). However, all insertion substrates tested so far are less efficient than the A6 substrate for ES1 deletion in both editing and photocrosslinking assays (Cifuentes-Rojas et al., 2007). The internally labeled substrate can also be used in parallel studies of full-round insertion and crosslinking activities of editing complexes. However, in contrast to standard full-round deletion reactions, the 3'-terminus of the insertion pre-mRNA must be blocked to prevent unwanted nucleotide extensions by the 3'-terminal U transferase activity of editing complexes. Such blockage can be accomplished by periodate oxidation of the pre-mRNA 3'-terminal ribose (Burgess et al., 1999) or by using a chemically synthesized thiolated donor fragment containing either a phosphoryl group or dideoxyribose at its 3' terminus (Golden and Hajduk, 2006; Igo et al., 2000).

2.3. Reaction conditions to perform parallel assays of full-round editing and crosslinking activities of editing complexes

Full-round deletion reactions are assembled in 20 μl mixtures containing 10 mM MRB buffer [10 mM Mg(OAc)$_2$, 10 mM KCl, 1 mM EDTA, pH 8, 25 mM Tris–HCl, pH 8, and 5% glycerol], 30 μM ATP, 3 mM ADP (or AMP-CP from Fluka). Full-round insertion reactions contain 3 μM ATP, and 150 μM UTP is used instead of ADP. The optional addition of a nonspecific protein at 1 mM such as hexokinase or bovine serum albumin (BSA) increases both full-round editing and crosslinking activities of editing complexes by at least 2-fold (Cruz-Reyes et al., 1998b). The assays are set up as follows:

1. Per assay, mix 1 μl of radiolabeled pre-mRNA ($\sim$10 fmol) with 1 μl of gRNA (1.2 pmol). RNA stocks and dilutions are prepared in buffer TE. Preanneal the transcripts by incubating this mixture at 37° for 10 min and then at room temperature for $\sim$10 min.
2. During the preannealing step, prepare a mixture containing reaction buffer, nucleotides, and water. It is recommended that a common ("master") mixture is prepared and replica assays are performed in all experiments.
3. Dispense identical aliquots of the master mix for each assay.
4. Add 2 μl of preannealed RNA substrate (prepared in step 1) to each tube in step 3. Gently mix the components and pulse-spin the tubes.
5. Add 2 μl of purified editing complexes, gently tap the tubes to homogenize all components, and incubate at 26° for 1 h. The complete mixture should contain 20 μl.

6. After incubation, deproteinize the reaction by phenol extraction, concentrate by ethanol precipitation, and resolve the samples on a 9% denaturing polyacrylamide gel.
7. Full-round deletion and insertion products are efficiently resolved in 60-cm and 40-cm–long (0.4-mm-thick) gels, respectively, using the substrates previously described. The former gels run for ~6 h and the latter for ~4 h at 1400 V.

Using native gels, we have determined that the preannealing step with a molar excess of gRNA ensures that virtually all radiolabeled pre–mRNA is incorporated into a hybrid (Cifuentes-Rojas et al., 2007). When needed, total pre–mRNA cleavage activity can be scored in the absence of RNA ligase activity. This is accomplished by using editing complexes pretreated with 10 mM PPi (Cruz-Reyes et al., 1998b). Briefly, 2 μl of editing complexes are incubated with 1 μl of 30 mM PPi (pH 8) on ice for 5 min. This mixture is added to complete the final 20 μl reaction, and the cleavage assay is incubated at 26° for 45 min. PPi (pH 8) at ~1.5 mM final should be used in the cleavage assay, as higher concentrations are inhibitory and unbuffered PPi causes RNA breakage (Cruz-Reyes et al., 1998b).

The standard photocrosslinking reaction is assembled as above except for the inclusion of nucleotides, which somewhat reduce the crosslinking efficiency. The specific steps are as follows:

1. Prepare complete reaction mixtures as in steps 1–5 of the deletion assay described above (except for the 1-h incubation).
2. Incubate the reactions for 10 min at 26° and then for 10 min on ice.
3. Place each sample on a 96-well plate (VRW no. 25227-308) on ice and ~5 cm below a Spectroline 150-V lamp.
4. Irradiate the samples with 365 nm UV light for 10 min.
5. Treat the samples with RNases A and T1 (50 μg/ml and 120 U/ml) for 10 min at 37°.
6. Add 7 μl of 4× Laemmli buffer (Laemmli, 1970) to each sample, boil, and analyze them by SDS–PAGE and autoradiography (overnight) or phosphorimaging.

If the insertion substrate is used, the reaction mixture can be scaled up a few fold to obtain autoradiography exposure times that are comparable to those with the deletion substrate.

2.4. Purification of native editing complexes for editing and crosslinking studies

Procyclic form (Pf) T. brucei is grown in Cunningham media and mitochondrial crude lysates are prepared as originally reported (Harris and Hajduk, 1992) with some modifications (Sollner-Webb et al., 2001). The lysates are then fractionated by consecutive ion-exchange chromatography in Q-Sepharose (Q1), DNA-cellulose (D), and Q-Sepharose (Q2) columns, as previously described (Rusche *et al.*, 1997). The elution fractions containing concentrated editing complexes are readily determined by either self-adenylylation of RNA ligases in the presence of $[\alpha\text{-}^{32}P]ATP$, 3000 $\mu Ci/mmol$ (Rusche *et al.*, 1997), *in vitro* editing activity, or Western blot analysis of editing complex subunits using available antibodies (Sacharidou *et al.*, 2006). The editing complexes in the Q2 peak fractions prepared in our laboratory are comparable in protein composition to the ~20 subunit complexes reported by other groups (Simpson *et al.*, 2004; Stuart *et al.*, 2005; Hernandez *et al.*, in preparation). Silver staining and editing activity of this material are shown in Fig. 5.3A. Importantly, editing complexes in Q2 and D peak fractions consistently exhibit the same proteinase K-sensitive photocrosslinking pattern of interactions with editing complex subunits (Fig. 5.3B and data not shown); thus both preparations appear equivalent for analyzing these interactions. The significantly less pure

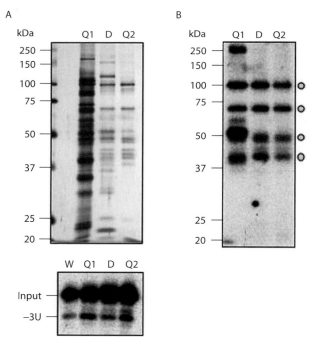

Figure 5.3 Comparison of representative Q1, D, and Q2 peak fractions of chromatography-purified editing complexes. (A) Silver staining (top) and full-round editing activity scored by formation of 3U deletion product (-3U; bottom). A lane with a whole mitochondrial lysate (W) is also shown. (B) UV photocrosslinking activity at ES1. Ten microliters of each peak fraction were silver stained, and 2 μl were used for the crosslinking and editing reactions. If necessary, the crosslinking assay can be scaled up 2-fold. One microliter of diluted (1/10) whole mitochondrial lysate was used for editing.

Q1 fractions produce additional photocrosslinks that are both less reproducible and do not precisely copurify with editing complexes (Sacharidou *et al.*, 2006). Furthermore, crude mitochondrial lysates and sedimentation fractions generate excessive nonspecific photocrosslinks and seem less suitable for studies. The specific protein-dependent crosslinking interactions in Pf trypanosomes are also observed in editing complexes from bloodstream form (Bf) trypanosomes collected from infected rats (Halbig *et al.*, 2004). This shows that these editing complex interactions are not affected by dramatically different growth conditions.

3. Assays to Assess the Specificity of RNA–Protein Crosslinking Interactions in Editing Complexes

3.1. Overview

The precise biochemical copurification of functionally active editing complexes and protein crosslinking interactions with thiolated substrates strongly suggests that such contacts occur in assembled complexes. However, additional tests should be performed to further demonstrate colocalization of any potentially relevant crosslink and editing complexes. Furthermore, it is necessary to evaluate the stability and specificity of the interactions since the associated editing subunits may exhibit a degree of nonspecific RNA binding, and the (^{32}P)4-thioU used in the methodology described here is highly reactive. The following immunoprecipitation and competition assays have been used to examine the specificity of RNA–protein associations in assembled editing complexes.

3.2. Coimmunoprecipitation of editing complexes and crosslinking interactions

Several monoclonal and polyclonal antibodies currently available from various laboratories are suitable for performing immunoprecipitation assays of editing complexes (McManus *et al.*, 2001; Panigrahi *et al.*, 2001; Sacharidou *et al.*, 2006). Our laboratory has used this technique to demonstrate copurification of RNA–protein photocrosslinks with assembled editing complexes from procyclic and bloodstream form trypanosomes (Halbig *et al.*, 2004; Sacharidou *et al.*, 2006) (Fig. 5.4).

Immunoprecipitation assays can be performed essentially as previously described (Panigrahi *et al.*, 2001) with minor modifications for the analysis of editing complex/substrate crosslinking interactions. For immunoprecipitation analyses of these interactions, editing reactions are usually scaled up between 5 and 10 times and crosslinked as described above. For each assay, the steps are as follows:

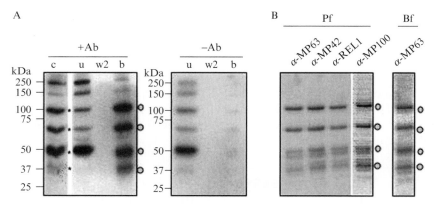

Figure 5.4 Coimmunoprecipitation (co-IP) of all major crosslinks at ES1 with purified editing complexes from procyclic and bloodstream trypanosomes. (A) Protein–RNA crosslinks in a peak Q1 fraction before (lane c) and after a co-IP assay with antibodies against the editing complex subunit MP63 (+Ab), including the unbound (u), second wash (w2), and bound immunoprecipitated (b) fractions. Note the enrichment of specific crosslinks in lane b. A parallel mock co-IP assay with no antibodies (-Ab) is shown. (B) Bound fractions of co-IP assays with antibodies to several subunits of editing complexes using the same assay conditions as (A). Immunoprecipitated complexes from procyclic (Pf) and bloodstream (Bf) trypanosomes are compared. Both types of complexes show the same major crosslinks described in Fig. 5.3.

1. Couple 100 μl of immunomagnetic beads (Dynabeads M–450; Dynal) with 225 μl of monoclonal antibody supernatants (kindly provided by the laboratory of Ken Stuart, SBRI, Seattle) and 1% BSA. As mentioned above, other antisera against editing subunits can also be used (McManus *et al.*, 2001; Sacharidou *et al.*, 2006).
2. Incubate the editing reaction with antibody-coated beads for 1 h at 4° using a biodirectional shaker and occasional tapping.
3. Wash the beads two times with 100 μl of immunoprecipitation buffer (10 mM Tris, pH 7.2, 10 mM MgCl$_2$, 200 mM KCl, 0.1% Triton-X 100).
4. Resuspend the beads with 100 μl of TE buffer.
5. Incubate the suspension in the presence of RNases A and T1 as previously described.
6. Add 30 μl of 4× Laemmli buffer, boil the bead suspension at 100° for 5 min, and analyze the supernatant by SDS–PAGE and autoradiography.

The entire 200-μl unbound fraction and 100-μl washes mixed with 60 μl and 30 μl of 4× Laemmli buffer, respectively, are boiled and also analyzed.

3.3. Parallel competition assays of crosslinking and catalytic activities

The specificity of editing complex interactions with RNA can also be analyzed by adapting the direct photocrosslinking reaction into competition assays. Importantly, the reaction conditions for competition can also be applied to functional editing assays performed in parallel (e.g., Figs. 5.5 and 5.6). In either case, the radiolabeled pre-mRNA/gRNA hybrid is mixed with a molar excess of homologous or heterologous transcript competitors (Fig. 5.5C). Editing complexes are then added to examine products of either UV radiation, full-round editing, or intermediate catalytic steps such as pre-mRNA cleavage (e.g., Fig. 5.6C) (Cifuentes-Rojas et al., 2007). Parallel inhibition of both photocrosslinking and enzymatic activities reflects titration of editing

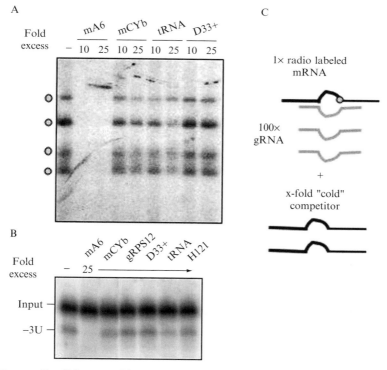

Figure 5.5 Parallel competition assays of (A) photocrosslinking and (B) full-round deletion activities of purified editing complexes. (C) The indicated fold excess of unlabeled ("cold") transcript competitors was used including homologous A6 pre-mRNA (mA6) and heterologous CYb pre-mRNA (mCYb), gRNA RPS12 (gRPS12), viral RNA H121, and tRNA. Other transcripts have also been tested (Cifuentes-Rojas et al., 2007). Note that additional cognate gRNA was also supplemented in some lanes (D33+) over the standard ~120-fold excess present in all lanes.

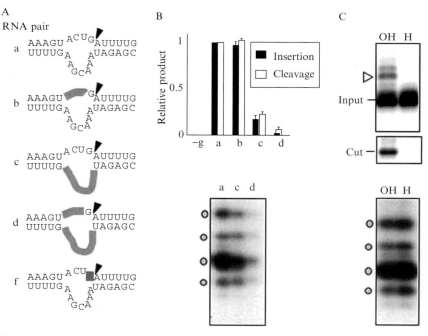

Figure 5.6 Ribose 2′ deoxy substitutions of internal loop residues around ES2 for full-round insertion. (A) Relevant residues of a minimal insertion substrate and modified derivatives with an arrowhead pointing at the scissile bond are shown. The complete sequence of the substrate has been published (Cifuentes-Rojas *et al.*, 2005). Boxes indicate multiple substitutions [RNA pairs (b–d)] or a single substitution [RNA pair (f)]. (B) Plots of full-round insertion and intermediate cleavage (top), and photocrosslinking (bottom) activities with the indicated substrates. (C) Full-round editing (top) and cross-linking (bottom) activities with the indicated substrates. OH and H indicate the specific 2′ functional group just 5′ of the scissile bond at ES2.

complexes by (unlabeled) bound competitor at the tested molar excess (Fig. 5.5C). Alternatively, conservation of crosslinking and editing activities indicates selective association of complexes to bona fide substrates upon dilution by a molar excess of unrelated transcripts. This technique in combination with standard functional assays can provide important insights into the mechanisms of substrate selectivity by purified editing complexes. For example, recent studies indicated that the initial complex/substrate association relies on secondary structure, not sequence-specific features, that is common to substrates for both full-round insertion and deletion (Cifuentes-Rojas *et al.*, 2007).

Specific recommendations to be taken into account when performing parallel competition assays for crosslinking and editing activities include the following: (1) a battery of homologous and heterologous potential competitors should be selected and preferably prepared with similar proto-cols; (2) the concentration of all RNA stocks may be determined simulta-neously and aliquots kept to perform replica experiments; (3) all RNAs should be subjected to quick heat and cooling in a small volume of reaction buffer to promote a proper folding immediately before performing the assays; and (4) all RNA competitors should be allowed to equilibrate along with the editing substrate for 5–10 min prior to addition of editing complexes. This is particularly relevant if a competitor exhibits potential complementarity to sequences in the preannealed pre-mRNA/gRNA hybrid or to free gRNA. The integrity of the labeled hybrid can be determined in native gels (Cifuentes-Rojas et al., 2007).

As previously mentioned, purified editing complexes evidently exhibit a preferential association with editing substrates, which depends on specific RNA secondary structure features. Interestingly, the crosslinking interac-tions with the deletion ES1 are more robust and stable than at the insertion ES2 in model A6 substrates. However, certain heterologous RNA compe-titors, particularly those with relatively stable structures, partially inhibit these associations at moderate molar excess (Cifuentes-Rojas et al., 2007). This suggests that purified editing complexes exhibit a level of nonspecific binding to unrelated RNA transcripts. In fact, a transcript that is ineffective as competitor (at 200-fold excess) exhibits low-stability crosslinking with editing complexes (Cifuentes-Rojas et al., 2007). While the competition assays described here can be used to compare the relative stability of RNA–protein crosslinking interactions in purified editing complexes under equi-librium conditions, complementary studies and assays will be needed to perform kinetic analysis (e.g., off-rate and on-rate measurements) and to unravel the precise mechanisms of stable association with bona fide editing substrates.

4. Additional Applications

4.1. Analysis of structural and biochemical requirements for catalysis and complex/substrate interactions

4.1.1. Overview

The analyses described above can be used to assess the relative specificity of interactions that affect both editing complex/substrate association and catal-ysis under optimal reaction conditions using purified native complexes. However, variations can also be introduced to the original assays to address specific biochemical and structural requirements that are either common to editing complex assembly and catalysis or specific to one of these processes. This is illustrated with the following examples.

4.1.2. Biochemical requirements of editing complexes

(1) Procyclic trypanosome cultures and editing activity are optimal at 26°; however, the photocrosslinking activity of editing complexes remains unaffected in reaction incubations kept on ice. This suggests that efficient assembly of the complexes does not require incubation at 26°. (2) ATP and ADP supplements (at μM and mM levels, respectively) significantly stimulate U-deletion editing (Cruz-Reyes *et al.*, 1998a), but photocrosslinking is efficient in their absence, suggesting that these nucleotide factors are required only for catalysis, not for docking of editing complexes on RNA substrates. (3) Mg^{2+} is essential for editing activity (Seiwert and Stuart, 1994), but editing complex crosslinking is efficient in reactions with 5 mM EDTA and no supplemented Mg^{2+}. Thus, Mg^{2+} ions also appear dispensable for editing complex docking. (4) Finally, it is known that addition of a nonspecific protein supplement, such as hexokinase or BSA (at 1 mM), markedly enhances the editing activity of very dilute purified complexes (Cruz-Reyes *et al.*, 1998a). Notably, supplemented hexokinase significantly enhances the crosslinking contacts with editing complexes. Thus, the increase in total protein concentration may stabilize diluted purified complexes and thereby enhance both their association with RNA substrates and catalysis. Reaction conditions tested so far that affect both crosslinking and editing activities, together with other conditions that impact editing but not crosslinking, suggest that the observed crosslinking interactions by editing complexes precede catalysis (P. Pavia *et al.*, unpublished observations).

4.1.3. Substrate structural requirement

The substrate determinants directly recognized by subunits of assembled editing complexes and the regulation of such interactions are unknown. The assays described in this chapter can be used to obtain important insights in this regard. For example, it has been shown that the single-strand character of residues adjoining individual ESs (i.e., 5′ to the paired thiolated U in Fig. 5.2B) stimulates crosslinking and catalysis by editing complexes (Sacharidou *et al.*, 2006). Related studies have further addressed the importance of secondary structure and specific chemical groups in substrates for both editing and crosslinking activities. For example, 2′-deoxyribonucleotide substitutions at either helix flanking an ES reduce both photocrosslinking and pre-mRNA cleavage activities of editing complexes (Cifuentes-Rojas *et al.*, 2007). The downstream "anchor" duplex is particularly sensitive to these modifications, suggesting a central role in docking of editing complexes. A similar approach defined specific hydroxyl groups in (potentially single-stranded) residues in the ES region that impact productive editing complex interactions. For example, 2′ hydroxyls in the gRNA strand are stimulatory for both pre-mRNA crosslinking and cleavage activities (Fig. 5.6A and B). Notably, a single 2′ H substitution just 5′ of the scissile bond obliterates catalysis of cleavage but has no effect on photocrosslinking interactions (Fig. 5.6C). This and other tested 2′ substituents suggest that the 2′ hydroxyl at the insertion ES2 may play a role in the catalytic mechanism of pre-mRNA endonuclease cleavage, rather than on docking of editing complexes (Cifuentes-Rojas et al., 2007).

5. Concluding Remarks

The detailed description of specific RNA–protein interactions and their specific roles in assembled editing complexes pose exciting challenges in the field of U-insertion/deletion editing in upcoming years. The use of site-specific radiolabeling and chemical modification to generate sensitive photoreagents has been extremely informative in similar studies of other complex ribonucleoprotein assemblies such as ribosomes and spliceosomes. The methodology and examples of applications described in this chapter illustrate the first steps in this direction using established in vitro systems that support robust full-round editing and direct photocrosslinking interactions with purified editing complexes. Our current studies began the identification of crosslinking subunits of editing complexes. As expected, those so far identified exhibit conserved RNA-binding domains, and previous genetic studies indicate that these editosome subunits must play central roles in the editing process (Hernandez et al., in preparation).

ACKNOWLEDGMENTS

We thank members of the Cruz-Reyes laboratory for comments on the manuscript and helpful discussions. This work was supported by a grant from the National Institutes of Health (GM067130) to JC-R.

REFERENCES

Allen, T. E., Heidmann, S., Reed, R., Myler, P. J., Goringer, H. U., and Stuart, K. D. (1998). Association of guide RNA binding protein gBP21 with active RNA editing complexes in Trypanosoma brucei. Mol. Cell. Biol. 18, 6014–6022.

Aphasizhev, R., Aphasizheva, I., Nelson, R. E., Gao, G., Simpson, A. M., Kang, X., Falick, A. M., Sbicego, S., and Simpson, L. (2003a). Isolation of a U-insertion/deletion editing complex from Leishmania tarentolae mitochondria. EMBO J. 22, 913–924.

Aphasizhev, R., Aphasizheva, I., Nelson, R. E., and Simpson, L. (2003b). A 100-kD complex of two RNA-binding proteins from mitochondria of Leishmania tarentolae catalyzes RNA annealing and interacts with several RNA editing components. RNA 9, 62–76.

Blom, D., Burg, J., Breek, C. K., Speijer, D., Muijsers, A. O., and Benne, R. (2001). Cloning and characterization of two guide RNA-binding proteins from mitochondria of Crithidia fasciculata: gBP27, a novel protein, and gBP29, the orthologue of *Trypanosoma brucei* gBP21. *Nucleic Acids Res.* **29,** 2950–2962.

Brecht, M., Niemann, M., Schluter, E., Muller, U. F., Stuart, K., and Goringer, H. U. (2005). TbMP42, a protein component of the RNA editing complex in African trypanosomes, has endo-exoribonuclease activity. *Mol. Cell* **17,** 621–630.

Burgess, M. L., Heidmann, S., and Stuart, K. (1999). Kinetoplastid RNA editing does not require the terminal 3′ hydroxyl of guide RNA, but modifications to the guide RNA terminus can inhibit *in vitro* U insertion. *RNA* **5,** 883–892.

Cifuentes-Rojas, C., Halbig, K., Sacharidou, A., De Nova-Ocampo, M., and Cruz-Reyes, J. (2005). Minimal pre-mRNA substrates with natural and converted sites for full-round U insertion and U deletion RNA editing in trypanosomes. *Nucleic Acids Res.* **33,** 6610–6620.

Cifuentes-Rojas, C., Pavia, P., Hernandez, A., Osterwisch, D., Puerta, C., and Cruz-Reyes, J. (2007). Substrate determinants for RNA editing and editing complex interactions at a site for full-round U insertion. *J. Biol. Chem.* **282,** 4265–4276.

Cruz-Reyes, J., Rusche, L. N., Piller, K. J., and Sollner-Webb, B. (1998a). *T. brucei* RNA editing: Adenosine nucleotides inversely affect U-deletion and U-insertion reactions at mRNA cleavage. *Mol. Cell* **1,** 401–409.

Cruz-Reyes, J., Rusche, L. N., and Sollner-Webb, B. (1998b). *Trypanosoma brucei* U insertion and U deletion activities co-purify with an enzymatic editing complex but are differentially optimized. *Nucleic Acids Res.* **26,** 3634–3639.

Cruz-Reyes, J., Zhelonkina, A., Rusche, L., and Sollner-Webb, B. (2001). Trypanosome RNA editing: Simple guide RNA features enhance U deletion 100-fold. *Mol. Cell. Biol.* **21,** 884–892.

Golden, D. E., and Hajduk, S. L. (2006). The importance of RNA structure in RNA editing and a potential proofreading mechanism for correct guide RNA:pre-mRNA binary complex formation. *J. Mol. Biol.* **359,** 585–596.

Halbig, K., De Nova-Ocampo, M., and Cruz-Reyes, J. (2004). Complete cycles of bloodstream trypanosome RNA editing *in vitro*. *RNA* **10,** 914–920.

Halbig, K., Sacharidou, A., De Nova-Ocampo, M., and Cruz-Reyes, J. (2006). Preferential interaction of a 25kDa protein with an A6 pre-mRNA substrate for RNA editing in *Trypanosoma brucei*. *Int. J. Parasitol.* **36,** 1295–1304.

Harris, M. E., and Hajduk, S. L. (1992). Kinetoplastid RNA editing: *In vitro* formation of cytochrome b gRNA-mRNA chimeras from synthetic substrate RNAs. *Cell* **68,** 1091–1099.

Igo, R. P., Jr., Palazzo, S. S., Burgess, M. L., Panigrahi, A. K., and Stuart, K. (2000). Uridylate addition and RNA ligation contribute to the specificity of kinetoplastid insertion RNA editing. *Mol. Cell. Biol.* **20,** 8447–8457.

Igo, R. P., Jr., Lawson, S. D., and Stuart, K. (2002). RNA sequence and base pairing effects on insertion editing in *Trypanosoma brucei*. *Mol. Cell. Biol.* **22,** 1567–1576.

Laemmli, U. K. (1970). Cleavage of structural proteins during the assembly of the head of bacteriophage T4. *Nature* **227,** 680–685.

Madison-Antenucci, S., and Hajduk, S. L. (2001). RNA editing-associated protein 1 is an RNA binding protein with specificity for preedited mRNA. *Mol. Cell* **7,** 879–886.

McManus, M. T., Shimamura, M., Grams, J., and Hajduk, S. L. (2001). Identification of candidate mitochondrial RNA editing ligases from *Trypanosoma brucei*. *RNA* **7,** 167–175.

Miller, M. M., Halbig, K., Cruz-Reyes, J., and Read, L. K. (2006). RBP16 stimulates trypanosome RNA editing *in vitro* at an early step in the editing reaction. *RNA* **12,** 1292–1303.

Milligan, J. F., Groebe, D. R., Witherell, G. W., and Uhlenbeck, O. C. (1987). Oligor-ibonucleotide synthesis using T7 RNA polymerase and synthetic DNA templates. *Nucleic Acids Res.* **15,** 8783–8798.

Missel, A., Souza, A. E., Norskau, G., and Goringer, H. U. (1997). Disruption of a gene encoding a novel mitochondrial DEAD-box protein in *Trypanosoma brucei* affects edited mRNAs. *Mol. Cell. Biol.* **17,** 4895–4903.

Moore, M. J., and Sharp, P. A. (1992). Site-specific modification of pre-mRNA: The 2′-hydroxyl groups at the splice sites. *Science* **256,** 992–997.

Muller, U. F., Lambert, L., and Goringer, H. U. (2001). Annealing of RNA editing substrates facilitated by guide RNA-binding protein gBP21. *EMBO J.* **20,** 1394–1404.

Panigrahi, A. K., Schnaufer, A., Carmean, N., Igo, R. P., Jr., Gygi, S. P., Ernst, N. L., Palazzo, S. S., Weston, D. S., Aebersold, R., Salavati, R., and Stuart, K. D. (2001). Four related proteins of the *Trypanosoma brucei* RNA editing complex. *Mol. Cell. Biol.* **21,** 6833–6840.

Panigrahi, A. K., Schnaufer, A., Ernst, N. L., Wang, B., Carmean, N., Salavati, R., and Stuart, K. (2003). Identification of novel components of *Trypanosoma brucei* editosomes. *RNA* **9,** 484–492.

Panigrahi, A. K., Ernst, N. L., Domingo, G. J., Fleck, M., Salavati, R., and Stuart, K. D. (2006). Compositionally and functionally distinct editosomes in *Trypanosoma brucei*. *RNA* **12,** 1038–1049.

Pelletier, M., and Read, L. K. (2003). RBP16 is a multifunctional gene regulatory protein involved in editing and stabilization of specific mitochondrial mRNAs in *Trypanosoma brucei*. *RNA* **9,** 457–468.

Rusche, L. N., Cruz-Reyes, J., Piller, K. J., and Sollner-Webb, B. (1997). Purification of a functional enzymatic editing complex from *Trypanosoma brucei* mitochondria. *EMBO J.* **16,** 4069–4081.

Sacharidou, A., Cifuentes-Rojas, C., Halbig, K., Hernandez, A., Dangott, L. J., De Nova-Ocampo, M., and Cruz-Reyes, J. (2006). RNA editing complex interactions with a site for full-round U deletion in *Trypanosoma brucei*. *RNA* **12,** 1219–1228.

Salavati, R., Ernst, N. L., O'Rear, J., Gilliam, T., Tarun, S., Jr., and Stuart, K. (2006). KREPA4, an RNA binding protein essential for editosome integrity and survival of *Trypanosoma brucei*. *RNA* **12,** 819–831.

Seiwert, S. D., and Stuart, K. (1994). RNA editing: Transfer of genetic information from gRNA to precursor mRNA *in vitro*. *Science* **266,** 114–117.

Seiwert, S. D., Heidmann, S., and Stuart, K. (1996). Direct visualization of uridylate deletion *in vitro* suggests a mechanism for kinetoplastid RNA editing. *Cell* **84,** 831–841.

Simpson, L., Aphasizhev, R., Gao, G., and Kang, X. (2004). Mitochondrial proteins and complexes in Leishmania and Trypanosoma involved in U-insertion/deletion RNA editing. *RNA* **10,** 159–170.

Sollner-Webb, B., Rusche, L. N., and Cruz-Reyes, J. (2001). Ribonuclease activities of trypanosome RNA editing complex directed to cleave specifically at a chosen site. *Methods Enzymol.* **341,** 154–174.

Stuart, K. D., Schnaufer, A., Ernst, N. L., and Panigrahi, A. K. (2005). Complex manage-ment: RNA editing in trypanosomes. *Trends Biochem. Sci.* **30,** 97–105.

Vanhamme, L., Perez-Morga, D., Marchal, C., Speijer, D., Lambert, L., Geuskens, M., Alexandre, S., Ismaili, N., Goringer, U., Benne, R., and Pays, E. (1998). *Trypanosoma brucei* TBRGG1, a mitochondrial oligo(U)-binding protein that co-localizes with an *in vitro* RNA editing activity. *J. Biol. Chem.* **273,** 21825–21833.

Vondruskova, E., van den Burg, J., Zikova, A., Ernst, N. L., Stuart, K., Benne, R., and Lukes, J. (2005). RNA interference analyses suggest a transcript-specific regulatory role for mitochondrial RNA-binding proteins MRP1 and MRP2 in RNA editing and other RNA processing in *Trypanosoma brucei*. *J. Biol. Chem.* **280,** 2429–2438.

CHAPTER SIX

Strategies of Kinetoplastid Cryptogene Discovery and Analysis

Dmitri A. Maslov* *and* Larry Simpson†

Contents

Abstract

The experimental approach to revealing the genetic information hidden in kinetoplastid cryptogenes and expressed through the posttranscriptional mRNA processing of U-insertion/deletion editing proceeds in reverse to the informational flow of the RNA editing process itself. While the editing integrates the informational content of maxicircle-encoded cryptogenes with that of minicircle-encoded gRNAs to produce functional edited mRNAs, the cryptogene analysis utilizes a comparison of the mature mRNA sequence with the cryptogene sequence to deduce the locations of edited sites and editing patterns, and a comparison of that mRNA sequence with the minicircle (or minicircle equivalent) sequences to identify the corresponding guide RNAs. Although a "direct" approach (prediction of a fully edited sequence pattern based on the analysis of cryptogene and minicircle sequences) seems to be theoretically possible, it proved to be not practically feasible. The major steps of the procedures utilized to decipher editing in a broad range of kinetoplastid species are presented in this chapter.

* Department of Biology, University of California, Riverside, California
† Department of Microbiology, Immunology and Molecular Genetics, David Geffen School of Medicine at UCLA, University of California, Los Angeles, California

Methods in Enzymology, Volume 424

ISSN 0076-6879, DOI: 10.1016/S0076-6879(07)24006-6

1. INTRODUCTION

Unicellular protists united in the class Kinetoplastea are characterized by the presence of a massive body of DNA, termed the "kinetoplast," situated near the basal body of the cell's flagellar apparatus (Vickerman, 1976). The group is well known for including a score of important human pathogens, as well as for several molecular and cellular mechanisms that are not observed in most eukaryotic systems (Campbell *et al.*, 2003; Lukeš *et al.*, 2005). This list includes the kinetoplastid U insertion/deletion RNA editing, which stands apart from other types of editing not only because of its unique guide RNA-based mechanism, but also by the pronounced genomic manifestations and genetic consequences of the process. While base-substitution/modification editing, as a rule, involves a relatively small alteration of an already-functional mRNA sequence, the insertion/deletion type of editing essentially creates a translationally competent template out of a preedited transcript that does not encode a functional product (Feagin *et al.*, 1988b; Shaw *et al.*, 1988). Accordingly, the sites of potential editing in kinetoplastid mitochondrial genomes become apparent by inspection of the reading frames displaying various defects (frameshifts or missing initiation codons), which could be corrected by U insertions/deletions at the mRNA level. Historically, this approach led to the original discovery of RNA editing in cytochrome *c* oxidase subunit II (COII) mRNA, wherein a −1 frameshift was corrected by the insertion of four Us (Benne *et al.*, 1986). The mechanism of editing involves an enzymatic cascade (Blum *et al.*, 1990); current information on this can be found in several recent reviews (Panigrahi *et al.*, 2006; Simpson *et al.*, 2004; Stuart *et al.*, 2005) as well as in other chapters of this book.

The trypanosomatids represent a subgroup of the kinetoplastids in which the original discovery of RNA editing was made and with which most subsequent work has been done. In these cells the organization of the mitochondrial genome is highly conserved. It consists of a network of interlocked minicircles and maxicircles (Simpson, 1972). The latter contain a set of genes usually found in mitochondria, including the genes with edited transcripts. The information for editing is provided by small (<50 nt) guide RNAs, most of which are encoded in minicircles (Blum *et al.*, 1990; Pollard *et al.*, 1990; Sturm and Simpson, 1990b). These guide RNAs are targeted to their respective editing sites by means of a short (10–15 nt) region of complementarity between the 5′ end of the gRNA (the anchor region) and an mRNA sequence immediately downstream of a preedited region. Upon completion of editing, the entire gRNA becomes complementary to a respective segment of the mRNA. In addition to canonical G–C and A–U base pairs, G–U base pairs frequently occur in the mRNA–gRNA duplex. A single gRNA mediates editing of only a few editing sites. Longer preedited regions in the mRNA are edited by a sequential involvement of two or more gRNAs. The process starts at the $3'$ end of a preedited transcript and gradually moves upstream, with the anchors for each upstream gRNA created by the editing mediated by an adjacent downstream gRNA (Maslov and Simpson, 1992).

The experimental and computational tools for the analysis of editing at the nucleotide sequence level were developed in the 1990s during the period of intensive investigation of RNA editing in the human pathogenic species, *Trypanosoma brucei* and *Trypanosoma cruzi*, and the nonpathogenic model organisms, such as *Leishmania tarentolae* and *Crithidia fasciculata*. Technically, the work consists of cloning and analysis of gene and mRNA sequences. Only limited attempts were made to develop a computational approach to predicting cryptogene editing patterns (Von Haeseler *et al.*, 1992). Most of the experimental methods employed are standard and have been described in detail previously, as were the methods of growing the trypanosomatid cultures, subcellular fractionation and isolation of the kinetoplast–mitochondrial fraction, and isolation of kinetoplast DNA (Simpson *et al.*, 1994, 1996). The most important computational methods employed include analysis of the DNA sequence for potential sites of editing and a search for a guide RNA gene by a complementary sequence match with an edited mRNA sequence. Most of the remaining kinetoplastids, collectively referred to as bodonids, remain uninvestigated, and their study may provide some important insight into the origin and evolution of this phenomenon. Thus, investigations of *Trypanoplasma borreli* and *Bodo saltans* have revealed that the kinetoplast DNA overall genomic organization, gene order, occurrence of preediting regions, and even structural features of guide RNAs are different from trypanosomatids (Blom *et al.*, 1998, 2000; Lukeš *et al.*, 1994; Maslov and Simpson, 1994; Yasuhira and Simpson, 1996). The main goal of this chapter is to complement earlier publications by describing general strategies applicable to analysis of RNA editing in diverse kinetoplastid species including bodonids.

2. Cell Growth

Besides trypanosomatids, only a few other members of the Kinetoplastea, such as a blood fish parasite, *T. borreli*, are cultivable axenically (Maslov *et al.*, 1993). Most bodonids are phagotrophic and need feeder bacteria for propagation in culture, as described below for *B. saltans* (Blom *et al.*, 2000) (J. Lukeš, personal communication).

2.1. Cultivation of *T. borreli*

1. Per 1 liter of the LIT medium, combine

Liver infusion	5 g
Tryptone	5 g
Na_2HPO_4	8 g
NaCl	4 g
KCl	0.4 g

Bring up the volume to 900 ml with distilled water, adjust the pH to 7.0, and autoclave. Add D-glucose to the final concentration of 2 g/liter (using 200 g/liter stock solution sterilized by filtration), hemin to 10 μg/ml (2 mg/ml stock solution in 0.5 N NaOH, sterilized by filtration), fetal bovine serum to 10% (sterile serum must be heat inactivated at 56° for 45–60 min).

2. Cells are grown at 15° in stationary T-flasks or in Erlenmeyer flasks with a very slow agitation. Transfers are made weekly by a 10-fold dilution with fresh medium. Cell density at the stationary phase is 7–9 × 10^6 cells/ml.

2.2. Cultivation of *B. saltans*

1. Per 1 liter of the 100-fold concentrated medium combine

Bactopeptone	9 g
Yeast extract	1 g

Bring the volume to 1 liter with distilled water and autoclave.

2. Maintain a culture of *Bodo* spp. in a 100-ml Erlenmeyer flask with a "seed" medium prepared by autoclaving two or three seeds of wheat in 50 ml of tap water (not distilled water). Keep this at room temperature and transfer biweekly by adding several milliliters of an old culture to a fresh medium. If a culture is kept at 16–17°, transfers can be done once a month.
3. Feeder bacteria (*Alcaligenes* spp.) are maintained using standard techniques in an LB medium (10 g/liter tryptone, 5 g/liter yeast extract, 10 g/liter NaCl) at 37°. To grow a large-scale *Bodo* culture, prepare a 5-ml feeder bacteria culture grown overnight.
4. Fill a 4-liter Erlenmeyer flask with 2–2.5 liters of tap water (not distilled water) and boil for 30 min. After cooling to room temperature, add 25 ml of the 100-fold concentrated medium stock solution and 0.3 ml of a stationary culture of the feeder bacteria. Leave this at room temperature for 24 h until a slightly whitish color appears. Transfer the flask to 16–17°, add 5 ml of a *Bodo* culture from the "seed" medium, and stir vigorously on a magnetic stirrer. After 4–6 days, the culture should become more transparent (most of the bacteria will be consumed) and the *Bodo* cell density should reach ~10^7 cells/ml.

5. The cells are pelleted by centrifugation at 4000×g for 10 min. The pellet is resuspended in 200 ml of fresh medium and the suspension is left at 4° in a cylinder to allow most of the remaining feeder bacteria to sediment.

3. DNA Isolation

The massive kinetoplast DNA networks of trypanosomatids are isolated by velocity sedimentation (Simpson and da Silva, 1971; Simpson et al., 1996). In general, this method is not applicable to other groups of kinetoplastids since many of them have a noncatenated structure of kinetoplast DNA (Blom et al., 2000; Lukeš et al., 1998, 2002; Maslov and Simpson, 1994; Štolba et al., 2001). In these cases, an equilibrium density centrifugation in CsCl gradients supplemented with ethidium bromide can be used to separate covalently closed molecules of kinetoplast–mitochondrial DNA from linear fragments of the nuclear chromosomal DNA (Maslov and Simpson, 1994). Alternatively, these components can be separated by the difference in their buoyant densities using CsCl-Hoechst 33258 gradients (Blom et al., 2000; Maslov and Simpson, 1994; Simpson, 1979). The same procedure allows for the removal of DNA of the feeder bacteria used to propagate the phagotrophic kinetoplastids (Blom et al., 2000). The starting material for both procedures is a total cell DNA preparation isolated by a standard detergent-phenol protocol or by a commercial kit.

3.1. Ethidium bromide method

1. Total cell DNA from 10–15 × 10^9 cells in 60 ml TE (10 mM Tris–HCl, pH 8.0, 1 mM EDTA) is combined with 63.8 g CsCl and 3.42 ml of 10 mg/ml ethidium bromide. The refractive index ($n_D^{25°}$) is adjusted to 1.3890.
2. The solution is centrifuged in two tubes in a Beckman type 50.2 Ti or equivalent fixed angle rotor at 45,000 rpm (RCF$_{avg}$ = 184,000×g) for 40 h at 20°.
3. The bands are detected by UV and the lower band is recovered by side puncture of the tube. To achieve a higher purity of the isolated covalently closed DNA, the volume of the recovered material is brought up to 9.5 ml with CsCl-ethidium bromide, the $n_D^{25°}$ is adjusted to 1.3850, and the solution is centrifuged in a single tube in a Beckman type 50 Ti or equivalent fixed angle rotor at 40,000 rpm (RCF$_{avg}$ = 106,000×g) for 40 h. The lower band is recovered as above.
4. The DNA of the lower band is extracted twice with an equal volume of isoamyl alcohol. The aqueous phase is diluted 3-fold with the TE buffer. DNA is precipitated by addition of three volumes of ethanol.

3.2. Hoechst 33258 method

1. The DNA in 52 ml of TE is mixed with 74 g CsCl and 4 ml of a 0.5 mg/ml solution of the dye. The refractive index is adjusted to 1.3950.
2. The material is centrifuged in two tubes in a Beckman VTi 50 or a similar vertical rotor at 40,000 rpm (132,000×g) for 40 h. A fixed angle rotor can also be used.
3. The upper band is collected and, after adjusting the refractive index to 1.3935 and the volume to 9.5 ml using a CsCl-Hoechst solution, the centrifugation is repeated in a Beckman Type 50 Ti rotor at 39,000 rpm (100,000×g) for 40 h.
4. The DNA of the upper band is extracted three times with isopropanol (readjusting the volume of the aqueous phase after each extraction to the initial value with TE), diluted 3-fold with TE, and precipitated with three volumes of ethanol.

4. SEARCH FOR CRYPTOGENES

The discovery of editing by Benne and coworkers in the insect trypanosomatid, *C. fasciculata* (Benne *et al.*, 1986) was followed by additional cases of frameshift repair or other types of a limited-scale editing in *L. tarentolae* and *T. brucei* (Feagin *et al.*, 1988b; Shaw *et al.*, 1988, 1989; Simpson and Shaw, 1989). The genomic organization of the maxicircle DNA was similar in both cases, with the exception that a few seemingly noncoding G-rich regions were present in *T. brucei* in place of some genes with 5' or internal editing in *L. tarentolae* (Simpson *et al.*, 1987). It was shown shortly thereafter that one of these regions in *T. brucei* represented a cryptogene that encoded an extensively edited (pan-edited) mRNA for NADH dehydrogenase subunit 7 (Feagin *et al.*, 1988a). Additional examples of pan-editing followed, including transcripts of the six conserved intergenic G-rich regions found in all trypanosomatid species (Corell *et al.*, 1994; Maslov *et al.*, 1992; Read *et al.*, 1992; Souza *et al.*, 1993; Thiemann *et al.*, 1994). The pan-editing process was asymmetrical, with U insertions more frequent than deletions, and thus the preedited mRNAs were largely devoid of genomically encoded U nucleotides, with most of the Us present in the mature transcripts acquired through editing. This explains why gene sequences for such extensively edited mRNAs appeared as G-rich (and T-poor) islands. This feature of preedited regions was particularly noticeable because maxicircle coding sequences that did not require editing (unedited or never edited) were highly T-rich (Simpson *et al.*, 1987). This bias is related to the highly hydrophobic nature of the corresponding polypeptides, with a frequent occurrence of Phe, Leu, Ile, and Val amino acids, which are usually encoded by the U-rich codons.

Sequence analysis of maxicircles, which represent a minor component of the kinetoplast DNA, is facilitated by their isolation using the CsCl-Hoechst procedure as described above (Simpson, 1979; Simpson et al., 1996). Once the more abundant minicircle component is removed, cloning of the maxicircle component is accomplished by a shotgun approach or by cloning of restriction fragments. For closely related species, an overlapping set of maxicircle fragments can be amplified using conserved primers derived from the known sequences (Simpson et al., 1998). In bodonids, identification of the maxicircles or their equivalents is not straightforward, because the mitochondrial genome organization can be completely different in these organisms. Successful approaches involved screening of a mitochondrial genomic library with a conserved trypanosomatid COI gene probe (Maslov and Simpson, 1994) or amplification of a COI or COII gene region with universally conserved sequence primers followed by library screening (Blom et al., 1998; Lukeš et al., 1994).

The G-rich regions can be found using a computer program that plots a distribution of nucleotide frequency in a sliding window along the DNA molecule. One of the best tools for this purpose is the GCG program WINDOW, which shows the difference between the observed and expected nucleotide frequency and, therefore, accounts for the compositional bias of the DNA. The observed G-rich regions (or C-rich regions for genes with an opposite strandedness) in all species correspond to the conserved pan-edited cryptogenes ND8 (G1), ND9 (G2), G3, G4, ND3 (G5), and RPS12 (G6), and in T. brucei to COIII, A6, and ND7. Additional G-rich regions observed in some but not all trypanosomatid species represent extensively edited gene segments, such as A6, which is 5' pan-edited in Leishmania, but only moderately edited in the 5' region in several other species (Maslov et al., 1994).

The same relative gene order is maintained in all investigated trypanosomatids with functional mitochondria, including the reduced mitochondrial genome of Phytomonas serpens, which has deletions of all cytochrome c oxidase and cytochrome b subunits (Arts et al., 1993; Nawathean and Maslov, 2000; Simpson et al., 1987). Thus, the identity of the cryptogene in question can be deduced from its position in relation to neighbors, even when the entire coding sequence is cryptic. However, this is not the case in the investigated bodonids, which showed a different gene order (Blom et al., 1998; Lukeš et al., 1994; Maslov and Simpson, 1994). Another difference with trypanosomatids was the presence of 3'-editing domains in some genes, a phenomenon that appears to contradict the theory of a gradual replacement of pan-edited cryptogenes by retroposition of partially edited mRNA copies (Maslov et al., 1994; Simpson and Maslov, 1994).

5. ELUCIDATION OF EDITED mRNA SEQUENCES

Once a putative preedited region is identified by inspection of a DNA sequence, the further analysis of editing requires cloning and sequence analysis of the transcripts. In most cases, this work involves a straightforward reverse-transcription PCR using oligonucleotide primers annealing upstream and downstream of the putative editing site(s). The products obtained usually represent a mixed population of molecules that have been edited to various degrees usually in a $3'$ to $5'$ gradient. A combination of two criteria is then used to derive the mature edited pattern: it usually represents a consensus of individual editing patterns and it contains an apparently functional reading frame. These criteria are not absolute. Thus, an editing consensus might be impossible to derive from the data obtained, while a few clones may exist that appear to represent a fully edited mRNA satisfying the functionality criterion. Yet, even this criterion may be difficult to apply in the absence of a clear homology between the derived polypeptide sequence and proteins from other organisms.

The effort required for elucidation of a mature edited sequence depends on the ratio of preedited and edited transcripts, which varies greatly for different genes, developmental stages, or species. In extreme cases, hundreds of cDNA clones have to be analyzed before a mature editing pattern will start to emerge, especially during analysis of long pan-edited transcripts with abundant partially edited intermediates. Since editing proceeds in a general $3'$-to-$5'$ direction, such partially edited molecules are edited at the $3'$ end and preedited at the $5'$ end. In order to enrich a collection of cDNA molecules with edited products, the $3'$ edited sequence found in the partially edited clones can be used to design an edited downstream PCR primer, which can be applied in combination with the same $5'$ primer. An additional caveat of this analysis is the presence of so-called junction regions between the $5'$ preedited and $3'$ edited sequences (Decker and Sollner-Webb, 1990; Sturm and Simpson, 1990a). Junction regions are also edited, although their editing patterns are diverse and do not match the final edited pattern. These regions represent sites of active editing, and their origin reflects the mechanism of the process, which can utilize noncognate guide RNAs with a successive reediting by correct guide RNAs ["misediting by misguiding" model (Sturm *et al.*, 1992)] and involves multiple rounds of editing and reediting until the best gRNA–mRNA match is achieved ["dynamic mRNA–guide RNA interactions" model (Koslowsky *et al.*, 1991)]. Moreover, some sections of extensively edited mRNA molecules may deviate from the majority of other edited molecules, which is apparently a consequence of reediting of the correctly edited sequence by a noncognate guide RNA.

6. Search for Guide RNA Genes

The last stage in cryptogene analysis is a search and assignment of guide RNAs, which becomes possible only after a fully edited mRNA sequence has been determined. A local similarity search algorithm (e.g., the BESTFIT program of the GCG package with a modified weight matrix) is employed to find a match between the mRNA and a minicircle segment or a putative gRNA transcript (Simpson et al., 1994).

There are two approaches to this task. The first approach works well with trypanosomatids. It is based on the observation that most guide RNAs are encoded in minicircles, and the position and polarity of the gRNA coding regions with respect to the minicircle CSB-3 region and the bent region are well conserved in each investigated species (Simpson, 1997). A minicircle library is constructed to include the entire anticipated complexity of the heterogeneous minicircle population. If possible, a rational negative selection and sequencing strategy is then developed to analyze most or all minicircle classes present without sequencing multiple reiterative clones. Thus, in various species of Leishmania, a single gRNA gene is localized at 150 bp from the CSB-3 region of the 600–900 bp minicircles (Sturm and Simpson, 1991). Full-length minicircle libraries were constructed by using single-cut restriction enzymes or incomplete digestions with multiple cutters (Gao et al., 2001; Maslov and Simpson, 1992). Clones with minicircle inserts are identified by hybridization with a CSB-3 probe. At first, positive clones are selected and sequenced at random and several minicircle classes are identified after the first round. For the second round of library screening, oligonucleotide probes are designed to represent each sequenced minicircle class, and a negative screening is performed to identify novel minicircle classes. In T. cruzi, minicircles have a tetrameric structure with a single gRNA gene per a 370-bp repeated unit flanked by CSB-3 sequences. A library of individual units was constructed instead of full-length minicircles to facilitate the acquisition of gRNA gene sequences (Avila and Simpson, 1995). Random cloning and analysis of full length minicircles were used in T. brucei, where it was not possible to devise a negative selection method (Hong and Simpson, 2003).

When the genomic localization of gRNA genes in unknown, as is the case in most kinetoplastids besides the Trypanosomatidae, guide RNA libraries are employed instead of minicircle libraries (Simpson et al., 1996). The method includes isolation of an RNA fraction enriched with gRNA by gel electrophoresis with subsequent amplification by 3' and 5' RACE procedures (Avila and Simpson, 1995; Simpson et al., 1996; Yasuhira and Simpson, 1996).

These transcripts (<50 nt) are smaller than tRNAs and migrate ahead of these molecules in polyacrylamide-urea gels. However, they are relatively low in abundance and their presence in gels is inconspicuous after staining with ethidium bromide. In order to visualize gRNA molecules, a method of selective capping gRNAs with [α-^{32}P]GTP and vaccinia virus guanylyl-transferase can be used. The method utilizes the fact that gRNAs possess 5′-triphosphate groups that are a substrate for the transferase. Purified mitochondrial RNA or total cell RNA can be used as a starting material.

1. Denature total cell RNA (10 μg) or mitochondrial RNA (1 μg) in 15 μl of water at 50° for 3 min. Place on ice.
2. To the RNA sample add
   2 μl of [α-^{32}P]GTP (800 Ci/mmol)
   2 μl 10× reaction buffer (600 mM Tris–HCl, pH 8.0, 60 mM MgCl$_2$, 100 mM DTT, supplied by the enzyme manufacturer)
   1 μl (5 U) vaccinia virus capping enzyme (available from Ambion)
   1 μl RNasin (20 U, optional)
3. Incubate at 37° for 30 min.
4. Add an equal volume of the standard formamide denaturing dye.
5. Resolve the RNA in 10% polyacrylamide 8 M urea gel.
6. Detect the labeled RNA by autoradiography.

Once the presence and gel position of gRNAs are determined, preparative isolation of gRNA is performed using at least 100 μg of total cell RNA or an equivalent amount of purified mitochondrial RNA. To facilitate detection of the gRNA bands in the gel, the content of the above GTP-labeling reaction can be added to unlabeled RNA prior to gel fractionation. The radioactive band is detected by autoradiography and the RNA is eluted. The rest of the procedure involves an oligo(A)-primed cDNA synthesis, a specific ligation of a 3′-modified "anchor" oligonucleotide to 3′ ends of the cDNA, PCR amplification of the products, cloning, and sequencing. This was described in detail previously (Simpson et al., 1996). For construction of a gRNA library in T. borreli, the cDNA synthesis step was performed after 3′ C-tailing the gel-isolated RNA with CTP and poly(A)-polymerase due to relatively short oligo(U)-tails of gRNAs in this organism (Yasuhira and Simpson, 1996). The procedure can be further modified to be used with commercially available 5′ and 3′ RACE systems.

REFERENCES

Arts, G. J., Van der Spek, H., Speijer, D., Van den Burg, J., Van Steeg, H., Sloof, P., and Benne, R. (1993). Implications of novel guide RNA features for the mechanism of RNA editing in Crithidia fasciculata. EMBO J. 12, 1523–1532.
Avila, H. A., and Simpson, L. (1995). Organization and complexity of minicircle-encoded guide RNAs in Trypanosoma cruzi. RNA 1, 939–947.

Benne, R., Van den Burg, J., Brakenhoff, J., Sloof, P., Van Boom, J., and Tromp, M. (1986). Major transcript of the frameshifted coxII gene from trypanosome mitochondria contains four nucleotides that are not encoded in the DNA. *Cell* **46**, 819–826.

Blom, D., De Haan, A., Van den Berg, M., Sloof, P., Jirků, M., Lukeš, J., and Benne, R. (1998). RNA editing in the free-living bodonid *Bodo saltans*. *Nucl. Acids Res.* **26**, 1205–1213.

Blom, D., De Haan, A., Van den Burg, J., Van den Berg, M., Sloof, P., Jirků, M., Lukeš, J., and Benne, R. (2000). Mitochondrial minicircles in the free-living bodonid *Bodo saltans* contain two gRNA gene cassettes and are not found in large networks. *RNA* **6**, 121–135.

Blum, B., Bakalara, N., and Simpson, L. (1990). A model for RNA editing in kinetoplastid mitochondria: "Guide" RNA molecules transcribed from maxicircle DNA provide the edited information. *Cell* **60**, 189–198.

Campbell, D. A., Thomas, S., and Sturm, N. R. (2003). Transcription in kinetoplastid protozoa: Why be normal? *Microbes Infect.* **5**, 1231–1240.

Corell, R. A., Myler, P., and Stuart, K. (1994). *Trypanosoma brucei* mitochondrial CR4 gene encodes an extensively edited mRNA with completely edited sequence only in bloodstream forms. *Mol. Biochem. Parasitol.* **64**, 65–74.

Decker, C. J., and Sollner-Webb, B. (1990). RNA editing involves indiscriminate U changes throughout precisely defined editing domains. *Cell* **61**, 1001–1011.

Feagin, J. E., Abraham, J., and Stuart, K. (1988a). Extensive editing of the cytochrome c oxidase III transcript in *Trypanosoma brucei*. *Cell* **53**, 413–422.

Feagin, J. E., Shaw, J. M., Simpson, L., and Stuart, K. (1988b). Creation of AUG initiation codons by addition of uridines within cytochrome b transcripts of kinetoplastids. *Proc. Natl. Acad. Sci. USA* **85**, 539–543.

Gao, G. G., Kapushoc, S. T., Simpson, A. M., Thiemann, O. H., and Simpson, L. (2001). Guide RNAs of the recently isolated LEM125 strain of *Leishmania tarentolae*: An unexpected complexity. *RNA* **7**, 1335–1347.

Hong, M., and Simpson, L. (2003). Genomic organization of *Trypanosoma brucei* kinetoplast DNA minicircles. *Protist* **154**, 265–279.

Koslowsky, D. J., Bhat, G. J., Read, L. K., and Stuart, K. (1991). Cycles of progressive realignment of gRNA with mRNA in RNA editing. *Cell* **67**, 537–546.

Lukeš, J., Arts, G. J., Van den Burg, J., De Haan, A., Opperdoes, F., Sloof, P., and Benne, R. (1994). Novel pattern of editing regions in mitochondrial transcripts of the cryptobiid *Trypanoplasma borreli*. *EMBO J.* **13**, 5086–5098.

Lukeš, J., Jirků, M., Avliyakulov, N., and Benada, O. (1998). Pankinetoplast DNA structure in a primitive bodonid flagellate, *Cryptobia helicis*. *EMBO J.* **17**, 838–846.

Lukeš, J., Guilbride, D. L., Votýpka, J., Zíková, A., Benne, R., and Englund, P. T. (2002). Kinetoplast DNA network: Evolution of an improbable structure. *Eukaryot. Cell* **1**, 495–502.

Lukeš, J., Hashimi, H., and Zíková, A. (2005). Unexplained complexity of the mitochondrial genome and transcriptome in kinetoplastid flagellates. *Curr. Genet.* **48**, 277–299.

Maslov, D. A., and Simpson, L. (1992). The polarity of editing within a multiple gRNA-mediated domain is due to formation of anchors for upstream gRNAs by downstream editing. *Cell* **70**, 459–467.

Maslov, D. A., and Simpson, L. (1994). RNA editing and mitochondrial genomic organization in the cryptobiid kinetoplastid protozoan, *Trypanoplasma borreli*. *Mol. Cell. Biol.* **14**, 8174–8182.

Maslov, D. A., Sturm, N. R., Niner, B. M., Gruszynski, E. S., Peris, M., and Simpson, L. (1992). An intergenic G-rich region in *Leishmania tarentolae* kinetoplast maxicircle DNA is a pan-edited cryptogene encoding ribosomal protein S12. *Mol. Cell. Biol.* **12**, 56–67.

Maslov, D. A., Elgort, M. G., Wong, S., Pecková, H., Lom, J., Simpson, L., and Campbell, D. A. (1993). Organization of mini-exon and 5S rRNA genes in the kinetoplastid *Trypanoplasma borreli*. *Mol. Biochem. Parasitol.* **61**, 127–136.

Maslov, D. A., Avila, H. A., Lake, J. A., and Simpson, L. (1994). Evolution of RNA editing in kinetoplastid protozoa. *Nature* **365,** 345–348.

Nawathean, P., and Maslov, D. A. (2000). The absence of genes for cytochrome *c* oxidase and reductase subunits in maxicircle kinetoplast DNA of the respiration-deficient plant trypanosomatid *Phytomonas serpens. Curr. Genet.* **38,** 95–103.

Panigrahi, A. K., Ernst, N. L., Domingo, G. J., Fleck, M., Salavati, R., and Stuart, K. D. (2006). Compositionally and functionally distinct editosomes in *Trypanosoma brucei. RNA* **12,** 1038–1049.

Pollard, V. W., Rohrer, S. P., Michelotti, E. F., Hancock, K., and Hajduk, S. L. (1990). Organization of minicircle genes for guide RNAs in *Trypanosoma brucei. Cell* **63,** 783–790.

Read, L. K., Myler, P. J., and Stuart, K. (1992). Extensive editing of both processed and preprocessed maxicircle CR6 transcripts in *Trypanosoma brucei. J. Biol. Chem.* **267,** 1123–1128.

Shaw, J., Feagin, J. E., Stuart, K., and Simpson, L. (1988). Editing of mitochondrial mRNAs by uridine addition and deletion generates conserved amino acid sequences and AUG initiation codons. *Cell* **53,** 401–411.

Shaw, J., Campbell, D., and Simpson, L. (1989). Internal frameshifts within the mitochondrial genes for cytochrome oxidase subunit II and maxicircle unidentified reading frame 3 in *Leishmania tarentolae* are corrected by RNA editing: Evidence for translation of the edited cytochrome oxidase subunit II mRNA. *Proc. Natl. Acad. Sci. USA* **86,** 6220–6224.

Simpson, L. (1972). The kinetoplast of the hemoflagellates. *Int. Rev. Cytol.* **32,** 139–207.

Simpson, L. (1979). Isolation of maxicircle component of kinetoplast DNA from hemoflagellate protozoa. *Proc. Natl. Acad. Sci. USA* **76,** 1585–1588.

Simpson, L. (1997). The genomic organization of guide RNA genes in kinetoplastid protozoa: Several conundrums and their solutions. *Mol. Biochem. Parasitol.* **86,** 133–141.

Simpson, L., and da Silva, A. M. (1971). Isolation and characterization of kinetoplast DNA from *Leishmania tarentolae. J. Mol. Biol.* **56,** 443–473.

Simpson, L., and Maslov, D. A. (1994). RNA editing and the evolution of parasites. *Science* **264,** 1870–1871.

Simpson, L., and Shaw, J. (1989). RNA editing and the mitochondrial cryptogenes of kinetoplastid protozoa. *Cell* **57,** 355–366.

Simpson, L., Neckelmann, N., de la Cruz, V., Simpson, A., Feagin, J., Jasmer, D., and Stuart, K. (1987). Comparison of the maxicircle (mitochondrial) genomes of *Leishmania tarentolae* and *Trypanosoma brucei* at the level of nucleotide sequence. *J. Biol. Chem.* **262,** 6182–6196.

Simpson, L., Simpson, A. M., and Blum, B. (1994). RNA editing in mitochondria. *In* "RNA Processing" (S. J. Higgins and B. D. Hames, eds.), Vol. II, pp. 69–105. IRL Press, Oxford.

Simpson, L., Frech, G. C., and Maslov, D. A. (1996). RNA editing in trypanosomatid mitochondria. *Methods Enzymol.* **264,** 99–121.

Simpson, L., Wang, S. H., Thiemann, O. H., Alfonzo, J. D., Maslov, D. A., and Avila, H. A. (1998). U-insertion/deletion Edited Sequence Database. *Nucleic Acids Res.* **26,** 170–176.

Simpson, L., Aphasizhev, R., Gao, G., and Kang, X. (2004). Mitochondrial proteins and complexes in *Leishmania* and *Trypanosoma* involved in U-insertion/deletion RNA editing. *RNA* **10,** 159–170.

Souza, A. E., Shu, H.-H., Read, L. K., Myler, P. J., and Stuart, K. D. (1993). Extensive editing of CR2 maxicircle transcripts of *Trypanosoma brucei* predicts a protein with homology to a subunit of NADH dehydrogenase. *Mol. Cell. Biol.* **13,** 6832–6840.

Štolba, P., Jirků, M., and Lukeš, J. (2001). Polykinetoplast DNA structure in *Dimastigella tryaniformis* and *Dimastigella mimosa* (Kinetoplastida). *Mol. Biochem. Parasitol.* **113,** 323–326.

Stuart, K. D., Schnaufer, A., Ernst, N. L., and Panigrahi, A. K. (2005). Complex management: RNA editing in trypanosomes. *Trends Biochem. Sci.* **30,** 97–105.

Sturm, N. R., and Simpson, L. (1990a). Partially edited mRNAs for cytochrome *b* and subunits III of cytochrome oxidase from *Leishmania tarentolae* mitochondria: RNA editing intermediates. *Cell* **61,** 871–878.

Sturm, N. R., and Simpson, L. (1990b). Kinetoplast DNA minicircles encode guide RNAs for editing of cytochrome oxidase subunit III mRNA. *Cell* **61,** 879–884.

Sturm, N. R., and Simpson, L. (1991). *Leishmania tarentolae* minicircles of different sequence classes encode single guide RNAs located in the variable region approximately 150 bp from the conserved region. *Nucleic Acids Res.* **19,** 6277–6281.

Sturm, N. R., Maslov, D. A., Blum, B., and Simpson, L. (1992). Generation of unexpected editing patterns in *Leishmania tarentolae* mitochondrial mRNAs: Misediting produced by misguiding. *Cell* **70,** 469–476.

Thiemann, O. H., Maslov, D. A., and Simpson, L. (1994). Disruption of RNA editing in *Leishmania tarentolae* by the loss of minicircle-encoded guide RNA genes. *EMBO J.* **13,** 5689–5700.

Vickerman, K. (1976). The diversity of the kinetoplastid flagellates. *In* "Biology of the Kinetoplastida" (W. H. R. Lumsden and D. A. Evans, eds.), pp. 1–34. Academic Press, London.

Von Haeseler, A., Blum, B., Simpson, L., Sturm, N., and Waterman, M. S. (1992). Computer methods for locating kinetoplastid cryptogenes. *Nucleic Acids Res.* **20,** 2717–2724.

Yasuhira, S., and Simpson, L. (1996). Guide RNAs and guide RNA genes in the cryptobiid kinetoplastid protozoan, *Trypanoplasma borreli. RNA* **2,** 1153–1160.

PHYSARUM INSERTION EDITING

CHAPTER SEVEN

RNA EDITING IN *PHYSARUM* MITOCHONDRIA: ASSAYS AND BIOCHEMICAL APPROACHES

Elaine M. Byrne,[†] Linda Visomirski-Robic,[‡] Yu-Wei Cheng,[§] Amy C. Rhee,* *and* Jonatha M. Gott*

Contents

Abstract

Mitochondrial RNAs in the myxomycete *Physarum polycephalum* differ from the templates from which they are transcribed in defined ways. Most transcripts contain nucleotides that are not present in their respective genes. These "extra" nucleotides are added during RNA synthesis by an unknown mechanism. Other differences observed between *Physarum* mitochondrial RNAs and the mitochondrial genome include nucleotide deletions, C to U changes, and the replacement of one nucleotide for another at the 5′ end of tRNAs. All of these alterations are remarkably precise and highly efficient *in vivo*. Many

* Center for RNA Molecular Biology, Case Western Reserve University, Cleveland, Ohio
† Trinity Centre for Engineering, Trinity College Dublin, Dublin, Ireland
‡ Department of Biological Sciences, Carnegie Mellon University, Pittsburgh, Pennsylvania
§ Department of Microbiology and Immunology, Weill Medical College of Cornell University, New York, New York

Methods in Enzymology, Volume 424
ISSN 0076-6879, DOI: 10.1016/S0076-6879(07)24007-8

of these editing events can be replicated *in vitro*, and here we describe both the *in vitro* systems used to study editing in *Physarum* mitochondria and the assays that have been developed to assess the extent of editing of RNAs generated in these systems at individual sites.

1. INTRODUCTION

Gene expression in *Physarum* mitochondria requires an incredible range of RNA editing activities, including the cotranscriptional insertion of nonencoded nucleotides, the deletion of encoded nucleotides, and posttranscriptional base changes (Gott and Visomirski-Robic, 1998; Gott *et al.*, 1993, 2005; Miller *et al.*, 1993). All classes of RNAs (mRNAs, tRNAs, and rRNAs) are subject to editing in *Physarum* mitochondria and each type of editing is both accurate and efficient. Indeed, nearly all mitochondrial transcripts present in the steady-state RNA pool are completely and correctly edited. The vast majority of the more than 500 known editing events involve the insertion of "extra" nucleotides, which occurs in a highly specific manner. Roughly 90% of the added nucleotides are single C insertions, but single U insertions and dinucleotide insertions (AA, UU, GU, GC, UA, and CU) are also observed at defined sites. The source of the information that specifies which nucleotide is to be added at a given site is currently unknown.

Two types of *in vitro* systems have been used to study insertional editing mechanisms: isolated mitochondria (Visomirski-Robic and Gott, 1995) and partially purified mitochondrial transcription elongation complexes (mtTECs) (Cheng and Gott, 2000). These mtTEC preparations consist of entire mitochondrial genomes (each ~63,000 bp) and their associated macromolecules, including elongating RNA polymerases and nascent transcripts. Transcripts made in isolated mitochondria or by elongation complexes are highly edited, which has made possible a range of mechanistic studies. Using these systems, it has been determined that the addition of nonencoded nucleotides is cotranscriptional, occurring at the $3'$ end of nascent transcripts (Cheng *et al.*, 2001; Visomirski-Robic and Gott, 1997a,b). Transcription of mitochondrial sequences using RNA polymerase isolated from *Physarum* mitochondria does not result in edited transcripts, however, and attempts to reconstitute editing on exogenous DNA templates have not been successful as yet. Thus, all of the currently available *Physarum in vitro* editing systems rely on run-on transcription by elongation complexes formed *in vivo*.

The DNA in mtTECs is accessible to enzymes such as restriction endonucleases and DNA ligase, allowing limited manipulation of the native template (Byrne and Gott, 2002). Digestion and religation in the presence or absence of exogenous DNA fragments result in the creation of chimeric templates that have yielded significant insights into the mechanism of insertional editing (Byrne and Gott, 2002; Byrne *et al.*, 2002). Experiments with chimeric templates have demonstrated that the region of the transcript that is made from the native portion of the DNA template is edited, but remarkably, parts of the same RNA transcript synthesized from exogenous DNA are unedited, indicating that insertion of nonencoded nucleotides requires some feature of the native template (Byrne and Gott, 2002). This is true even if deproteinized mitochondrial DNA is ligated to the DNA in mtTECs, indicating that the lack of editing is not simply due to a requirement for some covalent modification within the transcription/editing template. These findings have been exploited in recent experiments aimed at defining the regions of the template that are required for accurate editing (A. C. Rhee, unpublished observations).

To date, only nucleotide insertions have been systematically investigated in *Physarum* mitochondria. Consequently, the methods detailed in this chapter are primarily directed at dissecting insertional editing mechanisms, although many of the assays provide information regarding other forms of editing (Gott and Visomirski-Robic, 1998; Gott *et al.*, 1993, 2005; Rundquist and Gott, 1995; Visomirski-Robic and Gott, 1995). A number of other chapters in this volume [particularly those by Lohan and Gray (Chapter 10), Takenaka and Brennicke (Chapter 20), and Hayes and Hanson (Chapter 21)] contain assays that can be used to study other forms of editing that occur in *Physarum* mitochondria (i.e., tRNA deletion/insertion editing and C-to-U changes), and thus are not covered here.

2. Experimental Systems

2.1. Overview

The dissection of editing mechanisms in *Physarum* has been challenging because it is a cotranscriptional process that requires yet to be identified features of the native template (Byrne and Gott, 2002; Cheng *et al.*, 2001). Consequently, all of the current *in vitro* systems rely on the use of transcription elongation complexes formed *in vivo*, precluding the use of many "traditional" approaches. Both isolated mitochondria (Visomirski-Robic and Gott, 1995) and partially purified mtTECs (Cheng and Gott, 2000) can be used to study editing. Each system has its advantages and disadvantages, and these characteristics affect the choice of editing assays that can be used to analyze the resulting transcripts. While early mechanistic studies were predominantly performed with isolated mitochondria (Gott and Visomirski-Robic, 1998; Visomirski-Robic and Gott, 1995, 1997a,b), recent efforts have focused on the use of mtTECs (Byrne, 2004; Byrne and Gott, 2002, 2004; Cheng *et al.*, 2001) when possible, as they allow greater control and manipulation of transcription/editing conditions.

The efficiency and accuracy of editing differ somewhat in the two *in vitro* systems. Editing is more efficient in isolated mitochondria, reaching nearly 100% at most insertion sites under standard conditions (Visomirski-Robic and Gott, 1995). Overall levels of editing are typically lower when mtTECs are used (~50% on average), but the extent of editing at an individual site can reach nearly 100%, depending on the choice of conditions (Cheng and Gott, 2000; Cheng *et al.*, 2001). In both *in vitro* systems, the level of editing is sensitive to the concentration of the nucleotide to be added at an insertion site (Cheng *et al.*, 2001; Visomirski-Robic and Gott, 1997a). To achieve maximal editing at a given site when using mtTECs, the identity of the encoded nucleotide immediately downstream of the insertion site and the relative nucleotide triphosphate (NTP) concentrations in the reaction are critical. Lowering the concentration of the encoded nucleotide immediately downstream of a C insertion site enhances the extent of editing at that site (as well as other sites followed by the limiting nucleotide), while editing efficiency is reduced at sites not followed by the limiting nucleotide (Cheng *et al.*, 2001). The accuracy of *in vitro* editing is quite high. In isolated mitochondria, the extent of misediting falls below the level of detection of our standard assays (Visomirski-Robic and Gott, 1995, 1997a,b). In contrast, a low level (~5%) of misediting is observed in mtTECs, mostly involving the addition of a G at a C insertion site (Byrne *et al.*, 2002). Other misediting events that have been occasionally observed in transcripts made by mtTECs include dinucleotides added at C insertion sites, the addition of a single nucleotide at dinucleotide insertion sites, the deletion of one to three nucleotides immediately adjacent to editing sites, and intersite deletions, which involve the deletion of all encoded nucleotides that fall between two editing sites (Byrne and Gott, 2004; Byrne *et al.*, 2002).

A third difference is that mtTEC preparations contain mainly nascent transcripts, while isolated mitochondria contain, in addition to nascent RNAs, the full-length RNAs completed *in vivo* before isolation. Because editing efficiency is virtually 100% *in vivo* and nearly 100% in isolated organelles, RNAs synthesized in isolated mitochondria cannot be readily distinguished from those made *in vivo* unless the RNA made *in vitro* is radioactively labeled. This precludes the use of indirect techniques such as RT-PCR when analyzing RNAs made in isolated mitochondria. In contrast, because the mtTECs contain principally nascent transcripts, a wider range of techniques can be used to characterize the RNAs synthesized using mtTECs. Not only can radiolabeled populations of RNA be subjected to analysis, but unlabeled run-on RNA can be examined through RT-PCR. Individual transcripts can be characterized, as the presence of an unedited or misedited site can be used to the experimenter's advantage, as they indicate that, at a minimum, all downstream editing sites were made *in vitro*.

A fourth major difference between isolated mitochondria and mtTECs is endogenous nucleotide levels. Although isolated mitochondria are "leaky" enough to allow the incorporation of added radiolabeled nucleotides, they retain significant nucleotide pools (Visomirski-Robic and Gott, 1997a). Preincubation of mitochondria in Transcription buffer enhances incorporation of labeled nucleotides, but the absolute concentration of individual nucleotides cannot be ascertained. In contrast, mtTECs have negligible nucleotide pools (Cheng and Gott, 2000), making them the system of choice for most applications.

Finally, isolated mtTECs offer opportunities to manipulate the transcription/editing template that are currently lacking in isolated mitochondria (Byrne and Gott, 2002). The ability to cut and religate the DNA present in mtTEC preparations opens up a range of possible experiments, as described in more detail below. It should be noted that such experiments are not limited to the study of *Physarum* editing; conceivably any cotranscriptional process may be amenable to the application of these methods.

2.2. Isolation of mitochondria from *Physarum polycephalum*

2.2.1. Overview

Mitochondria prepared by differential centrifugation can be used for both the synthesis of labeled run-on transcripts and as a source of RNAs enriched in mitochondrial transcripts made *in vivo*. Reasonable quantities of cellular material can be obtained from either macroplasmodial or microplasmodial cultures. No significant differences in the extent of RNA synthesis or levels of editing have been observed between mitochondria isolated from macroplasmodia or microplasmodia. Macroplasmodia are generally more convenient for small-scale uses; microplasmodia are generally used for larger scale experiments. Additional purification steps (e.g., Percoll) can be used when more highly purified mitochondria are desired, but crude mitochondria are generally preferred due to the reduction in overall yield and activity during purification. Mitochondria should be freshly isolated when needed, as these preparations lose activity quickly, even when stored at $-80°$. All solutions and equipment are kept cold (on ice) unless otherwise specified; note that pH values are quoted for room temperature.

2.2.2. Isolation of mitochondria from macroplasmodia and microplasmodia

1. Culture *Physarum* as either macroplasmodia (on Schleicher & Schuell 576 filters underlain with media) or microplasmodia (in baffled flasks shaken at 150–200 rpm) at $26°$ in *Physarum* medium as described by Daniel and Baldwin (1964) for ~48 h. Note that the following protocol is written for 5–10 g of cells, but can be readily scaled up.
2. *To harvest macroplasmodia*: Using a sterile, flat spatula, transfer macroplasmodia into a tared conical tube containing 40 ml of ice-cold

BSS (250 mM sucrose/10 mM Tris–HCl, pH   7.5) and determine cell weight.

To harvest microplasmodia: Pour microplasmodia cultures into a large beaker, allow cells to settle, and decant the medium. Wash twice with sterile, distilled water, allowing cells to settle and decanting each wash. Resuspend cells in ~60 ml water and transfer to two preweighed Oak Ridge tubes. Pellet microplasmodia by centrifuging at 500×g for 5 min at 4°. Pour off the supernatant, rinse each pellet with ~15 ml water, vortex gently, and spin as above. Pour off supernatants, pipette off excess water, weigh tubes with pellets, and calculate cell weight. Resuspend cell pellets in 5 ml of BSS (250 mM sucrose/10 mM Tris–HCl, pH 7. 5) per gram of cells.

3. Pour the suspension into an appropriately sized Waring blender, adding sufficient BSS to cover the blades. Blend for 15 sec at ~50% power (a rheostat can be used to control blender speed) to lyse cells. Allow the mixture to settle for 30 sec (to reduce heat and foaming) and blend for an additional 15 sec as above. Visually determine the extent of lysis using a phase contrast microscope. Repeat the blending if necessary.

4. Filter the homogenate through a Schleicher & Schuell 520B1/2 or Whatman 113V filter prewet with BSS into either a 50-ml conical tube (small scale, gravity flow) or side arm flask (large scale, under low vacuum) embedded in ice.

5. Pour the homogenate into ice-cold Oak Ridge tubes and pellet the nuclei and cellular debris at ~700×g for 5 min at 4°. Decant the supernatant containing the mitochondria into fresh Oak Ridge tubes.

6. Pellet the mitochondria by centrifugation at ~6000×g for 5 min at 4°; decant and discard the supernatant.

7. Resuspend each mitochondrial pellet in 10 ml BSS, remove an aliquot, and determine the protein concentration using the Bradford reagent to estimate yield. This value is used to determine the volume of BSS used in step 9.

8. Pellet mitochondria at ~6000×g for 5 min at 4°; decant and remove as much of the supernatant as possible by draining the pellets upside down.

9. Resuspend the final mitochondrial pellet in BSS to a final protein concentration of 1–3 mg/ml.

2.3. Isolation of mitochondrial transcription elongation complexes

2.3.1. Overview

Although editing is less efficient in preparations enriched in elongation complexes than it is in isolated organelles, mtTECs offer a number of advantages as an experimental system, as noted above. In addition, unlike isolated mitochondria, mtTECs can be stored at −80° for up to a year without significant loss of transcription or editing activity. Reproducibility between aliquots is high, allowing for direct comparisons between experiments and reducing the number of variables that need to be examined in subsequent experiments with the same mtTEC preparation. Transcription elongation complexes are generally isolated from microplasmodia because it is easier to obtain 80–100 g of starting material from liquid culture, but macroplasmodia can also be used for this purpose.

2.3.2. Isolation of crude transcription elongation complexes from *Physarum* mitochondria

1. Culture *Physarum* as microplasmodia (3–4.5 liters in large baffled flasks shaken at 210 rpm) at 26° in *Physarum* medium as described by Daniel and Baldwin (1964) for ~45–46 h.

2. Isolate mitochondria as described above with the following exceptions. (a) Wash settled microplasmodia with ~500 ml ice-cold Sclerotia Salts (Daniel and Baldwin, 1964). (b) Centrifuge in 500-ml bottles rather than Oak Ridge tubes (the last wash of the mitochondrial pellets can be done in Oak Ridge tubes).

3. Remove as much of the supernatant as possible from the mitochondrial pellets, estimate the pellet volume, and resuspend well in ~1/2 volume of $2\times$ Buffer A [20% glycerol/20 mM Tris–HCl (pH 8)/10 mM $MgCl_2$/2 mM dithiothreitol (DTT)/1 mM EDTA] with gentle vortexing. Adjust to a final concentration of $1\times$ Buffer A and determine protein concentration (e.g., Bio-Rad DC protein assay). Note that the concentration of the mitochondrial suspension should be 10–20 mg/ml for optimal performance in the following steps, which should all be carried out at ~4°.

4. Add protease inhibitors directly to the mitochondrial suspension to a final concentration of 0.5 mM phenylmethylsulfonyl sulfate (PMSF) and 1 μg/ml leupeptin.

5. Add NP40 to a final concentration of 1%; vortex gently to lyse the mitochondria.

6. Add KCl to a final concentration of 250 mM, vortexing gently.

7. Clear lysate by centrifuging at ~130,000×g for 1 h at 4°. A table top ultracentrifuge (thick-walled polycarbonate tubes spun at 39 K rpm in a Beckman TLS 55 rotor) is useful for this purpose. Carefully pipette off the supernatants into a fresh tube, avoiding the pellet and any lipid layer at the top. Measure the volume and determine the protein concentration.

8. Add additional protease inhibitors to give a final concentration of 0.5 mM PMSF and 1 μg/ml leupeptin.

9. Load the cleared mitochondrial lysate onto a ~45-ml Sepharose 4B column (~1 cm diameter) equilibrated with $1\times$ Buffer A, then wash the column with $1\times$ Buffer A, collecting ~1-ml fractions for ~5 h.

10. Assay the fractions for *template-independent* (i.e., run-on) transcription as described below. Peak activity is generally found in fractions 10–15.

11. Pool active fractions and freeze in small (~50- to 100-μl) aliquots to avoid repeated freeze–thaw cycles, which reduce overall activity. Store at −80°. Active fractions can also be pooled and dialyzed into 10 mM Tris–HCl (pH 7.8)/0.5 mM MgCl$_2$/1 mM dithiothreitol (DTT)/ 0.1 mM EDTA/10% glycerol (3 changes, ~1 liter total) prior to freezing.

2.4. Generation of chimeric templates

2.4.1. Overview

The DNA present in active mtTECs can be efficiently digested by restriction enzymes (REs) (Byrne and Gott, 2002). This allows ligation of DNA sequences that are not normally adjacent to one another to create novel templates (Fig. 7.1), including both subgenomic circles and "scrambled" intermolecular chimeras (Byrne, 2004). Adding exogenous DNA to the ligation reaction results in chimeras between mtTEC and exogenous DNA (Fig. 7.1). While chimeras derived entirely from mtTEC DNA produce edited RNA, ligated exogenous DNA does not support editing, despite the fact that regions of the same transcript synthesized from the DNA derived from mtTECs are edited (Byrne and Gott, 2002). However, the addition of exogenous DNA has proven useful as a unique tag for RT-PCR, particularly in experiments aimed at defining template requirements for editing (Byrne and Gott, 2002) (A. C. Rhee, unpublished observations). For more details regarding the creation and use of chimeric templates, the reader is referred to Byrne (2004).

The methods described below outline the preparation of DNA cassettes for experiments involving ligation of exogenous DNA to fragmented mtTEC and production of chimeric templates. A number of interdependent parameters must be considered together when planning a chimeric template experiment. REs are chosen with the following considerations in mind. (1) Enzymes that generate staggered cuts in the duplex are preferable, as sticky ends are more efficiently ligated. (2) It is preferable to use pairs of nonisoschizomers to generate compatible sticky ends on the fragments targeted for ligation, if possible, due to the continued presence of the RE during subsequent steps. If a single RE is used and does not allow sufficient ligation product to accumulate, it may be possible to deplete the enzyme before ligation, using a BioSpin column (Bio-Rad). (3) Ligation of exogenous DNA should be more efficient if the targeted endogenous mtTEC fragment cannot circularize; in this case, it might be helpful to include an additional RE that generates an incompatible upstream end.

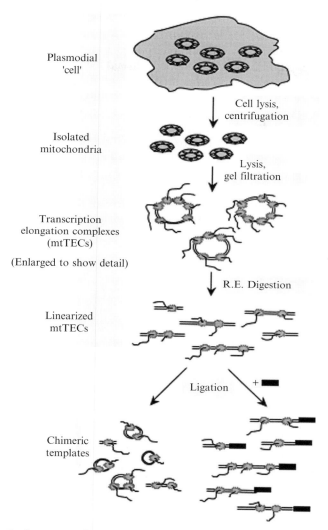

Plasmodial
'cell'

Cell lysis,
centrifugation

Isolated
mitochondria

Lysis,
gel filtration

Transcription
elongation complexes
(mtTECs)

(Enlarged to show detail)

R.E. Digestion

Linearized
mtTECs

Ligation

Chimeric
templates

Figure 7.1 *In vitro* systems for studying RNA editing in *Physarum* mitochondria. Logarithmically growing macroplasmodia or microplasmodia are broken open and the lysate is subjected to differential centrifugation to yield isolated mitochondria. Mitochondria are then lysed with detergent and the circular mitochondrial genome and associated macromolecules, including RNA polymerases and nascent trancripts (transcription elongation complexes, mtTECs), are partially purified from the cleared lysate via gel filtration chromatography. The DNA in mtTECs can be digested with restriction endonucleases (RE) to give linearized mtTECs (used for runoff experiments) or subsequently ligated in the absence or presence of added DNA cassettes (black boxes) to yield chimeric templates. See text for details.

2.4.2. Generation of DNA cassettes

Short cassettes can be made by annealing two suitably designed oligo-nucleotides, where one has a short unpaired terminus to act as a sticky end for ligation. The nonligating end should be blunt (and preferably without a 5' phosphate). Longer cassettes are most easily obtained by PCR amplification from suitably designed recombinant plasmids: if the cassette is to contain a section of mitochondrial sequence, it should include a downstream 3' terminal "tag" of a sequence not found in the mitochondrial genome to aid in the isolation/analysis of chimeric RNAs. The sticky end for ligation is then generated by RE digestion. If a preexisting RE site is not used, an appropriate recognition sequence can be engineered either when making the plasmid insert or when amplifying the cassette, by inclusion of the desired site at the 5' end of the upstream primer. After digestion, it is best to purify the cassette by agarose gel electrophoresis to prevent competition by other restriction fragment(s) in the subsequent ligation to mtTEC.

2.4.3. Restriction enzyme digestion of mtTECs

The protocols described below will act as a guide to setting up chimeric templates. Note that this digestion procedure can also be used to generate mtTEC cleaved near editing sites for "runoff" transcription to examine the role of downstream DNA; in this case, the ligation steps below are skipped and the digested mtTECs are used immediately for *in vitro* transcription (see Run-on transcription using mtTECs). When planning the details of the mtTEC cleavage-ligation-transcription series, the following considerations apply. (1) In order to reduce the chance of RE star activity, the final concentration of glycerol (contributed by the mtTEC preparation and added enzyme stocks) should not exceed 10%. (2) A high salt RE buffer (Roche "Buffer H") is the default digestion condition, as degradation of the mitochondrial DNA has been observed at lower salt concentrations. Therefore, if using REs that require a lower salt concentration, test digestions should first be carried out, monitoring the resulting fragment patterns by extracting the DNA and visualizing it via agarose gel electrophoresis. (3) If intermolecular ligation products are sought, it is best to keep the RE and ligation reaction volumes to a minimum; conversely, formation of intramolecular circles will be unaffected or facilitated by lower DNA concentrations. (4) For experiments involving labeled transcripts and direct RNA analysis, reactions can be scaled up by a factor of 2 or more. (5) It is desirable to check cleavage and ligation efficiency by agarose gel electrophoresis (followed by Southern blotting if necessary) and to include control reactions in which the REs, exogenous DNA, ligase, or the run-on transcription step are omitted (Byrne and Gott, 2002). The presence of ligated product can also be checked by PCR.

1. Thaw an aliquot of the mtTEC preparation on ice and remove a volume that contains ~5 μg protein. *Note*: mtTEC preparations may vary considerably in their protein and nucleic acid concentrations, and a range of 2–10 μg protein has been used successfully in this protocol.
2. Bring to 35 μl with 3.5 μl 10× Buffer H [500 mM Tris–HCl/100 mM MgCl$_2$/1 M NaCl/10 mM dithioerythritol (Roche); pH of 1× is 7. 5 at 37°], water and 15 U of each RE (use high concentration stocks, if available, to minimize glycerol addition). Combine water and buffer before mixing with mtTEC; add RE(s) last.
3. Incubate at 30° for 15–20 min.

2.4.4. Ligation reactions
After digestion of the DNA in mtTECs by one or more restriction endonucleases, the resulting fragments are then ligated to each other or to exogenous DNA prior to run-on transcription/editing reactions (as described in the section Run-on transcription using mtTECs). Detection of chimeric transcripts demonstrates success of the ligation reaction, which can also be evaluated directly by agarose gel electrophoresis with or without subsequent Southern blotting (Byrne and Gott, 2002) or by the PCR protocol below (PCR and Cloning).

1. Bring the reaction to 40 μl by adding, in the following order:
   a. ATP to 500 μM
   b. 2 pmol exogenous DNA, where appropriate. *Note*: The aim is to have an approximately 20-fold molar excess of exogenous cassette over mtTEC fragment ends. However, a reasonable proportion of the mtTEC fragment pool is ligated to the exogenous DNA at much smaller molar excesses of the latter; some ligation has been observed at a 2-fold excess.
   c. 1.5 U T4 DNA ligase (10 U/μl, Roche; can dilute in 10 mM Tris–HCl, pH 7.5, as necessary).
2. Incubate at 16° for 30 min.

2.5. RNA synthesis
2.5.1. Overview
Optimal RNA synthesis requires different conditions in isolated organelles and mtTECs due to both physical differences (e.g., the presence or absence of the mitochondrial membranes) and to differences in endogenous nucleotide pools. However, because both systems rely on run-on transcription rather than *de novo* initiation, transcription reactions can be carried out under a broad range of conditions. Neither transcription nor editing is particularly salt sensitive; both occur at KCl concentrations up to ~450 mM. Preincubation of isolated mitochondria in Transcription buffer

without added nucleotides depletes the nucleotide pools, allowing some control of nucleotide concentrations and greatly enhancing incorporation of labeled nucleotides during RNA synthesis. Mitochondrial pools of ATP and GTP are quite high and significant RNA synthesis is observed in isolated mitochondria even in the absence of added ATP or GTP, whereas only a limited amount of RNA is made when either UTP or CTP is omitted from the reaction (L. Visomirski-Robic, unpublished observations). In contrast, all four nucleotides are required for RNA synthesis in mtTECs, which have negligible nucleotide pools. It is important to add all of the nucleotides at the same time, so that each individual nucleotide is incorporated under the same conditions and labeling is uniform. Nucleotides are usually added as part of a master mix with other ingredients, but if the NTP mixture is added separately, it should be added last. To make labeled RNA for direct analysis, an $[\alpha\text{-}^{32}P]NTP$ (typically 30 μCi for mtTECs and 0.1–1 mCi for isolated mitochondria) is included in the run-on transcription reaction. If RNase T1 fragments are to be subjected to secondary analysis, the local sequence will determine which label is most useful; if radioactive GTP is used, all T1 fragments will be labeled with equal intensity. There are additional considerations regarding nucleotide concentrations. (1) Reducing the concentration of a given NTP decreases the efficiency with which it is inserted at editing sites and, conversely, increases the efficiency of editing at sites that immediately precede locations at which that nucleotide is conventionally encoded (Cheng *et al.*, 2001). (2) There are practical limits to reducing the concentrations of more than one nucleotide at a time; RNA yields must be adequate for subsequent analyses.

2.5.2. Run-on transcription in isolated mitochondria

When planning mitochondrial labeling experiments, the extent of RNA synthesis needs to be taken into account. Based on S1 protection experiments of pulse-labeled run-on RNAs, nascent transcripts synthesized in isolated mitochondria are extended only ~150–300 nucleotides when nucleotide concentrations are limiting, but can be extended at least 450 nucleotides at high nucleotide concentrations (Visomirski-Robic and Gott, 1997a). Typically these experiments are done using $[\alpha\text{-}^{32}P]GTP$. This choice has two advantages: first, the limiting nucleotide is GTP, and thus insertional editing remains unaffected, and second, all RNase T1 fragments (which end in G) are visible in subsequent gel electrophoresis or fingerprinting analyses. If experiments are done using $[\alpha\text{-}^{32}P]CTP$, care must be taken in choosing conditions that both allow editing to occur and produce sufficient amounts of labeled RNA for analysis.

1. Mix the crude mitochondrial preparation (at 1.0–2.4 mg/ml protein) with 0.25 volume 5× Transcription buffer (100 mM Tris–HCl,

pH 7.5/100 mM $MgCl_2$/50 mM KCl/10 mM DTT) and preincubate at 35° for 3 min to deplete endogenous NTP pools.

2. Bring to a final concentration of 1× Transcription buffer by the addition of exogenous NTPs to a final concentration of 0.15–0.45 μM [α-^{32}P] NTP (3.3 μM stock at 3000 Ci/mmol) and 100–200 μM nonlabeled NTPs, and incubate at 35° for ~15 min. Run-on RNAs can be extended further (chased) by the addition of 10–100 μM of the limiting NTP (unlabeled) and incubating at 35° for an additional ~5 min. Incubation conditions can be varied; for example, in some pulse–chase experiments, we have used a 3 min pulse labeling followed by a 12 min chase (Visomirski-Robic and Gott, 1997b).

3. Terminate the reaction by the addition of an equal volume of Stop Solution [10 mM Tris–HCl (pH 7.4)/20 mM EDTA/0.4% sodium dodecyl sulfate (SDS)] and isolate the nucleic acids as described in the section on RNA Isolation from Mitochondria.

2.5.3. Run-on transcription using mtTECs

1. *Synthesis of labeled transcripts* using isolated transcription elongation complexes is carried out using a final concentration of ~0.5 mg/ml transcription complexes in 20 mM Tris–HCl, pH 8/10 mM $MgCl_2$/0.1 mg/ml bovine serum albumin (BSA)/2 mM DTT/500 μM nonlimiting NTPs/ 2 μM limiting NTP [~0.37 μM contributed by the labeled NTP (3.3 μM stock at 3000 Ci/mmol)]. Under these conditions, 500 or more nucleotides are added to nascent RNAs (Cheng and Gott, 2000). The concentration of limiting nucleotide can be varied between 0.2 μM and 100 μM.
   Synthesis of unlabeled transcripts is done under the same conditions except that all NTPs are unlabeled (500 μM unless required at lower concentration to modulate editing efficiency).

2. Incubate at 30° for 40 min, then chase with 500 μM limiting NTP, incubating another 10 min at 30°.

3. Stop the reaction by adding one-ninth volume 10× Transcription Stop Solution (100 mM Tris pH 7.5/200 mM EDTA/1% SDS) and isolate the nucleic acids as described under RNA Isolation from Mitochondria.

Note: When *assaying column fractions* during isolation of mtTECs, use 5 μl of column fraction (or column buffer control) in a final volume of 25 μl under the conditions given in step 1, then incubate for 30 min at 30° prior to spotting 10-μl aliquots onto duplicate DE81 filters. Wash the filters four times with 10 mM sodium pyrophosphate/0.3 M ammonium formate, pH 7.8, by swirling in a large beaker, then do a final wash with 100% ethanol. Dry the filters about 1 min on a paper towel and count using a scintillation counter.

2.5.4. Transcription from chimeric templates

Conditions for run-on transcription from chimeric templates (and also from mtTEC that have been cleaved but not ligated) are similar to those described above for intact mtTECs, except for higher levels of Tris (approximately 40 mM, pH 7. 8) and NaCl (approximately 70 mM), which are contributed by the RE digestion buffer in addition to the mtTEC buffer; 40–50 mM Tris and 67–90 mM NaCl have been present in previous successful experiments. The protocol below describes an unlabeled transcription reaction with all nucleotides at the same concentration. Some ATP is consumed during the ligation reaction, but the amount has not been determined. If a defined low concentration is desired during transcription, most of the cold ATP can be removed after ligation by passing the reaction through a BioSpin-30 column (Bio-Rad; one per standard size reaction) prior to transcription (the columns should first be preequilibrated with $0.75\times$ mtTEC buffer/$1\times$ Buffer H/1 mM DTT).

1. Bring the freshly ligated chimeric template to a final concentration of 0.1 mg/ml BSA, 2.5 mM DTT, 500 μM CTP, GTP, and UTP, and 200 μM ATP (in addition to the ATP added to the ligation reaction) in a final volume of 50 μl. The final magnesium concentration should be at least 8 mM (taking into account free magnesium from the mtTEC preparation, buffer added for the RE digestion reaction, and chelation by the nucleotides).
2. Incubate at 30° for 30 min.
3. Stop the transcription reactions by adding one-ninth volume $10\times$ Transcription Stop Solution [100 mM Tris–HCl (pH 7.5)/200 mM EDTA/ 1% SDS; stored at room temperature].

2.6. Isolation of RNA

2.6.1. RNA isolation from mitochondria

1. Typically, mitochondrial pellets from 5–10 g of cells are resuspended in 3 ml TES [20 mM Tris–HCl (pH 7. 5)/1 mM EDTA/0.1% SDS] at room temperature and mixed thoroughly by vortexing. Adjust the volumes proportionately for the amount of mitochondria used in labeling experiments.
2. Extract the mitochondrial lysate by adding an equal volume of phenol (equilibrated to pH 7.6 with Tris base), vortexing, and centrifuging to yield two phases. Transfer the upper (aqueous) phase to a fresh tube.
3. Repeat the phenol extractions until the interface is clear.
4. Extract with an equal volume of CIA (25:1 chloroform:isoamyl alcohol) as above and remove the upper layer to a tube suitable for ethanol precipitation (e.g., Corex). Measure the volume.

5. Ethanol precipitate by adding 1/10 volume of 4 M ammonium acetate and 2 volumes of ethanol. Chill for at least 2 h at $-20°$ and pellet the nucleic acids by centrifuging at $13,000 \times g$ for 30 min at $4°$. Decant the supernatant.

6. Wash the pellet with 70% ethanol, spinning as above. Decant the supernatant.

7. Air dry the pellet (resuspension is more difficult if the pellet is overdried) and resuspend it in TE (10 mM Tris–HCl, pH 7.5/1 mM EDTA) on ice. Determine the nucleic acid concentration (OD_{260}); about one-third of the total nucleic acid will be RNA.

8. Digest the DNA in the mitochondrial nucleic acid sample by bringing it to 10 mM DTT, 0.4 units/μl RNase inhibitor, and 10 mM magnesium acetate (MgOAc), then adding 50 units of RNase-free DNase I. Incubate it on ice for 30 min, then for 30 min at $30°$.

9. Stop the reactions by adding acetic acid to 12 mM and SDS to 0.5%. Extract with an equal volume of equilibrated phenol, then twice with an equal volume of CIA as described in step 2 above.

10. Ethanol precipitate and wash as in steps 5 and 6.

11. For samples to be used for RT-PCR, repeat steps 8–10.

12. Resuspend in water or TE, determine the RNA concentration via OD_{260}, and store at $-20°$ until use.

2.6.2. RNA isolation from mtTECs and chimeric templates

1. Extract the nucleic acids once with an equal volume of equilibrated phenol, then once with an equal volume of CIA.

2. Ethanol precipitate by adding 1 μl 20 mg/ml glycogen (as carrier), 0.1 volume of 4 M ammonium acetate, and 2.5 volumes of ethanol, cooling on dry ice until the mixture is viscous (or at $-20°$ for at least 2 h). Pellet the nucleic acids by centrifuging at 12,000 rpm for 20 min in a microcentrifuge at $4°$. Pipette off the supernatant, then wash the pellet by adding 500 μl 70% ethanol and vortexing gently. Spin 5 min as above, pipette off the liquid, then air dry the pellet $\sim$15 min at room temperature.

3. Resuspend the pellets in 20 μl TE on ice for a minimum of 15 min. A small amount (e.g., 5%) should be taken as a pre-DNase sample for use in control PCR reactions.

4. Digest the DNA by bringing the suspension to 6.25 mM DTT, 0.25 units/μl RNase inhibitor, 12.5 mM MgOAc, 10 mM Tris–HCl (pH 7.5), and 1 mM EDTA; then add 20 units of RNase-free DNase I (10 U/μl, Roche) in a final volume of 40 μl. Incubate for 30 min on ice, then 30 min at $30°$.

5. Pass the reaction through a BioSpin P30 column (Bio-Rad), according to the manufacturer's instructions, to remove DNA digestion products that might interfere with subsequent S1 nuclease digestion or RT-PCR procedures.

6. To the effluent add SDS (to 0.6% final), acetic acid (to 12 m*M*), and EDTA (to 30 m*M*) in a final volume of 100 μl.

7. Extract the RNA once with an equal volume of a 1:1 mix of equilibrated phenol:CIA, then once with an equal volume of CIA. If the RNA is labeled, go directly to the section entitled S1 Nuclease Protection of Run-on RNA.

8. Otherwise, ethanol precipitate as in step 2.

9. RNA to be subjected to reverse transcription is resuspended as in step 3, DNase is treated for a second time by repeating steps 4–8, then resuspended as in step 3.

3. RNA EDITING ASSAYS

3.1. Overview

RNA made *in vivo*, i.e., RNA extracted from mitochondria or mtTECs prior to run-on transcription, can be characterized in bulk via primer extension sequencing with reverse transcriptase (Byrne and Gott, 2004; Gott *et al.*, 1993, 2005), or individual molecules can be analyzed by RT-PCR followed by cloning and sequencing (Byrne and Gott, 2002, 2004; Byrne *et al.*, 2002; Gott *et al.*, 1993, 2005) or, in cases where nucleotide insertion creates or destroys a recognition site, RE analysis (Byrne, 2004; Byrne and Gott, 2002; Rundquist and Gott, 1995). Unlabeled run-on RNAs from mtTEC can also be analyzed via RT-PCR, keeping in mind the potential pitfalls described in the sections below. In the case of chimeric templates, background signal from endogenous transcripts can be greatly reduced by using primers that span the ligation junction. Analysis of editing in labeled run-on transcripts can be performed directly by S1 nuclease protection of selected sequences followed by RNase T1 digestion. The resulting T1 fragments are then resolved by size (via gel electrophoresis) (Byrne and Gott, 2002; Cheng and Gott, 2000; Cheng *et al.*, 2001; Visomirski-Robic and Gott, 1997a,b) or by size and base content (via "RNA fingerprinting") (Cheng *et al.*, 2001; Visomirski-Robic and Gott, 1995, 1997a,b). Subsequent nearest-neighbor analysis may be employed to establish or confirm the identity of individual RNase T1 fragments and to determine the location and identity of sequence changes due to editing (Cheng and Gott, 2000; Visomirski-Robic and Gott, 1995, 1997b). The various analytic methods have different strengths and weaknesses, and their utility depends on the details of the sequence to be analyzed. If more than one technique can be applied for a given experiment, it may be advantageous to do so, as the different methods provide overlapping sets of information that complement each other.

Direct analysis of labeled transcripts has a number of advantages. For example, it examines only the portion of run-on transcripts that are made *in vitro*, as only these RNase T1 fragments are labeled. In contrast, only sequences downstream of an unedited or misedited insertion site or a chimeric junction can be defined as having been made *in vitro* when analyzing unlabeled transcripts using RT-PCR assays (which are not suitable for mitochondrial run-ons due to the preponderance of preexisting, fully edited RNA made *in vivo*). Direct analysis is also more suitable for quantifying editing efficiency at many sites within a population of RNAs simultaneously (Cheng and Gott, 2000; Cheng *et al.*, 2001; Visomirski-Robic and Gott, 1995, 1997a,b). In comparison, sequencing of cloned cDNAs on a logistically possible scale can be subject to sampling and cloning biases. Restriction analysis of bulk RT-PCR products shares the statistical power of direct analysis, but generally distinguishes between only two states at a time (e.g., correctly edited versus unedited/misedited) at a single editing site and its use is limited to only a small subset of editing sites (Byrne, 2004; Byrne and Gott, 2002; Rundquist and Gott, 1995). On the other hand, direct analysis is constrained by the details of the sequence of interest, which will determine whether individual RNase T1 fragments can be resolved from each other. When both the edited and unedited versions of a transcript yield diagnostic RNase T1 fragments separable by size alone, gel electrophoresis can be used to assess the extent of editing at that site (Byrne and Gott, 2002; Cheng and Gott, 2000; Cheng *et al.*, 2001; Visomirski-Robic and Gott, 1997a,b). However, gel electrophoresis does not yield information on the identity or position of added or removed nucleotides, whereas fingerprinting and secondary analyses can often provide this information (Cheng and Gott, 2000; Cheng *et al.*, 2001; Visomirski-Robic and Gott, 1995, 1997a,b).

RT-PCR-based analyses have different strengths. Importantly, cloning and sequencing of RT-PCR products provide a picture of what is occurring at every site within individual RNA molecules, yielding information on misediting and the succession of editing along single transcripts that cannot be gleaned from bulk analyses (Byrne and Gott, 2002, 2004; Byrne *et al.*, 2002). In addition, this technique may be the only option for analysis of very low abundance molecules, which is often the case when working with chimeric templates.

3.2. Direct analysis of labeled RNA

Sections of radiolabeled RNAs can be isolated by hybridizing them to complementary ssDNA probes (Fig. 7.2), then incubating with S1 nuclease, which digests the unprotected RNA. Polyacrylamide gel electrophoresis (PAGE) is used to resolve and purify the S1 digestion products. In the case of transcription from chimeric templates, the presence of the predicted

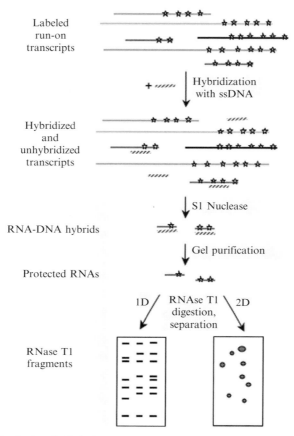

Figure 7.2 Analysis of labeled *in vitro* transcripts. Transcripts initiated *in vivo* are extended *in vitro* in the presence of radiolabeled nucleotides to yield labeled run-on transcripts (asterisks indicate incorporated radiolabeled nucleotides). These RNAs are derived from many different genes (represented here by different shades of gray), but only a subset of the labeled transcripts is analyzed in any given experiment. Note that even for transcripts from the same gene, the extent of labeling may be different, depending on the positions of RNA polymerases at the time the mitochondria are isolated (see short dark gray lines in this example). The total pool of transcripts is then annealed with single-stranded DNA (ssDNA) specific for a small region of a single gene (hybridized and unhybridized transcripts) prior to digestion with S1 nuclease. Single-stranded RNAs and DNAs are digested by S1 nuclease, whereas RNA–DNA hybrids are resistant to cleavage. The protected RNAs are then isolated by denaturing gel electrophoresis and digested by RNase T1, which cuts after G residues. The resulting RNase T1 fragments are then separated from one another in either one dimension (1D, high percentage denaturing polyacrylamide gel) or two dimensions (2D, RNA fingerprinting) and their patterns compared to those of edited and unedited control RNAs. See text for details.

novel S1 fragment is diagnostic of successful ligation and run-on transcription through the chimeric junction (Byrne and Gott, 2002). RNase T1, which cleaves after G residues, is then used to digest the protected RNA into small, defined fragments, some of which will encompass one, or occasionally more, insertion sites. In principle, any of the nucleotide-specific endonucleases could be used in this analysis, but RNase T1 is most suitable for *Physarum* mitochondrial sequences because (1) the transcripts are very A/U rich, so that cleavage by A- or U-specific enzymes would in general produce few fragments of distinctive size that could be resolved by PAGE; (2) many of the added Cs are inserted next to encoded Cs, so that correct editing at these sites would liberate only mononucleotides upon cytosine-specific digestion. High percentage PAGE can be used to separate the T1 fragments at single nucleotide resolution, allowing the efficiency of insertion at many sites to be measured. Alternatively, RNA fingerprinting can be used to resolve the T1 fragments in two dimensions (Branch *et al.*, 1989). Note that slightly different T1 digestion conditions are used for one-dimensional (1D) and two-dimensional (2D) analyses; these are described in separate sections below. Purified T1 fragments can then be subjected to nearest-neighbor analysis if required to establish sites of editing and the identity of inserted nucleotides (Cheng and Gott, 2000; Visomirski-Robic and Gott, 1995, 1997b). Protocols for analysis of single nucleotide insertion sites are first described, followed by a separate section that briefly discusses detection of editing at dinucleotide insertion sites.

3.2.1. Design and isolation of ssDNA probes

ssDNA probes are made using recombinant phagemids, which contain an origin of replication derived from M13 (or other single-stranded DNA phage) via standard methods. These are designed according to the following considerations. (1) It is generally preferable to protect RNA fragments of ~250–400 nucleotides; shorter fragments are likely to be contaminated by incompletely digested RNA fragments, while longer fragments generally give overly complex RNase T1 digestion patterns. (2) The size distribution of the fragments that will be produced upon T1 digestion of the S1 protected RNA: for resolution by single-dimension electrophoresis, relevant oligonucleotides should differ in size by at least one nucleotide from all others in the mix. (3) The sequence of the protecting ssDNA should correspond as closely as possible to that of the target or predicted transcript in the region to be protected. It is preferable to use the edited versions of the sequence; although single nucleotide insertion/deletions are largely resistant to nicking by S1 nuclease under the conditions used here, some nicking occurs at bulged bases, so it is better not to have these in the labeled RNA component of the hybrid. Two or more successive mismatches or insertions/deletions relative to the target RNA will constitute a substrate for S1 nicking. (4) The ssDNA should first be checked against labeled edited and unedited control RNAs following the protocol described in the section entitled S1 Nuclease Protection of Control RNA.

3.2.2. S1 nuclease protection of Run-on RNA

S1 nuclease protection experiments are done under conditions that allow individual *Physarum* mitochondrial RNA sequences to hybridize efficiently and stably enough for precise and efficient S1 nuclease protection, despite high A/T content and the presence or absence of inserted nucleotides (Byrne and Gott, 2002; Cheng and Gott, 2000; Cheng *et al.*, 2001; Visomirski-Robic and Gott, 1995, 1997a,b). A wide variation has been observed between the behaviors of S1 nuclease stocks from different suppliers, and even between different batches of enzyme from the same supplier. In our hands, the S1 nuclease supplied by Roche works best in the context of our experiments. Each new batch of nuclease is titrated to determine the number of units that generates a fragment of predicted size from labeled control RNAs mixed with unlabeled mitochondrial RNA, with minimal residual undigested RNA or overdigestion (cleavage within the hybridized region).

Two rounds of S1 nuclease protection are usually sufficient to obtain a reasonably clean protected fragment from RNAs derived from isolated mitochondria and native mtTECs and the pool of transcripts from exogenous DNA cotranscribed from multiple upstream genes in intergenic chimeras. Three rounds are usually necessary for less abundant individual chimeric transcripts and for RNAs that will be subjected to RNA fingerprinting. For the first round of S1 protection, the probe and target must interact in a complex RNA mixture, necessitating an overnight hybridization incubation; subsequently, the amount of nonspecific RNA is greatly reduced, so less time (~2 h) is needed for pairing between probe and target (however, the hybridizations can be left overnight without harm). Details of the ethanol precipitation, wash, and resuspension steps are designed to minimize loss and degradation of RNA and the accumulation of salt. These considerations are especially important for steps subsequent to the first S1 nuclease digestion, when the total quantity of intact RNA has been greatly reduced and the solution is high in salt. The quantity of labeled RNA present at each stage can be estimated by counting the dry pellets using a scintillation counter (although there may be significant coprecipitation of unincorporated label in the initial stages). Ethanol precipitations and 70% ethanol washes are carried out (as in step 2 of RNA isolation from mtTECs and chimeric templates), with modifications as noted; pellets are resuspended for at least 15 min on ice, unless otherwise indicated.

1. Mix labeled RNA with 4 µg of the appropriate antisense ssDNA, ethanol precipitate, wash twice with 70% ethanol, and dry the pellet for approximately 5 min.

2. Resuspend the pellet in 20 μl S1 Hybridization buffer [40 mM PIPES (pH 6.4)/1 mM EDTA/0.4 M NaCl/80% formamide; stock stored at −20°] for 15 min at room temperature. Heat at 85° for 10 min to denature the nucleic acids, spin briefly, and then incubate at 37° for a minimum of 16 h to hybridize.

3. At room temperature: spin briefly, add S1 nuclease (usually 200–800 U, determined in advance as described above) in 200 μl of ice-cold S1 Mapping buffer [0.75 M NaCl/50 mM sodium acetate (pH 4.5)/4.5 mM $ZnSO_4$; stored at 4°], then incubate at 26° for ∼1.5 h (range 1–2.5 h, experimentally determined for each region based on control titrations).

4. Stop the S1 reaction by adding 54 μl S1 Stop Solution [2 M sodium acetate/50 mM EDTA/50 μg/ml tRNA; stored at −20°], then extract once with a 1:1 mix of equilibrated phenol:CIA, then once with CIA. Ethanol precipitate without adding further salt; the pellet does not need to be dried. Resuspend the pellet in 100 μl TE.

5. Add 4 μg of the ssDNA and ethanol precipitate, without additional salt; glycogen is not needed unless the pellet in step 4 is very small. Wash twice with 70% ethanol, leaving 5 min on ice each time, and dry the pellet for 5 min.

6. Resuspend and denature the nucleic acids as in step 2, then hybridize at 37° for 2 h.

7. Perform S1 nuclease digestion as in step 3.

8. Stop the reaction, extract the nucleic acids, and ethanol precipitate as in step 4, but wash three times with 70% ethanol, then dry the pellet for approximately 5 min. Count the radioactive signal and resuspend in a small volume of TE (e.g., 20 μl, in case the RNA is ready for final electrophoresis).

9. At this point, samples of the RNA (e.g., 5–10%) can be examined by denaturing PAGE (see Separation and Purification of Protected RNAs below). If the expected S1 fragment(s) can be resolved above background, the third round of S1 protection (steps 10–13 below) can be omitted.

10. Add 4 μg of the ssDNA and ethanol precipitate, without additional salt; glycogen is not needed unless the pellet in step 8 is very small. Wash twice with 70% ethanol, leaving 5 min on ice each time, and dry pellet for approximately 5 min.

11. Resuspend and denature nucleic acids as in step 2, then hybridize at 37° for 2 h.

12. Perform S1 nuclease digestion as in step 3.

13. Stop reaction, extract nucleic acids and ethanol precipitate as in step 4, but wash three times with 70% ethanol, then dry pellet for approximately 5 min. Count the radioactive signal and resuspend in a small volume of TE (e.g., 20 μl).

3.2.3. S1 nuclease protection of control RNA

Labeled control RNAs (edited and unedited) should be made *in vitro* and S1 protected in separate reactions. These RNAs should be labeled with the same $[\alpha-^{32}P]$NTP used for the isolated mitochondria or mtTEC run-on transcription, and the full-length transcripts should be gel-purified. Their sequence should correspond as closely as possible to that of the run-on RNA in the region to be S1 protected. These RNAs make it possible to check S1 nuclease activity, ssDNA protection, and RNase T1 digestion, while the digestion products from successful procedures act as mobility markers for electrophoresis. To act as a positive control for S1 nuclease activity, the RNAs should extend for some length beyond the region of complementarity with the protecting ssDNA, at least at one end. When assaying a new batch of S1 nuclease, trace amounts of control RNAs should be mixed with unlabeled bulk mitochondrial RNA to determine the number of units of S1 nuclease that is required.

1. Mix labeled control RNA with 1 μg of the antisense ssDNA, then ethanol precipitate and dry the pellet for approximately 5 min. A substantial amount of labeled transcript can be used ($\sim$50,000 cpm) in order to provide ample protected fragment for subsequent analyses.
2. Resuspend the pellet in 20 μl Hybridization buffer for 15 min at room temperature. Heat at 85° for 10 min, spin briefly, then incubate at 37° for 2 h.
3. At room temperature: spin briefly, add S1 nuclease (the same amount as for S1 Nuclease Protection of Run-on RNA above) in 200 μl of ice-cold S1 Mapping buffer, then incubate at 26° for 1.5 h.
4. Stop the S1 reaction by adding 54 μl S1 Stop Solution, then extract once with a 1:1 mix of equilibrated phenol:CIA, then once with CIA. Ethanol precipitate without adding further salt, wash three times with 70% ethanol, dry the pellet for approximately 5 min, count the radioactive signal, and resuspend in TE.

3.2.4. S1 nuclease assay for editing at dinucleotide insertion sites

S1 nuclease can also be used to probe the extent of editing at dinucleotide insertion sites within labeled *in vitro* transcripts (Gott et al., 1993; Visomirski-Robic and Gott, 1995, 1997b). The key feature of these experiments is that an *unedited* ssDNA probe is hybridized to the pool of labeled RNAs, resulting in a dinucleotide bulge in the edited RNA that is highly susceptible to cleavage by S1 nuclease. This assay uses the same basic conditions as described above (S1 Nuclease Protection of Run-on RNA), but optimal conditions for each site must first be determined using edited and unedited control transcripts protected with edited and unedited ssDNAs. It is particularly critical to titrate salt and S1 nuclease concentrations to find conditions that yield complete cleavage at the dinucleotide bulge while leaving single nucleotide bulges (such as those resulting from single C insertions) intact. In the case of the GU insertion site located in the coI gene, three rounds of S1 nuclease protection were necessary to demonstrate efficient cleavage of edited RNAs (Visomirski-Robic and Gott, 1995). Importantly, S1 nuclease preparations from different suppliers behave differently in this assay and a number of different enzyme preparations should be assayed during optimization. For example, in previous side-by-side experiments done under identical conditions with the same labeled control RNAs and protecting ssDNAs, S1 nuclease purchased from Roche cleaved the dinucleotide bulge preferentially over single nucleotide bulges, S1 nuclease purchased from Pharmacia failed to cleave at the dinucleotide bulge, and S1 nuclease purchased from Sigma cleaved at both single and dinucleotide bulges (L. Visomirski-Robic, unpublished observations). The reasons for the differences between enzyme preparations are not known, but these findings emphasize the importance of optimization in this type of experiment.

3.2.5. Separation and purification of protected RNAs

1. Subject the S1 nuclease-protected RNAs to denaturing PAGE and autoradiography. Care should be taken to avoid loading too much of the control RNA samples on gels that also contain the products of run-on transcription experiments. Bands from run-on transcripts may be substantially less intense than predicted from the radioactivity in the final S1 protection pellet, as there may be a significant background of partially digested RNA and other by-products, as well as residual unincorporated label; an overnight exposure is usually required; 4% or 5% acrylamide gels are generally suitable.
2. Excise the bands of interest from the gel and extract the RNA fragments by soaking in elution buffer [10 mM Tris–HCl (pH 7.5)/0.25 M sodium acetate/1 mM EDTA/0.25% SDS] at room temperature for several hours (usually overnight). A second elution for a few hours in the same buffer is recommended, pooling the eluants. Extract the eluant once with equilibrated phenol and once with CIA, and then ethanol precipitate without additional salt. As a precaution against the presence of residual SDS, the RNA can be reprecipitated, and three 70% ethanol washes performed. Count the dry radioactive pellets and resuspend in a small volume of TE (as little as 5 μl); if there is more than one run-on sample, it is best to resuspend all at the same number of cpm per μl.

3.2.6. RNase T1 digestion for subsequent 1D gel electrophoresis

1. Denature the gel-purified S1 nuclease-protected RNA fragments in 5 μl TE by incubating at 95° for 2 min, then cool immediately on ice and spin briefly. Digest similar amounts of RNA in terms of radioactive counts

from each sample and control, so corresponding bands in different lanes have similar intensity.

2. Add 1 μl of 10 U/μl RNase T1 (Roche) and incubate for 45 min at 37°. *Note*: In earlier experiments we added 1 μl of 1 mg/ml tRNA (Sigma) to each RNA sample prior to digestion, using 10 times more enzyme (1 μl of 100 U/μl RNase T1) in each reaction. We now routinely use the more dilute enzyme in the absence of added tRNA to avoid gel migration problems due to the high levels of salt in the stock enzyme preparation.

3. Add 7 μl denaturing gel loading dye solution (7 M urea/1 × TBE/0.25% xylene cyanol/0.25% bromophenol blue) and heat at 85° for 2 min.

4. Separate the digestion products in one dimension by denaturing PAGE [20% acrylamide (19:1)/1 × TBE/7 M urea gels are generally suitable] and expose using a phosphoimager cassette.

3.2.7. RNA fingerprinting

Labeled RNase T1 fragments containing editing sites are often the same size as other fragments within the digest, making it difficult (or impossible) to quantify editing at a given site using one-dimension separation by denaturing PAGE. In many cases, this problem can be circumvented by separating the RNase T1 fragments by homochromatography ("RNA fingerprinting"), which involves separation based largely on size (charge) in one dimension and base composition in the other. An entire *Methods in Enzymology* chapter (Branch *et al.*, 1989) has been devoted to this powerful but technically demanding method, which truly is an art form; this article is an invaluable resource for those interested in using this technique. In that chapter, two different methods of separating oligonucleotides in the first dimension are detailed; we use the low pH denaturing polyacrylamide gel as described by Branch and colleagues (1989). When assaying labeled RNAs for editing, it is critical to determine ahead of time the fingerprint patterns for both edited and unedited control transcripts and to perform secondary analysis (see Nearest Neighbor Analysis described below) to confirm the identification of the resulting T1 fragments. Dry fingerprint plates can then be exposed to film or phosphoimager overnight to several weeks. When necessary, spots can be scraped off the fingerprint plates and eluted for nearest-neighbor analyses (Barrell, 1971).

3.2.8. RNase T1 digestion for subsequent RNA fingerprinting

1. Resuspend pellets of gel-purified S1 nuclease-protected RNA fragments in 10 μl of a 1 mg/ml solution of tRNA in water and lyophilize in silanized tubes.

2. Resuspend each sample in 2 μl of 10 mM Tris–HCl (pH 7.4)/1 mM EDTA containing RNase T1 (Sankyo) at 1 μg/μl and incubate for 35 min at 37°.

3. Separate digestion products in the first dimension using a 10% polyacrylamide gel containing 6 M urea and 25 mM sodium citrate, pH 3.5, followed by homochromatography as described by Branch et al. (1989).

3.2.9. Nearest neighbor analysis

In certain circumstances it is important to determine the identity of one or more specific nucleotides within an RNA molecule. This can often be accomplished via nearest neighbor analysis. In this technique, an α-^{32}P-labeled RNA is digested with enzymes that yield 5' OH and 3' P groups, resulting in the transfer of a labeled phosphate to the nucleotide immediately upstream. The resulting nucleotide 3' monophosphates are then separated via 2D chromatography and quantified (Keith, 1995; Krug et al., 1982). This technique has proven useful, for example, in demonstrating the accuracy of nucleotide insertion in both isolated mitochondria (Visomirski-Robic and Gott, 1997b) and mtTECs (Cheng and Gott, 2000).

1. Make up 50 ml of appropriate TLC solvents in separate chromatography tanks to equilibrate. Wear gloves and keep tanks in a fume hood (note that isobutyric acid is exceptionally noxious). Many different solvent systems can be used, but we routinely use Solvent A [isobutyric acid/concentrated NH_4OH/H_2O (66:1:33, v/v/v)] in the first dimension and Solvent C [0.1 M sodium phosphate (pH 6.8)/ammonium sulfate/n-propanol (100:60:2, v/w/v)] in the second dimension as described by Keith (1995).
2. Mix 2 μl labeled RNA sample in water and 1 μl supernuclease (3× stock containing 200 units/ml RNase T2, 250 units/ml RNase T1, and 50 μg/ml RNase A in 45 mM ammonium acetate).
3. Incubate at 37° for 30 min and then at 65° for ~15 min.
4. During the incubations, carefully cut the cellulose thin layer chromatography plates (e.g., Macherey-Nagel Polygram CEL400 UV$_{254}$) into 10 cm × 10 cm squares (wearing gloves).
5. Spot 1 μl Np standards (containing one or more unlabeled 3' nucleotide monophosphates at 2.5 mM) at origin and air dry.
6. Spot 1 μl of labeled RNA sample at the origin and let dry. Repeat until the entire sample is spotted.
7. For chromatography in the first dimension, put plates (with the origin at the bottom left) in the chromatography tank containing 50 ml solution A; allow the solvent to ascend until it is ~1 cm from the top; remove the plates and dry thoroughly in fume hood.
8. For chromatography in the second dimension, rotate the plates 90° and put them in the chromatography tank containing 50 ml solution C, allowing the solvent to ascend until ~1 cm from the top. Remove the plates and dry thoroughly in fume hood.

9. Use a hand-held UV lamp (254 nm) to visualize unlabeled standards; circle the standards with a pencil.
10. Wrap well, then expose to film or put up on a phosphoimager. Note that these solvents may damage phosphoimager screens if not well dried and wrapped.

3.3. Analysis of unlabeled transcripts

3.3.1. Sequencing of unlabeled RNA via primer extension

Primer extension sequencing of bulk RNA is generally used to validate sequences derived from RT-PCR products. For instance, the presence of both C to U changes (Gott *et al.*, 1993) and nucleotide deletions (Gott *et al.*, 2005) in steady-state mitochondrial RNAs was confirmed in this way.

1. *Primer-template annealing*: Denature 2–5 μg total mitochondrial RNA in 9 μl RT-Mg buffer (50 mM Tris–HCl (pH 8.5)/60 mM NaCl/10 mM DTT) by heating for 3 min at 65° in the presence of an appropriate end-labeled oligonucleotide primer.
2. Quickly cool the sample in a dry ice/ethanol bath, thaw on ice, and spin briefly to recover the condensate.
3. Add 1 μl 36 mM magnesium acetate in 50 mM Tris–HCl (pH 8.5)/60 mM NaCl/10 mM DTT.
4. *Sequencing reactions*: Aliquot 2 μl of an annealed primer-template mixture into each of five tubes (round-bottom 96-well microtiter plates work well if doing multiple reactions).
5. Add 1 μl 5× dNTP mix (1.875 mM of each dNTP) to each tube to yield a final concentration of 375 μM of each dNTP.
6. Add 1 μl appropriate 5× ddNTP (1 mM of a single ddNTP) to yield a final concentration of 200 μM ddNTP (ddATP, ddCTP, ddGTP, or dTTP) in each of four tubes; add 1 μl of RT buffer (50 mM Tris–HCl (pH 8.5)/60 mM NaCl/10 mM DTT/6 mM magnesium acetate) to the fifth tube to control for extraneous stops. Note that the ddNTP/dNTP ratio can be adjusted to sequence closer or further from the primer.
7. Dilute an appropriate amount of AMV-Reverse Transcriptase (AMV-RT, Life Sciences) to 3 units/μl using RT buffer. Add 1 μl diluted AMV-RT to each tube and incubate at 42° for 30 min.
8. Stop the reaction by the addition of 5 μl Formamide Stop Dye (95% formamide/20 mM EDTA/0.05% bromophenol blue/0.05% xylene cyanol) to each tube. Load 3 μl directly on 8% sequencing gel after heating samples to 95° for 3 min; store the remaining sample at −20°.

3.3.2. Reverse transcription and purification of cDNA

While RT-PCR is used in the initial characterization of editing sites, it is not generally used to assay *in vitro* editing of transcripts made in isolated mitochondria for the reasons outlined above. However, it is useful in the context of mtTECs, which edit less efficiently, and transcripts from chimeric templates (Byrne, 2004; Byrne and Gott, 2002, 2004; Byrne *et al.*, 2002). In the case of chimeric templates, RT-PCR is performed using pairs of primers that will not generate a product from the naturally occurring mtTEC nucleic acids, at least within the distance constraints over which RT and PCR operate. Thus, for example, (1) the primers that amplify the subgenomic circle transcripts are directed away from each other in natural sequences; (2) other chimeras are amplified using a primer from upstream of the chimeric junction in conjunction with one from the terminal tag sequence. A "nested" downstream primer can be used in the PCR step in order to decrease the probability of amplifying products based on spurious priming during RT. Before performing the RT reactions, it is useful to check for the presence of chimeric DNA by first performing PCR without RT on the pre-DNase sample (RNA Isolation from mtTECs and Chimeric Templates, step 3).

1. Anneal a sample of the RNA (e.g., 2 μl, i.e., 10%) from step 9 of Isolation from mtTECs and Chimeric Templates with 2.5 pmol RT primer in 10 μl by heating at 70° for 10 min, then placing on ice for 10 min; spin down any condensation. Scale-up 2-fold and split after annealing when including RT-minus reaction controls.

2. For a final reaction volume of 20 μl, add dNTPs to 0.5 mM, 100 U RNase H-minus M-MLV reverse transcriptase and reaction buffer (as supplied by the manufacturer).

3. Incubate at 48° for 60 min (preferably in an oven to minimize changes in concentrations due to condensation).

4. To digest the RNA, add 80 μl TE, heat at 95° for 5 min, then immediately add 1 μg RNase A and incubate at 37° for 15 min. Since use of an RNase H-minus reverse transcriptase is recommended, this step is included to prevent any spurious priming by RNA in the subsequent PCR reaction; the nucleic acids are first heated to free the RNA from hybrids with DNA.

5. Extract the cDNA once with equilibrated phenol and then once with CIA.

6. Ethanol precipitate by adding 1 μl 20 mg/ml glycogen, 0.1 volume of 4 M ammonium acetate, and 2 volumes of ethanol, cooling on dry ice until the mixture is viscous (or at −20° for at least 2 h). Pellet the nucleic acids by centrifuging at 12,000 rpm for 20 min at 4°. Pipette off the supernatant, then wash the pellet by adding 500 μl 70% ethanol and vortexing gently. Spin 5 min as above, pipette off the liquid, then air dry the pellet ~15 min at room temperature. Resuspend the pellets in 25 μl TE.

3.3.3. PCR and cloning

1. Using 5 μl of cDNA as template, perform 30 cycles of PCR (1 min at 94°, 1 min at 50°, and 2 min at 72°) with a 7 min final extension at 72°. For each reaction, add 2.5 U Taq DNA polymerase per 100 μl in buffer

supplied by the manufacturer, 200 μM dNTPs, and 100 pmol of each primer. If the PCR products are to be ligated directly to the vector, use primers with $5'$ phosphates.

2. Run the samples (e.g., 5–10%) on an agarose gel with size markers, stain with ethidium bromide, and look for bands of mobility consistent with the expected sequence.

3. Products of the expected size can be gel-purified from a preparative gel if necessary. Otherwise, purify the DNA by organic extraction and ethanol precipitation, the use of a spin-column, or a commercially available kit. Gel purification is indicated if there is a significant background of higher and/or lower molecular weight products that could reduce the efficiency of cloning or complicate the RE analysis (below). However, it might be worth investigating whether background bands represent novel sequence products of the transcription/editing process, e.g., deletions between editing sites, as have been found in the *Physarum* system (Byrne, 2002, 2004).

4. Clone into any convenient cloning vector using standard molecular biology techniques, then sequence.

3.3.4. Restriction enzyme analysis of labeled PCR fragments

At some editing sites, nucleotide addition generates recognition sequences for REs within the cDNA molecules that are not present in the original mitochondrial DNA sequence (Byrne, 2004; Byrne and Gott, 2002; Rundquist and Gott, 1995). The extent of correct editing at these sites can be estimated by incubating end-labeled RT-PCR products with the cognate REs, followed by PAGE (Rundquist and Gott, 1995). Radioactively labeled PCR products allow for more sensitive and quantitative RE analysis, and can be generated by having one of the primers $5'$ end labeled with ^{32}P (using polynucleotide kinase and $[\gamma\text{-}^{32}P]ATP$). It is best to have the unlabeled primer present in, e.g., 2-fold excess in order to minimize the chances of ending up with residual labeled single-stranded DNA, which could generate a false-negative signal during RE analyses (Byrne, 2004).

1. Using 5 μl of the cDNA as template, perform PCR in a 50 μl reaction over 30 cycles (1 min at 94°, 1 min at 50°, and 2 min at 72°) with a 7 min final extension. For each reaction, add 1.25 U Taq DNA polymerase in buffer supplied by the manufacturer, 200 μM dNTPs, 25 pmol of the unlabeled primer, and 12.5 pmol total of a ^{32}P-end-labeled primer.

2. Examine the samples (e.g., 5–10%) by denaturing PAGE (4%, unless the expected product is very small) and autoradiography, looking for bands of mobility consistent with the expected sequence.

3. Gel-purify the PCR product on a native acrylamide or agarose gel if necessary.

4. Incubate the samples of each PCR product with a series of REs that are diagnostic for editing sites and, if possible, one or two enzymes that cut at recognition sites not affected by editing (to confirm the identity of the DNA). PCR products can generally be digested without prior extraction and precipitation, provided they do not constitute more than, e.g., 10% of the reaction volume.

5. Examine all or a portion of each digest by denaturing PAGE and autoradiography. Choose an acrylamide percentage suitable for resolving the fragments that will result if all sites are edited; if the range of fragment sizes is very large, it may be beneficial to use several gels of differing percentages of acrylamide. If necessary, precipitate the DNA and resuspend it in a smaller volume before loading. Look for bands of mobility consistent with cleavage at editing sites (bearing in mind which terminus is labeled).

The methods described above have been used successfully to illuminate a number of aspects of *Physarum* editing. However, it is important to note that experiments examining run-on transcripts initiated *in vivo* and extended *in vitro* differ in significant ways from experiments that rely on added *in vitro* transcripts, and careful design is essential. In principle, many of the methods outlined here should be directly applicable to other cotranscriptional processes, such as RNA splicing, polyadenylation, and termination, and may therefore be of broad interest.

REFERENCES

Barrell, B. (1971). Fractionation and sequence analysis of radioactive nucleotides. *In* "Procedures in Nucleic Acid Research" (G. Cantoni and D. Davies, eds.), Vol. 2, pp. 751–779. Harper & Row, New York.

Branch, A. D., Benenfeld, B. J., and Robertson, H. D. (1989). RNA fingerprinting. *Methods Enzymol.* **180**, 130–154.

Byrne, E. M. (2004). Chimeric templates and assays used to study *Physarum* cotranscriptional insertional editing *in vitro*. *Methods Mol. Biol.* **265**, 293–314.

Byrne, E. M., and Gott, J. M. (2002). Cotranscriptional editing of *Physarum* mitochondrial RNA requires local features of the native template. *RNA* **8**, 1174–1185.

Byrne, E. M., and Gott, J. M. (2004). Unexpectedly complex editing patterns at dinucleotide insertion sites in *Physarum* mitochondria. *Mol. Cell. Biol.* **24**, 7821–7828.

Byrne, E. M., Stout, A., and Gott, J. M. (2002). Editing site recognition and nucleotide insertion are separable processes in *Physarum* mitochondria. *EMBO J.* **21**, 6154–6161.

Cheng, Y. W., and Gott, J. M. (2000). Transcription and RNA editing in a soluble *in vitro* system from *Physarum* mitochondria. *Nucleic Acids Res.* **28**, 3695–3701.

Cheng, Y. W., Visomirski-Robic, L. M., and Gott, J. M. (2001). Non-templated addition of nucleotides to the 3′ end of nascent RNA during RNA editing in *Physarum*. *EMBO J.* **20**, 1405–1414.

Daniel, J., and Baldwin, H. (1964). Methods of culture for plasmodial myxomycetes. *In* "Methods in Cell Physiology" (D. Prescott, ed.) pp. 9–41. Academic Press, New York.

Gott, J. M., and Visomirski-Robic, L. M. (1998). RNA editing in *Physarum* mitochondria. *In* "Modification and Editing of RNA" (H. Grosjean and R. Benne, eds.), pp. 395–411. ASM Press, Washington, DC.

Gott, J. M., Visomirski, L. M., and Hunter, J. L. (1993). Substitutional and insertional RNA editing of the cytochrome c oxidase subunit 1 mRNA of *Physarum* polycephalum. *J. Biol. Chem.* **268,** 25483–25486.

Gott, J. M., Parimi, N., and Bundschuh, R. (2005). Discovery of new genes and deletion editing in *Physarum* mitochondria enabled by a novel algorithm for finding edited mRNAs. *Nucleic Acids Res.* **33,** 5063–5072.

Keith, G. (1995). Mobilities of modified ribonucleotides on two-dimensional cellulose thin-layer chromatography. *Biochimie* **77,** 142–144.

Krug, M., de Haseth, P. L., and Uhlenbeck, O. C. (1982). Enzymatic synthesis of a 21-nucleotide coat protein binding fragment of R17 ribonucleic acid. *Biochemistry* **21,** 4713–4720.

Miller, D., Mahendran, R., Spottswood, M., Costandy, H., Wang, S., Ling, M. L., and Yang, N. (1993). Insertional editing in mitochondria of *Physarum. Semin. Cell Biol.* **4,** 261–266.

Rundquist, B. A., and Gott, J. M. (1995). RNA editing of the coI mRNA throughout the life cycle of *Physarum polycephalum. Mol. Gen. Genet.* **247,** 306–311.

Visomirski-Robic, L. M., and Gott, J. M. (1995). Accurate and efficient insertional RNA editing in isolated *Physarum* mitochondria. *RNA* **1,** 681–691.

Visomirski-Robic, L. M., and Gott, J. M. (1997a). Insertional editing in isolated *Physarum* mitochondria is linked to RNA synthesis. *RNA* **3,** 821–837.

Visomirski-Robic, L. M., and Gott, J. M. (1997b). Insertional editing of nascent mitochondrial RNAs in *Physarum. Proc. Natl. Acad. Sci. USA* **94,** 4324–4329.

CHAPTER EIGHT

COMPUTATIONAL APPROACHES TO INSERTIONAL RNA EDITING

Ralf Bundschuh

Contents

Abstract

Insertional RNA editing turns the task of locating edited genes and the editing sites within these genes into a challenge. Computational techniques can greatly simplify this task by providing specific predictions for the location of genes and the editing sites within these genes that can be directly used to design primers for the experimental verification of these predictions. To obtain good predictions from computational methods, a substantial amount of sequence analysis is necessary that takes into account the idiosyncrasies of the gene of interest and thus cannot be automated. Here, we describe step by step how to use our computational tool Predictor of Insertional Editing (PIE) to locate genes and

Department of Physics, The Ohio State University, Columbus, Ohio

Methods in Enzymology, Volume 424
ISSN 0076-6879, DOI: 10.1016/S0076-6879(07)24008-X

predict their editing sites using the nad4L genes of the mitochondrial genome of *Physarum polycephalum* as an example.

1. INTRODUCTION

The characteristic of RNA editing is that the RNA sequence of an edited gene differs from its genomic template (Bass, 2001; Keegan *et al.*, 2001). As reflected throughout this volume, the mechanisms by which the RNA is edited and by which the editing sites are determined are known for some types of editing and are a complete mystery for other types. To understand the mechanisms of RNA editing, it is very useful to have numerous examples of RNAs edited by a given RNA editing machinery as well as a complete understanding of the genomes under consideration. However, the very definition of RNA editing implies that these RNA editing sites can be identified only if both the genomic and the RNA sequence are known, i.e., significantly more work is needed than in understanding the genomes of organisms without RNA editing.

Determining RNA sequences is a rather challenging task in organisms with insertional RNA editing. The reasons for this are 2-fold. First, insertional RNA editing makes it rather difficult to even identify genes in a genome. While the RNA sequence for a gene has all the properties of a regular gene (such as distinct codon patterns and/or similarities to genes from other organisms), these patterns are absent from the underlying genomic sequence. Thus, it is not a coincidence that the very heavily edited genes in *trypanosomes* have been called "cryptogenes" (Simpson and Shaw, 1989)—they simply do not look like genes before their RNA is edited. The same situation holds true in the mitochondrion of the slime mold *Physarum polycephalum*. The most frequent editing event in this mitochondrion is the insertion of individual Cs, which occurs on average every 25 nucleotides in protein coding genes and every 40 nucleotides in structural RNAs (Gott, 2001; Horton and Landweber, 2002; Smith *et al.*, 1997). This is a much lower editing rate than in trypanosomes. Still, although the complete mitochondrial genome of this organism is known (Takano *et al.*, 2001), only 16 genes have been identified, a number that should be compared to the 44 genes annotated in the mitochondrial genome of *Dictyostelium discoideum* (Ogawa *et al.*, 2000), which is of comparable length to the mitochondrial genome of *P. polycephalum*. This lack of annotation is a direct consequence of the insertional RNA editing in *Physarum*.

The second challenge to finding the edited RNA sequences in organisms with RNA editing is primer selection. Even once the genomic location of a gene has been identified, it is necessary to design primers in order to determine the sequence of the edited RNA. These primers have to be complementary to regions of the edited RNA itself, the sequence of which is supposed to be determined by the very experiment. If editing is sufficiently frequent (such as the rate of 1 out of 25 nucleotides in *Physarum*), a primer chosen according to the genomic sequence alone is rather likely to overlap an editing site and thus not to work.

The key to overcoming the two challenges for finding edited RNA sequences is computation. In this chapter, we will present a detailed account of how to use computational methods to locate genes in the presence of insertional RNA editing and to predict the editing sites within these genes. Such a prediction can then be used to choose primers and to experimentally verify the predictions. We will focus on insertional RNA editing in *Myxomycota* and, more specifically, *P. polycephalum*. The reason for this choice is the high frequency of insertional editing paired with the availability of the complete mitochondrial genome and a general lack of understanding of the RNA editing machinery in *Physarum*. While our methods could in principle also be applied to trypanosomes that have an even higher frequency of insertional RNA editing (Benne *et al.*, 1986; Blum *et al.*, 1990; Simpson *et al.*, 2003; Stuart *et al.*, 2001), it is not very useful in this case since our knowledge of the guide RNAs being the information carriers that determine the RNA editing sites in trypanosomes can be turned into much more efficient computational tools for this system (von Haeseler *et al.*, 1992, see also Chapter 6, this volume).

2. OVERVIEW

Our tool for the computational prediction of genes and editing sites in organisms with insertional RNA editing is the Predictor of Insertional Editing (PIE). Its detailed computational structure, its performance in predicting editing sites on reference data (Bundschuh, 2004), and its usefulness in identifying new genes (Gott *et al.*, 2005) have been established in other publications. The focus of this chapter is to provide a detailed account and step-by-step instructions on how to successfully use PIE to identify genes and predict editing sites in organisms with insertional RNA editing. It is directed both at researchers who want to simply use PIE without worrying too much about how PIE actually works, and at more computationally interested researchers who want to understand how PIE works in detail in order to adapt it to other biological problems. We will achieve this dual purpose by marking paragraphs and footnotes that are more technical in nature with an asterisk. These sections can be safely skipped if there is no interest in the technical details.

Before explaining all the steps necessary to identify genes and predict editing sites in detail, we want to present an overview of the general idea behind PIE and the strategy of how to apply it. The main idea behind PIE is that given how little we understand about the mechanism of insertional RNA editing in *Myxomycota*, we have to use information external to these organisms in order to find genes and predict editing sites. Specifically, PIE uses the wealth of *protein sequence information* from many different species in order to find genes and editing sites in *Myxomycota*.

To be specific, we will use the example of the nad4L gene in *P. polycephalum* throughout this chapter. As a typical mitochondrial gene, protein sequences for nad4L genes from hundreds of organisms are known and can be found in GenBank (Benson *et al.*, 2006). All these sequences taken together capture the sequence features of a nad4L protein, or in slightly more technical terms, which amino acids are the most frequent ones in which position of a nad4L protein. Then, given a stretch of genomic sequence from the mitochondrion of *P. polycephalum*, one can in principle ask for every possible way of inserting extra Cs into this sequence how much the translation of this "putatively edited" sequence resembles a nad4L protein. PIE quantitates the answer to this question for *all* putatively edited sequences and identifies the method of inserting extra Cs that *maximizes* the similarity between the translation of the putatively edited sequence and a nad4L protein.

This leads to a natural sequence of steps in finding genes and editing sites. First, the information from protein sequences for a given gene from many organisms has to be collected and assembled into a quantitative model for this gene. Second, this model has to be compared to all regions of the genome of interest in order to identify the region most compatible with the protein model, i.e., the putative location of the gene. Third, after the general region in which the gene is located has been identified, the gene's precise location has to be found by determining the start and the stop codon of the gene. Lastly, the editing sites in the gene have to be predicted in order to allow for successful primer selection for experimental verification of the newly identified gene. In the remainder of this chapter, we will go through each of these steps in detail.

 ## 3. BUILDING GENE MODELS

3.1. Summary

Building a model for a gene involves identifying protein sequences for the gene in question, determining a multiple alignment of these protein sequences, and counting which amino acid occurs how often at which position in this multiple alignment. This is all done by the program PSI-BLAST (Altschul *et al.*, 1997; Schaffer *et al.*, 2001) offered by the National Center for Biotechnology Information (NCBI). The outcome of this procedure is a table of probabilities $p_i(a)$ that quantifies for each of the

```
99
# A      R      N      D      C      Q      E      G      H      I      L      K      M      F      P      S      T      W      Y      V
0.034  0.017  0.012  0.012  0.008  0.018  0.017  0.016  0.007  0.230  0.124  0.019  0.257  0.025  0.014  0.024  0.030  0.005  0.014  0.073
0.039  0.023  0.111  0.123  0.005  0.072  0.044  0.127  0.026  0.033  0.028  0.030  0.023  0.012  0.141  0.083  0.059  0.003  0.028  0.022
0.033  0.018  0.013  0.014  0.007  0.042  0.021  0.017  0.007  0.104  0.308  0.032  0.038  0.074  0.108  0.049  0.027  0.005  0.017  0.049
0.103  0.028  0.052  0.029  0.007  0.054  0.159  0.040  0.009  0.108  0.086  0.031  0.013  0.028  0.027  0.054  0.099  0.004  0.013  0.062
0.041  0.036  0.096  0.024  0.005  0.027  0.034  0.025  0.220  0.017  0.057  0.102  0.010  0.076  0.024  0.035  0.025  0.006  0.106  0.021
0.031  0.017  0.014  0.013  0.032  0.030  0.020  0.062  0.038  0.044  0.106  0.019  0.013  0.119  0.031  0.033  0.023  0.011  0.254  0.063
0.043  0.016  0.011  0.012  0.007  0.015  0.026  0.016  0.006  0.071  0.391  0.018  0.074  0.088  0.013  0.050  0.033  0.005  0.018  0.049
0.063  0.016  0.049  0.017  0.026  0.014  0.019  0.041  0.008  0.137  0.093  0.019  0.015  0.118  0.015  0.045  0.119  0.006  0.051  0.103
0.035  0.016  0.012  0.013  0.008  0.014  0.018  0.018  0.007  0.105  0.317  0.018  0.040  0.083  0.015  0.025  0.043  0.006  0.028  0.140
     .      .                                    .                           .
     .      .                                    .                           .
     .      .                                    .                           .
```

Figure 8.1 Example of a PSSM file describing a protein family. The first line specifies the number M of amino acid positions in the multiple alignment from which the protein family description has been constructed. The next line reminds readers of the order in which the probabilities for the 20 amino acids will be given. These two initial lines are followed by one line for each of the M positions of the protein family model. Each of these lines contains the 20 probabilities of finding a given amino acid at that position in the protein family model.

20 amino acids a at each of the M positions $i = 1, \ldots, M$ of the multiple alignment how likely this amino acid is to occur in a protein for the gene of interest. In practice, such a model is represented in a file like the one shown in Fig. 8.1.

3.2. Procedure

To build a model that is specific for the nad4L gene, we have to choose an arbitrary nad4L protein as a starting point. These proteins can be identified by searching in the protein section of GenBank for the keyword nad4L via NCBI Entrez. For the purpose of locating a gene, the organism from which we select the protein sequence really does not matter. Here, we choose the one from *Chaetosphaeridium globosum* (accession number 22417009).

Then the online version of PSI-BLAST (which can be reached from the BLAST home page http://www.ncbi.nlm.nih.gov/BLAST/) is run with this protein as the starting sequence and with default parameters except for the number of descriptions. It is important that the number of descriptions is larger than the number of nad4L proteins expected. Choosing a number of descriptions that is too large is by far not as bad as choosing a number that is too small, so the recommendation is to choose the largest number of descriptions available. In the first round, PSI-BLAST finds a list of close homologs of the *C. globosum* nad4L protein sequence and lists all of them. We can convince ourselves that all of these are indeed nad4L proteins and/or deselect any that are not before resubmitting the list of homologs for the next round of PSI-BLAST. In this next round, PSI-BLAST not only looks for homologs of the one sequence, but also uses the complete information on all homologs found so far in order to search for even more homologs. This cycle can be repeated several times; usually after three rounds, a good number of homologous proteins have been identified.

In the process of searching for more and more remote homologs of the original nad4L sequence, PSI-BLAST actually already internally constructs a multiple alignment of all the homologs found and computes probabilities $p_i(a)$ of finding amino acid a at position i of the gene from this multiple alignment. This information, which will be the basis of the following analysis, can be extracted by changing the "Alignment" choice for the display to "PSSM" on the formatting page and reformatting the output. The resulting output is a combination of numbers and letters that is not meant for human consumption.[1] However, after saving the page from the browser into a file (say `nad4Lgi22417009.ncbipssm`), it can be decoded into an explicit table of the $p_i(a)$ with the program pssmunpack, which is part of PIE via the command

```
pssmunpack nad4Lgi22417009.ncbipssm > nad4Lgi22417009.
pssm
```

The file nad4Lgi22417009.pssm looks like the one in Fig. 8.1 and describes the sequence features of a nad4L protein.

4. LOCATING GENES

4.1. Summary

To locate the nad4L gene in the mitochondrial genome of *Physarum*, we divide this genome into many regions and use PIE to determine for each of these regions which way of inserting Cs into the genomic region makes the putatively edited sequence most resemble a nad4L gene and how much it resembles a nad4L gene. While even for regions that do not contain the nad4L gene PIE will find *some* way of inserting Cs that makes them resemble a nad4L protein *most*, the score for these regions that captures how much this way of inserting Cs resembles a nad4L protein will be much lower than in the region that actually contains the nad4L gene. Figure 8.2 shows these scores as a function of the genomic position. It does not require much training or statistical analysis to recognize that the nad4L gene must be in the forward strand of the two fragments around genomic position 35000.

---

[1] The position-specific scoring matrix (PSSM) is represented by PSI-BLAST in ASN.1 format. This binary file is compressed with `bzip2` and then represented as a hexdump. `pssmunpack` undoes these encoding steps in the reverse order.

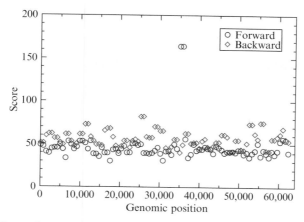

Figure 8.2   Scores for matches between a model for a nad4L protein and pieces of the *Physarum* mitochondrial genome of length 1120 nucleotides as a function of the genomic position of the piece. For each piece, the scores for the forward sequence (circle) and the reverse complement (diamond) are shown. It can be clearly seen that the scores for the forward sequences in the regions that contain the actual nad4L gene around genomic position 35000 are much higher than the scores for all the other regions. This allows the location and the orientation of the nad4L gene in the genome to be read off directly.

4.2. Scoring

While it is not strictly necessary when using PIE, it is helpful to understand how PIE quantifies how much the translation of some putatively edited sequence resembles a nad4L gene, i.e., how PIE assigns scores. To this end, we first have to define to what we actually assign scores. A configuration C that PIE scores consists of the following:

- The positions of the first and the last nucleotide in the genomic sequence that is part of the putative gene.
- The positions of all inserted Cs in the putative messenger RNA sequence. (PIE considers only insertions of Cs and at most one C can be inserted after every encoded nucleotide. However, there are in principle no restrictions on the number of inserted Cs in a configuration.)
- The first and the last amino acid position from the protein model that is aligned to the edited genomic sequence.
- The positions and lengths of amino acid gaps used in aligning the protein model to the translation of the genomic sequence with the inserted Cs.

An example of such a configuration C is shown graphically in Fig. 8.3. The score $S[C]$ of such a configuration is then calculated by adding the following quantities:

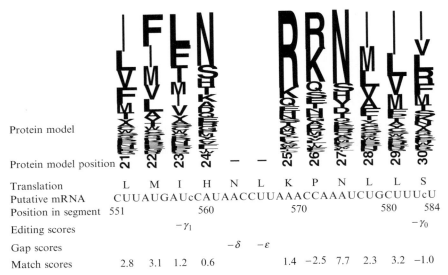

Protein model										
Protein model position	21	22	23	24	—	—	25	26	27	28	29	30
Translation	L	M	I	H	N	L	K	P	N	L	L	S
Putative mRNA	CUU	AUG	AUc	CAU	AAC	CUU	AAA	CCA	AAU	CUG	CUU	UcU
Position in segment	551				560			570			580	584
Editing scores				$-\gamma_1$							$-\gamma_0$	
Gap scores					$-\delta$	$-\varepsilon$						
Match scores	2.8	3.1	1.2	0.6			1.4	−2.5	7.7	2.3	3.2	−1.0

Figure 8.3 Example of a configuration for PIE scores and a score calculation for this configuration. The elements of such a configuration and their scores are as follows: (1) The positions of the first and the last nucleotide of the genomic sequence fragment, here 551 and 584. (2) The positions of inserted Cs (shown in lower case) in the putative mRNA sequence, here just upstream of genomic fragment positions 558 and 583. These obtain a score $-\gamma_1$ if they follow a purine–pyrimidine combination and a score $-\gamma_0$ if they do not follow a purine–pyrimidine combination. (3) The first and the last position of the protein sequence model used in the configuration, here 21 and 30. (4) The positions and lengths of amino acid gaps. This configuration has one gap of length 2 between protein model positions 24 and 25. The score for such a gap is $-\delta - (L - 1)$ ε where L is the length of the gap. Such gaps can also occur in the putative mRNA sequence (not shown in this configuration). Since they correspond to amino acid deletions, their length has to be a multiple of three, and for the purpose of scoring, their length is counted in codons and not in nucleotides. The remainder of the score is determined by the protein model. For example, the first codon in the putative mRNA sequence is a CUU, which translates into an L. Since this corresponds to protein model position 21 in this configuration, the score for this match is $1/\lambda \ln [p_{21} (L) / q_L]$ (see the text for an explanation of these quantities), which turns out to be 2.8. The total score for a configuration is the sum of all these contributions. For the optimal parameters $\delta = 12$, $\varepsilon = 1$, $\gamma_0 = 12$, and $\gamma_1 = 6$, the score of this configuration is −12.2. It is thus not a high-scoring configuration. This is largely due to the short length of this example configuration, which yields a good illustration but not a high score.

- A match score $\frac{1}{\lambda} \ln [p_i(a)/q_a]$ for each amino acid a generated by translating a codon from the edited genomic sequence and aligned to position i of the protein model. This number is positive if amino acid a occurs more frequently at protein model position i than the background probability q_a, i.e., than the frequency with which this amino acid occurs in protein sequences in general (Robinson and Robinson, 1991),[2] and negative otherwise.

- A penalty $-\gamma_0$ for each inserted C that does *not* follow a purine–pyrimidine combination.
- A penalty $-\gamma_1$ for each inserted C that *does* follow a purine–pyrimidine combination [this penalty is less severe than the general penalty γ_0 to reflect the fact that 70% of the experimentally observed editing sites follow a purine–pyrimidine combination (Miller *et al.*, 1993)].
- A penalty $-\delta - (L - 1)\,\varepsilon$ for each amino acid gap of length L. Here, δ is the gap opening penalty and ε is the gap extension penalty known from BLAST (Altschul *et al.*, 1997).

Figure 8.3 indicates for one specific example how the score of a configuration is calculated. The purpose of the PIE program is given a genomic DNA sequence and a protein model to find among all possible configurations the one with the highest score.

4.3. Procedure

To obtain the data shown in Fig. 8.2, the genomic sequence has to be cut into many short pieces. These pieces should be at least 1000 bases in length in order to cover a significant portion of the gene. Since the memory usage of PIE is quite extensive and depends linearly on the length of the DNA sequence, pieces much longer than 1000 bases should be used only if machines with memory resources of at least several GB are available. To ensure that genes that cross the boundaries between two pieces can be recognized, the pieces should overlap each other. For our study of *Physarum*, pieces of length 1120 bases were chosen. These can be easily created with a text editor since they contain 16 lines each of the FASTA file of the *Physarum* mitochondrial genome, which is formatted with 70 bases per line. The first piece then contains bases 1–1120, the second piece contains bases 561–1680, etc., and we call the files `physarum_1-1120.fasta`, `physarum_561-1680.fasta`, ..., respectively.

To compare each of the genomic pieces to the model for nad4L proteins, the model for nad4L proteins has to be converted into a model for DNA sequences coding for nad4L proteins. This is done by the command

---

[2] λ is a global scaling factor for the scores. It is chosen as the Gumbel parameter for gapless alignment statistics (Karlin and Altschul, 1990) for whatever protein scoring matrix is supplied to PIE in order to make gap insertion and extension costs comparable to what would usually be used in conjunction with that scoring matrix in protein sequence alignment programs such as BLAST (Altschul *et al.*, 1997).

```
hmmmake -p nad4Lgi22417009.pssm P_polycephalum_mit.cod\
BLOSUM62 12 1 > nad4Lgi22417009.hmm
```

The first argument in this command line is the protein model constructed in the first step described in the section on Building Gene Models. The second argument is a file with a codon usage table in GCG format (Devereux *et al.*, 1984). Such codon usage table files can be obtained for many organisms from the codon usage database (Nakamura *et al.*, 2000) (http://www.kazusa. or.jp/codon/). The purpose of this file is to specify the genetic code that is to be used for translating DNA sequences into protein sequences; the actual codon usage numbers are ignored. The next three arguments specify how to score comparisons between amino acids. The first of them is a scoring matrix in NCBI format. The next argument is the penalty δ for opening an amino acid gap, and the last argument is the penalty ε for extending an amino acid gap, introduced in the section on Scoring. For all three scoring arguments, the values given above are safe choices. The result of this command is the file nad4Lgi22417009.hmm.

The files generated by hmmmake specify hidden Markov models that are parsed against DNA sequences but that do not incorporate editing, as described in Fig. 8.2 of Bundschuh (2004). These files enumerate all nodes of the hidden Markov model. For each node, they identify which base it emits (or if it is silent), the amino acid to which it corresponds, the nodes to which transitions are allowed, and the weights of these transitions, and if a parse of the model is allowed to begin and/or end at that node.

The DNA sequence model generated by hmmmake can then be compared in all possible ways to insert Cs into a piece of the genomic sequence with the help of the program hmmrun. Specifically, to compare the model nad4Lgi22417009.hmm to the first piece physarum_1-1120.fasta of the *Physarum* genome the command

```
hmmrun -e C 12 -p 6 -r nad4Lgi22417009.hmm P_1-1120.fasta
```

is used. The first set of arguments, "-e C 12," specifies that the edited nucleotide is a C and that the score penalty γ_0 for every inserted C is 12. The next set of arguments, "-p 6," specifies that the editing penalty should be lowered by 6 if the editing site follows a purine–pyrimidine combination, i.e., that $\gamma_1 = \gamma_0 - 6 = 6$ in the language of the section on Scoring. Lastly, the "-r" demands that both the forward and the reverse complement of the genomic sequence are scanned for occurrences of the gene. These parameters have been found to give the best overall performance on experimentally verified genes (Bundschuh, 2004). The result will be two scores for the first 1120 bases of the *Physarum* mitochondrial genome, one for the forward sequence and one for its reverse complement. The data shown in Fig. 8.2 are then generated by repeating this last step for all pieces of the genome and collecting all scores.

The high scores indicate where in the genome the gene is located and if it is encoded in the forward or reverse strand.

5. DETERMINING START AND STOP CODONS

5.1. Summary

Once the genomic region in which the gene is located has been identified, the precise location of the gene, i.e., the positions of its start and stop codons, have to be found. In any type of gene prediction (with or without RNA editing), the determination of the relevant start codon for a gene is a challenge. Short of a stop codon appearing in between two putative start codons, there is little evidence from the sequence alone which start codon is the true start codon. In addition, terminal regions of proteins are typically not very strongly conserved between species and can vary significantly in length from species to species. Thus, even comparison with proteins from other species gives only limited information on the position of the start codon. This challenge also applies to organisms with RNA editing. However, while the position of the stop codon is trivially determined as the first in-frame stop codon in organisms without RNA editing, the challenges involving the start codon extend to the stop codon in organisms with RNA editing. In organisms with insertional RNA editing, it is not known a priori in which frame that part of the sequence is read, and thus several stop codons are candidates for the true stop codon.

Our approach to locating the start and stop codons is to build models for the nad4L gene starting from nad4L proteins from many different species and forcing PIE to match the identified piece of the genomic sequence to all these models under the constraint that the match has to start at a start codon and end at a stop codon. This yields one putative start codon and one putative stop codon for every nad4L model. A histogram of all these putative start and stop codons, as shown in Fig. 8.4, then guides the selection of the best candidate start and stop codons. It should be noted, though, that this method is far from being foolproof and the location of the predicted start and stop codon (as well as the predicted editing sites close to either end of the sequence) should not be trusted too much. For primer selection, it is definitely advisable to stay away from the ends of the predicted messenger RNA sequence since it may well be that the true end of the messenger RNA sequence is somewhere else and that this affects the prediction of the editing site locations in the vicinity of the end (indeed, in the example of nad4L in *Physarum*, the experimentally verified stop codon is the one at position 36064 and not the one at position 36053).

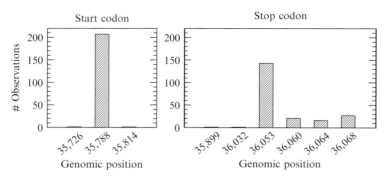

Figure 8.4 Histograms of putative genomic positions of the start and stop codon of the nad4L gene in *Physarum polycephalum*, respectively. The positions are obtained from matching 209 models built starting from nad4L proteins from different species to the genomic sequence of the *Physarum* mitochondrion using PIE.

5.2. Procedure

The first step to finding start and stop codon positions is to repeat the model building steps described in the section on Building Gene Models, starting from many different nad4L proteins. The choice of proteins should be as diverse as possible in order to obtain meaningful predictions. One should not build 100 models from 100 virtually identical nad4L proteins from different *Schistosoma* species that happen to be in the database because some study of the molecular phylogeny of *Schistosoma* was done with the nad4L gene (Morgan *et al.*, 2005). This is not only a waste of time but actually leads to incorrect conclusions, as most likely all 100 predictions of start and stop codon positions from these nearly identical models will be identical and will give a false impression of a preference for these positions in the histograms.

Next, some of the steps from the section on Locating Genes are repeated for all these models; however, there are some small differences to the procedure described in this section. First, an additional option "-s" is selected when creating the hidden Markov model, i.e., in the command

```
hmmmake -s -p nad4Lgi22417009.pssm P_polycephalum_mit.cod \
BLOSUM62 12 1 > nad4Lgi22417009_s.hmm
```

This will ensure that the generated model considers only configurations in which the initial nucleotide and the final nucleotide of the genomic sequence are at a start or stop codon, respectively.[3]

---

[3] In the hidden Markov models generated without the "-s" option, every node that describes the beginning of a codon is denoted a possible starting node and every node that describes the end of a codon is a possible final node. The option "-s" limits the starting nodes to only nodes at the beginning of start codons and the final nodes to nodes at the ends of stop codons.

Second, there is no need to compare each model to every genomic segment, but it is now enough to compare each model to only the genomic segment in which the nad4L gene has been located; in the case of nad4L, this is the segment in the file `physarum_35281-36400.fasta`. Third, we are now not only interested in the score of the best configuration of comparing a model with a genomic segment but in the actual configuration itself (at least in its first and last nucleotide). For these configurations to be reported, we add the option "-b" to the matching command, i.e., we use

```
hmmrun -b -e C 12 -p 6 -r nad4Lgi22417009_s.hmm \
physarum_35281-36400.fasta
```

(where the `-r` option is necessary only if the gene is located in the reverse strand).

The output of this command contains, in addition to the score for the forward and the reverse strand, a description of the configurations that generated these two scores. We choose the one in the strand in which we have identified the gene, i.e., in the case of nad4L the forward strand. The beginning and the end of this description of the configuration are shown in Fig. 8.5. The first column in this description of a configuration indicates the identity of the nucleotide in the mitochondrial genome of *Physarum*. The second column denotes the position of this nucleotide in the genome fragment given to PIE. This position can be converted into a genomic position by adding the genomic position of the first base in the file `physarum_35281-36400.fasta` minus one, i.e., 35280. The third column describes the amino acid of the protein predicted in this configuration. By looking at the very first (lower right in Fig. 8.5) and very last (upper left in Fig. 8.5) predicted codon, it can be read off that using the model built starting from the *C. globosum* nad4L protein sequence, the predicted locations of the start and stop codon are at positions 508 and 784 in the genomic fragment or 35788 and 36064 in the complete mitochondrial genome, respectively. Repeating the same analysis for models built starting from many different (in our example 209) initial nad4L protein sequences yields the histogram shown in Fig. 8.4.

6. PREDICTING EDITING SITE LOCATIONS

6.1. Summary

After the gene is located, the final step before an experimental verification is the prediction of the editing sites within the gene. In principle, the efforts described in the section on Determining Start and Stop Codons have already generated such predictions. For each of many protein models, the optimal configuration of matching this protein model to the genomic

```
A   786  *  (S94TAA3)                                    .
A   785  *  (S94TAA2)                                    .
T   784  *  (S94TAA1)                                    .
    784     (S94S)                       T   519  L  (S9CTT3)
    784     (S93E)                       T   518  L  (S9CTT2)
A   783  K  (S93AAA3)                    C   517  L  (S9CTT1)
A   782  K  (S93AAA2)                        516     (S9S)
A   781  K  (S93AAA1)                        516     (S8E)
    780     (S93S)                       A   516  T  (S8ACA3)
    780     (S92E)                       C   515  T  (S8ACA2)
T   780  V  (S92GTT3)                    A   514  T  (S8ACA1)
T   779  V  (S92GTT2)                        513     (S8S)
G   778  V  (S92GTT1)                        513     (S7E)
    777     (S92S)                       A   513  I  (S7ATA3)
    777     (S91E)                       T   512  I  (S7ATA2)
G   777  K  (S91AAG3)                    A   511  I  (S7ATA1)
A   776  K  (S91AAG2)                        510     (S7S)
A   775  K  (S91AAG1)                        510     (S6E)
    774     (S91S)                       G   510  M  (S6ATG3)
    774     (S90E)                       T   509  M  (S6ATG2)
      .                                  A   508  M  (S6ATG1)
      .
      .
```

Figure 8.5 Description of a configuration from which the locations of the start and stop codon can be determined. The left column shows the end of the predicted gene while the right column shows the beginning of the predicted gene (for technical reasons hmmrun outputs its results in reverse order, i.e., starting with the end of the sequences and finishing with the start codon). The nucleotides in the first column are the nucleotides of the *Physarum* mitochondrial genome fragment with their genomic position indicated in the second column. The third column represents the amino acid for which the corresponding codon codes. The lines without nucleotides and amino acids represent internal states of the computational model used to represent the protein sequences and can be ignored.

sequence of the *Physarum* mitochondrion has been determined. This optimal configuration contains, among other things, the positions of the putative editing sites. However, just because a configuration is optimal in the sense that it maximizes the score does not mean that it is biologically correct. Indeed, tests on six genes with experimentally known editing sites show that only about 70% of the editing sites are predicted correctly by PIE (Bundschuh, 2004).

There are two main reasons for this inaccuracy. First, the positions of the editing sites are determined such that they optimize the way the translated protein sequence fits the protein model. However, two (or more) competing putative positions for one editing site will have the same (or nearly the same) score if they either lead to the same protein sequence due to degeneracies in the genetic code or if they occur in regions of the protein model with very little conservation where the scores for the different amino acids at a model position are very similar. In this case, the editing site position chosen by PIE among several possible positions with the same or similar scores may not be the biologically correct one. Second, editing sites in regions with amino acid gaps have to be predicted without the benefit of knowing which amino acids are preferred at which position. This leads to low quality predictions inside (and close to) amino acid gaps.

We eliminate the second of these reasons for incorrect editing site predictions by constructing a protein model that matches the genomic sequence without amino acid insertions or deletions (indels). We achieve this by using the wealth of protein models generated in the section on Determining Start and Stop Codons. In the best possible case, one of these models already matches without amino acid indels and we simply use this specific model for the final editing site prediction. If no model matches without amino acid indels, we identify several models that have their amino acid indels in different regions of the predicted gene and combine the sections of these models without amino acid indels to a full length model without amino acid indels.

The first of the reasons for inaccuracy cannot be eliminated since these inaccuracies are due to an intrinsic lack of information about the editing site positions. However, we can address this issue by having PIE report how reliable each individual editing site prediction is, given the available information. This will allow regions with lower quality predictions (presumably due to low conservation in the protein model) for primer selection to be avoided. Specifically, instead of reporting just one (the optimally scoring one) configuration of editing sites, PIE will calculate an editing probability p (i) for every position i in the genomic sequence. If this probability is close to zero, it is very unlikely that there is a C insertion immediately downstream of position i in the genomic sequence. If $p(i)$ is close to one, it is very likely that there is a C insertion immediately downstream of position i in the genomic sequence. If $p(i)$ is neither close to one nor close to zero, this is an indication that the information available to PIE from the protein model and the genomic sequence is not enough to locate the position of the editing site. These editing site probabilities are graphically represented in Fig. 8.6. From these data, primers for experimental verification of the gene and its editing sites can be selected by focusing on regions in which there are no editing sites or only editing sites that are predicted with very high probability. For example, the long stretches without predicted editing sites from position 35821 to 35840 and from position 35993 to position 36018 are excellent candidates for primer design.

6.2. Procedure

To identify a protein model or a combination of protein models that can be matched against the genomic mitochondrial sequence of *Physarum* without amino acid indels, we have to identify the indels in the predictions generated in the section on Determining start and stop codons. Figure 8.7 shows how insertions and deletions can be identified in the optimal configurations

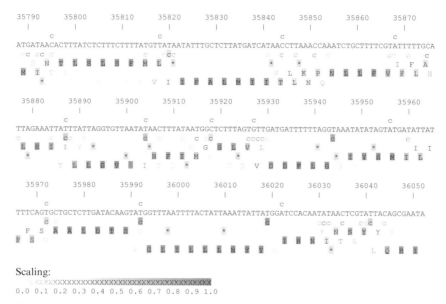

Scaling:

0.0 0.1 0.2 0.3 0.4 0.5 0.6 0.7 0.8 0.9 1.0

Figure 8.6 Prediction of editing sites for the nad4L gene in the mitochondrion of *Physarum polycephalum*. The middle of each of the three rows shows the genomic sequence from position 35788 to position 36052. The Cs in the line below indicate the probability of a C insertion immediately downstream of the corresponding position in the genomic DNA according to the scale given at the bottom of the figure. The three following lines show the amino acids encoded by the genomic sequence in the three different frames, also shaded by their probability of being encoded by the mRNA of the nad4L gene as well as all the stop codons in the genomic sequence, indicated by asterisks. The line above the genomic sequence indicates the locations of the experimentally determined editing sites (Gott *et al.*, 2005).

reported by hmmrun. Each line in a configuration report contains a label in parentheses.[4] This label starts with an S for substitutions, an I for insertions, and a D for deletions. If a report can be found that contains lines labeled with neither I nor D, the corresponding protein model only has to be truncated to the correct length to obtain a protein model that exactly fits the predicted nad4L protein in *Physarum*. For example, if the configuration, the beginning and end of which are shown in Fig. 8.5, did not contain any insertions and deletions, we would extract the lines for protein model positions 6 to 93 from the protein model file and make a new protein model file nad4Lnoindels.pssm that contains only these 88 positions.

---

[4] Each line in the configuration report corresponds to one node of the hidden Markov model in the optimal parse of the model. The labels denote the names of these nodes. The labels start with S for substitution, I for insertion, and D for deletion followed by the model position in the protein PSSM from which the hidden Markov model was built. Insertion and substitution nodes are further characterized by the codon and the codon position they represent.

```
      .                                              .
      .                                              .
      .                                              .
A   570 K (S25AAA3)                                661 (S58E)
A   569 K (S25AAA2)                         A     661 L (S58TTA3)
A   568 K (S25AAA1)                         T     660 L (S58TTA2)
    567   (I25E)                            T     659 L (S58TTA1)
    567   (S25S)                                  658 (S58S)
T   567 L (I25CTT3)                                658 (S57E)
T   566 L (I25CTT2)                         T     658 F (S57TTT3)
C   565 L (I25CTT1)                         T     657 F (S57TTT2)
    564   (I25S)                            T     656 F (S57TTT1)
    564   (I25E)                                  655 (S57S)
C   564 N (I25AAC3)                                655 (D56)
A   563 N (I25AAC2)                                655 (D55)
A   562 N (I25AAC1)                                655 (S54E)
    561   (I25S)                            T     655 D (S54GAT3)
    561   (S24E)                            A     654 D (S54GAT2)
T   561 H (S24CAT3)                         G     653 D (S54GAT1)
A   560 H (S24CAT2)                                652 (S54S)
C   559 H (S24CAT1)                                652 (S53E)
    558   (S24S)                            T     652 D (S53GAT3)
    558   (S23E)                            A     651 D (S53GAT2)
    558 I (edited) (S23ATC3 (edited))       G     650 D (S53GAT1)
T   558 I (S23ATC2)                                649 (S53S)
A   557 I (S23ATC1)                               .
    556   (S23S)                                  .
      .                                           .
      .
      .
```

Figure 8.7 Insertions and deletions in optimal configurations. Two excerpts from optimal configurations reported by PIE are shown. These have the same format as explained in the legend to Fig. 8.5. The left part corresponds to a subset of the configuration shown in Fig. 8.3. It shows a two amino acid insertion between protein model positions 24 and 25 that covers genome fragment positions 562–567. This insertion can be recognized from the labels in parentheses starting with I for insertion (not to be confused with the isoleucine at genomic positions 557 and 558). This configuration also shows how editing sites are represented with an inserted C in the third codon position for the codon at protein model position 23 just downstream of genomic fragment position 558. The right-hand side shows a fragment of a configuration with a two amino acid deletion indicated by the labels *in parentheses* that start with a D (again not to be confused with the aspartic acids at genomic positions 650–655). In this configuration, protein model positions 55 and 56 are deleted and this deletion is located between genomic fragment positions 655 and 656.

For most genes (including our example nad4L) there will be no protein model that matches the nad4L gene in *Physarum* without amino acid indels. This just means that actual amino acid insertion or deletion events occurred between the nad4L protein in *Physarum* and any nad4L protein in GenBank. In this case, we summarize the indels in all the configurations obtained in the section on Determining Start and Stop Codons by identifying them with the respective sequence position in the genomic fragment at which they occur. These positions are schematically indicated for a small but typical subset of the 209 predictions generated in Fig. 8.8.

The remaining task is to find a few models that can be "stitched together" to provide high-quality (and preferably indel-free) coverage of the whole *Physarum* nad4L protein. From Fig. 8.8 it can be seen that the

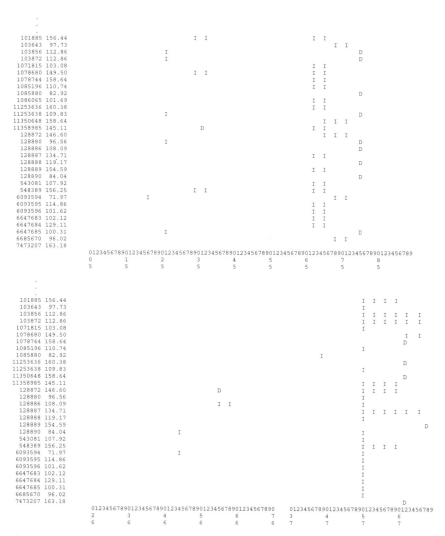

Figure 8.8 Schematic representation of positions of insertions and deletions in the various predictions of nad4L editing sites starting from different nad4L protein models. Each row corresponds to the results for the protein model built from the nad4L protein with the accession number listed at the beginning of the row. The number behind the accession number is the PIE score for the prediction. The Is and Ds in the row indicate by their locations with respect to the position numbers at the bottom of the figure at what positions in the genomic fragment amino acid insertions and deletions are predicted. These insertions and deletions correspond to insertions and deletions of complete amino acids in the course of evolution between *Physarum* and the organisms from which the protein sequences are taken; they do *not* have anything to do with editing. For clarity of presentation, some positions that do not show amino acid insertions and deletions in any of the predictions have been omitted and only a small but typical selection of the 209 predictions generated is shown.

(relatively high scoring) prediction generated by the protein model that was built starting from the nad4L protein with accession number 7473207 (last row in Fig. 8.8) is free of indels until position 761. Its prediction starts with the start codon at genomic position 35788 and ends with the stop codon at genomic position 36060, i.e., the start codon is the one we are looking for but the stop codon is not. Thus, this protein model is ideal for the beginning of the sequence up to the point of the deletion, but we need to use another model for the rest of the protein.

The search for a model for the rest of the sequence is somewhat harder. Figure 8.8 shows that all models have some indels toward the end of the sequence (which is a testimony of the weak sequence conservation at the end of genes). Inspection of the few models that have a late deletion rather than one or a few insertions reveals that none of the models with a late deletion has the stop codon we are looking for (this is still true if all models and not just the ones shown in Fig. 8.8 are taken into consideration). The best we can do is to choose one of the models with only a single insertion, e.g., the one built from the nad4L protein with accession number 6685670 (the penultimate line in Fig. 8.8), which has the stop codon we are looking for.

To stitch the two models together, we have to look at the predictions of the two models in the vicinity of the sequence position where we intend to do so. Figure 8.9 shows the predictions generated starting from the nad4L proteins with accession numbers 7473207 and 6685670 on the left and right, respectively. We want to use the model generated from 7473207 until downstream of the insertion in model 6685670 but still upstream of the deletion in model 7473207. For two models to be stitched together, it is mandatory for the predictions of the two models to be in the same frame at the stitching site. As Fig. 8.9 shows, this condition is met up to genomic fragment position 753; in the next codon, the frames are shifted due to a prediction of an editing site in model 6685670 that is not predicted in model 7473207. Genomic fragment position 753 barely covers the amino insertion at genomic fragment positions 751–753 in model 6685670. Thus, we will use the model built from the nad4L protein with accession number 7473207 up to model position 92 and continue from there with the model built from the nad4L protein with accession number 6685670 starting with model position 87 (as can be seen from Fig. 8.9, the nad4L proteins with accession numbers 7473207 and 6685670 differ in their length at the N-terminal end in such a way that model position 93 of the first and model position 87 of the second exactly correspond to each other, allowing us to continue after position 92 of the first model with position 87 of the second). In general, the stitching point should be chosen as far as possible from any indels and editing sites, but in the situation for the nad4L gene there is no choice in the position. Looking at the beginning and end of the predictions in the same way as described above for the case of a single model matching without

```
      .                                               .
      .                                               .
      .                                               .
      759 (S94E)                                      758 (S88E)
T     759  R  (S94CGT3)                          . G  758  S  (S88TCG3)
G     758  R  (S94CGT2)                            C  757  S  (S88TCG2)
C     757  R  (S94CGT1)                            T  756  S  (S88TCG1)
      756 (S94S)                                      755 (S88S)
      756 (S93E)                                      755 (S87E)
T     756  T  (S93ACT3)                               755  T  (edited)  (S87ACC3 (edited))
C     755  T  (S93ACT2)                            C  755  T  (S87ACC2)
A     754  T  (S93ACT1)                            A  754  T  (S87ACC1)
      753 (S93S)                                      753 (S87S)
      753 (S92E)                                      753 (I87E)
A     753  I  (S92ATA3)                            A  753  I  (I87ATA3)
T     752  I  (S92ATA2)                            T  752  I  (I87ATA2)
A     751  I  (S92ATA1)                            A  751  I  (I87ATA1)
      750 (S92S)                                      750 (I87S)
      750 (S91E)                                      750 (S86E)
T     750  N  (S91AAT3)                            T  750  N  (S86AAT3)
A     749  N  (S91AAT2)                            A  749  N  (S86AAT2)
A     748  N  (S91AAT1)                            A  748  N  (S86AAT1)
      747 (S91S)                                      747 (S86S)
      747 (S90E)                                      747 (S85E)
C     747  H  (S90CAC3)                            C  747  H  (S85CAC3)
A     746  H  (S90CAC2)                            A  746  H  (S85CAC2)
C     745  H  (S90CAC1)                            C  745  H  (S85CAC1)
      744 (S90S)                                      744 (S85S)
      744 (S89E)                                      744 (S84E)
C     744  I  (S89ATC3)                            C  744  I  (S84ATC3)
T     743  I  (S89ATC2)                            T  743  I  (S84ATC2)
A     742  I  (S89ATC1)                            A  742  I  (S84ATC1)
      741 (S89S)                                      741 (S84S)
      741 (S88E)                                      741 (S83E)
G     741  R  (S88CGG3)                            G  741  R  (S83CGG3)
G     740  R  (S88CGG2)                            G  740  R  (S83CGG2)
      739  R  (edited)  (S88CGG1 (edited))             739  R  (edited)  (S83CGG1 (edited))
      739 (S88S)                                      739 (S83S)
      .                                               .
      .                                               .
      .                                               .
```

Figure 8.9 Comparison of two editing site predictions in a region where they can be "stitched together." The prediction on the left is obtained using a protein model built starting from the nad4L protein with accession number 7473207, while the one on the right is obtained using a protein model built starting from the nad4L protein with accession number 6685670. It can be seen that these predictions agree over the range shown up to the codon starting at genomic fragment position 753. However, the one on the right has an amino acid insertion at genomic fragment positions 751–753. Thus, an indel free model can be constructed by using the left model up to model position 92 (covering the genomic fragment up to nucleotide position 753) and then switching to the right model starting at model position 87 (covering the genomic fragment from position 754 on). The apparent "jump" in the model position number from 87 to 92 reflects the fact that the proteins have different lengths at their N-terminal ends.

indels, we finally construct the file nad4Lnoindels.pssm from positions 7–92 of the model built from accession number 7473207 and 87–93 of the model built from accession number 6685670, yielding a model with a total of 93 positions.

The final step is the calculation of the editing probabilities $p(i)$ for each position in the genomic sequence shown in Fig. 8.6. These probabilities are calculated as

$$p(i) = \frac{\sum_{\text{configurations } C \text{ with an inserted C downstream of position } i} e^{\lambda S[C]}}{\sum_{\text{all configurations } C} e^{\lambda S[C]}} \qquad (1)$$

It is close to one if all high scoring configurations contain an inserted C immediately downstream of position i, close to zero if none of the high scoring configurations contain an inserted C immediately downstream of position i, and something in between if some of the high scoring configurations contain an inserted C immediately downstream of position i.

To perform the computation, we prepare a FASTA-file nad4Lgenomic.fasta that contains exactly the genomic region of the *Physarum* mitochondrion in which we predict the nad4L gene from the start codon to the last codon before the stop codon, i.e., according to our results in the section on Determining Start and Stop Codons from genomic position 35788 to genomic position 36052. Then we use the program globalign, which is part of PIE with the command line

```
globalign -o 35788 -P 6 -t 0.05 -h -n nad4Lnoindels.
pssm \ P_polycephalum_mit.cod nad4Lgenomic.fasta \
> nad4Leditingsites.html
```

The number after the -o is the genomic position of the first nucleotide in the file nad4Lgenomic.fasta and is used for display purposes only. The number after the -p is the weight for Cs inserted after a purine–pyrimidine combination[5] and the number after the -t is the minimal probability $p(i)$ above which an insertion site is displayed as a possible insertion site. The output nad4Leditingsites.html of the program is a file that can be viewed by a web browser with the result shown in Fig. 8.6 (of course without the line that indicates the experimentally verified editing sites).

7. CONCLUSIONS

The Predictor of Insertional Editing is a powerful tool for the location of genes and the prediction of editing sites in genomes of organisms with insertional RNA editing. Its probabilistic predictions, such as the one shown for the nad4L gene in Fig. 8.6, provide much information on the position of editing sites. As a comparison with the experimentally verified editing sites also shown in Fig. 8.6 demonstrates, these predictions are not perfect, but regions in which PIE is rather sure that there are no

---

[5] Consistent with the calculation shown in Eq. (1), the extra weight for a C after a purine–pyrimidine combination is $\exp[\lambda(\gamma_0 - \gamma_1)]$.

editing sites are indeed free of editing sites. Thus, these regions can be used to design primers, as has been done successfully for several genes including nad4L (Gott et al., 2005).

Obtaining a prediction such as the one shown in Fig. 8.6 requires a number of steps that cannot be automated since they depend on the specific circumstances of every single gene. In this chapter we explained these steps and the reasoning behind them in detail. Since the source code of PIE ready to be compiled on Linux and Mac OS X systems is freely available for noncommercial use upon request from the author, this will enable interested scientists to obtain editing site predictions for organisms with insertional RNA editing themselves.

ACKNOWLEDGMENTS

We acknowledge a very enjoyable collaboration with Jonatha Gott and her laboratory without whom none of this work would have come into existence and the Rustbelt RNA meeting supported by the National Science Foundation under Grant MCB0121758 at which this work was initiated. This work is supported by the National Science Foundation under Grant number DMR0404615.

REFERENCES

Altschul, S. F., Madden, T. L., Schaffer, A. A., Zhang, J., Zhang, Z., Miller, W., and Lipman, D. J. (1997). Gapped BLAST and PSI-BLAST: A new generation of protein database search programs. *Nucleic Acids Res.* **25,** 3389–3402.

Bass, B. L. (ed.), (2001). "RNA Editing." Oxford University Press, Oxford, UK.

Benne, R., Van den Burg, J., Brakenhoff, J. P., Sloof, P., Van Boom, J. H., and Tromp, M. C. (1986). Major transcript of the frameshifted coxII gene from trypanosome mitochondria contains four nucleotides that are not encoded in the DNA. *Cell* **46,** 819–826.

Benson, D. A., Karsch-Mizrachi, I., Lipman, D. J., Ostell, J., and Wheeler, D. L. (2006). GenBank. *Nucleic Acids Res.* **34,** 16–20.

Blum, B., Bakalara, N., and Simpson, L. (1990). A model for RNA editing in kinetoplastid mitochondria: "Guide" RNA molecules transcribed from maxicircle DNA provide the edited information. *Cell* **60,** 189–198.

Bundschuh, R. (2004). Computational prediction of RNA editing sites. *Bioinformatics* **20,** 3214–3220.

Devereux, J., Haeberli, P., and Smithies, O. (1984). A comprehensive set of sequence analysis programs for the VAX. *Nucleic Acids Res.* **12,** 387–395.

Gott, J. M. (2001). RNA editing in *Physarum polycephalum. In* "RNA Editing" (B. L. Bass, ed.), pp. 20–37. Oxford University Press, Oxford, UK.

Gott, J. M., Parimi, N., and Bundschuh, R. (2005). Discovery of new genes and deletion editing in Physarum mitochondria enabled by a novel algorithm for finding edited mRNAs. *Nucleic Acids Res.* **33,** 5063–5072.

Horton, T. L., and Landweber, L. F. (2002). Rewriting the information in DNA: RNA editing in kinetoplastids and myxomycetes. *Curr. Opin. Microbiol.* **5,** 620–626.

Karlin, S., and Altschul, S. F. (1990). Methods for assessing the statistical significance of molecular sequence features by using general scoring schemes. *Proc. Natl. Acad. Sci. USA* **87,** 2264–2268.

Keegan, L. P., Gallo, A., and O'Connell, M. A. (2001). The many roles of an RNA editor. *Nat. Rev. Genet.* **2,** 869–878.

Miller, D., Mahendran, R., Spottswood, M., Costandy, H., Wang, S., Ling, M. L., and Yang, N. (1993). Insertional editing in mitochondria of Physarum. *Semin. Cell Biol.* **4,** 261–266.

Morgan, J. A. T., Dejong, R. J., Adeoye, G. O., Ansa, E. D. O., Barbosa, C. S., Bremond, P., Cesari, I. M., Charbonnel, N., Correa, L. R., Coulibaly, G., D'Andrea, P. S., De Souza, C. P., *et al.* (2005). Origin and diversification of the human parasite Schistosoma mansoni. *Mol. Ecol.* **14,** 3889–3902.

Nakamura, Y., Gojobori, T., and Ikemura, T. (2000). Codon usage tabulated from international DNA sequence databases: Status for the year 2000. *Nucleic Acids Res.* **28,** 292.

Ogawa, S., Yoshino, R., Angata, K., Iwamoto, M., Pi, M., Kuroe, K., Matsuo, K., Morio, T., Urushihara, H., Yanagisawa, K., and Tanaka, Y. (2000). The mitochondrial DNA of Dictyostelium discoideum: Complete sequence, gene content and genome organization. *Mol. Gen. Genet.* **263,** 514–519.

Robinson, A. B., and Robinson, L. R. (1991). Distribution of glutamine and asparagine residues and their near neighbors in peptides and proteins. *Proc. Natl. Acad. Sci. USA* **88,** 8880–8884.

Schaffer, A. A., Aravind, L., Madden, T. L., Shavirin, S., Spouge, J. L., Wolf, Y. I., Koonin, E. V., and Altschul, S. F. (2001). Improving the accuracy of PSI-BLAST protein database searches with composition-based statistics and other refinements. *Nucleic Acids Res.* **29,** 2994–3005.

Simpson, L., and Shaw, J. (1989). RNA editing and the mitochondrial cryptogenes of kinetoplastid protozoa. *Cell* **57,** 355–366.

Simpson, L., Sbicego, S., and Aphasizhev, R. (2003). Uridine insertion/deletion RNA editing in trypanosome mitochondria: A complex business. *RNA* **9,** 265–276.

Smith, H. C., Gott, J. M., and Hanson, M. R. (1997). A guide to RNA editing. *RNA* **3,** 1105–1123.

Stuart, K. D., Panigrahi, A. K., and Salavati, R. (2001). RNA editing in kinetoplastid mitochondria. *In* "RNA Editing" (B. L. Bass, ed.), pp. 1–19. Oxford University Press, Oxford, UK.

Takano, H., Abe, T., Sakurai, R., Moriyama, Y., Miyazawa, Y., Nozaki, H., Kawano, S., Sasaki, N., and Kuroiwa, T. (2001). The complete DNA sequence of the mitochondrial genome of Physarum polycephalum. *Mol. Gen. Genet.* **264,** 539–545.

von Haeseler, A., Blum, B., Simpson, L., Sturm, N., and Waterman, M. S. (1992). Computer methods for locating kinetoplastid cryptogenes. *Nucleic Acids Res.* **20,** 2717–2724.

CHAPTER NINE

Evolution of RNA Editing Sites in the Mitochondrial Small Subunit rRNA of the Myxomycota

Uma Krishnan, Arpi Barsamian, *and* Dennis L. Miller

Contents

Department of Molecular and Cell Biology, The University of Texas at Dallas, Richardson, Texas

Methods in Enzymology, Volume 424
ISSN 0076-6879, DOI: 10.1016/S0076-6879(07)24009-1

Abstract

Because of their unique and unprecedented character, it is often difficult to imagine how and why the different, diverse types of RNA editing have evolved. Information about the evolution of a particular RNA editing system can be obtained by comparing RNA editing characteristics in contemporary organisms whose phylogenetic relationships are known so that editing patterns in ancestral organisms can be inferred. This information can then be used to build models of the origins, constraints, variability, and mechanisms of RNA editing. As an example of the types of information that can be obtained from these analyses, we describe how we have used cDNA, covariation, and phylogenetic analyses to study the evolution of the variation in RNA editing site location in the core region of the small subunit rRNA gene in the mtDNA of seven myxomycetes, including *Physarum polycephalum*. We find that the unique type of insertional RNA editing present in mitochondria of *P. polycephalum* is also present in the mitochondrial small subunit (SSU) rRNA of the other six myxomycetes. As in *Physarum*, this editing predominantly consists of cytidine insertions, but also includes uridine insertions and certain dinucleotide insertions such that any of the four canonical ribonucleotides can be inserted. Although the characteristics of RNA editing in these organisms are the same as in *Physarum*, the location of the insertion sites varies among the seven organisms relative to the conserved primary sequence and secondary structure of the rRNA. Nucleotide insertions have been identified at 29 different sites within this core region of the rRNA, but no one organism has more than 10 of these insertion sites, suggesting that editing sites have been created and/or eliminated since the divergence of these organisms. To determine the order in which editing sites have been created or eliminated, the sequences of the mitochondrial SSU rRNA have been aligned and this alignment has been used to produce phylogenetic trees showing the sequence relationship of these organisms. These phylogenetic trees are congruent with phylogenetic trees predicted by alignment of nuclear rDNA sequences. These trees indicate that editing sites change rapidly relative to mtDNA sequence divergence and suggest that some editing sites have been created more than once during the evolution of the Myxomycota.

1. INTRODUCTION

The processes that are, by definition, grouped together under the term RNA editing are diverse in function and mechanism, and, as such, are thought to be individually derived rather than being inherited from one or a few common ancestral progenitors within an "RNA World" (Gray, 2001). This implies that RNA editing has been "invented" by evolution numerous times and it is likely that each diverse type of RNA editing has a separate origin and evolutionary pathway leading to contemporary RNA editing systems.

Understanding these origins and evolutionary pathways provides clues to the purpose and mechanism for the particular RNA editing system. The classical approach to understanding evolutionary origins and pathways is to develop phylogenies of the organisms possessing a particular type of RNA editing and to use these phylogenies to infer the order of evolutionary events in the development of the RNA editing system. Here we describe these and related techniques that were used to obtain information about the mechanism and the *raison d'être* of the insertional RNA editing found in the myxomycetes.

A unique type of insertional RNA editing occurs in mitochondria of *Physarum polycephalum* (Gott, 2001; Gott and Visomirski-Robic, 1998; Mahendran *et al.*, 1991; Miller *et al.*, 1993a,b,c). This type of RNA editing is characterized by insertions of one or two nucleotides at sites separated on average by about 25 nucleotides. Single nucleotide insertions are predominantly cytidines, but uridines are added exclusively at a small number of specific sites. Dinucleotide insertions are also observed. They are less common and appear to be limited to a subset of the possible dinucleotide sequences. This subset includes AA, UU, CU (or UC), GU (or UG), GC (or CG), and AU (or UA). These insertions are often quite numerous (30–50 sites per mRNA) and constitute about 2–4% of the RNA. Insertion of these nucleotides creates the genetic information in mRNAs that is missing from the mitochondrial (mt) DNA template for the RNAs. These insertions also create conserved primary sequences and secondary structures in mitochondrial tRNAs (Antes *et al.*, 1998) and rRNAs (Mahendran *et al.*, 1994). The mechanism whereby sites are specified for insertion to create the genetic information is unknown.

It is not clear how this unique type of RNA editing has become established in the Myxomycota and how widely it is distributed throughout this phylum, nor is it generally known how RNA editing becomes established in organisms. Gray *et al.* (Covello and Gray, 1993; Gray, 2001; Price and Gray, 1998) have proposed a general model for the establishment of RNA editing. They proposed that first, RNA editing activity is produced as an alteration or duplication of a preexisting activity, i.e., a preadaptation. Second, a deleterious mutation occurs in a gene that can be corrected at the RNA level by this activity, so that this activity is able to restore essentially wild–type function. If this now essentially neutral mutation becomes fixed in a population by random genetic drift, then the activity necessary to make it neutral also becomes fixed by selection. Third, this activity may also be able to compensate for other mutations that occur in other genes and as the number of mutations neutralized by the RNA editing activity increases, the probability of these mutations reverting by classical mechanisms becomes very small and the organism becomes increasingly dependent on the activity. This model predicts that the number of mutations corrected by this activity, i.e., the number of editing sites in the RNA, will tend to increase over time until some limit of density is reached, and once established, editing sites should be retained if they can be eliminated only by classical mechanisms of reversion, which for deletions would be a rare event. If this type of editing was present in the common ancestor of the myxomycetes, this would mean that editing sites established before the divergence of these organisms would be expected to be present in all contemporary myxomycetes. If editing sites are selected for their ability to compensate for deletion mutations, then any mutation that can be neutralized by the editing activity could result in the creation of an editing site. In *Physarum* editing, only single cytidines, single uridines, and certain dinucleotides are inserted at editing sites, and so this would imply that the editing activity in *Physarum* could correct these types of deletion only in the mtDNA.

P. *polycephalum* is a member of the phylum Myxomycota, which is anciently diverged from the eukaryotic kingdoms of plants, animals, and fungi. The exact point of divergence of the Myxomycota from the main eukaryotic line is not completely clear. Phylogenies based on rRNA sequences indicate an ancient divergence from the eukaryotic line, prior to the divergence of plants, animals, and fungi, and place the myxomycetes among the first mitochondriate eukaryotes (Barns *et al.*, 1996; De Rijk *et al.*, 1995; Henricks *et al.*, 1991; Hinkle and Sogin, 1993; Krishnan *et al.*, 1990). However, Baldauf and Doolittle (1997) and Baldauf *et al.* (2000) have argued for a more recent divergence based on analysis of non-rRNA sequences, which would place the divergence of the myxomycete branch after the divergence of plants but before the divergence of fungi from the line leading to animals. Organisms within the phylum Myxomycota (Margulis and Schwartz, 1988) have been classified within two classes (Frederick, 1990; Spiegel, 1990), four subclasses, and seven orders based on morphological characteristics such as fruiting body morphology, spore color and ornamentation, and capillitium structure (Martin *et al.*, 1983; Stephenson and Stempen, 1994). The phylogenetic relationship of some of these organisms has only recently been determined by criteria of nuclear DNA sequence comparison (Fiore-Dunno *et al.*, 2005).

We wished to determine whether RNA editing was required for mitochondrial gene expression in myxomycetes related to *Physarum* and, if so, whether it was the same type of RNA editing as that characterized in *Physarum*. Antes *et al.* (1998) have previously shown that RNA editing of the *Physarum* type is necessary to produce mature tRNAs in mitochondria of *Didymium nigripes*, a myxomycete related to *Physarum*. Additionally, they have shown that these tRNAs were edited at different sites in analogous tRNAs of these two organisms. This implies that editing sites have been created, eliminated, or both since the divergence of *Physarum* and *Didymium*. Horton and Landweber (2000) have identified *Physarum*-like editing in the *co*I mRNA of three additional myxomycetes (*Stemonitis flavogenita*, *Arcyria cineria*, and *Clastoderma debaryanum*). The *co*I mRNA of both *A. cineria* and *C. bebaryanum* have a dramatically decreased density of insertional editing sites (five sites identified in *A. cineria* and four sites in

C. debaryanum) and a preference for uridine over cytidine insertion (nine U to one C insertion) with a complete absence of dinucleotide insertions.

To determine the distribution of this type of editing among the Myxomycota and to understand how this type of editing has evolved in the Myxomycota, we have surveyed a region of the mitochondrial small subunit (SSU) rRNA for editing sites and have compared the location of these sites to determine how much variation in editing site location is present among these organisms. To compare editing site variation to evolutionary sequence divergence, the phylogenetic relationships of these organisms based on sequence divergence have also been determined.

Here we report the identification of RNA editing sites of the *Physarum* type in the mitochondrial SSU rRNA of seven organisms of the Myxomycota. The location of nucleotide insertions varies among 29 sites. No site is used in all seven organisms. To determine how editing site location varies relative to phylogenetic relationships, we have determined the sequence of the core region of the SSU rRNA from both mitochondria and nucleus and have used alignments of the sequences to produce phylogenetic trees showing the sequence relationship of these organisms.

2. ISOLATION AND AMPLIFICATION OF MYXOMYCETE DNA FOR SEQUENCE ANALYSIS

2.1. Source of myxomycetes and culture conditions

D. nigripes and *S. flavogenita* were obtained as plasmodia from Carolina Biologicals. Macroplasmodia were grown at 27° in the dark on half-strength cornmeal agar (CMA/2) plates sprinkled with crushed oats. Cultures were initiated from macroplasmodia grown with attenuated bacteria. Macroplasmodia were subcultured by cutting out a 1-cm^2 agar plug from the growing plasmodial front and transferring it to a new plate.

Echinostelium minutum was obtained as plasmodia provided by Edward Haskins (University of Washington, Seattle, WA). Macroplasmodia were grown at 22° in the dark on CMA/2 plates (Hinchee and Haskins, 1981).

Physarum didermoides was obtained as lyophilized amebas with bacteria from the American Type Culture Collection (22485). Amebas were plated with heat-killed bacteria on full-strength cornmeal agar (CMA) plates, and grown at 23° in the dark. Amebas were subcultured by transferring amebas to a 1-ml tube of heat-killed bacteria, mixing well, and plating the mixture on two or three CMA plates.

Lycogala epidendrum aethalia from a single plasmodium were isolated from the wild. Aethalia were washed several times by centrifugation in water prior to DNA isolation.

2.2. Isolation of total nucleic acids

Total nucleic acids were isolated directly from aethalia or from macroplasmodia or ameba growing on plates. Macroplasmodia were scraped from plates and resuspended in BEST (0.1% bovine serum albumin, 60 mM EDTA, 300 mM sucrose, 20 mM Tris–HCl, pH 7.4). The plasmodia were pelleted by centrifugation at 8000×g for 10 min at 4° and the wash with BEST was repeated. Macroplasmodia were then lysed by the addition of sodium dodecyl sulfate (SDS) to a final concentration of 2%. After centrifugation, the supernatant was extracted with phenol and then with chloroform. Total nucleic acids were precipitated by the addition of 0.2 volumes of 3 M sodium acetate, pH 7.8, and 3 volumes of 95% ethanol. *Lycogala* fruiting bodies were washed a couple of times with BEST. Fruiting bodies were disrupted by blending for about 15–25 sec in 2% SDS. This lysate was then extracted with phenol/chloroform and the nucleic acids were precipitated with 95% ethanol.

2.3. Oligonucleotide primers

Primers used for the polymerase chain reaction (PCR), reverse transcription PCR (RT-PCR), and DNA sequence analysis were commercially synthesized (GIBCO-BRL and Fisher-Genosys).

2.3.1. Mitochondrial SSU primers

mt SSU *Bam*HI   5′ CGGGATCCAGCAGCCGCGGTAA 3′
mt SSU *Kpn*I   5′ CGGGTACCTCGAATTAAACCACAT3′
mt SSU Internal   5′ CAAGGTCTCTAATCC 3′
Lycogala mt SSU Internal   5′ GGACTTCGAAGGCGAAA 3′
Stemonitis SSU Internal   5′ GTTTCACGCGT 3′
Stemonitis Internal SSU 3′   5′ CTCCGTCTTGATTCAC 3′
mt SSU Eco 3′   5′ GGAATTCGAATTAAACCACAT 3′

2.3.2. Nuclear SSU primers

nuc SSU *Bam*HI   5′ CGGGATCCAGCACCCGCGGTAA 3′
nuc SSU *Eco*RI   5′ GGAATTCGTCAAATTAAGCCGCAGG 3′

2.4. Polymerase chain reaction

Total nucleic acids were resuspended in TE (10 mM Tris–HCl, pH 8, 1 mM EDTA), and RNA was removed by digestion with DNase-free RNase A (Sigma Chemical Co.) at 37° for 30 min. The DNA was used as the template for the PCR (Saiki *et al.*, 1988) with the set of either SSU mitochondrial primers or the SSU nuclear primers to amplify the core region of the SSU rRNA from the mitochondria and the nucleus, respectively.

The amplification was performed in 10 mM Tris–HCl, pH 8.3, 50 mM KCl, 2.5 mM MgCl$_2$, 200 μM deoxynucleotide triphosphates, and 2.5 units Taq DNA polymerase (Fisher Biotech). The thermal cycle regimen consisted of 94° for 1 min, 55° for 2 min, and 72° for 2 min for 35 cycles. During the final cycle, an extension of 10 min at 72° was performed and the reaction products were slowly cooled to 4° for complete annealing.

2.5. cDNA synthesis (RT-PCR)

Total nucleic acids were treated with RNase-free RQ1 DNaseI (Promega) in 40 mM Tris–HCl, pH 7.9, 10 mM NaCl, 6 mM MgCl$_2$, and 10 mM CaCl$_2$. Approximately 1 μg of 3′ mitochondrial SSU primer was annealed with the remaining RNA in 250 mM KCl, 10 mM Tris–HCl, pH 8.3, by heating to 80° and slowly cooling to 45° over the course of 30 min. After annealing, the primer–RNA hybrid was precipitated with 0.5 M sodium acetate and ethanol. The RNA–primer pellet was resuspended in 50 mM Tris–HCl, pH 8.3, 40 mM KCl, 10 mM dithiothreitol, 7 mM MgCl$_2$, 0.1 mg/ml bovine serum albumin, 25 mM deoxynucleotide triphosphates, and 1 unit of Prime RNase Inhibitor (5′–3′, Inc.) to a final volume of 50 μl. cDNA synthesis was accomplished using MMLV reverse transcriptase (Promega) for 45 min at 37°. After cDNA synthesis, RNA was digested using RNase A (Sigma Chemical Co.) for 5 min at 37°. cDNAs were recovered by three phenol–chloroform extractions and precipitated with ethanol in 2.5 M ammonium acetate. A double-stranded cDNA amplification product was produced using the original 3′ cDNA primer and an appropriate mitochondrial SSU 5′ primer as described above.

2.6. Cloning and sequencing of mitochondrial and nuclear core SSU rDNA amplification products

Restriction sites on the ends of the primers (*Eco*RI and *Bam*HI for mitochondrial SSU primers) were used for cloning. Amplification products of approximately 480 bp were produced using the mitochondrial SSU primers and were cloned into the *Eco*RI and *Bam*HI sites of pUC18. Digested amplification products were ligated with digested pUC18 in 30 mM Tris–HCl, pH 7.8, 10 mM dithiothreitol, 10 mM MgCl$_2$, and 10 mM ATP with 3 units T4 DNA ligase (Promega) at 16° for 18 h. The ligase reaction was precipitated with 0.5 M sodium acetate and ethanol. An aliquot was used to transform *Escherichia coli* XL1-blue (Stratagene) and recombinant DNA clones were identified and isolated using conventional methods. The clones were screened with restriction enzyme digests to identify plasmids with inserted amplification products. Amplification products produced from *Lycogala* mitochondrial RNA and from *Echinostelium* mitochondrial DNA and RNA were sequenced directly without cloning. Sequences of the amplification products were determined using the dideoxynucleotide chain-termination method with synthesized primers (Sanger *et al.*, 1977; SequenaseVersion 2.0; Amersham Life Sciences) or the cycle sequencing, dye-labeled terminator procedure (ABI PRISM, Applied Biosystems). Thirteen picomoles of primer in 4 μl of water was extended on a template of 80 ng of PCR amplification product in 4 μl of water using ampliTaq DNA Polymerase FS in the presence of big dye terminator mix. Thermocycling conditions were 90° for 30 sec, 50° for 15 sec, and 60° for 3 min. Sequence data were analyzed using the MicroGenie sequence analysis package (Queen and Korn, 1984).

2.6.1. Cloning and sequencing of cDNA amplification products

The cDNA amplification products were cloned using the restriction enzyme recognition sequences for *Eco*RI or *Bam*HI at the ends of the primers. Digested amplification products were ligated to the *Eco*RI and *Bam*HI sites of pUC18 as described above. The clones were screened and sequenced as described above. cDNA amplification products produced from *Lycogala* and *Echinostelium* mitochondrial RNA were sequenced directly, without cloning.

2.7. Phylogenetic analysis

DNA sequences were initially aligned using the multiple alignment algorithm of MicroGenie sequence software (Queen and Korn, 1984) and then manually adjusted. Regions of length variation were omitted in the alignment. Maximum parsimony trees were produced from alignments using the DNAPARS algorithm from the PHYLIP package, version 3.57c (Felsenstein, 1995). Distance matrices were produced from pairwise comparisons of aligned sequences or by using the maximum likelihood (ML) option of the DNA-DIST algorithm of PHYLIP on multiple alignments. Trees were produced from these matrices using the FITCH algorithm of PHYLIP. Bootstrap values were determined using the SEQBOOT algorithm of PHYLIP.

3. ANALYSIS OF MITOCHONDRIAL DNA AND cDNA SEQUENCES OF THE SSU rRNA CORE REGION

3.1. Rationale for choice of myxomycete; amplification product size and location

To survey RNA editing among the organisms of the Myxomycota, total nucleic acids were isolated from six organisms at varying evolutionary distances from *P. polycephalum* based on their classification (Frederick, 1990; Stephenson and Stempen, 1994). The organisms used in the study (with increasing classification distance from *P. polycephalum*) were *P. didermoides* (same genus, different species), *D. nigripes* and *Didymium iridis* (same order, different family), *L. epidendrum* (different order, same subclass), and

S. *flavogenita* (different subclass). These seven organisms represent five families (Enterdiaceae, Echinosteliaceae, Physaraceae, Didymiaceae, and Stemonita-ceae), four orders (Liceales, Echinosteliales, Physarales, and Stemonitales), and two subclasses (Myxogastromycetidae and Stemonitomycetidae) of the class myxomycetes. PCR was used to amplify a specific region of the mitochondrial SSU rRNA gene and the corresponding cDNA of the mitochondrial SSU rRNA. The sequences of these amplification products were compared to identify editing sites. The primers used to amplify these core regions were chosen so as to specifically amplify the mitochondrial core SSU rRNA in preference to the nuclear SSU rRNAs. A region corresponding to coordinates 676–1127 of the mitochondrial small SSU rRNA of *P. polycephalum* (Mahendran *et al.*, 1994) was amplified. This region contains 10 editing sites in *P. polycephalum*, 8 single cytidine insertions, and 2 dinucleotide insertions (AA at position 729 and CU at position 1040). Total DNA and total RNA were isolated from these organisms and used to produce mtDNA and cDNA amplification products, respectively. Mitochondrial DNA amplification products varied from 436 to 520 nucleotides depending on the size of an internal variable region in the amplified core region.

3.2. Covariation analysis of the core region of the mitochondrial SSU rRNA

All of the sequences from the mitochondrial SSU rRNA core region were consistent with the secondary structure of the rRNA deduced for *P. polycephalum* (Mahendran *et al.*, 1994). Numerous sites of nucleotide covariation were observed that served to maintain the secondary structure (Fig. 9.1). Covariation analysis could be preformed on nine coaxial secondary structure elements (coaxial elements 19, 20, 21, 22, 23, 24, 25, 26a, and 27; Noller, 2005) completely included within the core SSU rRNA sequence. A coaxial element analogous to element number 26 of the 16S rRNA of *E. coli* (Noller, 2005) could not be analyzed because the region variable in length and sequence in myxomycetes is contained within this element. Covariation analysis confirmed that the secondary structure within the SSU rRNA in myxomycete mitochondria is similar to that identified for the 16S rRNA of *E. coli*. In addition to the Watson–Crick and U–G wobble base pairs that make up the secondary structure, several non-Watson–Crick base pairing interactions are preserved in myxomycete SSU rRNA. A–C base pairs are indicated at several positions and can invert or exchange with U–G, G–C, A–U, or pyrimidine–pyrimidine base pairing interactions. Twenty-three editing sites are contained within the regions available for covariation analysis. All of the nucleotides inserted by RNA editing, with the possible exception of sites U and P, are involved in base pairing interactions as determined by covariation analysis of the secondary structure. This indicates that these nucleotides are important to the

secondary structure of the SSU rRNA and are presumably important
to SSU rRNA function.

3.3. Identification of RNA editing sites by alignment and comparison of cDNA and mtDNA sequences of the core region of the mitochondrial SSU rRNA

Comparison of cDNAs and mtDNAs revealed 8–10 editing sites of the type
characterized in *P. polycephalum* with the insertion of 9–12 nucleotides in
the RNA of each organism. Alignment of the edited and unedited
sequences from the seven organisms revealed 29 editing site locations
(each indicated by a separate letter in Fig. 9.2) only 8–10 of which were

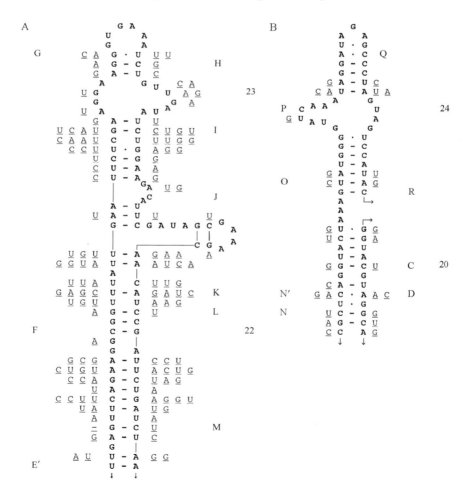

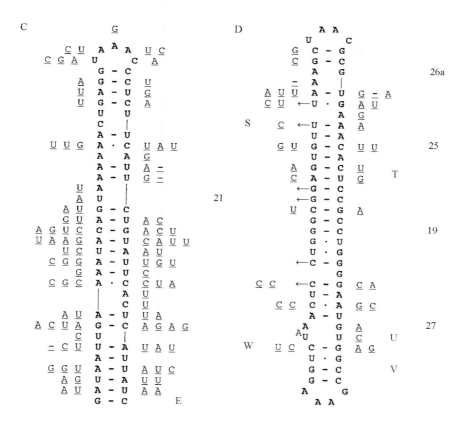

Figure 9.1 Secondary structures within the core mitochondrial SSU rDNA used for covariation analysis. The four panels show conserved coaxial secondary structure elements within the mitochondrial SSU rRNA core region. The numbering of the coaxial elements reflects the numbering system used by Noller (2005) to indicate the analogous elements in the *E. coli* 16S rRNA; 23 and 22 for (A), 24 and 20 for (B), 21 for (C), and 26a, 25, 19, and 27 for (D). RNA editing sites are indicated by the bold letters used to label the editing sites in Fig. 9.2. In each panel the sequence of the SSU rDNA from *Physarum polycephalum* is written in bold letters that are used to define the secondary structure. Covariation in the sequences from the other organisms are indicated on both sides of the base pair interaction as underlined nucleotide sequences. Paired covariant nucleotides at a base pairing site are read outward from the secondary structure.

used in any one organism. The sequences of *D. iridis* are 96.5% identical to *D. nigripes* and are not shown in Fig. 9.2. The location of the editing sites and the identity of the inserted nucleotides are identical in these two organisms. Fifteen of the 29 sites are present in more than one organism, but no site is used in all seven organisms. Fourteen sites are unique to a single organism. As in *P. polycephalum,* most of the RNA editing sites

A Pp mtDNA   1   CAGCAGCCGCGGTAAAACGGGGGGGGT TAGTGTTATTCGTGATGACTGGGCGT   AGGG
   Pp   cDNA   1   CAGCAGCCGCGGTAAAACGGGGGGGGT TAGTGTTATTCGTGATGACTGGGCGTAAAGGG
   Dn mtDNA   1   CAGCAGCCGCGGTAAAACGGGGGGGGC TAGTGTTATTCGTGATGACTGGGCGT   AGGG
   Dn   cDNA   1   CAGCAGCCGCGGTAAAACGGGGGGGGC TAGTGTTATTCGTGATGACTGGGCGTAAAGGG
   Pd mtDNA   1   CAGCAGCCGCGGTAAAACGGGGGGGG   TAGTGTTATTCGTGATGACTGGGCGT   AGGG
   Pd   cDNA   1   CAGCAGCCGCGGTAAAACGGGGGGGGC TAGTGTTATTCGTGGTGACTGGGCGTAAAGGG
   Sf mtDNA   1   CAGCAGCCGCGGTAAAACGGGGGGGGCTTAGTGTTATTCGTGATGACTGGGCGT   AGGG
   Sf   cDNA   1   CAGCAGCCGCGGTAAAACGGGGGGGGC TAGTGTTATTCGTGATGACTGGGCGTAAAGGG
   Le mtDNA   1   CAGCAGCCGCGGTAAAACGGGGGGGAGC TAGTGTTATTCGTGATGACTGGGTGTATAGGA
   Le   cDNA   1   CAGCAGCCGCGGTAAAACGGGGGGAGC TAGTGTTATTCGTGATGACTGGGTGTATAGGA
   Em mtDNA   1   CAGCAGCCGCGGTAAAACGGGGGGTAC TGGTGTTATTCGTGATGACTGGGCGTAAAGGG
   Em   cDNA   1   CAGCAGCCGCGGTAAAACGGGGGGTAC TGGTGTTATTCGTGATGACTGGGCGTAAAGGG
                             AA'   B

   Pp mtDNA 58   TACGTAGGCAG ATAATTGAAAATACAGTAAAAAACTGAGGTAAACCCTCTTCATTCTGT
   Pp   cDNA 60   TACGTAGGCAG ATAATTGAAAATACAGTAAAAAACTGAGGTAAACCCTCTTCATTCTGT
   Dn mtDNA 58   TACGTAGGCGG ATAATCAAAAATGTAGTAAAAGACTGAGGATAACCCTCTTAAGCTAC
   Dn   cDNA 60   TACGTAGGCGG ATAATCAAAAATGTAGTAAAAGACTGAGGATAACCCTCTTTAAGCTAC
   Pd mtDNA 57   TACGTAGGTGG AGTATTATCAATACTGTAAAAGACTGAGGTTATCCCTCTTTAATCAGA
   Pd   cDNA 60   TACGTAGGTGG AGTATTATCAATACTGTAAAAGACTGAGGTTATCCCTCTTTAATCAGA
   Sf mtDNA 59   TATGTAGGCGG TAAATTTAAAGTAGAGTAAAAGACTGAAGATGACCTTCTTTAAGCTCT
   Sf   cDNA 60   TATGTAGGCGG TAAATTTAAAGTAGAGTAAAAGACTGAAGATGACCTTCTTTAAGCTCT
   Le mtDNA 60   TA GTAGGTGGTAAGACTAAGGGCTAAAAAAAATACTTTGGGAACACCGATTTA   CTTT
   Le   cDNA 60   TACGTAGGTGGTAAGACTAAGGGCTAAAAAAAATACTTTGGGAACACCGATTTA   CTTT
   Em mtDNA 60   TATGT GGTGGTAGGA TCTCACTGGGTATAAATACTGTAGCCAAACTACTTAGTTCCCC
   Em   cDNA 60   TATGTGGTGGTAGGA TCTCACTGGGTATAAATACTGTAGCCAAACTACTTAGTTCCCC
                   C   D   ■

   Pp mtDNA 117   ATTCACTCAT TAT TTGAGTTCAGAAGGCGGTTTATTGAATTCTCGAAGGAAGGGTGAA
   Pp   cDNA 119   ATTCACTCAT TATCTTGAGTTCAGAAGGCGGTTTATTGAATTCTCGAAGGAAGGGTGAA
   Dn mtDNA 117   ATTCACTAAT TAT TTGAGTTTTGAGGGCGATCAAATGAATCCTTTAAGTAAGGGTTAA
   Dn   cDNA 119   ATTCACTAAT TATCTTGAGTTTTGAGGGCGATCAAATGAATCCTTTAAGTAAGGGTTAA
   Pd mtDNA 116   ATTCACTGTT ATTCTTGGGTATTATAGG GGTTTATTGAATTCCAAAAGGAAGGGTGAA
   Pd   cDNA 119   ATTCACTGTT ATTCTTGGGTATTATAGGCGGTTTATTGAATTCCAAAAGGAAGGGTGAA
   Sf mtDNA 118   ATTCACTGAT TTAC TGAGTTCTAGAGGCGAGGT TTGAATTCCTTAAGTAAGAGTAAA
   Sf   cDNA 119   ATTCACTGAT TTACTTGAGTTCTAGAGGCGAGGT TTGAATTCCTTAAGTAAGAGTAAA
   Le mtDNA 117   AGCTACAGATCTTTCTAGA ATCTCTCGA GGTAT ATCTCCACATGTAGACGTAAA
   Le   cDNA 118   AGCTACAGATCTTTCTAGA ATCTCTCGACGGTAT GGGTACTCCACATGTAGACGTAAA
   Em mtDNA 117   TTTATTTATTCCTAC TGAGTTTTCCGGGCGATGT GTGAATTTCCTGTGTAGA GTGAA
   Em   cDNA 119   TTTATTTATTCCTACTTGAGTTTTCCGGGCGATGT GTGAATTTCCTGTGTAGACGTGAA
                 ■   EE'   ■   F   ■   G

   Pp mtDNA 175   AT TGTTGATAT TGGAAGACATTCGATAGCGAAAGCAACATCCGATTCTGATCTAACGC
   Pp   cDNA 178   ATCTGTTGATATCTGGAAGACATTCGATAGCGAAAGCAACATCCGATTCTGATCTAACGC
   Dn mtDNA 175   AT TGTTGATAT TAGGAGACGTTCGAAGGCGAAAGCAATGATCGACTCAAATCTGACGT
   Dn   cDNA 178   ATCTGTTGATATCTAGGAGACGTTCGAAGGCGAAAGCAATGATCGACTCAAATCTGACGT
   Pd mtDNA 174   ATCTGTTGATATTTGGAAGACATTCAAAAGTGAAAACATTATCCGATATAGTT CGACAC
   Pd   cDNA 178   ATCTGTTGATATTTGGAAGACATTCAAAAGTGAAAACATTATCCGATATAGTTCCGACAC
   Sf mtDNA 175   ATCTGTTGATAT TGGAAGACATTCAAAGGCGAAAGCGATAATCGATCTAGATCTGACGC
   Sf   cDNA 177   ATCTGTTGATATCTGGAAGACATTCAAAGGCGAAAGCGATAATCGATCTAGATCTGACGC
   Le mtDNA 174   ATGCGCAGATATGGCGAGGAC TTCGAAGGCGAAAGCACGTG CGACTAATGATTGACA
   Le   cDNA 176   ATGCGCAGATATGGCGAGGACCTTCGAAGGCGAAAGCACGTGCCGACTAATGATTGACAC
   Em mtDNA 175   ATGCGAGAATATTGGGAAGTCATTCAGCGGCGAAAGCAAT TTCGATGGAAAT TGACAC
   Em   cDNA 178   ATGCGAGAATATTGGGAAGTCATTCAGCGGCGAAAGCAATCTTCGATGGAAATCTGACAC
               H   I   J   K L   M   N

```
B Pp mtDNA 233  T AGGTACTAAAGTATGGGGAT AAATAGGATTAGAGA CCTAGTAGTCCATACCTTAAA
  Pp cDNA  238  TCAGGTACTAAAGTATGGGGATCAAATAGGATTAGAGACCCTAGTAGTCCATACCTTAAA
  Dn mtDNA 233  T AGGTACTAAAGCATGGGTAT GAAAAGGATTAGAGA CCTTGTAGTCCATGCTGTAAA
  Dn cDNA  238  TCAGGTACTAAAGCATGGGTATCGAAAAGGATTAGAGACCCCTTGTAGTCCATGCTGTAAA
  Pd mtDNA 233  TAAGGTACGAAAGTATGGGGAT AAATGGGATTAGAGACCCCAGTAGTCCATA CTTAAA
  Pd cDNA  238  TAAGGTACGAAAGTATGGGGATCAAATGGGATTAGAGACCCCAGTAGTCCATACCTTAAA
  Sf mtDNA 234  TAAGGTACTAAAGCATGGGTAT GAAAAGGATTAGAGA CCTTGTAGTCCATGCCTTAAA
  Sf cDNA  237  TAAGGTACTAAAGCATGGGTATCGAAAAGCATTAGAGACCCCTTGTAGTCCATGCCTTAAA
  Le mtDNA 231  TGAGGTATGAAGGTATGGGTATCGATCGGGATTAGAGACCCCAGTAGTCCATA AGTAAA
  Le cDNA  236  TGAGGTATGAAGGTATGGGTATCGATCGGGATTAGAGACCCCAGTAGTCCATACAGTAAA
  Em mtDNA 233  TACGGTACGAAAG GTGGGGAGCAAAAGGGATTAGAGACCCCTGTAGTCCATG CTTAAA
  Em cDNA  238  TACGGTACGAAAGCGTGGGGAGCAAAAGGGATTAGAGACCCCTGTAGTCCATGCCTTAAA
                N'        O        P              Q           R
```

```
  Pp mtDNA 290  CAATGAG  T G    TTCAACG       TCTTT    AA TA   TAGTCAA
  Pp cDNA  298  CAATGAG  T G    TTCAACG       TCTTT    AA TA   TAGTCAA
  Dn mtDNA 290  CCATGAG  TAG    TTCAACG       TCCTTTTGA GA TA   TGGCTTAT
  Dn cDNA  298  CCATGAG  TAG    TTCAACG       TCCTTTTGA GA TA   TGGCTTAT
  Pd mtDNA 291  CAATGAG  T G    T CAAC        ATCTT       TT   GACACTAGT
  Pd cDNA  298  CAATGAG  T G    TCCAAC        ATCTT       TT   GACACTAGT
  Sf mtDNA 292  CAATGAG  T G    TTCAACGTCTTAAAACATT        CT TA   GATATCATT
  Sf cDNA  297  CAATGAG  T G    TTCAACGTCTTAAAACATT        CT TA   GATATCATT
  Le mtDNA 290  CGCTGCA  TAT    T ACTA        TTGGAT    GA TA   TA AAAACT
  Le cDNA  296  CGCTGCA  TAT    TCACTA        TTGGAT    GA TA   TA AAAACT
  Em mtDNA 291  CGATGAGTATTGAATTTTTTTATTTTCTCCTCTTTCCGCGGACTAGGGTCTAAAGTTATG
  Em cDNA  298  CGATGAGTATTGAATTTTTTTATTTTCTCCTCTTTCCGCGGACTAGGGTCTAAAGTTATG
                ----------S--------variable region-------------------
```

```
  Pp mtDNA 322              TATA  TT   T                 AG         GG
  Pp cDNA  330              TATA  TT   T                 AG         GG
  Dn mtDNA 329        AT    CAAC  AT   AACATAAGACACAAAAG            GG
  Dn cDNA  337        AT    CAAC  AT   AACATAAGACACAAAAG            GG
  Pd mtDNA 321        TG    T CA  TT                     G          GG
  Pd cDNA  329        TG    T CA  TT                     G          GG
  Sf mtDNA 333        TT    TATA  GTGAATCAAGACGGAGTTTTAG           GG
  Sf cDNA  338        TT    TATA  GTGAATCAAGACGGAGTTTTAG           GG
  Le mtDNA 323        TT    GCTT  TGCGAAGGAAAAAAGAAGTTCA          GT
  Le cDNA  330        TT    GCTT  TGCGAAGGAAAAAAGAAGTTCA          GT
  Em mtDNA 351  TGAGCCCCCTTGGGGTTCCTATACCTGCCCCCGAAACAAAGGGAGAGGAAGAGAAATAGG
  Em cDNA  358  TGAGCCCCCTTGGGGTTCCTATACCTGCCCCCGAAACAAAGGGAGAGGAAGAGAAATAGG
                ----------------------variable region----------------------
```

```
  Pp mtDNA 333  G  GTTTAAAGCTAACGCGTGAAACA  CCGCCTGGGGAATGTGGCCGCAAGGT TAAAC
  Pp cDNA  341  G  GTTTAAAGCTAACGCGTGAAACACTCCGCCTGGGGAATGTGGCCGCAAGGTCTAAAC
  Dn mtDNA 356  G  GTTTGAAGCTAACGCGTGAAACA  CCGCCTGGGGAATGTGGCCGCAAGGT TAAAC
  Dn cDNA  364  G  GTTTGAAGCTAACGCGTGAAACACTCCGCCTGGGGAATGTGGCCGCAAGGTCTAAAC
  Pd mtDNA 331  G  GTTTAAAGCTAACGCGTAGAACA  CCGCCTGGGGAATGTGGCCGCAAGGT TAAAC
  Pd cDNA  339  G  GTTTAAAGCTAACGCGTAGAACACTCCGCCTGGGGAATGTGGCCGCAAGGTCTAAAC
  Sf mtDNA 363  G  GTTTAAACGTAACGCGTGAAACA  CCGCCTGGGGAATGTGGCCGCAAGGT TAAAC
  Sf cDNA  368  G  GTTTAAACGTAACGCGTGAAACACTCCGCCTGGGGAATGTAGCCGCAAGGTCTAAAC
  Le mtDNA 353  G  GTTC  AGCTAACGCGTTAAATATGCCACCTGGGCAGTA GGCCGCAAGGTTAAAAC
  Le cDNA  360  G  GTTC  AGCTAACGCGTTAAATATGCCACCTGGGCAGTACGGCCGCAAGGTTAAAAC
  Em mtDNA 411  GACGATTAAAGCTAACGCGAGAAATA  CCGCCTGGGAACTATGG CGCAAGGTTGAAAC
  Em cDNA  418  GACGATTAAAGCTAACGCGAGAAATACTCCGCCTGGGAACTATGGCCGCAAGGTTGAAAC
                ---   —                T            U  V        W
```

Figure 9.2 (*continued*)

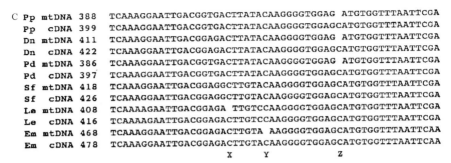

C Pp mtDNA 388 TCAAAGGAATTGACGGTGACTTATACAAGGGGTGGAG ATGTGGTTTAATTCGA
  Pp cDNA 399 TCAAAGGAATTGACGGTGACTTATACAAGGGGTGGAGCATGTGGTTTAATTCGA
  Dn mtDNA 411 TCAAAGGAATTGACGGAGACTTATACAAGGGGTGGAG ATGTGGTTTAATTCGA
  Dn cDNA 422 TCAAAGGAATTGACGGAGACTTATACAAGGGGTGGAGCATGTGGTTTAATTCGA
  Pd mtDNA 386 TCAAAGGAATTGACGGTGACTTATACAAGGGGTGGAG ATGTGGTTTAATTCGA
  Pd cDNA 397 TCAAAGGAATTGACGGTGACTTATACAAGGGGTGGAGCATGTGGTTTAATTCGA
  Sf mtDNA 418 TCAAAGGAATTGACGGAGGCTTGTACAAGGGGTGGAGCATGTGGTTTAATTCGA
  Sf cDNA 426 TCAAAGGAATTGACGGAGGCTTGTACAAGGGGTGGAGCATGTGGTTTAATTCGA
  Le mtDNA 408 TCAAAAGAATTGACGGAGA TTGTCCAAGGGGTGGAGCATGTGGTTTAATTCGA
  Le cDNA 416 TCAAAAGAATTGACGGAGACTTGTCCAAGGGGTGGAGCATGTGGTTTAATTCGA
  Em mtDNA 468 TCAAAGGAATTGACGGAGACTTGTA AAGGGGTGGAGCATGTGGTTTAATTCAA
  Em cDNA 478 TCAAAGGAATTGACGGAGACTTGTACAAGGGGTGGAGCATGTGGTTTAATTCAA
                                            X   Y   Z

Figure 9.2 Alignment of mtDNA and cDNA sequences from the mitochondrial SSU rRNA from five myxomycetes. The sequences corresponding to a core region of the mitochondrial rRNA of five different myxomycetes are aligned, *Physarum polycephalum* (Pp, lines 1 and 2), *Didymium nigripes* (Dn, lines 3 and 4), *Physarum didermoides* (Pd, lines 5 and 6), *Stemonitis flavogenita* (Sf, lines 7 and 8), *Lycogala epidendrum* (Le, lines 9 and 10), and *Echinostelium minutum* (Em, lines 11 and 12). Even numbered lines are mtDNA sequences; odd numbered lines are cDNA sequences. Editing sites are labeled with letters below the alignments. Solid squares indicate gaps in the alignment that are not compensated by editing.

are single cytidine insertions. However, occasional sites of uridine insertion (site E′) and dinucleotide insertions (sites B and T) were observed. In addition to these previously characterized types of editing site, a novel type of site (site A′) was observed in *S. flavogenita*. At this position in the mtDNA, an extra T residue is present relative to the other mtDNAs. This T residue is absent in the cDNA sequence from *Stemonitis* and is interpreted as a uridine deletion in the mitochondrial SSU rRNA of *Stemonitis*. This is the first documentation of a uridine deletion editing site in the RNA of the Myxomycota, although Gott *et al.* (2005) have recently described a triple A deletion editing site in the *nad2* gene of *P. polycephalum*. Although editing sites consist of nucleotides present in RNAs relative to their mtDNAs, mtDNAs from other organisms that lack RNA editing at that site already have a nucleotide at that position. Because of this, there are generally no gaps in the alignment of edited RNA sequences with mtDNA sequences from organisms that lack editing at a particular site.

Editing sites present in the different myxomycetes are summarized in Fig. 9.3. *P. polycephalum* and *D. nigripes* have the same set of editing sites at the same locations. This identity in editing site location is in contrast to the variation in the location of editing sites in tRNAs between *P. polycephalum* and *D. nigripes* (Antes *et al.*, 1998). *Stemonitis*, *P. didermoides*, *Echinostelium*, and *Lycogala* have editing patterns that differ from those observed in *P. polycephalum* and the *Didymium* species. Some sites are present in more than one organism, e.g., *Stemonitis* has six sites in common with *P. polycephalum* and *D. nigripes*, and *P. didermoides* has five sites in common. However, *Lycogala* has no sites in common with *P. polycephalum* and *D. nigripes*. *Stemonitis*, *P. didermoides*,

	Dn	Sf	Pp	Pd	Em	Le	
A	–	–	–	+	–	–	C
A′	–	+	–	–	–	–	(U)
B	+	+	+	+	–	–	AA
C	–	–	–	–	–	+	C
D	–	–	–	–	+	–	C
E	+	–	+	–	–	–	C
E′	–	+	–	–	+	–	U
F	–	–	–	+	–	+	C
G	–	–	–	–	+	–	C
H	+	–	+	–	–	–	C
I	+	+	+	–	–	–	C
J	–	–	–	–	–	+	C
K	–	–	–	–	+	–	C
L	–	–	–	–	–	+	C
M	–	–	–	+	+	–	C
N	–	–	–	–	–	+	C
N′	+	–	+	–	–	–	C
O	–	–	–	–	+	–	C
P	+	+	+	+	–	–	C
Q	+	+	+	–	–	–	C
R	–	–	–	+	+	+	C
S	–	–	–	+	–	+	C
T	+	+	+	+	+	–	CU
U	–	–	–	–	–	+	C
V	–	–	–	–	+	–	C
W	+	+	+	+	–	–	C
X	–	–	–	–	–	+	C
Y	–	–	–	–	+	–	C
Z	+	–	+	+	–	–	C

Figure 9.3 Distribution of editing sites among six myxomycetes. The letters on the left column indicate the editing sites defined in Fig. 9.2. Myxomycetes [*Didymium nigripes* (Dn), *Stemonitis flavogenita* (Sf), *Physarum polycephalum* (Pp), *Physarum didermoides* (Pd), *Echinostelium minutum* (Em), and *Lycogala epidendrum* (Le)] in which the nucleotides at these positions in the RNA are created by editing are designated with a +; those in which the nucleotide is specified by the mtDNA are designated with a –. Nucleotides indicated in the right-hand column are the nucleotides inserted by editing; the U in parentheses is a nucleotide deleted by editing.

Echinostelium, and *Lycogala* have unique editing sites that are not present in the other organisms.

Despite differences in editing site location, the overall distribution of editing sites is retained (Fig. 9.4). In general, editing sites are distributed throughout the entire core sequence in each case. However, editing sites are enriched in conserved sequences relative to variable sequence regions resulting in gaps between sites D and E and between sites S and T (Fig. 9.4). The average spacing between editing sites ranges from 41 to 46 nucleotides for the organisms studied. However, eliminating the large gaps associated with these variable regions reduced the average distances between editing sites to 30–38 nucleotides.

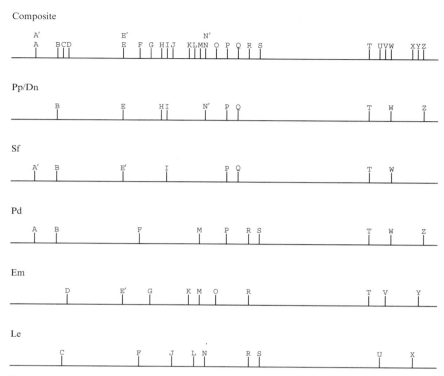

Figure 9.4   Distribution of editing site locations in the core region of the mitochondrial SSU rRNA. The distribution of editing sites in the core conserved region of SSU rRNA from six different myxomycetes is shown to scale. The editing sites are named as defined in Fig. 9.2. The letters above each line at the left indicate the organism containing the rRNA using the abbreviations defined in Figs. 9.2 and 9.3. The top line is a composite showing the relative position of all 29 editing sites.

Although several of the editing sites (A and A′; C and D; E and E′; G and H; M, N, and N′; U, V, and W; and X and Y) are located closer together than the nine nucleotide minimum distance between editing sites observed in *P. polycephalum* mtRNAs (Miller *et al.*, 1993a,b,c), these sites are not used in the same organism, so that the distance between editing sites is nine nucleotides or more in each case. This minimum distance between editing sites has been observed in all myxomycetes studied and is probably a constraint on editing site location in the Myxomycota.

3.4. The two nucleotides preceding editing sites show a sequence bias in all the myxomycetes analyzed

Miller *et al.* (1993b) identified a purine–pyrimidine bias in the two nucleotides prior to an editing site in 165 sites in *P. polycephalum*. Gott *et al.* (2005) have also detected this bias as a purine–U bias at positions −2, −1 in 277 sites in *P. polycephalum*. They also detected an underrepresentation of AA, UU, and GG at positions −2 and −1. Analysis of the 45 C-insertion editing sites detected in the myxomycetes of this study gives a similar but not identical result. GU and AU precede editing in 48% of the unambiguous examples, and so the purine-U bias is maintained in other myxomycetes. The observation that the number of times that AU precedes editing sites is about double the number of times that GU precedes editing sites is also maintained. No examples of AA, UC, or CG preceding editing sites were observed. However, GG (sites A, F, and V) and UU (sites M, N′, and S) are each observed preceding three sites in myxomycetes other than *P. polycephalum*.

3.5. Phylogenetic analysis of myxomycete mitochondrial core SSU rDNA sequences

Variation in the location of the editing sites in these myxomycetes implies that editing sites have been created or eliminated or both since the divergence of these organisms from a common ancestor. If the gain or loss of editing sites is rare relative to the general rate of sequence divergence in the mtDNA, then this pattern of editing site variation would be expected to reflect the pattern of sequence divergence in the organisms. Since the pattern of sequence divergence had not been determined for these organisms, the sequence variation of the core SSU rRNA was used to generate phylogenetic trees using both distance matrix and maximum parsimony methods. To determine the pattern of divergence of these organisms (i.e., their phylogenetic relationships), the edited core rRNA sequences were aligned and 335 nucleotides of unambiguously aligned sequence were used to infer the phylogenetic relationship of these organisms. Analogous sequences from *E. coli* (Brosius *et al.*, 1978) and *Oenothera* mitochondria (Brennicke *et al.*, 1985) were included as outgroups. A distance matrix of percentage dissimilarity of the sequences of these organisms was generated by comparing all possible pairwise alignments of the core SSU rRNA sequences and adjusting using the ML option of the DNADIST algorithm of the PHYLIP package of Felsenstein (1995). The phylogenetic tree shown in Fig. 9.5 is derived from this matrix using the FITCH algorithm of PHYLIP. The aligned sequences were also used to produce a single most parsimonious tree relating the sequences of all nine organisms using the DNAPARS algorithm of the PHYLIP package. This tree has the same topology as the tree shown in Fig. 9.5. Branch lengths shown in Fig. 9.4 are proportional to the composite percentage dissimilarity from the distance matrix that is indicated on each branch. Also indicated on each branch is the number of variations between sequences at each node of the most parsimonious tree. Similar sequence analysis of core regions of the nuclear SSU rRNAs also gives similar tree topologies (data not shown).

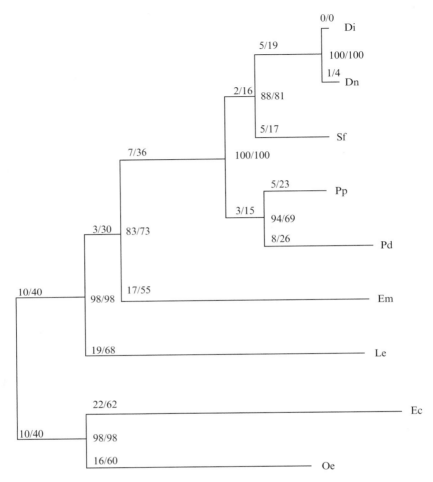

Figure 9.5 Phylogenetic tree of mitochondrial SSU rRNA sequences of six myxomycetes. *Oenothera* (Oe) mitochondrial SSU rRNA and *E. coli* (Ec) SSU rRNA sequences were used as outgroups to "root" the tree. The FITCH algorithm of the PHYLIP sequence analysis package was used to derive tree topology. The distance matrix of the percentage dissimilarity of pairwise comparisons of sequence alignments was produced using the DNADIST algorithm with the maximum likelihood (ML) option. The branch lengths shown are proportional to the composite percentage dissimilarity between sequences. The topology of the one most parsimonious tree derived from the sequence alignments using the DNAPARS algorithm of PHYLIP is the same as that determined using the distance matrix approach. Branch lengths are shown on each branch as *a/b* where *a* is the composite percentage dissimilarity derived from distance matrix analysis and *b* is the number of changes in the sequences between nodes in the most parsimonious tree. Bootstrap resampling values (100 replicates) are shown at branching sites as *x/y* where *x* is the bootstrap values for distance matrix analysis and *y* is the bootstrap values for maximum parsimony analysis. Organisms are identified at the right using the abbreviations defined in Figs. 9.2 and 9.3.

The tree produced from these analyses does not have a topology consistent with trees predicted by the classification system. *Stemonitis* would be expected to be more deeply divergent than the divergence of *Physarum* and *Didymium* species, which are members of the same order, since *Stemonitis* is a member of a separate subclass. On the other hand, *Lycogala* and *Echinostelium*, members of the same subclass as *Physarum* and *Didymium*, are more deeply divergent than any of the other myxomycetes relative to the outgroups, *E. coli* and *Oenothera*. Also, *P. didermoides* and *P. polycephalum* are much more deeply divergent than expected based on their classification in the same genus. A similar tree topology has been predicted by Fiore-Donno *et al.* (2005) using nuclear DNA sequences for elongation factor 1-A and the SSU rRNA, although the branching order of *Echinostelium* and *Lycogala* is reversed in the rooted tree based on elongation factor 1-A sequences.

To determine the pattern of variation in editing site location in the mitochondrial SSU rRNA of these organisms, the editing sites present in contemporary organisms have been superimposed on the topology of the phylogenetic tree determined using mtDNA sequence variation (Fig. 9.6). The most parsimonious assignment of editing site variation consistent with the sites at the external nodes and the topology of the phylogenetic tree is shown in Fig. 9.6. A minimum of 45 events (defined as either the creation or elimination of editing sites) is required to produce the pattern of editing sites at the external nodes (contemporary organisms). Several different patterns of variation are equally parsimonious, but all patterns include both editing site creation and editing site elimination. At least 19 different editing sites (A, A', B, D, E, E', G, H, I, K, M, N', O, P, Q, V, W, Y, Z) are unambiguously predicted to have been independently created since the divergence from the common ancestor. Only one site (R) is unambiguously predicted to be present in the common ancestor and eliminated in some organisms.

3.6. Implications of RNA editing site distribution and variation

Comparison of cDNAs and mtDNA sequences from organisms representing the major classifications of the Myxomycota reveals that RNA editing of the type characterized in *P. polycephalum* is distributed throughout the Myxomycota and is present in at least four of the five orders within the subclass Myxogastromycetidae. Although the type of editing is conserved, the location of editing sites varies dramatically among the different organisms. This implies that the mechanism of nucleotide insertion or deletion at RNA editing sites is separate from the process in which editing sites are specified. *P. polycephalum*, *D. nigripes*, and *Stemonitis* have a completely separate set of RNA editing sites from *Lycogala*. Only *P. didermoides* and *E. minutum* have some editing sites in common with all of the other organisms.

The single most parsimonious tree relating the editing site differences in these five organisms predicts 24 events where events are defined as the loss or

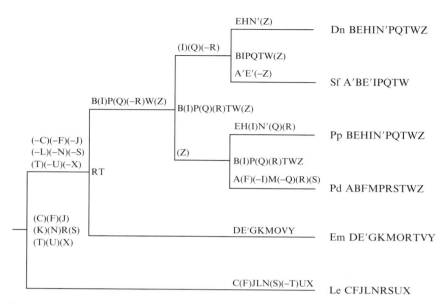

Figure 9.6 Editing site variations superimposed on phylogenetic tree topology. The phylogenetic tree topology is derived from Fig. 9.5. To the right of the external nodes are the abbreviations for the contemporary organisms and the letters indicating the editing sites present in that organism. At the internal nodes are the letters for the editing sites inferred to be present in ancestral organisms using the most parsimonious assignment of editing site changes. Parentheses indicate sites that may be present or may be created or eliminated depending on which of several equally parsimonious scenarios of editing site variation is used. Minus signs (–) indicate the potential elimination of an editing site.

gain of editing sites (data not shown). Depending on where this tree is rooted, these events could be either the loss or gain of editing sites. However, this most parsimonious tree relating editing site differences is not congruent with the topology of the most parsimonious tree based on overall sequence divergence within these organisms (Fig. 9.5). Sequence divergence using either nuclear or mitochondrial SSU rRNA core sequences and either distance matrix or maximum parsimony algorithms gives trees with the same topology and similar branch lengths. The topology of these trees based on sequence variation is different from the most parsimonious treebased on editing site variation. This topology implies that identical editing sites must have been independently created in different lineages. This could mean that certain mtDNA locations have a propensity to become editing sites, or that there is a constraint on editing site location, or that a determinant that defines the location of an editing site can be transferred horizontally.

Analysis of editing site distribution within the SSU rRNA core sequence of these five organisms indicates a general conservation in the number of editing sites and in their uniform distribution, even though the editing sites are at different locations. Although some editing sites are closer than the nine nucleotide minimum distance between editing sites observed in *P. polycephalum* RNAs, these sites are not used in the same organism and so are still consistent with this limitation. The fact that this distribution and density of editing sites are conserved among these anciently diverged organisms may indicate a constraint that maintains these features. For these features to have been maintained in ancestral as well as contemporary organisms, editing sites would have to be both created and lost at a rate sufficient to produce a completely nonoverlapping set of sites since the divergence of *Lycogala* and *P. polycephalum*.

How insertional editing sites could be gained or lost in the Myxomycota is not known. The fact that most inserted nucleotides in RNAs correspond to nucleotides already present in the mtDNA of organisms that lack editing sites at that position implies a two-step process in which a nucleotide is removed from the mtDNA and then is replaced at the RNA level to maintain the genetic information. Conversely, to eliminate an editing site, the mtDNA sequence would have to be restored, and the activity acting to insert a nucleotide at that site in the RNA must be eliminated. In both editing site creation and editing site elimination, alteration of mtDNA and RNA must be coordinated. We have developed a model for the evolution of RNA editing in myxomycetes based on the general model for the evolution of RNA editing systems proposed by Gray *et al.* (Covello and Gray, 1993; Gray, 2001; Price and Gray, 1998), which would explain this coordination and the observed editing site distribution. In this model, editing sites are created by random deletion mutation of nucleotides in mtDNA. The large number of observed editing site locations indicates that editing sites can be essentially at any location within an mtDNA sequence. On the other hand, the fixation of an editing site requires an activity that includes detection of the deletion as well as correction of the deletion by insertion of a nontemplated nucleotide in the RNA. This fixation seems to be constrained by at least three criteria: the preference for C nucleotide insertions, the weak bias for a purine–U to precede the editing site, and an absolute constraint preventing editing sites from being closer to each other than nine nucleotides. These constraints on the fixation of random deletions would produce the distribution of editing sites observed in the various myxomycetes. As editing sites accumulate through additional mutation, the target sites for permissible fixation under these constraints decrease until editing sites reach a maximum density and the rate of fixation approaches zero. For example, two editing sites separated by 19 nucleotides would present only a one-nucleotide target for establishment of an editing site between them. Since the probability of a deletion occurring at exactly that site would be low, the rate of fixation would also be low.

Another constraint on editing site fixation seems to be the functional importance of the nucleotide deleted. Regions that diverge rapidly, presumably because they are under little selection constraint, are depleted in editing sites, while conserved regions, presumably under strong selection constraints, have essentially the maximum density of editing sites.

When editing sites reach a maximum density, the fixation of new sites would cease, maintaining the existing distribution of editing sites. Because these distribution patterns have been observed to change, some mechanism must exist to eliminate RNA editing sites. Landweber (1992) and Simpson *et al.* (Maslov *et al.*, 1994; Simpson and Maslov, 1994) have proposed retrotransposition as a mechanism of eliminating editing sites in trypanosomes. Integration of cDNAs produced by reverse transcription of edited RNA would remove accumulated mutations, cleansing the region of editing sites. This would allow new patterns of editing site distribution to develop as new mutations accumulate during myxomycete divergence. Analysis of additional sequences from other organisms of the Myxomycota will be necessary to test this model.

REFERENCES

Antes, T. A., Mahendran, R., Spottswood, M. S., and Miller, D. L. (1998). Insertional editing in the mitochondria of *Physarum polycephalum* and *Didymium nigripes*. *Mol. Cell Biol.* **18,** 7521–7527.

Baldauf, S. L., and Doolittle, W. F. (1997). Origin and evolution of the slime molds (Mycetozoa). *Proc. Natl. Acad. Sci. USA* **94,** 12007–12012.

Baldauf, S. L., Roger, A. J., Wenk-Siefert, I., and Doolittle, W. F. (2000). A kingdom-level phylogeny of eukaryotes based on combined protein data. *Science* **290,** 972–976.

Barns, S. M., Delwiche, C. F., Palmer, J. D., and Pace, N. R. (1996). Perspectives on archael diversity, thermophily and monophyly from environmental rRNA sequences. *Proc. Natl. Acad. Sci. USA* **93,** 9188–9193.

Brennicke, A., Möller, A., and Blanz, P. A. (1985). The 18S and 5S ribosomal RNA genes in *Oenothera* mitochondria: Sequence rearrangements in the 18S and 5S rRNA genes of higher plants. *Mol. Gen. Genet.* **198,** 404–410.

Brosius, J., Palmer, M. L., Kennedy, P. J., and Noller, H. F. (1978). Complete nucleotide sequence of a 16S ribosomal RNA gene from *Escherichia coli. Proc. Natl. Acad. Sci. USA* **75,** 4801–4805.

Covello, P. S., and Gray, M. W. (1993). On the evolution of RNA editing. *Trends Genet.* **9,** 265–268.

De Rijk, P., Van de Peer, Y., Vanden Broeck, I., and De Wachter, R. (1995). Evolution according to large ribosomal subunit RNA. *J. Mol. Evol.* **41,** 366–375.

Felsenstein, J. (1995). PHYLIP: Phylogeny inference package, version 3.57c. University of Washington, Seattle, WA.

Fiore-Dunno, A., Berney, C., Pawlowski, J., and Baldauf, S. L. (2005). High-order phylogeny of plasmodial slime molds (Myxogastria) based on elongation factor 1-A and small subunit rRNA gene sequences. *J. Eukaryot. Microbiol.* **52,** 201–210.

Frederick, L. (1990). Phylum plasmodial slime molds, class Myxomycota. *In* "Handbook of Protoctista" (L. Margulies, J. O. Corliss, M. Melkonian, and D. J. Chapman, eds.), Chapter 27a,pp. 467–483. Jones & Bartlett, Boston, MA.

Gott, J. M. (2001). RNA editing in *Physarum polycephalum. In* "RNA Editing" (B. L. Bass, ed.), pp. 20–37. Oxford University Press, Oxford.

Gott, J. M., and Visomirski-Robic, L. M. (1998). RNA editing in *Physarum* mitochondria. *In* "Modification and Editing of RNA" (H. Grosjean and R. Benne, eds.), pp. 395–411. ASM Press, Washington, DC.

Gott, J. M., Parimi, N., and Bundschuh, R. (2005). Discovery of new genes and deletion editing in *Physarum* mitochondria enabled by a novel algorithm for finding edited mRNAs. *Nucleic Acids Res.* **33**, 5063–5072.

Gray, M. W. (2001). Speculations on the origin and evolution of editing. *In* "RNA Editing" (B. L. Bass, ed.), pp. 160–184. Oxford University Press, Oxford.

Henricks, L., DeBaere, R., Van de Peer, Y., Neefs, J., Goris, A., and De Wachter, R. (1991). The Evolutionary position of the rhodophyte *Porphyra umbilicalis* and the basidiomycete *Leucosporidium scottii* among other eukaryotes as deduced from complete sequences of small subunit ribosomal RNA. *J. Mol. Evol.* **32**, 167–177.

Hinchee, A. A., and Haskins, E. F. (1981). A light microscopical investigation of amoebal and plasmodial mitosis in the myxomycete, *Echinostelium minutum*. *Protoplasma* **108**, 361–367.

Hinkle, G., and Sogin, M. L. (1993). The evolution of the Vahlkampfiidae as deduced from 16S-like ribosomal RNA analysis. *J. Eukaryot. Microbiol.* **40**, 599–603.

Horton, T. L., and Landweber, L. F. (2000). Evolution of four types of RNA editing in myxomycetes. *RNA* **6**, 1339–1346.

Krishnan, S., Barnabas, S., and Barnabus, J. (1990). Interrelationships among major protistan groups based on a parsimony network of 5S rRNA sequences. *BioSystems* **24**, 135–144.

Landweber, L. F. (1992). The evolution of RNA editing in kinetoplastid protozoa. *BioSystems* **28**, 41–45.

Mahendran, R., Spottswood, M. S., and Miller, D. L. (1991). RNA editing by cytidine insertion in mitochondria of *Physarum polycephalum*. *Nature* **349**, 434–438.

Mahendran, R., Spottswood, M. S., Ghate, A., Ling, M. L., Jeng, K., and Miller, D. L. (1994). Editing of the mitochondrial small subunit rRNA in *Physarum polycephalum*. *EMBO J.* **13**, 232–240.

Margulis, L., and Schwartz, K. V. (1988). "Five Kingdoms, an Illustrated Guide to the Phyla of Life on Earth." W. H. Freeman & Co., New York.

Martin, G. W., Alexopoulos, C. J., and Farr, M. L. (1983). "The Genera of Myxomycetes." University of Iowa Press, Iowa City, IA.

Maslov, D. A., Avila, H. A., Lake, J. A., and Simpson, L. (1994). Evolution of RNA editing in kinetoplastid protozoa. *Nature* **368**, 345–348.

Miller, D. L., Mahendran, R., Spottswood, M. S., Ling, M., Wang, S., Yang, N., and Costandy, H. (1993a). RNA editing in mitochondria of *Physarum polycephalum*. *In* "RNA Editing: The Alteration of Protein Coding Sequences of RNA" (R. Benne, ed.), pp. 87–103. Ellis Horwood, New York.

Miller, D., Mahendran, R., Spottswood, M., Costandy, H., Wang, S., Ling, M., and Yang, N. (1993b). Insertional editing in mitochondria of *Physarum*. *Semin. Cell Biol.* **4**, 261–266.

Miller, D., Mahendran, R., Spottswood, M., Ling, M., Wang, S., Yang, N., and Costandy, H. (1993c). RNA editing in mitochondria of *Physarum polycephalum*. *In* "Plant Mitochondria" (A. Brennicke and U. Kuck, eds.), pp. 53–62. VHC, Weinheim, Germany.

Noller, H. F. (2005). RNA structure: Reading the ribosome. *Science* **309**, 1508–1514.

Price, D. H., and Gray, M. W. (1998). Editing of tRNA. *In* "Modification and Editing of RNA" (H. Grosjean and R. Benne, eds.), pp. 289–305. ASM Press, Washington, DC.

Queen, C., and Korn, L. J. (1984). A comparative sequence analysis program for the IBM personal computer. *Nucleic Acids Res.* **12**, 581–599.

Saiki, R. K., Gelfand, D. H., Stoffel, S., Scharf, S. J., Higuchi, R., Horn, G. T., Mullis, K. B., and Erlich, H. A. (1988). Primer-directed enzymatic amplification of DNA with a thermostable DNA polymerase. *Science* **239**, 487–491.

Sanger, F., Nicklen, S., and Coulson, A. R. (1977). DNA sequencing with chain-terminating inhibitors. *Proc. Natl. Acad. Sci. USA* **74**, 5463–5467.

Simpson, L., and Maslov, D. A. (1994). Ancient origin of RNA editing in kinetoplastid protozoa. *Curr. Opin. Genet. Dev.* **4,** 887–894.

Spiegel, F. W. (1990). Phylum plasmodial slime molds, class Protostelida. *In* "Handbook of Protoctista" (L. Margulis, J. O. Corliss, M. Melkonian, and D. J. Chapman, eds.) Chapter 27b,pp. 484–497. Jones & Bartlett, Boston, MA.

Stephenson, S. L., and Stempen, H. (1994). "Myxomycetes: A Handbook of Slime Molds." Timber Press, Portland, OR.

5′ DELETION/INSERTION EDITING OF tRNAs

ANALYSIS OF 5'- OR 3'-TERMINAL tRNA EDITING: MITOCHONDRIAL 5' tRNA EDITING IN *ACANTHAMOEBA CASTELLANII* AS THE EXEMPLAR

Amanda J. Lohan *and* Michael W. Gray

Contents

Abstract

Editing processes that result in the structural retailoring of the aminoacyl acceptor stems of mitochondrial tRNAs are the focus of this chapter. This type of tRNA editing is the most frequently observed and widely distributed and involves nucleotide replacement within the 5' or 3' half of the aminoacyl acceptor stem in either a template-directed or a template-independent fashion. We provide a detailed protocol that allows demarcation of 5'-terminal tRNA editing events from those occurring on the 3' side of the acceptor stem. We present the mitochondrial 5' tRNA editing system in *Acanthamoeba castellanii* as the exemplar of terminal tRNA editing. The methodology involves RNA ligase-mediated circularization of tRNAs, cDNA synthesis primed by tRNA-specific oligonucleotides,

Department of Biochemistry and Molecular Biology, Dalhousie University, Halifax, Nova Scotia, Canada

Methods in Enzymology, Volume 424
ISSN 0076-6879, DOI: 10.1016/S0076-6879(07)24010-8

amplification of cDNA via polymerase chain reaction, and cloning and sequencing of multiple products. This approach permits (1) simultaneous determination of 5′ and 3′ acceptor stem sequences, (2) delineation of 5′ or 3′ editing, (3) identification of mature tRNAs (characterized by having a 3′-CCA$_{OH}$ motif), (4) identification of processing/editing intermediates, and (5) mechanistic insights.

1. INTRODUCTION

The term "RNA editing," originally coined to define the phenomenon of uridine insertion in mitochondrial transcripts of trypanosomatids (Benne *et al.*, 1986), now encompasses a much broader range of processes acting on a wide variety of both nuclear and organellar RNA substrates (Gott and Emeson, 2000; Gray, 2003; Koslowsky, 2004). A favored definition, especially pertinent to the mitochondrial editing events addressed in this chapter, is "any programmed alteration of RNA primary structure to generate a sequence that could have been directly encoded at the DNA (gene) level" (Price and Gray, 1999a).

Transfer RNA (tRNA) editing was first discovered in the mitochondrion of the amoeboid protist *Acanthamoeba castellanii* (Lonergan and Gray, 1993a,b). The *Acanthamoeba* mitochondrial genome (mtDNA) encodes 15 bona fide tRNA species (Burger *et al.*, 1995). Secondary structure modeling revealed that of the 15 predicted tRNA structures, 12 had one or more mismatches in the first three base pairs of the acceptor stem, a situation deemed incompatible with normal tRNA function. The initial investigation of a subset of the mitochondrial tRNAs revealed that the inferred acceptor stem mismatches were, in fact, "corrected" by a 5′-terminal editing process (Fig. 10.1) (Lonergan and Gray, 1993a). The nucleotide incorporation activity (putative tRNA editing activity) has been partially purified from *A. castellanii* mitochondria (A. J. Lohan and M. W. Gray, unpublished observations; Price and Gray, 1999b) and was shown to be capable of excising the first three nucleotides (editable sites) at the 5′ end of the acceptor stem, prior to catalyzing their reincorporation (in a 3′-to-5′ direction) utilizing the corresponding 3′ half as a *cis*-acting template guide (Price and Gray, 1999b) (Fig. 10.1). As a consequence of this editing, mismatched acceptor stem positions are converted to canonical Watson–Crick pairings (A–U, U–A, G–C, or C–G). Edits involve purine-to-purine, pyrimidine-to-purine, and pyrimidine-to-pyrimidine replacements; notably the majority of evident edits represent replacement of U or A by G.

In the intervening years, various forms of mitochondrial tRNA editing have been reported for a wide variety of eukaryotes (Gray, 2001; Hopper and Phizicky, 2003; Price and Gray, 1998). These forms include C-to-U

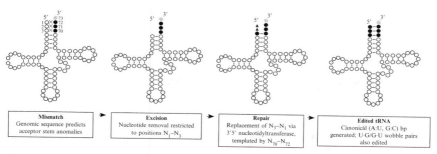

Mismatch	Excision	Repair	Edited tRNA
Genomic sequence predicts acceptor stem anomalies	Nucleotide removal restricted to positions N_1–N_3	Replacement of N_3–N_1 via 3′5′ nucleotidyltransferase, templated by N_{70}–N_{72}	Canonical (A:U, G:C) bp generated; U·G/G·U wobble pairs also edited

Figure 10.1 Proposed mechanism of 5′ tRNA editing in *Acanthamoeba castellanii* mitochondria. Secondary structure modeling predicts that 12 of 15 mtDNA-encoded tRNAs have one or more mismatches in the first three base pairs of their acceptor stems. The tRNA editing activity in *Acanthamoeba* mitochondria is capable of excising the first three nucleotides at the 5′ end of the acceptor stem, prior to catalyzing their replacement in a 3′-to-5′ direction, utilizing the corresponding 3′ half of the stem as a *cis*-acting template guide. Discriminator nucleotide (position N_{73}) is denoted by a gray circle; 3′-terminal-CCA extension is omitted. (Adapted from Lohan and Gray, 2004.)

substitution editing in plants (Fey *et al.*, 2000) and animals (Janke and Pääbo, 1993), C or U insertional editing in myxomycete protozoa (Antes *et al.*, 1998; Miller *et al.*, 1993), a type of template-independent editing observed at the 3′ ends of mitochondrial tRNAs in many metazoans (Hatzoglou *et al.*, 1995; Tomita *et al.*, 1996; Yamazaki *et al.*, 1997; Yokobori and Pääbo, 1995a,b), and the template-dependent processes evidenced as 5′ editing in the mitochondria of amoeboid protozoa (Angata *et al.*, 1995; Cole and Williams, 1994; Lonergan and Gray, 1993a,b) and several chytridiomycete fungi (Bullerwell and Gray, 2005; Forget *et al.*, 2002; Laforest *et al.*, 1997), as well as the template-directed 3′ mitochondrial tRNA editing first characterized in centipede (Lavrov *et al.*, 2000) and more recently demonstrated in the jakobid flagellate *Seculamonas ecuadoriensis* (Leigh and Lang, 2004).

This chapter focuses on editing processes that result in the structural retailoring of the aminoacyl acceptor stems of mitochondrial tRNAs. This is the most frequently observed and widely distributed type of tRNA editing and involves nucleotide replacement within the 5′ or 3′ half of the aminoacyl acceptor stem in either a template-directed or a template-independent fashion. Below, we detail a protocol that allows delineation of 5′-terminal tRNA editing events from those occurring on the 3′ side of the acceptor stem.

Originally, Lonergan and Gray (1993a) used oligonucleotide-primed reverse transcriptase sequencing to identify 5′ editing in a subset of *Acanthamoeba* mitochondrial tRNAs. Utilizing a version of the method detailed in this chapter [reverse transcriptase polymerase chain reaction (RT-PCR)-mediated amplification primed off circularized tRNAs], Price and Gray (1999a) confirmed the previously characterized tRNA edits and expanded the study to encompass the remaining predicted mitochondrial tRNA edits. The ability of T4 RNA ligase to catalyze the joining of a 5′-phosphoryl moiety to a 3′-hydroxyl group between or within single-stranded regions of RNA or DNA (Sugino et al., 1977) allows circularization of the characteristically 5′-monophosphate-, 3′-hydroxyl-terminated tRNAs (Fig. 10.2B). Use of such a circularized template for cDNA synthesis offers several advantages compared to other methods for analysis of terminal editing in tRNA. Primarily, cDNA synthesis on a circularized tRNA template spans the entire region of the acceptor stem, thus allowing simultaneous determination of 5′ and 3′ acceptor stem sequences. Depending on where the forward and reverse primers are situated within the tRNA sequence, regions of the tRNA flanking the acceptor stem may also be analyzed (Figs. 10.2B and 10.3). A further benefit of the circularization method is that it reveals the presence of the 3′-CCA_{OH} motif, which mature tRNAs require for aminoacylation. Unlike their eubacterial counterparts, the 3′-CCA sequence is generally not encoded in mitochondrial tRNA genes, but is added posttranscriptionally by the ATP(CTP):tRNA nucleotidyltransferase (Hopper and Phizicky, 2003; Schürer et al., 2001). The presence of a 3′-terminal CCA therefore indicates that the corresponding tRNA gene is functional.

In this chapter, we present the mitochondrial 5′ tRNA editing system in Acanthamoeba as the exemplar of terminal tRNA editing. The procedures outlined herein describe in detail the preparation of Acanthamoeba mitochondrial RNA and tRNA fractions, RT-PCR, and expected results. Analysis of Acanthamoeba tRNA^{Leu2} (Price and Gray, 1999a), Seculamonas tRNA^{Glu} (Leigh and Lang, 2004), and Euhadra tRNA^{Tyr} (Yokobori and Pääbo, 1995b) is presented as an illustration of the three principal manifestations of tRNA editing, namely, cis-guided 5′ editing, cis-guided 3′ editing, and nontemplated 3′ editing (Fig. 10.4).

2. METHODS

The methods outlined in the following sections cover the growth and fractionation of purified mitochondria from the amoeboid protozoon A. castellanii. Detailed procedures for isolating both total mitochondrial RNA and tRNA-enriched fractions are given. Conditions for circularization of tRNA species and subsequent RT-PCR methodology for generation of DNA sequences for cloning are provided. Where relevant, note segments address alternate techniques, tips, explanations, or general considerations with regard to methodology or equipment.

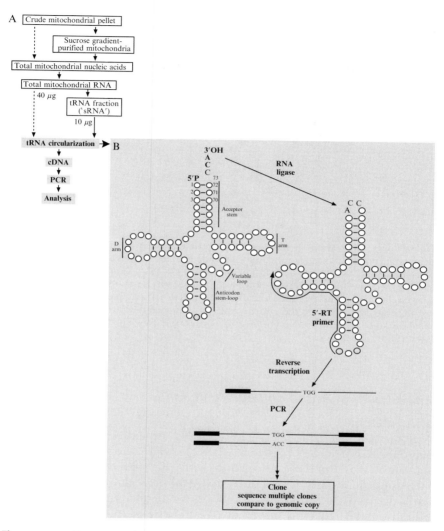

Figure 10.2 Overview of procedures for analysis of 5′- or 3′-terminal tRNA editing detailed in this chapter. (A) Flow chart representation of the methods outlined for analysis of 5′ tRNA editing in highly purified *Acanthamoeba* mitochondria. Boxed titles refer to subsections in the Methods section. Gray-boxed titles are Methods sections further illustrated in (B). Solid arrows refer to methodology described in detail. Dotted lines with a solid arrowhead signify alternate steps, or abbreviated methodology applied to other organisms, or when material is not abundant. (B) Illustration of tRNA secondary structure, and a schematic of steps for tRNA circularization, cDNA synthesis, and PCR analysis. The tRNA secondary structure, numbering, and labels are according to convention (Dirheimer *et al.*, 1995; Sprinzl and Vassilenko, 2005). (Adapted from Lohan and Gray, 2004.)

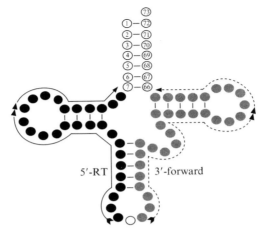

Figure 10.3   Diagrammatic illustration of optimal regions within a typical tRNA for primer placement. Dotted or solid line and shaded circles delineate the regions of primer placement for cDNA synthesis (5′-RT primer) and PCR (5′-RT and 3′-forward primers). 5′-RT primer positioning is indicated by solid black circles and 3′-forward primer by gray circles. The arrowhead indicates direction of synthesis. Primers may lie within but not necessarily encompass the entire stretch of sequence indicated in the figure. The acceptor stem is numbered according to convention (see Fig. 10.2). (Adapted from Lohan and Gray, 2004.)

2.1. Growth of *Acanthamoeba castellanii*

2.1.1. Materials

Neff base medium (pH 7): 1-liter: 0.75% yeast extract, 0.75% proteose peptone, 2 mM KH$_2$PO$_4$, 1 mM MgSO$_4$. Adjust the volume to 890 ml and autoclave. Prior to inoculation with the starter culture, the base medium requires aseptic addition of 100 ml 15% glucose, 10 ml 10 mM ferric citrate, 50 μl 1 M CaCl$_2$, 100 μl 10 mg/ml thiamine, 100 μl 2 mg/ml D-biotin, and 2 μl 0.5 mg/ml vitamin B$_{12}$ from sterile stock solutions. Four-liter Erlenmeyer or 2.8-liter Fernbach flasks are the growth vessels of choice. Neff base medium may be stored at 4°.

2.1.2. Method

Five-milliliter starter cultures are maintained on a shaking platform in 50-ml tissue culture tubes at room temperature. A 4-day starter culture is used to inoculate a 1-liter culture. The required growth conditions are 30° with moderate shaking until an OD$_{550}$ reading of 0.9–1.0 is achieved, typically 4—5 days. The doubling time of a healthy culture of *Acanthamoeba* trophozoites is approximately 18—20 h.

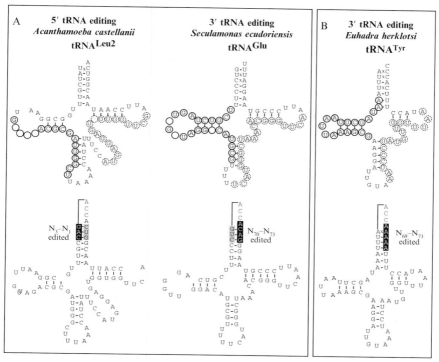

Figure 10.4 Schematic of products of 5′-templated, 3′-templated, and 3′-nontemplated mitochondrial tRNA editing. tRNA secondary structures inferred from (1) DNA sequence (upper portion of the panel) and (2) cDNA sequence (lower portion of the panel). Regions encompassing primer binding sites are indicated by solid circles (5′-RT) or dashed circles (3′-forward). Crosses denote mismatched nucleotide pairings. Lines connecting 5′ and 3′ ends of tRNAs indicate bonds formed by RNA ligase during tRNA circularization. Edited nucleotides are denoted as white characters boxed in black. The stem portion acting as the *cis* guide is boxed in gray. (A) Illustration of products generated by template-directed editing: 5′ editing of *Acanthamoeba* tRNA^{Leu2} (Price and Gray, 1999a) and 3′ editing of *Seculamonas* tRNA^{Glu} (Leigh and Lang, 2004). (B) Polyadenylation-mediated 3′ editing of *Euhadra* tRNA^{Tyr} (Yokobori and Pääbo, 1995b); note the $A_1 \times A_{72}$ mismatch that is generated from the mtDNA-encoded $A_1 \times C_{72}$ mismatch by editing.

Note: A healthy culture will have a slightly granular appearance. Individual *Acanthamoeba* cells are readily visible by conventional light microscopy (40× magnification). Healthy cells have slender hair-like projections (acanthapodia), prominent vacuoles, and a prominent nucleolus. No staining is required to visualize any of these features of a growing cell. To permit optimal air exchange during growth when scaling up or down, culture volumes should not exceed a depth of 4–5 cm in the chosen vessel.

2.2. Isolation of crude mitochondrial pellet

While the mitochondrial isolation protocol is specific for *Acanthamoeba*, with minor alterations it has been used successfully by us for other protists.

2.2.1. Materials

Phosphate-buffered saline (PBS) (pH 7.4): 4.3 mM Na_2HPO_4, 1.4 mM KH_2PO_4, 137 mM NaCl, 2.7 mM KCl (1 liter of PBS is required for every liter of culture processed).

10% Bovine serum albumin (BSA): (a) BSA fraction V, (b) BSA fraction V essentially fatty acid free—used only for the gradient solutions (may be stored at $-20°$).

Phenylmethylsulfonyl fluoride (PMSF): 100 mM solution (1.74 mg/ml), dissolved in isopropanol. This can be stored at $-20°$.

Homogenization buffer: 10 mM Tris–HCl (pH 7.6), 1 mM $ZnCl_2$, 0.25 M sucrose, 1 mM dithiothreitol (DTT), 0.1% BSA—from sterile stock solutions. Prepare this fresh on the day of use, with the addition of the BSA and DTT immediately prior to use.

Mitochondrial wash buffer: 10 mM Tris–HCl (pH 7.6), 10 mM EDTA, 0.35 M sucrose, 1 mM DTT, 0.1% BSA—from sterile stock solutions. Prepare this fresh on the day of use, with the addition of the BSA and DTT just prior to use.

Dounce tissue grinder (Wheaton): 15 or 40 ml, with supplied large (A) and small (B) pestles.

Pipette controller: Motorized or battery-operated.

2.2.2. Method

All manipulations should be carried out on ice, all buffers and tubes are stored on ice, and all centrifugation steps should be at $4°$.

1. Divide the culture between two chilled 750-ml plastic centrifuge bottles and centrifuge at $900 \times g$ for 5 min. A swinging bucket rotor is preferable but is not an absolute requirement. Wash the cell pellets twice with 0.5 volume cold PBS; after each wash, centrifuge for 5 min at $900 \times g$. The pellet should be resuspended either by gentle swirling or flushing the buffer over the pellet using a pipette (the use of a vortex mixer is not recommended).

2. Resuspend each PBS-washed pellet in 25 ml of homogenization buffer (5 ml of buffer is required for every 100 ml of initial culture). Disrupt the cells in a chilled Dounce tissue grinder—35 strokes with pestle A followed by 5 strokes with pestle B. A stroke is a single down–up movement. Greater than 90% cell breakage is achieved with this method.

3. A low-speed centrifugation (900×g for 10 min) sediments the nuclei and any intact cells. A swinging bucket rotor is preferable but is not an absolute requirement.

4. The supernatant from the low-speed spin is divided between two cooled 30-ml tubes. When transferring the supernatant, make sure to avoid touching the pellet–supernatant interface. A small volume (<1 ml) may be left at the interface in order to avoid any carryover contamination. The gray-brown-colored pellets (nuclei, large membrane fragments and intact cells) are discarded.

5. Centrifuge the postnuclear supernatants for a further 20 min at 9000×g to yield the crude mitochondrial pellet (typically caramel-brown colored). Discard the supernatant (crude cytoplasmic fraction) from this spin. Use a fixed-angle rotor for this and all subsequent centrifugations.

6. Wash the mitochondrial pellets with 10 ml of wash buffer. For efficient washing, resuspend the pellet in the wash buffer by repeated pipetting using the motorized pipette controller. Resediment the mitochondria with a 10-min centrifugation at 9000×g. Repeat the wash but increase the centrifugation time to 15 min.

7. Final pellets constitute the crude mitochondrial yield.

8. The crude mitochondria can be (a) further purified by sucrose gradient centrifugation or (b) directly used for RNA extraction (see flowchart Fig. 10.2A).

Note: Substitution of the Dounce (step 2) with one or two passages at 1500 psi in a French press has yielded crude mitochondrial fractions from a variety of different protists, including *Hartmannella vermiformis, Physarum polycephalum,* and *Polysphon-dylium pallidum.* The most efficient method of resuspension of the mitochondrial pellets utilizes the motorized or battery-operated pipette. Although repeated pipetting is generally sufficient to resuspend *Acanthamoeba* mitochondrial pellets, for other organisms use of a tissue grinder was sometimes required. Briefly, pellets are partially resuspended in the relevant buffer and transferred to the chilled Potter–Elvehjem Teflon tissue grinder (Wheaton); 5–10 strokes of the pestle are usually sufficient to produce a uniform suspension, detailed for wheat mitochondria in Spencer *et al.* (1992).

2.3. Purification of mitochondria on sucrose gradients

The crude mitochondria are further purified by ultracentrifugation on a two-step (1.3 M, 1.55 M) sucrose gradient. Each sucrose gradient consists of a 7.5-ml lower layer of 1.55 M sucrose, on which is carefully layered 15 ml of 1.3 M sucrose solution. The protocol outlined below is specifically for use with a Beckman SW25.1 swinging bucket rotor and 1 × 3-in. (25 × 76-mm) centrifuge tubes. The Beckman SW25.1 rotor requires three tubes; therefore, three step gradients must be prepared for each run.

2.3.1. Materials

Sucrose gradient solutions: 1.55 M sucrose and 1.3 M sucrose in 50 mM Tris–gHCl (pH 8), 3 mM EDTA (pH 8), 1 mM DTT, 0.1% BSA (fatty acid-free stock)—from sterile stock solutions. Prepare fresh on the day of use, with addition of the BSA and DTT just prior to use.

Presucrose gradient buffer: 50 mM Tris–HCl (pH 8), 3 mM EDTA (pH 8), 0.25 M sucrose, 1 mM DTT, 0.1% BSA—from sterile stock solutions. Prepare fresh on the day of use, with addition of the BSA and DTT just prior to use.

Postsucrose gradient/storage buffer: 20 mM Tris–HCl (pH 8), 0.5 mM EDTA, 0.25 M sucrose, 15% (v/v) glycerol, 1 mM DTT, 1 mM PMSF—from sterile stock solutions. Prepare fresh on the day of use, with the addition of the DTT and PMSF just prior to use.

Ultracentrifuge: Required rotor: Beckman SW25.1 swinging bucket rotor. Tubes: 1 × 3 in. (25 × 76 mm).

2.3.2. Method

1. Pipette 7.5 ml of 1.55 M sucrose into each centrifuge tube. To add the upper sucrose layer, place the pipette tip on the side of the tube just above the lower 1.55 M layer and, with the motorized pipette at the lowest setting, carefully layer on the sucrose mix. A correctly poured gradient will have a visible sharp interface.

2. Resuspend the crude mitochondrial pellets in 1.5 ml presucrose gradient buffer by repeated pipetting (employing the automatic pipette aid). The suspension should have no visible clumps. Each gradient will hold 3–5 ml of applied material.

3. Pool the two pellet suspensions and load onto a single gradient. Gently layer the mitochondrial suspension onto the gradient, taking care to avoid mixing with the 1.3 M shelf. To act as balance tubes, the remaining two gradients are blind-loaded with an equivalent volume of pregradient buffer.

4. Centrifuge the gradients at 4° for 1 h at 22,500 rpm. The brake should be on the gentlest setting allowed by the manufacturer.

5. The mitochondria should form a compact caramel-brown-colored band at the interface of the 1.3 M and 1.55 M layers. A cloudy lipid band may be visible near the top of the gradient, at the interface between the gradient and the applied sample. If there was carryover nuclei contamination, a small pellet may be visible. A large pellet with little material at the 1.3 M–1.55 M interface is indicative of loss of

mitochondrial integrity and/or the loading of incompletely resuspended pellets.

6. Remove the bulk of the upper sucrose layer by aspiration, being especially careful to avoid disturbing the mitochondrial band.

7. Using an 18-gauge needle (bent at a 90° angle) attached to a 5-ml syringe, slowly recover the mitochondrial band. The bent syringe allows the needle to be placed across the top of the mitochondrial band. Once the needle tip is in the band, pull the contents into the syringe. The mitochondrial material has a viscous (clot-like) consistency. Take note of the volume, generally 3–4 ml, and transfer to a chilled 15-ml snap cap culture tube.

8. Dilute the mitochondria slowly (over a 15–20 min period) with 2 volumes postsucrose gradient/storage buffer. This introduces both glycerol as a cryoprotectant and PMSF and DTT as protease inhibitors.

9. Centrifuge at $12,000 \times g$ for 30 min (4°).

10. Decant the supernatant and discard. The mitochondrial pellet should be firm, caramel-brown colored, and 0.6–0.7 g in weight (typical yield from a 1-liter culture).

11. The mitochondrial pellet may be frozen in liquid nitrogen and stored at −70° for periods in excess of 2 years.

Note: There is no "one-size-fits-all" approach for gradient purification of mitochondria. Applying the latter method, mitochondria from other organisms may not be as efficiently fractionated; where gradient fractionation methods are not optimized or mitochondrial yields are low (<0.2 g), the crude mitochondrial pellet may be directly used for RNA preparation.

2.4. Preparation of mitochondrial RNA fractions

In this section detailed procedures for the preparation of total mitochondrial nucleic acids, and subsequent manipulations to yield both total RNA and tRNA fractions, are provided. Note that volumes are small enough so that most centrifugations may be carried out in microfuges. To minimize potential RNase contamination, disposable gloves should be worn throughout all procedures. Mitochondrial lysis and phenolic extraction protocols are based on procedures outlined in Parish and Kirby (1966) and Spencer *et al.* (1992).

2.4.1. Materials

Phenol-cresol mix (500:70:0.5): to 500 g of phenol crystals dissolved in 55 ml distilled water (dH$_2$O) or alternatively 500 ml liquified phenol add 70 ml *m*-cresol and 0.5 g 8-hydroxyquinoline. Mix and equilibrate against 10 mM–50 mM Tris–HCl (pH 7.6).

Sodium dodecyl sulfate (SDS): 10% solution.

TE: 10 mM Tris–HCl (pH 7.6), 1 mM EDTA (pH 8).

NaOAc: 3 M stock (pH 5.2).
Ethanol: 100% and 80% stocks.
Nuclease-free dH$_2$O.
DNase I buffer (10×): 400 mM Tris–HCl (pH 7.6), 60 mM MgCl$_2$.

2.4.2. Total Mitochondrial Nucleic Acids
2.4.2.1. Method

1. Resuspend either the crude or sucrose gradient-purified mitochondrial pellets in 2 ml chilled TE. Lyse the mitochondria by sequential addition of SDS to 1%, NaOAc (pH 5.2) to 0.3 M, and an equal volume phenol-cresol mix.
2. Vortex the lysis mixture briefly, place on ice for 5 min, and centrifuge at 10,000×g for 10 min (4°). Since RNA is the final goal, possible shearing of DNA by vortexing is not a major concern.
3. Transfer the aqueous phase to clean tubes, add NaOAc (pH 5.2) to 0.3 M and reextract with an equal volume of phenol-cresol. An additional phenol-cresol extraction is generally sufficient to remove any contaminating material.
4. Precipitate the nucleic acids from the aqueous phase with 2.5 volumes of 100% ethanol (1 h; −70° or −20°).
5. Centrifuge at 12,000×g for 15 min. Wash the final pellet once in 70–80% ethanol, dry, and redissolve in 200 μl nuclease-free dH$_2$O. This constitutes the total yield of mitochondrial nucleic acids.

Note: The phenol-cresol mix used for phenolic extraction is based on protocols previously outlined (Parish and Kirby, 1966; Spencer *et al.*, 1992). The 8-hydroxyquinoline acts as both an antioxidant and a mild RNase inhibitor. Addition of the salt [0.1 volume 3 M NaOAc (pH 5.2)] to the mitochondrial nucleic acid mix prior to phenol extraction acidifies the phenolic lysate. Under acid conditions, the DNA will collect at the organic–aqueous interface, whereas the RNA will remain in the aqueous phase. An alternate method for nucleic acid extraction involves the use of the commercially available TRIzol Reagent[TM], a phenol/guanidine isothiocyanate mix (GIBCO-BRL).

2.4.3. Preparation of total RNA fraction

Due to the acidic condition of the phenolic extractions, the RNA preparation is minimally contaminated with DNA. The residual contaminating DNA is eliminated by incubation with DNase I to yield the total RNA fraction.

2.4.3.1. Method

1. Final reaction volume 300 μl: mix 180 μl of the total mitochondrial nucleic acids with 150–200 units of RNase-free DNase I, 30 μl 10× DNase buffer, with the remaining volume made up with nuclease-free dH$_2$O.
2. Incubate the DNase reaction at 37° for 30 min.

3. Following the incubation period, add NaOAc (pH 5.2) to 0.3 M to the mix and extract once with an equal volume of phenol-cresol.

4. Recover the RNA by two rounds of ethanol precipitation. For precipitation: to the aqueous phase add 2.5 volumes ethanol (1 h; $-70°$ or $-20°$). On addition of the ethanol the precipitating RNA will turn the solution "milky." Two rounds of precipitation are recommended in order to ensure complete removal of any contaminating phenol.

5. Centrifuge for 15 min at $12,000 \times g$ (4°).

6. Decant the alcohol, redissolve the pellet in 50 μl nuclease-free dH$_2$O, and add 5 μl 3 M NaOAc (pH 5.2) and 125 μl 100% ethanol (1 h; $-70°$ or $-20°$). Centrifuge again at $12,000 \times g$ for 15 min.

7. Wash the final pellet with 70–80% ethanol and dry. Do not overdry the RNA pellet, as this will substantially reduce its subsequent solubility.

8. Redissolve the RNA in 100 μl nuclease-free dH$_2$O; the concentration should be approx 5 $\mu g/\mu l$. The estimated yield is 300–500 μg total RNA per 0.4–0.6 g sucrose gradient-purified mitochondria.

9. Assess the purity and concentration of the total RNA preparation before proceeding to (a) salt fractionation (to yield tRNA-enriched fraction—sRNA) or (b) directly to the circularization protocol.

Note: The salt fractionation procedure is based on the observation (Crestfield *et al.*, 1955) that high-molecular-weight rRNA is insoluble in aqueous solution at elevated ionic strength, whereas tRNA (and 5S rRNA) remain soluble under the same conditions. Quality control: using standard methodology (Sambrook *et al.*, 1989), a small amount of the RNA may be electrophoresed on a 3–4% agarose or 10% polyacrylamide gel. Intact RNA should band reproducibly; intense bands representative of the rRNAs with minimal background smearing are a good sign that the RNA is undegraded. Transfer RNA and other small RNA species will appear as a diffuse blob at the leading edge of the gel. Lack of definition and excessive smearing are usually indicative of degradation of the RNA sample. As an additional quality control, an ultraviolet absorbance spectrum of the RNA can be determined over the range of 200–300 nm. A pure sample should generate the characteristic skewed bell-shaped curve, with an ideal A_{260}/A_{280} ratio of ~2 for pure RNA (Sambrook *et al.*, 1989).

2.4.4. Salt fractionation to recover a tRNA-enriched fraction

For salt precipitation to work effectively, the initial RNA concentration must be in excess of 4–5 $\mu g/\mu l$.

2.4.4.1. Method

1. To the total mitochondrial RNA, add NaCl to a concentration of 1.2 M.
2. Store the RNA–salt solution overnight at 4° (or not less than 8 h).

3. Centrifuge for 10 min at 12,000×g (4°). Draw off the supernatant; this constitutes the bulk of the salt-soluble (mostly transfer) RNAs (sRNA). Store at −70° or −20°.

4. Resuspend the pellet in 20 μl of nuclease-free dH$_2$O and repeat the salt fractionation (solution adjusted to 1.2 M NaCl, 4° overnight or not less than 8 h).

5. Once again, centrifuge the salt–RNA mixture at 10,000×g for 10 min. The pellet constitutes the salt-insoluble (mostly ribosomal and messenger) RNAs (iRNA).

6. Combine the supernatants from steps 3 and 5 and precipitate sRNA with 2.5 volumes of ethanol (−70°; 1 h). Centrifuge for 15 min at 12,000×g (4°). Wash the pellet with 80% ethanol, dry, and dissolve in 200 μl of nuclease-free dH$_2$O.

7. Measure the absorbance at 260 nm and calculate the concentration, assuming that a 40 μg/ml RNA solution gives an A_{260} of 1.0.

8. Add 0.1 volume 3 M NaOAc (pH 5.2) and extract once with phenol-cresol.

9. To the aqueous phase, add 2.5 volumes ethanol (−70°; 1 h) and centrifuge for 15 min at 12,000×g (4°).

10. Decant the ethanol, resuspend the pellet in 50 μl nuclease-free dH$_2$O, and add 0.1 volume 3 M NaOAc (pH 5.2) and 125 μl 100% ethanol; store for 1 h at −70° or −20°. Centrifuge for 15 min at 12,000×g (4°). Wash the final pellet once in 70−80% ethanol and dry.

11. Store the sRNA at −70° at a concentration of 1 μg/μl.

2.5. tRNA circularization and generation of tRNA clones

This section outlines the conditions required for intramolecular ligation of individual tRNAs to yield circular products that serve as substrates for cDNA synthesis. The cDNA is then subjected to PCR and the final product is visualized. For a schematic of the methods in this section, refer to Fig. 10.2.

2.5.1. Materials

Reverse transcriptase: avian myeloblastosis virus (AMV) with supplied buffer.
Dimethyl sulfoxide (*DMSO*).
BSA: commercial nuclease-free.
T4 RNA ligase.
ATP: 2 mM stock solution, stored at −20°.
tRNA-specific oligonucleotide primers: For each tRNA to be analyzed, 1 pmol and 10 pmol stocks of 5′-RT primer and a 10 pmol stock of 3′-forward primer are required.
Cloning kit: Invitrogen's TOPO TA Cloning^{TM}.

2.5.2. Selection of primers

Two primers appropriately flanking the acceptor stem are required for each tRNA analyzed (Fig. 10.3). The primers should be minimally 17 nucleotides long and situated within the regions indicated by the solid or dotted lines and shaded circles (Fig. 10.3).

2.5.3. tRNA circularization

The ligation conditions are a combination of those found in the references (Lohan and Gray, 2004; Price and Gray, 1999a; Yokobori and Pääbo, 1995a). This combined methodology has been optimized for the use of either total RNA or tRNA (salt-solubilized fraction—sRNA) preparations in the ligation reaction. As only 10–20% of total RNA comprises tRNA, 40 μg of total RNA is substituted for the recommended 10 μg sRNA in order to ensure that there is sufficient target in the circularization reaction.

2.5.3.1. Method

1. Denature 10 μg sRNA or 40 μg total RNA by heating to 90° for 5 min. Cool on ice and centrifuge briefly.
2. Include several controls during the first use of this protocol:
   a. RNase-treated DNA
   b. DNA no RNase treatment—both (a) and (b) should be treated under the same conditions as the RNA
   c. Unligated RNA

3. Ligation conditions: 20 μl total reaction volume; 50 mM HEPES (pH 7.5), 15 mM $MgCl_2$, 10% DMSO, 0.1 mg/ml nuclease-free commercial BSA, 3.3 mM DTT, 100 μM ATP, 15 units T4 RNA ligase, and the denatured cooled RNA. Allow ligation to proceed for 8–10 h at 37°.
4. Subsequent to ligation, dilute the mix to 50 μl, add 0.1 volume 3 M NaOAc (pH 5.2), and extract once with an equal volume of phenol-cresol.
5. The circularized tRNAs are precipitated twice: add 0.1 volume 3 M NaOAc (pH 5.2) and 2.5 volume 100% ethanol (−70° for 1 h). Centrifuge for 15 min at 12,000×g (4°).
6. Decant the ethanol, redissolve the pellet in 50 μl nuclease-free dH_2O, and add 0.1 volume 3 M NaOAc (pH 5.2) and 125 μl 100% ethanol; store for 1 h at −70° or −20°.
7. Centrifuge at 12,000×g for 15 min.
8. Resuspend the pellet in 10 μl nuclease-free dH_2O; the final concentration should be approximately 1 μg/μl if sRNA is used or approximarely 4 μg/μl if total RNA is the substrate. Store the circularized tRNA solution at −70°.

2.5.4. cDNA synthesis from circularized tRNA templates
2.5.4.1. Method

1. In a total volume of 21.5 μl, combine 1 μl of the circularized tRNA solution with 1 pmol 5'-RT primer, adding dH_2O to bring the volume to the required 21.5 μl.
2. Heat the tRNA–primer mix to 90° for 3 min, allow it to cool to room temperature over a 15- to 20-min period, and finally place it on ice for a further 15 min.
3. To the cooled primer–tRNA mix, add 0.1 volume AMV buffer, 1.8 μl 1 mM dNTP mix (60 μM final concentration of each of dATP, dCTP, dGTP, and dTTP), 15 units AMV reverse transcriptase, and dH_2O to a final reaction volume of 30 μl.
4. Allow cDNA synthesis to proceed at 45° for 45 min.
5. Store the resultant cDNA at $-20°$.

Note: AMV is preferred over other commonly used reverse transcriptases (e.g., Moloney murine leukemia virus, MMLV); its higher optimal reaction temperature (42°) may help to reduce possible tRNA secondary structure issues.

2.5.5. PCR amplification of transcribed circularized tRNAs
2.5.5.1. Method

1. 50 μl total volume: 1.0 μl cDNA, 10 pmol each of the 5'-RT and 3'-forward tRNA-specific primers, 0.1 volume PCR buffer, 0.2 mg/ml commercial BSA or gelatin, 2.5 units Taq DNA polymerase, and 250 μM of each of the four dNTPs.
2. Cycle parameters: denaturation (94°; 3 min), followed by 30–35 cycles of denaturation (94°; 40 sec), annealing (50°/55°; 40 sec), extension (72°; 60 sec), and a final 12-min extension step at 72°. Preheating the block to 94° prior to introducing the reaction tubes minimizes potential aberrant annealing or primer extension during the initial denaturation step.

2.5.6. Analysis of reverse-transcribed circularized tRNAs

PCR products may be visualized using standard agarose or polyacrylamide electrophoresis techniques (Sambrook *et al.*, 1989). Cloning of the PCR products is recommended (1) for sequencing by automated means and (2) so that multiple clones may be generated to quantify editing and identify possible editing intermediates. There are a plethora of commercially available kits for cloning PCR products. In our hands, Invitrogen's TOPO TA Cloning[TM] technology has proved to be one of the quicker, more reliable, and more efficient means of cloning the PCR products for analysis. Multiple clones from each circularized tRNA should be analyzed. Sequencing is carried out with either the specific tRNA primers or preferably vector primers if vector cloning is utilized.

2.6. Data interpretation

During data analysis special attention should be paid to the following.

1. The presence of a $3'\text{-CCA}_{OH}$ tail. For a tRNA to undergo aminoacylation and participate in protein biosynthesis the presence of the $3'\text{-CCA}_{OH}$ is required. Most mitochondrial tRNAs require posttranscriptional addition of this motif; therefore, maturation and functionality are inferred when a sequence corresponding to the $3'\text{-CCA}_{OH}$ is present in the cDNA copy.

2. Lack of $3'\text{-CCA}_{OH}$ motif. The absence of the expected CCA may be indicative of an editing/processing intermediate. In these instances, care must be taken when comparing the "corrected" cDNA to its corresponding genomic copy. Extension at the 5′ and/or 3′ ends might correspond to mtDNA-encoded flanking gene or intergenic sequences. This possibility is of special interest in cases where tRNA genes are clustered in the mtDNA or overlap other genes.

3. Identity of replacements. Does the editing restore canonical Watson–Crick base pairing (A–U, U–A, G–C, or C–G)? Are all four bases used in mismatch repair? Of special note is conversion of G·U and U·G pairs to Watson–Crick pairings. The observation of both the latter and former nucleotide replacements is indicative of template-directed editing. In the case of the 5′ editing described in *Acanthamoeba*, one of the unexpected discoveries was the replacement of the uridine residue in the U·G pairings in both tRNA^{Met} ($U_1 \cdot G_{72}$) and tRNA^{Leu2} ($U_3 \cdot G_{70}$) with a cytidine (Price and Gray, 1999a). The generation of the canonical G–C base pair emphasized the templated nature of the nucleotide replacement in the *Acanthamoeba* system (Lonergan and Gray, 1993a; Price and Gray, 1999a) (Figs. 10.1 and 10.4A). It should be noted that both U·G and G·U base pairs are frequently observed and tolerated in tRNA helical regions (Dirheimer *et al.*, 1995; Sprinzl and Vassilenko, 2005) and, in some cases, may serve as identity determinants (Beuning *et al.*, 1997). Similar nucleotide replacement patterns were revealed in the analysis of mitochondrial 3′ editing systems. In the case of the centipede (*Lithobius forficatus*) editing restores canonical base pairing, including conversion of U·G pairs in tRNA^{Gln} and tRNA^{His} to Watson–Crick U–A pairings (Lavrov *et al.*, 2000). The mt-3′ tRNA editing mechanism in the jakobid flagellate *Seculamonas ecuadoriensis* superficially resembles that elucidated in centipede (Leigh and Lang, 2004) (Fig. 10.4A). The third editing variant, nontemplated 3′ editing, was first revealed in the land snail (*Euhadra herklotsi*), where it was demonstrated to be unidirectional—i.e., all changes were to an A residue, including conversion of the mtDNA-predicted $A_1 \times C_{72}$ mismatch to the "corrected" $A_1 \times A_{72}$ pairing (Fig. 10.4B) (Yokobori and Pääbo, 1995b).

4. Partial editing: The sequencing of multiple clones from the same circularization reaction may allow identification of partially edited intermediates. In the case of *Acanthamoeba*, no editing intermediates—i.e.,
sequences lacking the CCA tail or incompletely edited—were found
for any of the 15 bona fide tRNAs analyzed (Price and Gray, 1999a).
Conversely, both partial and fully edited tRNAs were identified in the
Seculamonas 3′ editing system. For both tRNA^{Ser} (26%) and tRNA^{Glu}
(28%), the furthest upstream nucleotide (G$_{70}$ and A$_{70}$, respectively) was
found to have the mtDNA-encoded identity (G or A) as opposed to the
expected edited identity (U or G): as a consequence, the authors posit a
mechanism of 3′ exonucleolytic degradation, followed by resynthesis
using the 5′ side of the acceptor stem as a template (Leigh and Lang, 2004).

3. Concluding Remarks

The relatively simple method outlined here allows generation of a
large dataset for each potentially edited tRNA. When significant multiples
of circularized products are analyzed, the characteristics of the editing in
addition to its position on the 5′ or 3′ side of the acceptor stem may be
assessed. Possible mechanisms and the temporal order of processing and
editing events may be inferred from these types of data.

Recent literature provides an alternate use for the circularization method
detailed above. Analysis of the genome sequence of the archaeal parasite
Nanoarchaeum equitans revealed a surprising arrangement for four of the
encoded tRNA species (Randau *et al.*, 2005). Computational analysis identified nine tRNA halves spread throughout the chromosome. In this organism,
creation of functional tRNAs for glutamate, histidine, tryptophan, and methionine (initiator) requires splicing of their widely separated 5′ and 3′ halves. The
presence of six tRNA half-transcripts was confirmed by RT-PCR and
sequence analysis. Functionality of the tRNAs was then inferred by utilizing
the circularization method detailed above to identify the 5′ and 3′ ends of the
matured tRNAs, as evidenced by the presence of a posttranscriptionally added
3′-CCA$_{OH}$ motif to the spliced 5′ and 3′ halves (Randau *et al.*, 2005).

REFERENCES

Angata, K., Kuroe, K., Yanagisawa, K., and Tanaka, Y. (1995). Codon usage, genetic code
and phylogeny of *Dictyostelium discoideum* mitochondrial DNA as deduced from a 7.3-kb
region. *Curr. Genet.* **27**, 249–256.
Antes, T., Costandy, H., Mahendran, R., Spottswood, M., and Miller, D. (1998). Insertional
editing of mitochondrial tRNAs of *Physarum polycephalum* and *Didymium nigripes*. *Mol.
Cell. Biol.* **18**, 7521–7527.

Benne, R., Van den Burg, J., Brakenhoff, J. P. J., Sloof, P., Van Boom, J. H., and Tromp, M. C. (1986). Major transcript of the frameshifted *coxII* gene from trypanosome mitochondria contains four nucleotides that are not encoded in the DNA. *Cell* **46**, 819–826.

Beuning, P. J., Yang, F., Schimmel, P., and Musier-Forsyth, K. (1997). Specific atomic groups and RNA helix geometry in acceptor stem recognition by a tRNA synthetase. *Proc. Natl. Acad. Sci. USA* **94**, 10150–10154.

Bullerwell, C. E., and Gray, M. W. (2005). *In vitro* characterization of a tRNA editing activity in the mitochondria of *Spizellomyces punctatus*, a chytridiomycete fungus. *J. Biol. Chem.* **280**, 2463–2470.

Burger, G., Plante, I., Lonergan, K. M., and Gray, M. W. (1995). The mitochondrial DNA of the amoeboid protozoon, *Acanthamoeba castellanii*: Complete sequence, gene content and genome organization. *J. Mol. Biol.* **245**, 522–537.

Cole, R. A., and Williams, K. L. (1994). The *Dictyostelium discoideum* mitochondrial genome: A primordial system using the universal code and encoding hydrophilic proteins atypical of metazoan mitochondrial DNA. *J. Mol. Evol.* **39**, 579–588.

Crestfield, A. M., Smith, K. C., and Allen, F. W. (1955). The preparation and characterization of ribonucleic acids from yeast. *J. Biol. Chem.* **216**, 185–193.

Dirheimer, G., Keith, G., Dumas, P., and Westof, E. (1995). Primary, secondary and tertiary structures of tRNAs. *In* "tRNA: Structure Biosynthesis and Function" (D. Söll and U. L. RajBhandrary, eds.), pp. 93–126. American Society for Microbiology, Washington, DC.

Fey, J., Tomita, K., Bergdoll, M., and Maréchal-Drouard, L. (2000). Evolutionary and functional aspects of C-to-U editing at position 28 of tRNA^{Cys}(GCA) in plant mitochondria. *RNA* **6**, 470–474.

Forget, L., Ustinova, J., Wang, Z., Huss, V. A. R., and Lang, B. F. (2002). *Hyaloraphidium curvatum*: A linear mitochondrial genome, tRNA editing, and an evolutionary link to lower fungi. *Mol. Biol. Evol.* **19**, 310–319.

Gott, J. M., and Emeson, R. B. (2000). Functions and mechanisms of RNA editing. *Annu. Rev. Genet.* **34**, 499–531.

Gray, M. W. (2001). Speculations on the origin and evolution of RNA editing. *In* "RNA Editing" (B. L. Bass, ed.), pp. 160–184. Oxford University Press, Oxford.

Gray, M. W. (2003). Diversity and evolution of mitochondrial RNA editing systems. *IUBMB Life* **55**, 227–233.

Hatzoglou, E., Rodakis, G. C., and Lecanidou, R. (1995). Complete sequence and gene organization of the mitochondrial genome of the land snail *Albinaria coerulea*. *Genetics* **140**, 1353–1366.

Hopper, A. K., and Phizicky, E. M. (2003). tRNA transfers to the limelight. *Genes Dev.* **17**, 162–180.

Janke, A., and Pääbo, S. (1993). Editing of a tRNA anticodon in marsupial mitochondria changes its codon recognition. *Nucleic Acids Res.* **21**, 1523–1525.

Koslowsky, D. J. (2004). A historical perspective on RNA editing: How the peculiar and bizarre became mainstream. *In* "RNA Interference, Editing, and Modification" (J. M. Gott, ed.), Vol. 265, pp. 161–197. Humana Press, Totowa, NJ.

Laforest, M.-J., Roewer, I., and Lang, B. F. (1997). Mitochondrial tRNAs in the lower fungus *Spizellomyces punctatus*: tRNA editing and UAG 'stop' codons recognized as leucine. *Nucleic Acids Res.* **25**, 626–632.

Lavrov, D. V., Brown, W. M., and Boore, J. L. (2000). A novel type of RNA editing occurs in the mitochondrial tRNAs of the centipede *Lithobius forficatus*. *Proc. Natl. Acad. Sci. USA* **97**, 13738–13742.

Leigh, J., and Lang, B. F. (2004). Mitochondrial 3′ tRNA editing in the jakobid *Seculamonas ecuadoriensis*: A novel mechanism and implications for tRNA processing. *RNA* **10**, 615–621.

Lohan, A. J., and Gray, M. W. (2004). Methods for analysis of mitochondrial tRNA editing in *Acanthamoeba castellanii*. In "RNA Interference, Editing, and Modification" (J. M. Gott, ed.), Vol. 265, pp. 315–331. Humana Press, Totowa, NJ.

Lonergan, K. M., and Gray, M. W. (1993a). Editing of transfer RNAs in *Acanthamoeba castellanii* mitochondria. *Science* **259**, 812–816.

Lonergan, K. M., and Gray, M. W. (1993b). Predicted editing of additional transfer RNAs in *Acanthamoeba castellanii* mitochondria. *Nucleic Acids Res.* **21**, 4402.

Miller, D., Mahendran, R., Spottswood, M., Costandy, H., Wang, S., Ling, M. L., and Yang, N. (1993). Insertional editing in mitochondria of *Physarum*. *Semin. Cell Biol.* **4**, 261–266.

Parish, J. H., and Kirby, K. S. (1966). Reagents which reduce interactions between ribosomal RNA and rapidly labelled RNA from rat liver. *Biochim. Biophys. Acta* **129**, 554–562.

Price, D. H., and Gray, M. W. (1998). Editing of tRNA. In "Modification and Editing of RNA" (H. Grosjean and R. Benne, eds.), pp. 289–305. ASM Press, Washington, DC.

Price, D. H., and Gray, M. W. (1999a). Confirmation of predicted edits and demonstration of unpredicted edits in *Acanthamoeba castellanii* mitochondrial tRNAs. *Curr. Genet.* **35**, 23–29.

Price, D. H., and Gray, M. W. (1999b). A novel nucleotide incorporation activity implicated in the editing of mitochondrial transfer RNAs in *Acanthamoeba castellanii*. *RNA* **5**, 302–317.

Randau, L., Münch, R., Hohn, M. J., Jahn, D., and Söll, D. (2005). *Nanoarchaeum equitans* creates functional tRNAs from separate genes for their $5'$- and $3'$-halves. *Nature* **433**, 537–541.

Sambrook, J., Fritsch, E. F., and Maniatis, T. (1989). "Molecular Cloning: A Laboratory Manual." Cold Spring Harbor Laboratory Press, Cold Spring Harbor, NY.

Schürer, H., Schiffer, S., Marchfelder, A., and Mörl, M. (2001). This is the end: Processing, editing and repair at the tRNA $3'$-terminus. *Biol. Chem.* **382**, 1147–1156.

Spencer, D. F., Gray, M. W., and Schnare, M. N. (1992). The isolation of wheat mitochondrial DNA and RNA. In "Modern Methods of Plant Analysis New Series" (H. F. Linskens and J. F. Jackson, eds.), Vol. 14, (Seed Analysis) pp. 347–360. Springer-Verlag, Berlin.

Sprinzl, M., and Vassilenko, K. S. (2005). Compilation of tRNA sequences and sequences of tRNA genes. *Nucleic Acids Res.* **33**, D139–D140.

Sugino, A., Snopek, T. J., and Cozzarelli, N. R. (1977). Bacteriophage T4 RNA ligase. Reaction intermediates and interaction of substrates. *J. Biol. Chem.* **252**, 1732–1738.

Tomita, K., Ueda, T., and Watanabe, K. (1996). RNA editing in the acceptor stem of squid mitochondrial tRNA^{Tyr}. *Nucleic Acids Res.* **24**, 4987–4991.

Yamazaki, N., Ueshima, R., Terrett, J. A., Yokobori, S., Kaifu, M., Segawa, R., Kobayashi, T., Numachi, K. I., Ueda, T., Nishikawa, K., Watanabe, K., and Thomas, R. H. (1997). Evolution of pulmonate gastropod mitochondrial genomes: Comparisons of gene organizations of *Euhadra*, *Cepaea* and *Albinaria* and implications of unusual tRNA secondary structures. *Genetics* **145**, 749–758.

Yokobori, S., and Pääbo, S. (1995a). Transfer RNA editing in land snail mitochondria. *Proc. Natl. Acad. Sci. USA* **92**, 10432–10435.

Yokobori, S. I., and Pääbo, S. (1995b). tRNA editing in metazoans. *Nature* **377**, 490.

A-TO-I EDITING

Comparative Genomic and Bioinformatic Approaches for the Identification of New Adenosine-to-Inosine Substrates

Jaimie Sixsmith *and* Robert A. Reenan

Contents

Abstract

Adenosine-to-inosine (A-to-I) RNA editing is the enzymatic deamination of A-to-I catalyzed by ADAR (adenosine deaminase acting on RNA). Adenosine is read by ribosomes as guanosine causing a codon change and potentially protein recoding. A-to-I RNA editing can be either promiscuous, where the editing is nonspecific, or site specific, which requires a complex target RNA secondary structure formed by intramolecular base pairings between editing sequences and intronic or exonic editing site complementary sequences (ECSs). The most numerous

Department of Molecular and Cellular Biology, Brown University, Providence, Rhode Island

Methods in Enzymology, Volume 424
ISSN 0076-6879, DOI: 10.1016/S0076-6879(07)24011-X

editing sites have been found in noncoding regions containing Alu repeats, such as 3′ untranslated regions, while specific editing sites are mostly found in transcripts involved in the transmission of neuronal signals. Previously A-to-I RNA editing sites were discovered by chance, but recently investigators have used comparative genomic and bioinformatics methods to identify novel sites. In this chapter, we discuss these approaches to identifying new editing sites.

1. ADENOSINE-TO-INOSINE RNA EDITING

Proteins differing at the primary sequence level can be synthesized from the same DNA sequence by two separate posttranscriptional processes: alternative splicing and RNA editing. Following the synthesis of premessenger RNA (pre-mRNA) from the genomic DNA template, RNA editing can lead to protein recoding. One method of RNA editing involves the chemical conversion of the base adenosine (A) to the base inosine (I) in premRNA and is thus referred to as adenosine-to-inosine (A-to-I) RNA editing (Valente and Nishikura, 2005). The process is performed by the enzyme ADAR (adenosine deaminase acting on RNA) (Schaub and Keller, 2002; Seeburg, 2002). Inosine is recognized by the translation machinery as guanosine (G) (Basilio et al., 1962), causing a codon change, which may lead to an alternate amino acid being incorporated in the resulting protein. One consequence of incomplete editing is that both forms of the RNA and the protein can exist at the same time in the same cell.

Different RNA editing systems are diverse in their actions and phylogenetic distribution; however, A-to-I RNA editing is highly conserved in metazoan animals. A-to-I editing, first reported in 1988, was discovered as a result of its promiscuous editing activity, an RNA duplex unwinding/modifying activity (Bass and Weintraub, 1988). A number of different ADARs have since been discovered in various species. For example, three are present in humans (ADAR1, ADAR2, and ADAR3) (Valente and Nishikura, 2005), two in *Caenorhabditis elegans* (ADR1 and ADR2) (Tonkin et al., 2002), and one in *Drosophila melanogaster* (dADAR) (Palladino et al., 2000a).

There are two classes of A-to-I RNA editing. First, there is promiscuous nonspecific editing in which there are many edited positions. This is involved in gene regulation, such as in defense against viruses where the editing disrupts open reading frames (Bass and Weintraub, 1988). This type of A-to-I RNA editing requires extended and complete complementarity of double-stranded RNA (dsRNA) as its target (Lehmann and Bass, 1999). Promiscuous editing results in about half of the adenosines present being converted to inosine. When more than 50% of the adenosines in the transcript have been converted to inosine, the process halts, since the double-stranded RNA loses structural stability (Cattaneo et al., 1988).

It was originally believed that extensive A-to-I RNA editing might be prevalent and that gene targets of editing might be found in functionally diverse ontological classes. However, numerous discoveries of A-to-I editing were shown to possess only a single or a few edited adenosines in coding regions of genes, mostly those encoding nervous system gene products. A-to-I RNA editing resulting in a single or a few deamination events per target has been termed site-specific editing.

Specific editing requires an imperfect dsRNA duplex secondary structure (Lehmann and Bass, 1999). ADAR recognizes partially base paired dsRNA structures (more than 20 base pairs in length), which form by intramolecular base pairing between exonic sequences destined for editing and an (often intronic) editing site complementary sequence (ECS) in the pre-mRNA (Keegan et al., 2001). An ECS can be positioned hundreds of nucleotides away from the A-to-I RNA editing site (Herb et al., 1996). Mouse models lacking the ECS for the GluR-B Q/R site element have been shown to lack editing and die postnatally of an epilepsy-like syndrome; this is in an animal lacking editing at only one site. Also mutagenesis in the nucleotides surrounding the editing site has been shown to prevent ADAR catalysis (Maas et al., 2001).

Depending on which ADAR is acting upon the RNA, there is a 5' preference for particular neighboring bases at the potential editing site. In humans, if ADAR1 is acting, then the preference in 5' neighboring bases is uracil followed by adenosine, cytosine, and guanosine (Polson and Bass, 1994), while ADAR2 prefers a 5' neighbor to be uracil or adenosine then cytosine or guanosine equally and a 3' neighbor to be uracil or guanosine then cytosine or adenosine equally (Lehmann and Bass, 2000). Also ADAR1 will not edit an adenosine that is closer than three nucleotides to the 5' end of a duplex or eight nucleotides from the 3' end (Polson and Bass, 1994). ADAR2 appears to lack this limitation (Lehmann and Bass, 2000).

ADARs from all species studied contain two distinct sets of domains. First, there are dsRNA binding domains, the number of which varies between ADARs. For instance, there are three present in human ADAR1, while only two are present in human ADAR2 (Keegan et al., 2004). There is also a C-terminal deaminase domain responsible for the catalytic activity of the enzyme (Kim et al., 1994; O'Connell et al., 1995). Certain structures identify certain ADAR families such as the zinc binding domain in humans in ADAR1 (Herbert et al., 1997) and the arginine/lysine-rich region in ADAR3 (Chen et al., 2000).

ADARs bind to RNA as a dimer in both animals and flies (Gallo et al., 2003). Dimerization may allow for the correct alignment of the active site for its access to the target adenosine. In mammals, ADAR1 and ADAR2 homodimerize rather than forming heterodimers with each other, but ADAR3 possesses no detectable catalytic activity in mammals (even for nonspecific editing), most likely due to the fact that it does not appear to dimerize (Cho et al., 2003).

In mammals, the most abundant A-to-I RNA editing sites are located in noncoding regions, such as untranslated regions, introns, and noncoding RNAs, where editing could alter processes such as alternative splicing (Rueter et al., 1999). Mostly, these include areas of repetitive sequences, such as Alu repeats, which are abundant and homologous in humans (Athanasiadis et al., 2004; Lander et al., 2001). Alu regions can fold back into an extensive hairpin structure, which acts as the double-stranded substrate for ADAR activity (Hess et al., 1985). This was emphasized in mouse studies, which lack Alu elements and display greatly reduced numbers of A-to-I RNA editing sites (Kim et al., 2004).

In many organisms, including humans and D. melanogaster, A-to-I RNA editing sites have frequently been found to be present in transcripts expressed in the central nervous system (CNS) and, in particular, in genes encoding proteins necessary for the transmission of rapid electrical and chemical signals. The transcripts targeted for A-to-I RNA editing in Drosophila encode ligand and voltage-gated ion channels and presynaptic release proteins (Hoopengardner et al., 2003). The "model" example of A-to-I RNA editing in the CNS is the site-specific editing of the mammalian GluR–B glutamine (Q)/arginine (R) site by ADAR2. The editing leads to a glutamine codon being replaced by an arginine, which decreases the calcium permeability of the ion channel. Lack of editing is extremely rare with this editing site and generates a channel more permeable to calcium ions (Sommer et al., 1991). Also, one receptor for the neurotransmitter serotonin is altered by A-to-I editing, leading to a reduction in G protein coupling to the receptor (Burns et al., 1997).

There are often serious consequences when ADAR is deficient. Mouse embryos lacking ADAR1 are embryonic lethal and manifest liver disintegration (Hartner et al., 2004). A deficiency in ADAR2 has also been shown to lead to seizures and higher postnatal mortality in mice (Higuchi et al., 2000). In D. melanogaster, lack of editing has been shown to result in tremors, lack of coordination, mating defects, and neurodegeneration (Palladino et al., 2000b). In humans, abnormal levels of editing have also been linked to suicidal depression (Yang et al., 2004).

Although a number of editing sites had been identified by the early 2000s, these were often discovered by chance. Therefore, methods were necessary to directly search for these sites. We believe that bioinformatics and comparative genomics are powerful tools for the identification of new A-to-I RNA editing sites.

2. BIOINFORMATICS

Bioinformatics has been one of the most rapidly growing areas of scientific investigation in recent years. Specifically, bioinformatics involves the formulation of algorithms plus statistical and computational methods, which are used to answer questions created by the collection and analysis of biological data. It is important to separate bioinformatic approaches from those of computational biology, even though the two terms are frequently used interchangeably. Bioinformatics involves technique–driven research, while computational biology is hypothesis driven for the analysis of biological data using computers. Computational biology utilizes both experimental and simulated data, with the primary aim of discovery of novel biological information.

Much bioinformatic software is now available to perform these tasks, of which the best documented and widely used is Basic Local Alignment Search Tool (BLAST). The National Center for Biotechnology Information (NCBI) provides the BLAST service to allow comparison of new unknown sequences with previously sequenced genes via rapid searching of nucleotide and protein databases. This may provide important clues to the function of uncharacterized proteins. The NCBI BLAST website is available at http://www.ncbi.nlm.nih.gov/BLAST/. NCBI also provides ENTREZ, which is a system that allows access to numerous NCBI databases for retrieval of relevant sequence and order information.

Another example of available software is EMBOSS at http://emboss. sourceforge.net/. EMBOSS stands for "The European Molecular Biology Open Software Suite," and the software automatically handles data in a number of formats and can even retrieve sequence data from the Internet. EMBOSS provides extensive libraries, thus allowing other scientists to develop and release software in an open spirit. EMBOSS also combines a number of currently available packages and tools for sequence analysis into its setup. Many more are available, which can be accessed by a web search.

A theoretical bioinformatic pipeline is displayed in Fig. 11.1. The ability to search for areas of high conservation or difference between genes and species that bioinformatics has presented has led to great advancements in areas such as the study of evolutionary biology, biodiversity, drug discovery, and prediction of protein structure. One of the most important areas on which bioinformatics has had an impact is the field of comparative genomics.

3. COMPARATIVE GENOMICS

Although there is a large diversity of species, they have one commonality, DNA. In an ideal world, the complete sequence of every existing genome would be known and available, with every polymorphism documented. However, this is not realistic, so methods are needed to identify and classify the genes present in organisms. One such method is comparative genomics, which is the study of the similarities and differences between genes from separate species with the goal of estimating how closely species are related on the evolutionary scale.

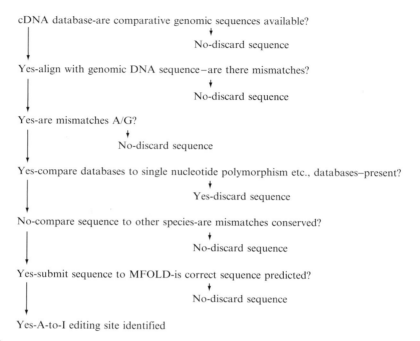

cDNA database-are comparative genomic sequences available?

No-discard sequence

Yes-align with genomic DNA sequence – are there mismatches?

No-discard sequence

Yes-are mismatches A/G?

No-discard sequence

Yes-compare databases to single nucleotide polymorphism etc., databases – present?

Yes-discard sequence

No-compare sequence to other species-are mismatches conserved?

No-discard sequence

Yes-submit sequence to MFOLD-is correct sequence predicted?

No-discard sequence

Yes-A-to-I editing site identified

Figure 11.1 An example of a simplified bioinformatics pipeline to identify A-to-I RNA editing sites. A database of cDNAs is subjected to a number of parameters to determine if it contains any A-to-I RNA editing sites. Each parameter narrows the list of candidates for A-to-I RNA editing sites, until all of the criteria are met and a site can be confirmed to the investigator's satisfaction.

Comparative genomics can provide information on the control of gene function, which, in a narrow view, could lead to development of novel medical treatments. On a wider, less human-biased scale, the goal of comparative genomics is to discover the true evolutionary relationship between different species and to resolve different versions of evolutionary trees. However, for comparative genomics to be most helpful as a technique, the data about genomes should be available in large, continuous, and sequenced pieces. Computer software must also be available to allow easy access and analysis.

One of the most important uses of comparative genomics has been to understand the vast data derived from the Human Genome Project, a 13-year project headed by the U.S. Department of Energy and the National Institutes of Health, incorporating the British Wellcome Trust and Celera Genomics, as well as being aided by worldwide efforts from countries such as Japan, France, Germany, and China. The sequencing part of this project was completed in 2003.

However, the success of comparative genomics has been hindered by the vast number of different and complicated events that have occurred throughout evolution. Thus, the use of smaller organisms, where numerous species have been sequenced, has taken on much prominence in the field of comparative genomics, for example, the fruit fly *D. melanogaster.*

Because of the success of comparative genomics, the demand for sequences from various species has increased multifold. For this reason, the National Human Genome Research Institute (NHGRI) has set up a priority setting process as of 2001, which evaluates the benefits (medical, agricultural, and biological opportunities) that sequencing a particular species would bring.

In 2000, it was discovered that fruit flies share nearly 60% of human genes. The importance of *D. melanogaster* in comparative genomics, plus other similar fields, cannot be underestimated. For instance, its use in the study of A-to-I RNA editing has been crucial in the understanding of the process. ADAR-deficient flies have been created and the major phenotypic consequences, such as locomotor and mating defects, have indicated the importance of A-to-I editing in the *Drosophila* nervous system (Palladino *et al.,* 2000b).

There have recently been a number of novel A-to-I RNA editing sites identified using bioinformatic and comparative genomic methods. In this chapter, we review what we see as the most interesting and important papers describing this work with regard to sound scientific methodology and results. We cover each article in the same manner, by highlighting the reasoning behind the work and the methods that were used, as well as the results that were gained. We hope that this will be useful as a reference to any investigator who is proposing future work to identify new A-to-I RNA editing sites.

4. NERVOUS SYSTEM TARGETS OF RNA EDITING IDENTIFIED BY COMPARATIVE GENOMICS (HOOPENGARDNER *ET AL.,* 2003)

The formation of the structure required for ADAR binding has already been described. It was hypothesized that intronic *cis* elements, which were involved in forming the editing site duplex structure, would be conserved between species if A-to-I RNA editing was also conserved. An A-to-I editing site of the *paralytic* Na^+ channel gene was investigated in various *Drosophila* species (*D. melanogaster Canton-S* and *Oregon-R, pseudoobscura, virilis, melanica, mercatorum,* and *Chymomyza procnemis*).

The coding sequence flanking the editing site was found to be identical in all these species. These highly conserved exonic region sequences near to sites of A-to-I editing were identified and recognized as a signature of ADAR substrates. This conservation was used to potentially identify novel editing sites in *Drosophila*. A comparative genomic analysis between *D. melanogaster* and *D. pseudoobscura* examined 914 genes using TraceBlast bioinformatics (http://www.ncbi.nlm.nih.gov/blast/tracemb.html). Of these genes, 135 were annotated as ion channels, 178 as G protein–coupled receptors (GPCRs), 102 as synaptic transmission proteins, and 499 as transcription factors. The algorithm for the study looked primarily for exonic regions with 98% identity or greater. Large areas of conservation (50 base pairs plus) were given a priority. Sequences were ranked on percentage of sequence identity.

It was discovered that 41 of these genes had a high potential for A-to-I editing because of the high sequence conservation they displayed, which was similar to known targets of ADAR. The relevant region of each gene was sequenced in wild-type and ADAR-deficient backgrounds following reverse transcriptase polymerase chain reaction (RT-PCR) and a mixed peak of adenosine and guanosine signals was searched for in the ADAR positive backgrounds. When these signals were found (in 16 of the 41 genes), a pure adenosine peak was always seen in the corresponding position in sequenced products from ADAR-lacking animals. Examples of the use of direct sequencing to identify A-to-I RNA editing sites are displayed in Figs. 11.2 and 11.3.

The 16 genes targeted by ADAR were all involved in the transmission of neuronal signals and were classified as either ligand-gated ion channels [including a γ-aminobutyric acid (GABA) receptor and nicotinic acetyl-choline receptor α and β subunits], voltage-gated ion channels (sodium, calcium, and potassium channels), or components of presynaptic release machinery (including calcium sensors and adaptor proteins, such as synaptotagmin).

The orthologue of the *Shaker* gene (hKv1.1) was found to be edited in human, rat, and mouse by using similar comparative genomic techniques, demonstrating the conserved ADAR bases mechanism. A wide range of editing levels was displayed across human brain regions (17–77%). Results from rat and mouse brain displayed similar spatial regulation of A-to-I RNA editing to humans, confirming that the regulation of editing site was conserved between these species.

5. MOLECULAR DETERMINANTS AND GUIDED EVOLUTION OF SPECIES-SPECIFIC RNA EDITING (REENAN, 2005)

Most A-to-I RNA editing, which causes recoding, occurs in genes that affect neuronal signaling. *Drosophila synaptotagmin I* (*dsytI*) is the Ca^{2+} sensor for neurotransmitter release (Yoshihara and Littleton, 2002) and a

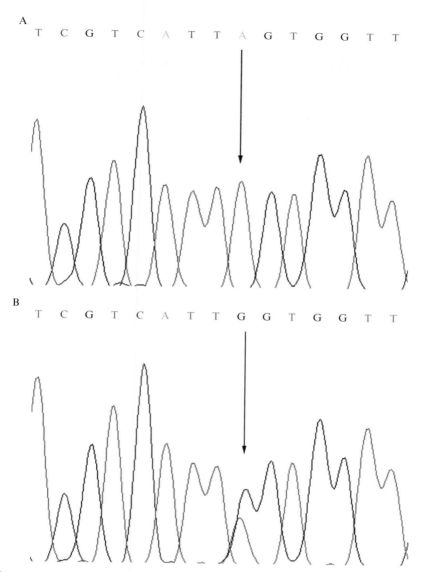

A

T C G T C A T T A G T G G T T

B

T C G T C A T T G G T G G T T

Figure 11.2 Comparison of electropherograms to display A-to-I RNA editing. (A) Electropherogram of ADAR null *Drosophlia* displaying a sequence without A-to-I RNA editing. The arrow indicates only a pure A signal (green). (B) Electropherogram of ADAR^{+} *Drosophila* of the same region of the sequence with A-to-I RNA editing. The arrow indicates a mixed A/G signal (green/black). (See color insert.)

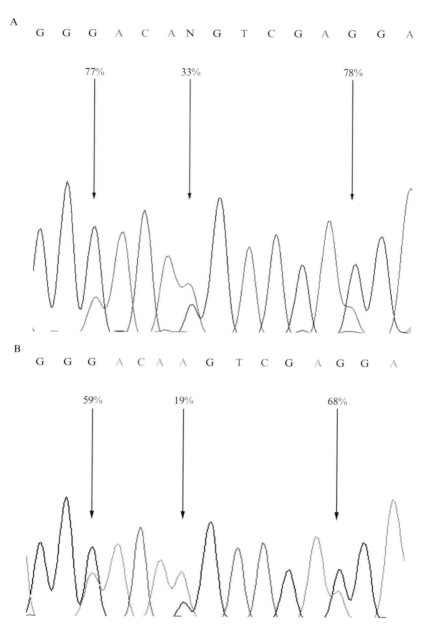

Figure 11.3 Direct sequencing method for quantification of A-to-I RNA editing sites. The presence of more than one editing site in close proximity allows comparisons of electropherograms to measure changes in levels of editing between species (A) *Drosophila hydei* and (B) *Drosophila eohydei*. The percentage of editing at each site is indicated above the relevant peak. This figure is the area of the G peak (black) as a percentage of the combined areas of both mixed peaks, including the A peak (green). (See color insert.)

target for ADAR (Hoopengardner *et al.*, 2003), with one exon containing four A-to-I RNA editing sites. These editing sites affect the *SytI* C2B domain, which is essential for synchronous neurotransmitter release (Craxton, 2001). Studies reported that A-to-I RNA editing differed between arthropod species (Grauso *et al.*, 2002; Hanrahan *et al.*, 2000). To determine any variation in editing between species in *sytI*, a number of insects were studied: fruit fly (*D. melanogaster*), malaria mosquito (*Anopheles gambiae*), tobacco hawkmoth (*Manduca sexta*), honeybee (*Apis mellifera*), red flour beetle (*Tribolium castaneum*), and German cockroach (*Blatella germanica*).

Following cDNA synthesis and RT-PCR, the DNA products of each of these were gel purified and directly sequenced to identify the editing sites. The electropherograms confirmed that *SytI* was not edited in all species (honeybees, cockroaches, beetles) and even when it was, no two species shared the same set of four editing sites. Among the species that displayed editing, there was one common site in all species, while each species also had its own unique editing sites. To confirm the unique editing sites, close relations from the same taxonomic group were examined.

Conserved specific ADAR modification between species is indicative of conservation of the *cis* elements that control RNA structure formation. To explore this, the genomic regions around the editing sites of 10 of the Drosophilidae family were cloned and sequenced. This identified two conserved intronic sequences for all of the family members. This intron and its two neighboring exons were expressed as a minigene in *Drosophila* S2 cells along with *dADAR*. This minigene transcript underwent RT-PCR and the level of RNA editing in the products was determined by the presence or absence of certain restriction sites. The intensity of bands on an agarose gel was quantified using Kodak 1D v.3.6 image analysis software to determine the level of editing. A strong level of editing in these cells at two of the editing sites (the universal site and a site shared by fruit fly and mosquito) was detected. There was no editing at other unexpected adenosines or when there was no coexpressed *dADAR* present (data not presented).

MFOLD software (http://mfold.burnet.edu.au/) was used to generate potential structures incorporating the two new elements. A pseudoknot could be predicted with the intronic elements pairing with separate exonic regions to create two domains. This structure was confirmed when mutations were introduced in each domain that prevented editing at the relevant sites independently. Structurally compensatory double mutations restored editing. This suggested that the structure had two areas of separate ADAR action.

The universal and unique A-to-I RNA editing sites of lepidopterans were examined in 10 species of moths and butterflies and revealed a single highly conserved region in the intronic region downstream of the editing sites. A similar experiment was undertaken with S2 cells, *dADAR,* and the genomic region containing the conserved intronic sequence with its flanking exons from *M. sexta sytI*. Strong editing was present in both sites, showing that the intron could direct species-specific A-to-I RNA editing (rather than ADAR enzymes). MFOLD predicted a structure with conserved intronic sequences binding to the region of the editing sites. This displayed two domains with separate areas of editing, which was different from the pseudoknot of the Drosophilidae. This same structure was consistent for all of the moths and butterflies examined, despite differences in intron sequence. Two separate mutational analysis experiments altered the editing efficiency of each site individually but not of the other, confirming that they were separable.

Two of the editing sites shared by *A. gambiae* and *Aedes aegypti* (mosquito) were the same sites that were in the original pseudoknot, but with an additional mosquito specific site between them. Sequence analysis of five mosquito species revealed two conserved intronic regions similar to *Drosophila*. Structural analysis predicted another pseudoknot with some minor differences compared to the *Drosophila* species. In expression analysis, *dADAR* recognized the third editing site in *A. gambiae* reporter constructs.

It was suggested that the common editing site seen in all of the species was ancestral to the other editing sites, beginning in the common ancestor to the lepidopteran and dipteran lines. This was either through differences in the intron regions causing the creation of new conserved *cis* domains or by editing sites progressing through slight alterations of sequence to generate new editing sites by structural changes.

6. SYSTEMATIC IDENTIFICATION OF ABUNDANT ADENOSINE-TO-INOSINE EDITING SITES IN THE HUMAN TRANSCRIPTOME (LEVANON *ET AL.*, 2004)

Previous use of bioinformatic methods to search for new A-to-I RNA editing sites was hindered by sequencing mistakes existing in available databases, plus the risk of falsely identifying single nucleotide polymorphisms (SNPs) as editing sites. However, investigators addressed this problem by designing a fresh bioinformatic approach, which used an algorithm with certain regulations for sequence alignment and comparison.

Human expressed sequence tags (ESTs) and cDNAs (from NCBI GenBank-http://www.ncbi.nlm.nih.gov/dbEST) were aligned to the genome (from Human Genome Build-http://www.ncbi.nlm.nih.gov/genome/guide/human) and were ordered into genes or parts of genes if they displayed at least 94% similarity to their genome. As dsRNA is a target for ADAR, the algorithm was automated to align the expressed part of the gene with its corresponding genome and search only for reverse complement alignments in excess of 32 base pairs with more than 85% bases matching. The algorithm also removed SNPs, random mismatches, and mutations, plus repeats of more than four base pairs; the 150 base pairs at each end of the sequence and 50 nucleotide stretches with five or higher mismatches (unless four or more were present consecutively) were ignored because of potential sequencing errors. Any mismatches that were present in fewer than 5% of the sequences or less than three times in an exon were ignored.

The algorithm initially revealed 12,723 potential A-to-I RNA editing sites in 1637 genes. To confirm that the predicted editing sites were actual ADAR substrates, experimental validation investigated 30 of these genes. For these genes, total RNA and genomic DNA were obtained from the same tissue (a variety of areas, including brain, liver, and lung). RT was used to produce cDNA before both this and the genomic DNA underwent PCR to produce short sequences of about 200 base pairs. Following gel electrophoresis, bands of correct size were sequenced. These sequences were aligned and screened for A-to-I editing sites. This confirmed 26 new editing sites, while editing was confirmed in all of the tissues analyzed for the majority of the genes. The unedited signal mainly dominated the edited signal.

Two of the genes were validated using HEK293 and HeLa cell lines, as these possessed varying ADAR activity levels. Poly(A)$^+$ RNA was isolated from tissue culture cells and collected using magnetic beads, before it underwent RT, as was genomic DNA. First strand cDNAs and their equivalent genomic sequences were amplified and then had a polyadenosine tail attached. These were purified by gel electrophoresis and extraction before they were cloned for transformation in *Escherichia coli*. Individual plasmids were sequenced and aligned. Alignment utilized ClustalW (http://www.ebi.ac.uk/clustalw/), a sequence alignment program for DNA and proteins, which produces multiple sequence alignments of divergent sequences. It calculated the best match for the selected sequences (sequencing was performed from both ends) and aligned them for comparison. There was a strong correlation between the levels of A-to-I editing and the known ADAR activity of the cell lines.

A-to-I RNA editing was seen to occur mainly in noncoding regions (12% in 5′ UTR, 54% in 3′ UTR, 33% in introns), in particular in Alu regions. By reviewing the best complimentary alignment of all of the predicted editing sites, it was suggested that A-to-I RNA editing would occur only in a position where it could affect double-strand RNA stability, in particular make it more stable.

7. EVOLUTIONARY CONSERVED HUMAN TARGETS OF ADENOSINE-TO-INOSINE RNA EDITING (LEVANON *ET AL.*, 2005)

This study identified novel targets for A-to-I RNA editing in humans using comparative genomics and expressed sequence analysis. Because the few mammalian A-to-I RNA editing sites already discovered could not explain all of the phenotypes exhibited in ADAR1-deficient mice, it was hoped that new targets would result (Hartner *et al.*, 2004; Higuchi *et al.*, 2000). A large bank of potential A-to-I RNA editing sites was created for both humans and mice (human ESTs and cDNAs from NCBI GenBank) by alignment of expressed sequences with their corresponding genomes (from the Human Genome Build). When a mismatch was found between expressed and genomic, a specially designed algorithm was used to identify whether it was actually due to an editing site or an SNP.

In brief, the algorithm aligned all the expressed sequences against the genome and arranged them into genes with overlapping expressed sequences and corresponding genomic sequences multiply aligned. This multiple alignment model was designed in a previous study (Sorek *et al.*, 2002). It was a spliced alignment model allowing long gaps and only sequences having >94% identity to the genome. Sequences that mapped to two or more chromosomes and sequences with inferred introns longer than 400,000 base pairs were discarded, as were low-quality sequence ends that disagreed with the genomic DNA.

For each gene, mismatches between expressed and genomic sequences were identified and the probability of each being due to RNA editing/ SNPs or sequencing/alignment errors was estimated (which is described in depth in the paper). As long as this probability of the event being due to RNA editing/SNP exceeded the designated cutoff level, while the probability of it being due to either sequencing error/alignment mistakes did not, then the nucleotide difference was declared to be the result of RNA editing or an SNP. The algorithm discarded any sequences that were rated as unimportant for consideration, such as regions of low alignment quality, low multiple alignment quality, or repetitive sequences of single bases/mismatches.

This algorithm yielded potential A-to-I RNA editing sites in both humans and mice, which then were aligned against each other. Alignments were discarded if they were less than 50 nucleotides in length, displayed less than 85% identity, or lacked the mismatches at the same positions. Duplicated genomic sequences were also ignored. Four new substrates for A-to-I RNA editing were identified in humans. These sites were the bladder cancer associated protein (BLCAP), filamin A protein (FLNA), cytoplasmic FMR1 interacting protein 2 (CYFIP2), and insulin-like growth factor binding protein 7 (IGPFB7).

To confirm these results, DNA and RNA were obtained from the same specimens of human and mouse from up to six tissues. For human RNA, RT-PCR produced cDNA, and both this and the genomic DNA underwent PCR to produce sequences of approximately 200 nucleotides. Following gel electrophoresis, only bands of the correct predicted size were sequenced. For mouse tissue, poly(A)$^+$ was isolated and collected on magnetic beads before it underwent RT-PCR. The PCR products had adenosine tails added and were cloned and transformed into *E. coli* before being sequenced and aligned with ClustalW. The three tissues that expressed the highest level of editing were sequenced for both human and mouse.

Each of the four A–to–I RNA editing sites was confirmed in humans, while only IGFBP7 was not identified in mouse. Tissues were obtained from chicken (treated by same protocol as mouse) and editing in both CYFIP2 and FLNA was verified. MFOLD was used to predict the structures of each new editing site.

8. IS ABUNDANT ADENOSINE-TO-INOSINE RNA EDITING PRIMATE SPECIFIC? (EISENBERG *ET AL.*, 2005)

Previous work indicated that many human A–to–I RNA editing sites were present in Alu elements (Kim *et al.*, 2004; Levanon *et al.*, 2004). These Alu regions were found only in primates (Kapitonov and Jurka, 1996; Schmid, 1996), so it was therefore hypothesized that the number of editing sites might be related to some characteristic of the Alu region and primate specific. Therefore, similar editing levels would not be seen in other organisms.

Human and mouse expressed sequences (from NCBI GenBank) were aligned with their corresponding genomes (from Human Genome Build) and screened for A–to–I RNA editing sites using the algorithm detailed previously (Levanon *et al.*, 2004). Terminal vector, low complex, and repeat sequences in expressed sequences were masked and greater than 94% alignment was required. This identified 12,723 potential ADAR substrates in 1637 human genes, but only 302 similar sites in 87 mouse genes.

A second screen for editing sites used a different sequence alignment method, where 128,068 human and 102,895 mouse RNAs (from UCSC genome browser–http://genome.ucsc.edu) were aligned against their respective genomes. Again, a large difference was displayed in potential editing site numbers, approximately 50,000 in humans to 3000 in mice. As clustering of ADAR substrates was common, regions were screened for, with three or more editing sites grouped locally. In humans, only about 4000 such RNA were identified, but only about 220 were present in mice.

Humans had displayed an order of magnitude higher level of A–to–I RNA editing than mice, so the single and multiple editing site screens were repeated with rat, chicken, and fly models (from UCSC genome browser) to determine if this was either primate or rodent specific. These three species revealed results similar to the mouse model, which indicated that there was, indeed, a primate increase in repeat effect on A–to–I RNA editing. To confirm that these results were not due to tissue bias, the same tissues (brain, thymus, testis) from both human and mouse were screened as in the primary experiment. Significant editing was detected in all of the human tissues, while it was undetectable in mouse tissues. Most of the editing in humans was detected in the Alu elements, so these were possibly responsible for the higher order of magnitude of A-to-I RNA editing between humans and the other species.

9. A Bioinformatics Screen for Novel Adenosine-to-Inosine RNA Editing Sites Reveals Recoding Editing in BC10 (Clutterbuck *et al.*, 2005)

By reviewing the known characteristics of A-to-I RNA editing sites, a bioinformatic protocol was designed to identify new editing substrates. The protocol contained seven predictive features to screen mismatches identified between human and mouse expressed and genomic sequences (Table 11.1). Human data were used only to confirm the mouse data.

Exon sequences were obtained from Ensembl (http://www.ensembl.org/index.html) to be used as reference, while GluR–6 was obtained from GenBank to make computational methods simpler. These contained nonredundant sequences for most of the known human and mouse genes, but included most of the known recoding A-to-I RNA editing sites. Orthologous mouse–human pairs (Ensembl) were obtained and in the case of multiple homologues, the best was chosen on the basis of sequence conservation and mismatches.

MegaBlast (http://www.ncbi.nlm.nih.gov/blast/megablast.shtml) was used to identify any mismatches between the Ensembl mouse and human genes and available expressed sequences and their genomes. This was a refined BLAST search, which included an algorithm for a DNA sequence gapped

Table 11.1   Seven predictive features for identifying potential A-to-I RNA editing sites

Predictive feature	Description
1	Number of putatively edited mouse or cDNAs or ESTs with conserved mismatch
2	Number of unedited mouse cDNAs/ESTs and publicly available genomic sequences for a mismatch
3	Is mouse editing site conserved at same position in humans?
4	Does editing site alter amino acid sequence?
5	What is most conserved 120 base pair window across editing site?
6	Inverted repeats present?
7	Cluster of editing sites?

alignment search. Parameters were included to ignore any BLAST matches of fewer than 100 base pairs or that shared less than 98% homology. Sequences were obtained from the databases of dbEST, EMBL (http://www.ebi.ac.uk/embl/), GenBank, and Ensembl. An editing region was defined as an exon containing at least one editing site, which occurred with an Ensembl transcript. Known recoding A-to-I RNA editing sites acted as positive controls for the experiment.

Analysis of the known recoding A-to-I RNA editing sites confirmed that sequence conservation and cluster analysis were good predictors of this type of site. Mouse ECS elements were successfully identified and could be scored by the quality of the best predicted inverted repeat within 2000 bases of the potential editing site.

Initially, 28,992 A-to-G mismatches were identified and these were all analyzed using the seven predictive features. The results of these were combined to produce a log-odds (LOD) score for each mismatch, by which they could all be ranked. Also included were filters, which removed SNPs, as well as errors because of sequencing and alignment.

Most of the positive controls were successfully identified by LOD score as well as sequence conservation, clustering, and ECS score. The 10 top ranked potential recoding A-to-I RNA editing sites where further studied by PCR in mouse brain and heart tissue for the presence of editing. Whole RNA and genomic DNA were extracted and underwent PCR. The correct bands identified by gel electrophoresis were extracted and sequenced. The top four ranked were also cloned and sequenced. Only one, BC10, met the criteria of the screen.

BC10 had an ECS element upstream and it was 99% conserved between mouse and human. It contained three recoding editing sites, all of which were in the 5′ UTR and the N-terminus of the protein. Finally, there was a high level of editing in the brain tissue. The editing sites were conserved through a number of species. BC10 has been implicated in bladder and renal cancer (Gromova et al., 2002; Rae et al., 2000). The results indicated that using a number of features is a powerful tool to identify new A-to-I RNA editing sites.

10. CONCLUSION

The papers discussed have highlighted the recent advances in both comparative genomics and bioinformatics, which have led to the successful identification of novel substrates for A-to-I RNA editing. The list of ADAR substrates will undoubtedly keep growing as databases keep expanding with new biological information and bioinformatics tools become more powerful.

REFERENCES

Athanasiadis, A., Rich, A., and Maas, S. (2004). Widespread A-to-I RNA editing of Alu-containing mRNAs in the human transcriptome. *PLoS Biol.* **2**, e391.

Basilio, C., Wahba, A. J., Lengyel, P., Speyer, J. F., and Ochoa, S. (1962). Synthetic polynucleotides and the amino acid code. V. *Proc. Natl. Acad. Sci. USA* **48**, 613–616.

Bass, B. L., and Weintraub, H. (1988). An unwinding activity that covalently modifies its double-stranded RNA substrate. *Cell* **55**, 1089–1098.

Burns, C. M., Chu, H., Rueter, S. M., Hutchinson, L. K., Canton, H., Sanders-Bush, E., and Emeson, R. B. (1997). Regulation of serotonin-2C receptor G-protein coupling by RNA editing. *Nature* **387**, 303–308.

Cattaneo, R., Schmid, A., Eschle, D., Baczko, K., ter Meulen, V., and Billeter, M. A. (1988). Biased hypermutation and other genetic changes in defective measles viruses in human brain infections. *Cell* **55**, 255–265.

Chen, C. X., Cho, D. S., Wang, Q., Lai, F., Carter, K. C., and Nishikura, K. (2000). A third member of the RNA-specific adenosine deaminase gene family, ADAR3, contains both single- and double-stranded RNA binding domains. *RNA* **6**, 755–767.

Cho, D. S., Yang, W., Lee, J. T., Shiekhattar, R., Murray, J. M., and Nishikura, K. (2003). Requirement of dimerization for RNA editing activity of adenosine deaminases acting on RNA. *J. Biol. Chem.* **278**, 17093–17102.

Clutterbuck, D. R., Leroy, A., O'Connell, M. A., and Semple, C. A. (2005). A bioinformatic screen for novel A-I RNA editing sites reveals recoding editing in BC10. *Bioinformatics* **21**, 2590–2595.

Craxton, M. (2001). Genomic analysis of synaptotagmin genes. *Genomics* **77**, 43–49.

Eisenberg, E., Nemzer, S., Kinar, Y., Sorek, R., Rechavi, G., and Levanon, E. Y. (2005). Is abundant A-to-I RNA editing primate-specific? *Trends Genet.* **21**, 77–81.

Gallo, A., Keegan, L. P., Ring, G. M., and O'Connell, M. A. (2003). An ADAR that edits transcripts encoding ion channel subunits functions as a dimer. *EMBO J.* **22**, 3421–3430.

Grauso, M., Reenan, R. A., Culetto, E., and Sattelle, D. B. (2002). Novel putative nicotinic acetylcholine receptor subunit genes, Dalpha5, Dalpha6 and Dalpha7, in *Drosophila melanogaster* identify a new and highly conserved target of adenosine deaminase acting on RNA-mediated A-to-I pre-mRNA editing. *Genetics* **160**, 1519–1533.

Gromova, I., Gromov, P., and Celis, J. E. (2002). bc10: A novel human bladder cancer-associated protein with a conserved genomic structure downregulated in invasive cancer. *Int. J. Cancer* **98**, 539–546.

Hanrahan, C. J., Palladino, M. J., Ganetzky, B., and Reenan, R. A. (2000). RNA editing of the *Drosophila para* Na(+) channel transcript. Evolutionary conservation and developmental regulation. *Genetics* **155**, 1149–1160.

Hartner, J. C., Schmittwolf, C., Kispert, A., Muller, A. M., Higuchi, M., and Seeburg, P. H. (2004). Liver disintegration in the mouse embryo caused by deficiency in the RNA-editing enzyme ADAR1. *J. Biol. Chem.* **279**, 4894–4902.

Herb, A., Higuchi, M., Sprengel, R., and Seeburg, P. H. (1996). Q/R site editing in kainate receptor GluR5 and GluR6 pre-mRNAs requires distant intronic sequences. *Proc. Natl. Acad. Sci. USA* **93**, 1875–1880.

Herbert, A., Alfken, J., Kim, Y. G., Mian, I. S., Nishikura, K., and Rich, A. (1997). A Z-DNA binding domain present in the human editing enzyme, double-stranded RNA adenosine deaminase. *Proc. Natl. Acad. Sci. USA* **94**, 8421–8426.

Hess, J., Perez-Stable, C., Wu, G. J., Weir, B., Tinoco, I., Jr., and Shen, C. K. (1985). End-to-end transcription of an Alu family repeat. A new type of polymerase-III-dependent terminator and its evolutionary implication. *J. Mol. Biol.* **184**, 7–21.

Higuchi, M., Maas, S., Single, F. N., Hartner, J., Rozov, A., Burnashev, N., Feldmeyer, D., Sprengel, R., and Seeburg, P. H. (2000). Point mutation in an AMPA receptor gene rescues lethality in mice deficient in the RNA-editing enzyme ADAR2. *Nature* **406**, 78–81.

Hoopengardner, B., Bhalla, T., Staber, C., and Reenan, R. (2003). Nervous system targets of RNA editing identified by comparative genomics. *Science* **301,** 832–836.

Kapitonov, V., and Jurka, J. (1996). The age of Alu subfamilies. *J. Mol. Evol.* **42,** 59–65.

Keegan, L. P., Gallo, A., and O'Connell, M. A. (2001). The many roles of an RNA editor. *Nat. Rev. Genet.* **2,** 869–878.

Keegan, L. P., Leroy, A., Sproul, D., and O'Connell, M. A. (2004). Adenosine deaminases acting on RNA (ADARs): RNA-editing enzymes. *Genome Biol.* **5,** 209.

Kim, D. D., Kim, T. T., Walsh, T., Kobayashi, Y., Matise, T. C., Buyske, S., and Gabriel, A. (2004). Widespread RNA editing of embedded alu elements in the human transcriptome. *Genome Res.* **14,** 1719–1725.

Kim, U., Wang, Y., Sanford, T., Zeng, Y., and Nishikura, K. (1994). Molecular cloning of cDNA for double-stranded RNA adenosine deaminase, a candidate enzyme for nuclear RNA editing. *Proc. Natl. Acad. Sci. USA* **91,** 11457–11461.

Lander, E. S., Linton, L. M., Birren, B., Nusbaum, C., Zody, M. C., Baldwin, J., Devon, K., Dewar, K., Doyle, M., FitzHugh, W., Funke, R., Gage, D., *et al.* (2001). Initial sequencing and analysis of the human genome. *Nature* **409,** 860–921.

Lehmann, K. A., and Bass, B. L. (1999). The importance of internal loops within RNA substrates of ADAR1. *J. Mol. Biol.* **291,** 1–13.

Lehmann, K. A., and Bass, B. L. (2000). Double-stranded RNA adenosine deaminases ADAR1 and ADAR2 have overlapping specificities. *Biochemistry* **39,** 12875–12884.

Levanon, E. Y., Eisenberg, E., Yelin, R., Nemzer, S., Hallegger, M., Shemesh, R., Fligelman, Z. Y., Shoshan, A., Pollock, S. R., Sztybel, D., Olshansky, M., Rechavi, G., *et al.* (2004). Systematic identification of abundant A-to-I editing sites in the human transcriptome. *Nat. Biotechnol.* **22,** 1001–1005.

Levanon, E. Y., Hallegger, M., Kinar, Y., Shemesh, R., Djinovic-Carugo, K., Rechavi, G., Jantsch, M. F., and Eisenberg, E. (2005). Evolutionarily conserved human targets of adenosine to inosine RNA editing. *Nucleic Acids Res.* **33,** 1162–1168.

Maas, S., Patt, S., Schrey, M., and Rich, A. (2001). Underediting of glutamate receptor GluR-B mRNA in malignant gliomas. *Proc. Natl. Acad. Sci. USA* **98,** 14687–14692.

O'Connell, M. A., Krause, S., Higuchi, M., Hsuan, J. J., Totty, N. F., Jenny, A., and Keller, W. (1995). Cloning of cDNAs encoding mammalian double-stranded RNA-specific adenosine deaminase. *Mol. Cell. Biol.* **15,** 1389–1397.

Palladino, M. J., Keegan, L. P., O'Connell, M. A., and Reenan, R. A. (2000a). dADAR, a Drosophila double-stranded RNA-specific adenosine deaminase is highly developmentally regulated and is itself a target for RNA editing. *RNA* **6,** 1004–1018.

Palladino, M. J., Keegan, L. P., O'Connell, M. A., and Reenan, R. A. (2000b). A-to-I pre-mRNA editing in *Drosophila* is primarily involved in adult nervous system function and integrity. *Cell* **102,** 437–449.

Polson, A. G., and Bass, B. L. (1994). Preferential selection of adenosines for modification by double-stranded RNA adenosine deaminase. *EMBO J.* **13,** 5701–5711.

Rae, F. K., Stephenson, S. A., Nicol, D. L., and Clements, J. A. (2000). Novel association of a diverse range of genes with renal cell carcinoma as identified by differential display. *Int. J. Cancer* **88,** 726–732.

Reenan, R. A. (2005). Molecular determinants and guided evolution of species-specific RNA editing. *Nature* **434,** 409–413.

Rueter, S. M., Dawson, T. R., and Emeson, R. B. (1999). Regulation of alternative splicing by RNA editing. *Nature* **399,** 75–80.

Schaub, M., and Keller, W. (2002). RNA editing by adenosine deaminases generates RNA and protein diversity. *Biochimie* **84,** 791–803.

Schmid, C. W. (1996). Alu: Structure, origin, evolution, significance and function of one-tenth of human DNA. *Prog. Nucleic Acid Res. Mol. Biol.* **53,** 283–319.

Seeburg, P. H. (2002). A-to-I editing: New and old sites, functions and speculations. *Neuron* **35,** 17–20.

Sommer, B., Kohler, M., Sprengel, R., and Seeburg, P. H. (1991). RNA editing in brain controls a determinant of ion flow in glutamate-gated channels. *Cell* **67,** 11–19.

Sorek, R., Ast, G., and Graur, D. (2002). Alu-containing exons are alternatively spliced. *Genome Res.* **12,** 1060–1067.

Tonkin, L. A., Saccomanno, L., Morse, D. P., Brodigan, T., Krause, M., and Bass, B. L. (2002). RNA editing by ADARs is important for normal behavior in *Caenorhabditis elegans. EMBO J.* **21,** 6025–6035.

Valente, L., and Nishikura, K. (2005). ADAR gene family and A-to-I RNA editing: Diverse roles in posttranscriptional gene regulation. *Prog. Nucleic Acid Res. Mol. Biol.* **79,** 299–338.

Yang, W., Wang, Q., Kanes, S. J., Murray, J. M., and Nishikura, K. (2004). Altered RNA editing of serotonin 5-HT2C receptor induced by interferon: Implications for depression associated with cytokine therapy. *Brain Res. Mol. Brain Res.* **124,** 70–78.

Yoshihara, M., and Littleton, J. T. (2002). Synaptotagmin I functions as a calcium sensor to synchronize neurotransmitter release. *Neuron* **36,** 897–908.

CHAPTER TWELVE

GENETIC APPROACHES TO STUDYING ADENOSINE-TO-INOSINE RNA EDITING

James E. C. Jepson *and* Robert A. Reenan

Contents

Abstract

Increasing proteomic diversity via the hydrolytic deamination of adenosine to inosine (A-to-I) in select mRNA templates appears crucial to the correct functioning of the nervous system in several model organisms, including *Drosophila*, *Caenorabditis elegans*, and mice. The genome of the fruitfly, *Drosophila melanogaster*, contains a single gene encoding the enzyme responsible for deamination, termed ADAR (for adenosine deaminase acting on RNA). The mRNAs that form the substrates for ADAR primarily function in neuronal

Department of Molecular and Cellular Biology, Brown University, Providence, Rhode Island

Methods in Enzymology, Volume 424
ISSN 0076-6879, DOI: 10.1016/S0076-6879(07)24012-1

signaling, and, correspondingly, deletion of ADAR leads to severe nervous system defects. While several ADAR enzymes are present in mice, the presence of a single ADAR in *Drosophila*, combined with the diverse genetic toolkit available to researchers and the wide range of ADAR target mRNAs identified to date, make *Drosophila* an ideal organism to study the genetic basis of A-to-I RNA editing. This chapter describes a variety of methods for genetically manipulating *Drosophila* A-to-I editing both in time and space, as well as techniques to study the molecular basis of ADAR–mRNA interactions. A prerequisite for experiments in this field is the ability to quantify the levels of editing in a given mRNA. Therefore, several commonly used methods for the quantification of editing levels will also be described.

1. INTRODUCTION

A puzzle arising from large-scale genome sequencing projects was the realization that gene number does not appear to be directly correlated to phenotypic complexity. For example, the genome of the fruitfly, *Drosophila melanogaster*, contains several thousand fewer genes than that of the nematode *Caenorhabditis elegans* (~14,000 compared to ~20,000), and conversely the number of human genes is only one-third larger than this simple worm and of a similar size to *Arabidopsis thalia*. The solution to this so-called N- or g-value paradox (Claverie, 2001; Hahn and Wray, 2002) may in part be due to the fact that proteomic diversity, and thus the genomic information content ("I-value"; Hahn and Wray, 2002), can be exponentially increased in complex organisms by a variety of posttranscriptional and posttranslational mechanisms.

The most widely studied form of posttranscriptional modification is alternative splicing. The consequences of alternative splicing of mRNAs include the inclusion/exclusion of alternative exons, intron inclusion, and the generation of alternative promoters and initial exons (Blencowe, 2006). In addition to alternative splicing, a more subtle mechanism of mRNA modification, known as RNA editing, also occurs within a diverse array of phyla. RNA editing was first documented in trypanosomal kinetoplast *coxII* mRNA, where the insertion of several nucleotides restores the mRNA open reading frame, allowing the translation of functional CoxII protein (Benne *et al.*, 1986). In addition to the insertion/deletion of nucleotides, RNA editing also encompasses the enzymatic conversion of bases within RNA (Keegan *et al.*, 2001). This modification may consist of either a C-to-U or A-to-I transition. In this chapter, we focus on the most universal type of editing in metazoans, namely, A-to-I RNA editing, a process that results in genetic recoding. We illustrate how the unique genetic toolkit of the model organism *D. melanogaster* may be used to further our knowledge of this intriguing biological process.

2. ADARs: Structure and Function

The term "A-to-I RNA editing" refers to the hydrolytic deamination of adenosine to inosine, catalyzed by ADARs (adenosine deaminases acting on RNA). Since inosine is interpreted by the ribosome as guanosine (Basillo *et al.*, 1962), A-to-I editing may alter the coding potential of the target mRNA transcripts. In fact, more than half of the amino acids of the genetic code may be posttranscriptionally reassigned by RNA editing. Interestingly, the mRNAs recoded by the A-to-I editing machinery in *Drosophila*, squid, and mammals appear to largely function in the transmission of neuronal signals (Hoopengardner *et al.*, 2003; Patton *et al.*, 1997). This suggests that the primary function of A-to-I editing is to increase the diversity of neuronally expressed proteins, which may in turn be required for the correct functioning of the nervous system. In support of this hypothesis, null mutants in both the *Drosophila adar* and the mouse *adar2* genes exhibit neurological defects (Higuchi *et al.*, 2000; Palladino *et al.*, 2000a).

The domain organization of ADAR enzymes is conserved throughout a diverse range of phyla, including nematodes, *Drosophila*, mice, and humans, although the number of genomically encoded ADARs varies between species. For example, the genome of *Drosophila* contains a single *adar* locus (Palladino *et al.*, 2000b), while two ADAR genes (*adr-1* and *adr-2*) are present in *C. elegans* (Tonkin *et al.*, 2002) and three (ADAR1, ADAR2, and ADAR3/RED2) in mammalian genomes (Keegan *et al.*, 2001). ADAR proteins contain two or three double-stranded RNA (dsRNA) binding domains and a catalytic deaminase domain. This catalytic domain shares distant homology to the catalytic domains of APOBEC-1, a mammalian cytidine deaminase (Teng *et al.*, 1993) and ADATs (adenosine deaminases acting on tRNA), which are likely to be the evolutionary precursors of ADARs (Gerber and Keller, 1999; Gerber *et al.*, 1998). The substrate requirements for ADAR activity have been well documented. In the nervous system of *Drosophila* and mammals, ADARs target imperfect dsRNA duplexes formed by the base pairing of the target mRNA and a complementary RNA strand known as the editing site complementary sequence (ECS). In the majority of cases documented to date, the ECS element resides in intronic sequences adjacent to the exon containing the editing site. Examples of this include the mammalian GluR-B subunit (Higuchi *et al.*, 1993), *Drosophila synaptotagmin 1* (*syt 1*; Reenan, 2005), and several sites in the *Drosophila paralytic* (*para*) sodium channel mRNA (Hanrahan *et al.*, 2000) (Fig. 12.1.) However, for other editing sites, such as the *para Fsp* site (Hanrahan *et al.*, 2000; Fig. 12.1) and an editing site within the mammalian Kv1.1 potassium channel (Bhalla *et al.*, 2004), the ECS element has been shown to reside within exonic sequences, providing a double

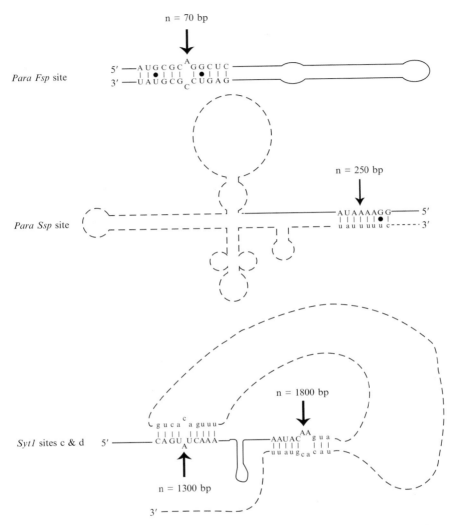

Figure 12.1 Schematic diagrams illustrating the diversity of exon–ECS structures. Three experimentally deduced structures formed by the binding of the ECS element to the region around the targeted adenosine are shown (not to scale), ranging from a simple exonic hairpin loop (the *para Fsp* site; Hanrahan *et al.*, 2000), an exonic region pairing with an intronic ECS sequence to form a more complex hairpin structure (the *para Ssp* site; Hanrahan *et al.*, 2000), and the highly complex pseudoknot that is required for editing at *Syt1* sites c and d (Reenan, 2005). The approximate number of nucleotides separating the edited adenosine and the ECS element is given above the target adenosine (shown by arrows). Exonic sequence stretches are in capital letters or full lines, while intronic sequences are lower case or doted. The sequences shown surrounding the edited adenosine highlight the lack of a consensus sequence for the binding of ADAR.

constraint on the evolution of these nucleotides. The case of *para* also illustrates that a single pre-mRNA may contain both intronic and exonic ECS elements that can direct editing.

3. THE *DROSOPHILA* ADAR IS REQUIRED FOR NORMAL ADULT BEHAVIOR AND UNDERGOES TRANSCRIPTIONAL AND POSTTRANSCRIPTIONAL REGULATION

The *Drosophila adar* locus is situated on the 1B6–7 region of the X-chromosome, and encodes a protein containing two dsRNA-binding domains (DRBD1 and 2) and an adenosine deaminase catalytic domain (Palladino *et al.*, 2000b). ADAR functions as a dimer to bind stretches of dsRNA, a process that requires the N-terminus and DRBD1 (Gallo *et al.*, 2003). Functional expression of ADAR is complex and is regulated both transcriptionally and posttranscriptionally (Palladino *et al.*, 2000b). The *adar* locus has two alternative transcription start sites, located in the furthermost 5' pair of adjacent exons (termed –4a and –4b, respectively). Transcription from these alternative exons is under developmental control, with transcripts produced constitutively throughout development from the –4a start site, and transcripts initiated from the –4b start site detected solely from the pupal stage onward (Palladino *et al.*, 2000b). Several alternatively spliced exons can be detected in pre-mRNA transcripts: exons –3, –2, –1, and 0 contain alternative starting methionines, with translation predicted to begin at either exons –2, –1, 0, and 1. Of these, –1 and 1 appear to be the most common starting sites utilized throughout development (Palladino *et al.*, 2000b). Alternative splicing is also observed in an exon located between DRBD1 and 2 (exon 3a), the inclusion or exclusion of which alters the distance between the two DRBDs. Intriguingly, ADAR proteins can edit *adar* transcripts lacking the 3a exon at a site that causes an S-to-G transition six amino acids downstream of motif II of the catalytic deaminase domain (Palladino *et al.*, 2000b). Editing at this site increases throughout development to a final level of approximately 40% at the adult stage (Palladino *et al.*, 2000b) and has been shown to decrease the ability of the ADAR enzyme to edit its own transcript and an editing site in the *cacophony* (*cac*) calcium channel mRNA (Keegan *et al.*, 2005). Therefore, self-editing appears to act as a form of negative feedback regulation.

The consequences of abolishing RNA editing in *Drosophila* are profound. Deletion of the *adar* locus through site-directed *P*-element mutagenesis followed by transposase-mediated *P*-element excision produced morphologically wild-type *adar*-deficient-(*adar⁻*) flies that exhibited a range of neurological defects, including locomoter defects, severe tremors, circling behavior, extreme mating defects in males, and brain lesions and abnormalities in retinal anatomy that increase with age (Palladino *et al.*, 2000a). These phenotypes were accompanied by complete loss of editing at 25 sites in three distinct mRNA targets (*para, cac,* and the glutamate-gated chloride channel, *DrosGluCl-α*). Indeed, all editing sites demonstrated to date are unedited in *adar⁻* flies (Hoopengardner *et al.*, 2003; Palladino *et al.*, 2000a; Stapleton *et al.*, 2006). Surprisingly, *adar⁻* flies exhibit normal life spans even though the loss of editing activity is accompanied by progressive brain degeneration. It should be noted that due to the difficulty in obtaining *adar⁻* female progeny, most experiments to date have been performed on *adar⁻* males.

At present, it is not clear precisely when editing activity is required during development for normal adult behavior. In mammals, global levels of inosine incorporation appear to be maximal in adult brains (Paul and Bass, 1998). Correspondingly, in *Drosophila* the editing of numerous target editing sites [such as those in *adar, para,* and *cac* (Hanrahan *et al.*, 2000; Keegan *et al.*, 2005; Palladino *et al.*, 2000b)] peak in the adult stage, and the pupal and adult stages exhibit the highest degree of *adar* expression (Palladino *et al.*, 2000b). These findings suggest that the ADAR-mediated increase in the diversity of the neuronal proteome is primarily required at the adult stage for correct functioning of the nervous system.

4. IDENTIFYING THE TARGETS OF ADAR

What are the target mRNA substrates for the ADAR enzyme in *Drosophila*? The initial discoveries of editing in four neuronal transcripts, *DrosGluCl* (Semenov and Pak, 1999), *para* (Hanrahan *et al.*, 2000; Reenan *et al.*, 2000), *cac* (Smith *et al.*, 1998), and the Dα6 nicotinic acetylcholine receptor subunit (Grauso *et al.*, 2002), were serendipitous, based on the observation of mixed A/G peaks in cDNA sequence chromatograms and pure A peaks in the corresponding genomic DNA. More recently, a comparative genomics approach based on the high levels of homology between orthologous exons containing edited sites in distinct *Drosophila* species revealed a further 53 editing sites in 16 target mRNAs (Hoopengardner *et al.*, 2003). All of the new targets were involved in the transmission of neuronal signals, with seven classified as voltage-gated ion channels, four as ligand-gated ion channels, and five as components of the presynaptic release machinery (see Table 12.1 for details).

In addition, a recent large-scale screen of over 10,000 isogenic *Drosophila* cDNAs revealed a further 58 sites in 27 genes exhibiting mixed A/G peaks (Stapleton *et al.*, 2006). Many of these potential new targets, which include the

Table 12.1 Targets of A-to-I editing in *Drosophila*[a]

Gene	Protein function	Number of sites edited	Conservation of altered residue?
Voltage-gated ion channels			
Paralytic(para)	Na^+ channel	12	\\
DCS1	Na^+ channel	1	\\
Cacophony(cac)	Ca^{2+} channel	11	\\
Ca-alpha1T	Ca^{2+} channel	1	\
DmCa1D	Ca^{2+} channel	5	\\
$\alpha_2\delta$	Ca^{2+} channel Accessory subunit	3	\
Shaker (Sh)	K^+ channel	6	\\
Shab	K^+ channel	6	?
Ether-a-go-go (eag)	K^+ channel	6	\\
Slowpoke (slo)	Ca^{2+}-activated K^+ channel	2	\
Ligand-gated ion channels			
Dα5	nAChR α-subunit	7	\\
Dα6	nAChR α-subunit	7	\\
Dβ1	nAChR β-subunit	4	?
Dβ2	nAChR β-subunit	2	\\
Rdl	GABA receptor	6	\\
DrosGluCl	Glutamate-gated Cl^- channel	4	?
Presynaptic release machinery			
Synaptotagmin (syt)	Ca^{2+} sensor	4	\\
Complexin (cpx)	SNARE "clamp"	3	?
Like AP180 (lap)	Adaptor protein	1	?
Stoned B (stnB)	Interacts with Syt	1	\
Dunc-13	SNARE "primer"	1	\\
Other			
Adar	Editing enzyme	1	\\

[a] The proteins altered by A-to-I editing fall into three distinct classes, all of which function in electrical and chemical neurotransmission. The residues affected are mostly highly conserved between species and phyla (\\, sequence invariant; \, sequence identity conserved; ?, degree of conservation unknown).

Mob1 and *boss* G-protein-coupled receptors, the *CG31116* chloride channel, and the *SK* calcium-activated potassium channel, exhibit neuronal expression, in concordance with the previous editing targets identified to date.

With the exception of ADAR itself (as detailed above), the functional consequences of editing of these target genes, both at the protein and whole-organism level, are largely unknown. However, high sequence conservation, combined with functional studies of orthologous sites in other model organisms, suggests that A-to-I editing may lead to significant alterations in the properties of the resulting protein. For example, in the *DSC1* voltage-gated sodium channel, editing alters a residue in the functionally crucial IFM motif within the channel inactivation gate (Eaholtz *et al.*, 1994; Hoopengardner *et al.*, 2003), and editing at several sites in *Syt1* alters residues in the calcium-sensing C2B domain, which also plays a role in the binding of voltage-gated calcium channels and other presynaptic release proteins such as Stoned B, which is also a target for ADAR (Chapman *et al.*, 1998; Hoopengardner *et al.*, 2003; Phillips *et al.*, 2000).

Several neuronal targets have also been defined for the mammalian ADAR2 enzyme, including the GluR2 AMPA receptor subunit, the GluR5 and GluR6 kainate receptor subunits, the G-protein-coupled serotonin ($5HT_{2C}$) receptor, and the Kv1.1 potassium channel (Bhalla *et al.*, 2004; Burns *et al.*, 1997; Hoopengardner *et al.*, 2003; Sommer *et al.*, 1991). As with *Drosophila*, the mammalian ADAR2 enzyme also edits its own transcript (Rueter *et al.*, 1999). However, in this case, autoediting appears to regulate the alternative splicing of the ADAR2 transcript by converting two intronic AA sequences to AG splice acceptor sites, leading to the inclusion of 47- and 30-bp sequence stretches in the $5'$ region and the deaminase catalytic domain, respectively. Inclusion of these additional sequences has been shown to modify the enzymatic properties of the ADAR2 protein (Rueter *et al.*, 1999).

In contrast to *Drosophila*, the effects of editing at several mammalian sites have been well documented, in some cases at both the protein and whole-organism levels. For example, editing at the Q/R site within the transmembrane II segment of the GluR2 AMPA receptor subunit (Sommer *et al.*, 1991) affects numerous aspects of channel function and biogenesis, including ionic conductance (Verdoorn *et al.*, 1991), exit of the GluR2 transcript from the endoplasmic reticulum, and subunit tetramerization (Greger *et al.*, 2002, 2003). The physiological importance of editing at this site is highlighted by the fact that expression of a constitutively edited isoform of the GluR2 subunit is sufficient to rescue lethality in the mouse ADAR2 null background (Higuchi *et al.*, 2000). Editing in the $5HT_{2C}R$ significantly affects receptor–G-protein coupling, and thus the relationship between extracellular serotonin concentration and downstream signaling (Burns *et al.*, 1997). Intriguingly, not only have editing levels in the $5HT_{2C}R$ been shown to be responsive to neuronal serotonin levels in mice (Gurevich *et al.*, 2002a), but misregulation of this potentially activity-dependent editing in human brains has been shown to correlate with severe depression and suicide (Gurevich *et al.*, 2002b). These data suggest that A-to-I editing in the nervous system may be physiologically regulated, and that incorrect regulation of particular target sites may have significant behavioral consequences.

5. THE USE OF GENETIC TECHNIQUES TO STUDY ADENOSINE-TO-INOSINE EDITING IN *DROSOPHILA*

The quest to understand the roles of A-to-I editing is still at an early stage. Although an increasing variety of ADAR targets have been revealed, the most significant questions remain unanswered: at what time during development is A-to-I editing required for correct functioning of the nervous system? Do the alternative isoforms of ADAR bind distinct mRNA substrates, and if so, what are they? Where are the ECS elements involved in directing RNA editing for each target mRNA and, what structures do they form? How is editing developmentally regulated, and is the process of catalytic deamination capable of being modified by non-ADAR protein factors or RNA? Finally, how does the editing of particular neuronal mRNAs alter the physiology of the nervous system and thus the output behaviors? We consider *Drosophila* to be an ideal organism to study these diverse aspects of RNA editing due to the multitude of genetic techniques that can be used to modify the editing process. In this section, we will briefly describe the basic techniques for creating transgenic flies, and then move on to detail methods for expressing transgenes of choice in defined cell types, at particular developmental stages, or both. The coexpression of mRNA substrates and ADAR enzymes in S2 cells—a method used to study the substrate requirements for ADAR–mRNA binding (Reenan, 2005)—will be discussed. In addition, we will also review the principles of quantifying the extent of A-to-I editing at particular sites.

To answer such fundamentals as when and where A-to-I editing is required for wild-type behavioral outputs requires the ability to express ADAR isoforms in particular cell types and at defined developmental stages. This, in turn, requires the generation of transgenic lines that can express ADAR isoforms under the control of spatially and temporally specific promoters (described below) in an *adar* null background. This technique also allows us to study how the overexpression of ADAR isoforms in a wild-type background may affect normal development and behavior. Using these approaches, it has recently been shown that an altered ADAR isoform that is incapable of being edited (thus mimicking the genome-encoded isoform) can rescue adult *adar* ⁻ locomotor defects when expressed constitutively in cholinergic neurons (using a *Cha*-GAL4 driver; Keegan *et al.*, 2005). Approaches such as these may be used to determine the neuronal loci in which A-to-I editing is most crucial.

The principle of creating transgenic *Drosophila* lines is based on the use of *P*-element vectors containing the desired transgene, which in the presence of a transposase source can stably insert into germ line DNA (Rubin and Spradling, 1982). Typically, dechorionated embryos that have yet to form pole cells are injected under oil with 1–100 pl of fluid containing the plasmid-transgene construct at a concentration of 300–1000 μg/ml, using a sharp-tipped needle with a tip diameter of <2 μm (Ashburner et al., 1989). This procedure requires a suitable microinjection apparatus. Generally, transgenes are cloned into vectors containing the UAS promoter sequence, allowing expression to be controlled by the presence or absence of the GAL4 activator protein (described below). The transposase required for P-element insertion may be either coinjected with the plasmid or injected in transgenic embryos expressing the Δ2–3 transposase. Several markers, contained within the plasmid, may be used to test for successful transgene integration. These include rosy (ry^+), alcohol dehydroganse (Adh^+), and white (w^+) (Ashburner et al., 1989). Plasmids carrying wild-type copies of these markers are injected into embryos lacking the corresponding gene (e.g., w^1). Thus, stable rescue of the marker gene phenotype implies successful integration of the plasmid. In the case of injecting a plasmid containing both the transgene of choice and the w^+ allele, w^1; + ; Δ2–3 embryos (which at the adult stage have white eyes) are injected and the resulting adults (G_0) are mated individually to w^1 flies. The offspring (G_1) containing fully integrated transgenes will have a uniform nonwhite eye color that will vary from light yellow to dark red depending on the transgene insertion site and copy number. The location of the transgene(s) may then be mapped (see Ashburner et al., 1989) and subsequently either homozygosed or balanced over a chromosome such as Fm7, CyO, or Tm3/Tm6 depending on the site of insertion. Since G_0 flies may be gonadal mosaics for the transgene, it is important to separate the individual G_1 offspring of single G_0 adults prior to mapping, since G_1 siblings may have independent transgene insertion sites.

6. SPATIAL AND TEMPORAL CONTROL OF TRANSGENES USING THE UAS-GAL4, GENE-SWITCH, AND TARGET SYSTEMS

Once a stable transgenic line containing an ADAR isoform under the control of a promoter such as the upstream activating sequence (UAS) is generated, how can its expression be regulated? Several genetic systems have been developed in Drosophila that allow the expression of a desired transgene to be regulated in time, space, or both. In this section, we will review three such methods: the UAS-GAL4 system (spatial control) and the Gene-Switch and TARGET systems (spatial and temporal control). Other systems, such as the heat shock promoter (Hsp) system (temporal control) and the Hsp-FLP-FRT and tetracycline-transactivator systems (spatial and temporal control), have been reviewed elsewhere (for example, see McGuire et al., 2004a).

6.1. Spatial control using the bipartite UAS-GAL4 system

GAL4 is a protein from the yeast *Saccharomyces cerevisiae* known to regulate the transcription of galactose-induced genes such as *GAL1* and *GAL10* (Laughon and Gesteland, 1982, 1984). The DNA sequence motif required for GAL4 binding has been defined and consists of four homologous 17-bp repeats termed the UAS (Giniger *et al.*, 1985). This regulatory system can be recreated in *Drosophila* to drive transgenes in a spatially controlled fashion by generating separate lines containing (1) a transgene of interest under the control of a UAS-promoter sequence and (2) a *GAL4* gene under the control of an endogenous tissue-specific promoter (Brand and Perrimon, 1993). By crossing these two lines, offspring can be generated that express the transgene in a defined set of cell types (for a detailed review, see Duffy, 2002; Fig. 12.2). The ability to separate transgene and driver lines allows the generation of transgene responder lines that encode lethal or toxic products, since the transgene is transcriptionally inactive until crossed with the appropriate driver line. For example, the constitutively unedited iso-form of ADAR is lethal when ubiquitously expressed throughout the fly using a *tubulin* driver (Keegan *et al.*, 2005). Experiments using the UAS-GAL4 system have suggested that the locus of this lethality is in fact the muscle, since expression of the unedited isoform using a muscle-specific *Mef2* driver recapitulated the observed lethality (although at a later developmental stage) while neuronal expression using a *Cha* driver (specific to cholinergic neurons) was not lethal (Keegan *et al.*, 2005).

Using the UAS-GAL4 system, ADAR transgenes can be driven in a wide range of different cell types, allowing the physiological and behavioral consequences of overexpression of these enzymes to be studied (for a list of GAL4-driver lines, see http://flystocks.bio.indiana.edu/Browse/misc-browse/gal4.htm). Alternatively, RNAi constructs that may knock down ADAR expression (and thus the levels of A-to-I editing) can be expressed in a cell-specific fashion, allowing roles of A-to-I editing in individual cell types to be examined. From a methodological point, the induction of a transgene in a cell-type-specific manner is accomplished simply by crossing a virgin female containing the transgene of choice under control of a UAS promoter to a male carrying the GAL4 gene under control of a tissue-specific promoter (see Fig. 12.2), or vice versa, and collecting the offspring expressing both the transgene and driver constructs. However, it should be noted that the induction of GAL4 activity in *Drosophila* is temperature dependent, being minimal at 16° and optimal at approximately 29° (Duffy, 2002). Therefore, flies expressing GAL4–transgene combinations should either be kept at room temperature or in a controlled incubator at 25–29°. It is also important when making physiological or behavioral studies of GAL4–transgene flies to compare the experimental offspring to both the transgene and driver lines alone, since the insertion of both

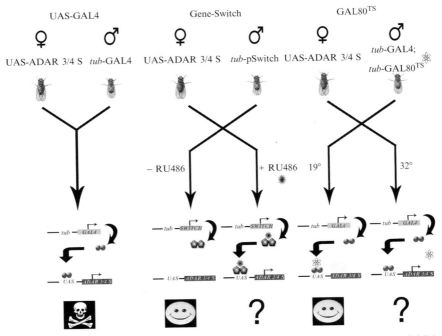

Figure 12.2 Spatial and temporal control of ADAR expression using the UAS–GAL4, Gene–Switch, and TARGET systems. For each genetic system, the following are indicated: the parental genotypes, the molecular events occurring in the F1 progeny containing driver and transgene lines, and the resulting F1 phenotypes. Constitutive ubiquitous expression of an uneditable ADAR isoform (UAS-ADAR 3/4 S) using a *tubulin* (*tub*) driver via the standard UAS-GAL4 system results in early lethality (Keegan *et al.*, 2005). This lethality can potentially be circumvented using either the Gene-Switch or TARGET systems. In the case of Gene-Switch, females containing the UAS-ADAR 3/4 S transgene are crossed to males carrying the *tub*-Switch construct. In the absence of RU486, the chimeric GAL4 is inactive, and thus the ADAR isoform is absent in the F1 progeny, resulting in wild-type larval and adult flies. However, if transferred to food or a sucrose solution containing RU486, the chimeric GAL4 will be activated and drive ADAR transcription. Similarly, in the case of the TARGET system, females containing the UAS-ADAR 3/4 S transgene are crossed to males carrying the *tub*-GAL4 and *tub*-GAL80^TS constructs. In the progeny containing all three transgenes, GAL4 activity will be inhibited at 19° by the temperature-sensitive GAL80. However, when transferred to a 32° incubator, GAL80^TS is no longer able to inhibit GAL4, thus allowing GAL4-driven expression of the ADAR construct at the desired developmental time point. The phenotypic consequences of ubiquitous UAS-ADAR 3/4 S expression solely at the larval or adult stages of *Drosophila* using these approaches are currently unknown.

types of construct may have unintended physiological consequences. Furthermore, for a correct functional comparison of wild-type and overexpression phenotypes, it is helpful to determine whether the transgene encoding an endogenous protein mimics the correct expression pattern that occurs *in vivo*.

6.2. Spatiotemporal control using UAS-GAL4 upgrades

Two genetic techniques have recently been developed that allow a temporal component to be added to the spatial specificity of the UAS-GAL4 system: the Gene-Switch and the TARGET systems (McGuire *et al.*, 2004a). The Gene-Switch system (Osterwalder *et al.*, 2001; Roman *et al.*, 2001) makes use of a GAL4 chimera formed of the GAL4 DNA-binding domain and the ligand-binding domain of the human progesterone receptor (Burcin *et al.*, 1998). In this system, the GAL4 chimera is inactive in the absence of a progesterone-receptor ligand, such as the antiprogestin RU486. Upon RU486 ingestion by the fly, when either in solution or mixed with fly food, GAL4 DNA-binding and subsequent transgene transcription are activated. Since the GAL4 chimera can be expressed under a tissue-specific promoter, both temporal and spatial control of transgene expression can be accomplished simply by controlling the time of exposure to RU486. Addition of RU486 to food or sucrose solution has been shown to induce GFP expression at both the larval (Osterwalder *et al.*, 2001) and adult stages (Fig. 12.3). In addition, embryonic transgene expression can be activated by maternal feeding of RU486. Therefore, transgene expression can be temporally controlled throughout the *Drosophila* life cycle.

The protocols for expressing transgenes using the Gene-Switch system are summarized below, and have also been detailed previously (McGuire *et al.*, 2004b).

1. A 10 mM stock solution of RU486 is produced by dissolving a set amount of RU486 in the appropriate volume of 80% ethanol. This can be stored at 4° for as long as a few months.

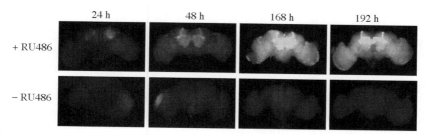

Figure 12.3 Expression of GFP in adult *Drosophila* brains using the Gene-Switch system. Females containing the UAS-GFP transgene were crossed with males carrying the RU486-activated GAL4-chimera under the control of the pan-neuronal *elav* (*embryonic lethal abnormal vision*) driver. The resulting adult offspring (UAS-GFP/*elav*-Switch) were placed on food ±200 μM RU496. GFP expression above background levels was detected in dissected adult brains after 24–48 h, and peaked after approximately 7 days on RU486-containing food (J. E. C. Jepson and R. A. Reenan, unpublished results). No neuronal GFP expression above background was detected in flies fed food without RU486. (See color insert.)

2. The stock solution may then be either diluted to 500 μM in a 2% (w/v) sucrose solution or to 200 μM in standard molten fly food (for recipes, see http://fly.bio.indiana.edu/media-recipes.htm). RU486 food can be stored for approximately a month at $4°$.

3. Mate virgin females of the transgene line to males of the driver line (or vice versa). The time of placement of the resulting offspring on RU486 food/sucrose solution depends on the developmental stage in which the transgene is required to be activated. As stated above, activation at the embryonic stage can be accomplished by allowing mating and subsequent egg-laying to occur on RU486 food. Larval and adult expression is achieved by placing larvae/adults onto RU486 food or soaking larvae in an RU486 sucrose solution.

The kinetics of transgene expression have to be determined for each individual transgene at each developmental stage. Green fluorescent protein (GFP) expression has been detected via Western blotting 5 h after exposure of larvae to RU486 (Osterwalder et al., 2001). However, the rate of expression will depend on the site of transgene insertion, the stability of the protein, and also the developmental stage of the fly; since larvae have a higher feeding rate than adults, the kinetics of expression are likely to be faster. In the case of adults, the rate of transgene induction can be increased by briefly starving adults prior to placement on RU486 food or sucrose/solution.

One drawback of the Gene-Switch system is that there is currently minimal promoter diversity to drive the RU486-activated GAL4 chimera, thus limiting the range of cell types that can be studied in this fashion. An alternative strategy is to use the TARGET system (temporal and regional gene expression targeting; McGuire et al., 2003), which in addition to the standard UAS-GAL4 combination, incorporates a temperature-sensitive yeast GAL4 repressor—GAL80^{TS}. Expression of this protein has been shown to successfully repress GAL4 at all stages of the *Drosophila* life cycle. This repression is optimal at $19°$ and minimal above $30°$ (McGuire et al., 2004a). Simple two generation mating crosses can be designed to produce offspring that carry a standard driver-GAL4, UAS-transgene combination in addition to a GAL80^{TS} transgene under the control of a ubiquitous driver such as *tubulin* or *actin*, in which activation of the transgene is repressed by storing the flies at $19°$. Repression may then be alleviated by transferring flies to a $32°$ incubator.

The Gene-Switch and TARGET systems are particularly useful for studying the overexpression of transgenes that incur lethal phenotypes when expressed in the early stages of development. In Fig. 12.2 we illustrate how these techniques might be employed to overcome the lethality observed in flies ubiquitously overexpressing the constitutively unedited ADAR isoform, thus allowing further study of the physiological and behavioral consequences at later developmental stages.

6.3. Molecular genetic approaches to determine ECS element location

In the *Drosophila* nervous system, ADAR binding and subsequent enzymatic activity require the formation of an imperfect dsRNA structure, generally involving the exonic sequences surrounding the editing site and a complementary RNA sequence (the ECS) often located within an adjacent intron (Dawson *et al.*, 2004; Herb *et al.*, 1996; Higuchi *et al.*, 1993). However, for the vast majority of *Drosophila* editing sites, the location of the ECS and the nature of the exon–ECS structure are unknown.

Candidate ECS elements for a given editing site can be determined using comparative genomics approaches (Reenan, 2005) and their structures predicted using the MFOLD program of the MacFarlane Burnet Centre MFOLD server (http://mfold.burnet.edu.au/). However, further experiments using molecular and genetic techniques must be used to determine whether the predicted structures accurately reflect those occurring *in vivo*. In this section, we describe one such method: an *in vitro* approach using *Drosophila* S2 cells.

6.4. Mimicking *in vivo* editing in *Drosophila* S2 cells

S2 (Schneider line 2) cells are an immortalized cell line derived from primary cultures of late-stage *Drosophila* embryos (Schneider, 1972) and are widely used for *in vitro* expression, functional studies of exogenously expressed ion channels and receptors, and large-scale RNAi studies (Towers and Sattelle, 2002). In the context of RNA editing, S2 cells allow the experimental determination of ECS element sequences via simultaneous *in vitro* expression of (1) a reporter construct expressing an editable target RNA sequence in tandem with the hypothetical ECS element and (2) an inducible *adar* transgene under the control of a metallothionein promoter. If the correct ECS element is present *in cis* to the target adenosine, the ADAR protein produced following the addition of copper sulfate may correctly edit the adenosine. The extent of editing at the target site can then be examined by amplifying the reporter transgene using primers to both the cDNA in question and the plasmid flanking sequence and subsequently quantifying the levels of A-to-G conversion in the resulting cDNA (see below). This method has recently been used to determine the ECS elements and structures required for editing of *syt1* in a wide range of divergent insect species (Reenan, 2005).

Here we describe a simple protocol for S2 cell transfection with reporter and *adar* constructs. The prerequisite for these experiments is the generation, using standard molecular techniques, of the appropriate plasmids containing the experimental reporter construct (for example, an exon containing an edited adenosine, followed by a short linker sequence of 100–200 bp, followed by the hypothesized intronic ECS element; if the ECS element is predicted to occur within the same exon as the editing site, then only the complete exon is required) and the *adar* transgene of choice under the control of a metallothionein promoter (±3a; genomic, edited or constitutively uneditable if the 3/4 isoform).

1. Grow S2 cells in six-well plates in a CO_2-free incubator at room temperature (~25°) in serum-free media (SFM, Invitrogen) for 2 days. SFM is generally supplemented with L-glutamine and penicillin/streptomycin. Prior to transfection, remove SFM and replace with new media.
2. Mix 250 μl of SFM and 7 μl of Gene Juice transfection reagent (Novagen) via pipette in a test tube for 5 min.
3. Add 0.2 μg of reporter plasmid, 2 μg of *adar*-containing plasmid, and mix via pipetting. Allow to stand for 15 min.
4. Remove fresh SFM from S2 cells and replace with the above mix. Leave for 4–5 h.
5. Add 15 μl of 1 μM copper sulfate to activate *adar* transcription.
6. Whole RNA can be isolated from transfected S2 cells 24–48 h after the initiation of *adar* transcription, and the reporter construct amplified using standard reverse transcriptase polymerase chain reaction (RT-PCR) protocols. Since endogenous copies of the reporter transgene may also be expressed in S2 cells, it is important to use a plasmid-specific primer in addition to the primer designed to the reporter sequence.
7. Quantify the levels of A-to-I editing present in the reporter construct using restriction digests, clone counting, or poisoned primer extension (described below).

7. QUANTIFYING ADENOSINE-TO-INOSINE EDITING

How is the enzymatic activity of ADAR measured in the context of specific editing sites? When examining the molecular correlates of, for example, overexpressing ADAR isoforms in the nervous system of *Drosophila*, or coexpressing editable substrates with ADAR isoforms in S2 cells (as detailed above), methods to quantify the extent of A-to-I conversion are required. In this section, we describe four such techniques. We begin with the least quantitative (comparing the ratio of A-to-G peaks in sequence chromatograms) and then move on to detail fully quantitative methods such as restriction enzyme digests, plasmid clone-counting, and poisoned primer extension.

7.1. Ratiometric A/G measurement from sequence chromatograms

Editing sites may be easily identified by the presence of a mixed A/G (or in rarer cases a full G peak) in a cDNA sequence chromatogram, where the genomically encoded base is adenosine. The ratio of the A-to-G peaks provides a qualitative indication of the level of editing, since the relative area of each A or G peak is proportional to the amount present in the sequenced cDNA (see Fig. 12.4 for an example of low, medium, and high levels of A-to-I editing at distinct sites within Dα6 nicotinic acetylcholine receptor cDNA isolated from Canton-S *Drosophila*, illustrated using sequence chromatograms). Certain studies have shown a reasonable correlation between the level of editing estimated via the A/G peak height ratio and more quantitative techniques such as clone counting (for example, see Grauso *et al.*, 2002). However, it should be kept in mind that this method is not quantitative, and can be used only as an estimate of the editing levels at a given site.

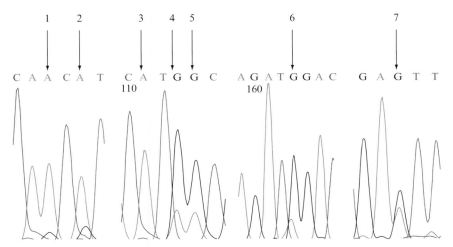

Figure 12.4 Mixed A/G peaks in the cDNA of a *Drosophila* ADAR target. Sequence chromatograms of cDNA encoding the *Drosophila* Dα6 nicotinic acetylcholine receptor amplified by RT-PCR from Canton-S flies exhibit several sites of mixed A/G peaks in exons 5 and 6 (J. E. C. Jepson and R. A. Reenan, unpublished results), corresponding to the seven known editing sites within this gene (Grauso *et al.*, 2002). The ratio of A/G peaks has been shown to agree approximately with the ratio of A- and G-containing cDNA clones for these seven sites (Grauso *et al.*, 2002). Low, medium, and high levels of editing are observed at sites 1 and 2, 7, and 4–6, respectively. Site 3 shows no detectable editing. (See color insert.)

7.2. Restriction digest analysis

A second, fully quantitative method takes advantage of the fact that editing at particular sites may either create or destroy consensus sites for restriction digest enzymes. Examples of this include the *Drosophila syt1* sites C and D, where A/G transitions create PshAI and HpyCH4V cutting sites, respectively (Reenan, 2005; see Fig. 12.5). Alternatively, the A/G cDNA transition may result in the destruction of a cutting site. For example, A/G conversion at site 7 of cDNA coding for the Dα6 acetylcholine receptor (Grauso *et al.*, 2002) results in the alteration of a GA*A*TTC *Eco*RI cutting site to a GAGTTC sequence, which is no longer amenable to restriction digest. In cases such as these, quantification becomes straightforward. The sequence surrounding the editing site may be amplified via standard RT-PCR. The resulting PCR product containing edited and unedited cDNA strands may then either be digested directly or cloned into a suitable plasmid and subsequently digested. Preferably, this process is performed with three or more independent primer sets in order to discount primer-specific artifacts and to produce data amenable to statistical analysis. Once the cDNA template of interest has been obtained and purified, the following simple digestion procedure may be used (In this example, the volume of purified cDNA is 10 μl. If lower or higher, the volume of water may be adjusted accordingly to make a final volume of 30 μl.):

1. Mix together in a tube 10 μl cDNA, 1 μl cutting enzyme, 3 μl buffer, and 16 μl H$_2$O.
2. Incubate at 37° for 1 h.
3. Purify DNA using standard techniques.

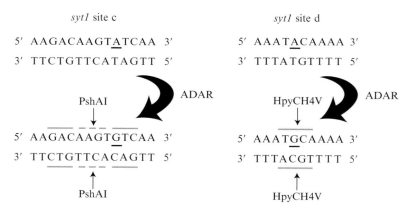

Figure 12.5   Editing at *syt1* sites c and d creates PshAI and HpyCH4V cutting sites. The edited adenosines and resulting guanosines in the *syt1* cDNA sequence are underlined in black. Gray underlined areas indicate the consensus site required for restriction enzyme binding that are created following the conversion of A→G in the cDNA of interest by ADAR activity. Arrows indicate the cutting points of the enzymes.

4. Run the sample on a 1–1.5% polyacrylamide gel. For quantification of band strength, an imaging system that is able to record the fluorescent intensity of DNA bands is required. In cases where a cutting site is created by editing, the percentage of A/G conversion is equal to the sum of the subsaturation fluorescent intensities of the two lower (cut) bands divided by the intensity of the upper, full length, band multiplied by 100. In cases where editing destroys a cutting site, the opposite applies. In both instances, the fluorescent band intensity is normalized to band size before ratiometric calculation. Alternatively, radiolabeled primers (e.g., ^{32}P-labeled) can be used to generate end-labeled PCR products that, following restriction digest, may be visualized and quantified using a phosphoimaging system.

It should be noted that the time required for complete cutting may vary between enzymes and depending on the amount of cDNA. Also, bovine serum albumin (BSA) is required for certain enzymes. For further information on restriction enzymes, see http://www.neb.com/nebecomm/tech_reference/restriction_enzymes/default.asp.

7.3. Quantifying editing levels by counting clones

In many cases, A–to–G conversions neither create nor destroy restriction sites. In these instances, distinct methods are required for quantification. A technique that is simple in principle, though time-consuming and relatively expensive in practice, is to clone cDNA PCR products containing the edited adenosine in question into a suitable vector [such as the TOPO-PCRII plasmid (Invitrogen); see http://www.invitrogen.com/content.cfm?pageid=4073 for information]. By transforming competent bacteria with the cDNA-containing vector, individual vector-containing colonies may be grown up and the plasmid vector subsequently isolated from each colony. These plasmids, which will contain the cDNA product with either an adenosine or guanosine at the edited site, can then be sequenced en masse using standard methods and the number of plasmids containing either A or G at the editing site determined. For each sample, a large number of plasmids must be sequenced to give an accurate ratio of A:G (e.g., 3 × 30 plasmids).

There are several advantages of this method compared to restriction digest. For example, if a cDNA sequence contains several editing sites (for example, the four edited sites in exon 9 of *syt1* are within 117 bp of each other), then the complete editing profile of a gene under different conditions may be determined. Also, since clones are sequenced individually, it is possible to determine whether editing at distinct sites are linked or unlinked, and if particular combinations of editing occur within individual transcripts.

7.4. Quantification via primer extension

Poisoned primer extension (PPE) assays using PCR products as a template have been widely used to measure editing efficiencies at particular sites (Burns et al., 1997; Keegan et al., 2005; Peeters and Hanson, 2002; Schiffer and Heinemann, 1999). The principle of PPE assays involves the use of a ^{32}P-labeled primer to bind a cDNA region just 5' of the editing site in question. The primer is extended via the addition of polymerase in the presence of three deoxynucleotide triphosphates (dATP, dTTP, dCTP) and one dideoxynucleotide triphosphate (ddGTP), which acts as the reaction terminator. If the cDNA bound by the polymerase is unedited, the polymerase will fall off the DNA at the target adenosine following incorporation of ddGTP. However, if the adenosine has been deaminated to an inosine, resulting in a guanosine in the cDNA, the polymerase will carry on until the next adenosine, thus creating a larger product. The ratio of these radiolabeled products when run out on a high-concentration acrylamide gel can be visualized and measured using a phosphoimaging system. Recently, the accuracy of this technique has been improved by using fluorescently labeled instead of radiolabeled primers (Roberson and Rosenthal, 2006). Visualization and quantification may then be performed using software similar to that required for measuring band intensity following restriction digests.

REFERENCES

Ashburner, M., Golic, K. G., and Hawley, R. S. (1989). In "Drosophila: A Laboratory Handbook," 2nd Ed. Academic Press, San Diego, CA.

Basillo, C., Wahba, A. J., Lengyel, P., Speyer, J. F., and Ochoa, S. (1962). Synthetic polynucleotides and the amino acid code. Proc. Natl. Acad. Sci. USA 48, 613–616.

Benne, R., Van den Burg, J., Brakenhoff, J. P., Sloof, P., Van Boom, J. H., and Tromp, M. C. (1986). Major transcript of the frameshifted coxII gene from trypanosome mitochondria contains four nucleotides that are not encoded in the DNA. Cell 46, 819–826.

Bhalla, T., Rosenthal, J. J., Holmgren, M., and Reenan, R. (2004). Control of human potassium channel inactivation by editing of a small mRNA hairpin. Nat. Struct. Mol. Biol. 11, 950–956.

Blencowe, B. J. (2006). Alternative splicing: New insights from global analyses. Cell 126, 37–47.

Brand, A. H., and Perrimon, N. (1993). Targeted gene expression as a means of altering cell fates and generating dominant phenotypes. Development 118, 401–415.

Burcin, M. M., O'Malley, B. W., and Tsai, S. Y. (1998). A regulatory system for target gene expression. Front. Biosci. 3, c1–c7.

Burns, C. M., Chu, H., Rueter, S. M., Hutchinson, L. K., Canton, H., Sanders-Bush, E., and Emeson, R. B. (1997). Regulation of serotonin-2C receptor G-protein coupling by RNA editing. Nature 387, 303–308.

Chapman, E. R., Desai, R. C., Davis, A. F., and Tornehl, C. K. (1998). Delineation of the oligomerization, AP-2 binding, and synprint binding region of the C2B domain of synaptotagmin. *J. Biol. Chem.* **273**, 32966–32972.

Claverie, J. M. (2001). Gene number. What if there are only 30,000 human genes? *Science* **291**, 1255–1257.

Dawson, T. R., Sansam, C. L., and Emeson, R. B. (2004). Structure and sequence determinants required for the RNA editing of ADAR2 substrates. *J. Biol. Chem.* **279**, 4941–4951.

Duffy, J. B. (2002). GAL4 system in *Drosophila*: A fly geneticist's Swiss army knife. *Genesis* **34**, 1–15.

Eaholtz, G., Scheuer, T., and Catterall, W. A. (1994). Restoration of inactivation and block of open sodium channels by an inactivation gate peptide. *Neuron* **12**, 1041–1048.

Gallo, A., Keegan, L. P., Ring, G. M., and O'Connell, M. A. (2003). An ADAR that edits transcripts encoding ion channel subunits functions as a dimer. *EMBO J.* **22**, 3421–3430.

Gerber, A. P., and Keller, W. (1999). An adenosine deaminase that generates inosine at the wobble position of tRNAs. *Science* **286**, 1146–1149.

Gerber, A., Grosjean, H., Melcher, T., and Keller, W. (1998). Tad1p, a yeast tRNA-specific adenosine deaminase, is related to the mammalian pre-mRNA editing enzymes ADAR1 and ADAR2. *EMBO J.* **17**, 4780–4789.

Giniger, E., Varnum, S. M., and Ptashne, M. (1985). Specific DNA binding of GAL4, a positive regulatory protein of yeast. *Cell* **40**, 767–774.

Grauso, M., Reenan, R. A., Culetto, E., and Sattelle, D. B. (2002). Novel putative nicotinic acetylcholine receptor subunit genes, $D\alpha5$, $D\alpha6$, and $D\alpha7$, in *Drosophila melanogaster* identify a new and highly conserved target of adenosine deaminase acting on RNA-mediated A-to-I pre-mRNA editing. *Genetics* **160**, 1519–1533.

Greger, I. H., Khatri, L., and Ziff, E. B. (2002). RNA editing at arg607 controls AMPA receptor exit from the endoplasmic reticulum. *Neuron* **34**, 759–772.

Greger, I. H., Khatri, L., Kong, X., and Ziff, E. B. (2003). AMPA receptor tetramerization is mediated by Q/R editing. *Neuron* **40**, 763–774.

Gurevich, I., Englander, M. T., Adlersberg, M., Siegal, N. B., and Schmauss, C. (2002a). Modulation of serotonin 2C receptor editing by sustained changes in serotonergic neurotransmission. *J. Neurosci.* **22**, 10529–10532.

Gurevich, I., Tamir, H., Arango, V., Dwork, A. J., Mann, J. J., and Schmauss, C. (2002b). Altered editing of serotonin 2C receptor pre-mRNA in the prefrontal cortex of depressed suicide victims. *Neuron* **34**, 349–356.

Hahn, M. W., and Wray, G. A. (2002). The g-value paradox. *Evol. Dev.* **4**, 73–75.

Hanrahan, C. J., Palladino, M. J., Ganetzky, B., and Reenan, R. A. (2000). RNA editing of the *Drosophila para* Na^+ channel transcript. Evolutionary conservation and developmental regulation. *Genetics* **155**, 1149–1160.

Herb, A., Higuchi, M., Sprengel, R., and Seeburg, P. H. (1996). Q/R site editing in kainate receptor GluR5 and GluR6 pre-mRNAs requires distant intronic sequences. *Proc. Natl. Acad. Sci. USA* **93**, 1875–1880.

Higuchi, M., Single, F. N., Kohler, M., Sommer, B., Sprengel, R., and Seeburg, P. H. (1993). RNA editing of AMPA receptor subunit GluR-B: A base-paired intron-exon structure determines position and efficiency. *Cell* **75**, 1361–1370.

Higuchi, M., Maas, S., Single, F. N., Hartner, J., Rozov, A., Burnashev, N., Feldmeyer, D., Sprengel, R., and Seeburg, P. H. (2000). Point mutation in an AMPA receptor gene rescues lethality in mice deficient in the RNA-editing enzyme ADAR2. *Nature* **406**, 78–81.

Hoopengardner, B., Bhalla, T., Staber, C., and Reenan, R. (2003). Nervous system targets of RNA editing identified by comparative genomics. *Science* **301**, 832–836.

Keegan, L. P., Gallo, A., and O'Connell, M. A. (2001). The many roles of an RNA editor. *Nat. Rev. Genet.* **2,** 869–878.

Keegan, L. P., Brindle, J., Gallo, A., Leroy, A., Reenan, R. A., and O'Connell, M. A. (2005). Tuning of RNA editing by ADAR is required in *Drosophila. EMBO J.* **24,** 2183–2193.

Laughon, A., and Gesteland, R. F. (1982). Isolation and preliminary characterization of the GAL4 gene, a positive regulator of transcription in yeast. *Proc. Natl. Acad. Sci. USA* **79,** 6827–6831.

Laughon, A., and Gesteland, R. F. (1984). Primary structure of the *Saccharomyces cerevisiae* GAL4 gene. *Mol. Cell. Biol.* **4,** 260–267.

McGuire, S. E., Le, P. T., Osborn, A. J., Matsumoto, K., and Davis, R. L. (2003). Spatiotemporal rescue of memory dysfunction in *Drosophila. Science* **302,** 1765–1768.

McGuire, S. E., Roman, G., and Davis, R. L. (2004a). Gene expression systems in *Drosophila*: A synthesis of time and space. *Trends Genet.* **20,** 384–391.

McGuire, S. E., Mao, Z., and Davis, R. L. (2004b). Spatiotemporal gene expression targeting with the TARGET and gene-switch systems in *Drosophila. Sci. STKE* **2004,** pl6.

Osterwalder, T., Yoon, K. S., White, B. H., and Keshishian, H. (2001). A conditional tissue-specific transgene expression system using inducible GAL4. *Proc. Natl. Acad. Sci. USA* **98,** 12596–12601.

Palladino, M. J., Keegan, L. P., O'Connell, M. A., and Reenan, R. A. (2000a). A-to-I pre-mRNA editing in *Drosophila* is primarily involved in adult nervous system function and integrity. *Cell* **102,** 437–449.

Palladino, M. J., Keegan, L. P., O'Connell, M. A., and Reenan, R. A. (2000b). dADAR, a *Drosophila* double-stranded RNA-specific adenosine deaminase is highly developmentally regulated and is itself a target for RNA editing. *RNA* **6,** 1004–1018.

Patton, D. E., Silva, T., and Bezanilla, F. (1997). RNA editing generates a diverse array of transcripts encoding squid Kv2 K^+ channels with altered functional properties. *Neuron* **19,** 711–722.

Paul, M. S., and Bass, B. L. (1998). Inosine exists in mRNA at tissue-specific levels and is most abundant in brain mRNA. *EMBO J.* **17,** 1120–1127.

Peeters, N. M., and Hanson, M. R. (2002). Transcript abundance supercedes editing efficiency as a factor in developmental variation of chloroplast gene expression. *RNA* **8,** 497–511.

Phillips, A. M., Smith, M., Ramaswami, M., and Kelly, L. E. (2000). The products of the *Drosophila* stoned locus interact with synaptic vesicles via synaptotagmin. *J. Neurosci.* **20,** 8254–8261.

Reenan, R. A. (2005). Molecular determinants and guided evolution of species-specific RNA editing. *Nature* **434,** 409–413.

Reenan, R. A., Hanrahan, C. J., and Barry, G. (2000). The mle^{napts} RNA helicase mutation in *Drosophila* results in a splicing catastrophe of the *para* Na^+ channel transcript in a region of RNA editing. *Neuron* **25,** 139–149.

Roberson, L. M., and Rosenthal, J. J. (2006). An accurate fluorescent assay for quantifying the extent of RNA editing. *RNA* **12,** 1907–1912.

Roman, G., Endo, K., Zong, L., and Davis, R. L. (2001). P[Switch], a system for spatial and temporal control of gene expression in *Drosophila melanogaster. Proc. Natl. Acad. Sci. USA* **98,** 12602–12607.

Rubin, G. M., and Spradling, A. C. (1982). Genetic transformation of *Drosophila* with transposable element vectors. *Science* **218,** 348–353.

Rueter, S. M., Dawson, T. R., and Emeson, R. B. (1999). Regulation of alternative splicing by RNA editing. *Nature* **399,** 75–80.

Schiffer, H. H., and Heinemann, S. F. (1999). A quantitative method to detect RNA editing events. *Anal. Biochem.* **276,** 257–260.

Schneider, I. (1972). Cell lines derived from late embryonic stages of *Drosophila melanogaster*. *J. Embryol. Exp. Morphol.* **27,** 353–365.

Semenov, E. P., and Pak, W. L. (1999). Diversification of *Drosophila* chloride channel gene by multiple posttranscriptional mRNA modifications. *J. Neurochem.* **72,** 66–72.

Smith, L. A., Peixoto, A. A., and Hall, J. C. (1998). RNA editing in the *Drosophila* DMCA1A calcium-channel alpha 1 subunit transcript. *J. Neurogenet.* **12,** 227–240.

Sommer, B., Kohler, M., Sprengel, R., and Seeburg, P. H. (1991). RNA editing in brain controls a determinant of ion flow in glutamate-gated channels. *Cell* **67,** 11–19.

Stapleton, M., Carlson, J. W., and Celniker, S. E. (2006). RNA editing in *Drosophila melanogaster*: New targets and functional consequences. *RNA* **12,** 1922–1932.

Teng, B., Burant, C. F., and Davidson, N. O. (1993). Molecular cloning of an apolipoprotein B messenger RNA editing protein. *Science* **260,** 1816–1819.

Tonkin, L. A., Saccomanno, L., Morse, D. P., Brodigan, T., Krause, M., and Bass, B. L. (2002). RNA editing by ADARs is important for normal behavior in *Caenorhabditis elegans*. *EMBO J.* **21,** 6025–6035.

Towers, P. R., and Sattelle, D. B. (2002). A *Drosophila melanogaster* cell line (S2) facilitates post-genome functional analysis of receptors and ion channels. *BioEssays* **24,** 1066–1073.

Verdoorn, T. A., Burnashev, N., Monyer, H., Seeburg, P. H., and Sakmann, B. (1991). Structural determinants of ion flow through recombinant glutamate receptor channels. *Science* **252,** 1715–1718.

CHAPTER THIRTEEN

A METHOD FOR FINDING SITES OF SELECTIVE ADENOSINE DEAMINATION

Johan Ohlson *and* Marie Öhman

Contents

Abstract

Single sites of selective adenosine (A) to inosine (I) RNA editing with functional consequences on the proteome are rarely found in mammals. Here we describe a method that can be used to detect novel site-selective A-to-I editing in various tissues as well as species. The method utilizes immunoprecipitation of intrinsic RNA–protein complexes to extract substrates subjected to site-selective *in vivo* editing. We show that known single sites of A-to-I editing are enriched utilizing an antibody against the ADAR2 protein. We propose that this method is suitable for identification of novel substrates subjected to site-selective A-to-I editing.

Department of Molecular Biology and Functional Genomics, Stockholm University, Stockholm, Sweden

Methods in Enzymology, Volume 424
ISSN 0076-6879, DOI: 10.1016/S0076-6879(07)24013-3

1. INTRODUCTION

The family of adenosine deaminase acting on RNA (ADAR) proteins binds to double-stranded RNA without a stringent sequence specificity. Within the structured RNA, adenosine is converted to inosine by a hydrolytic deamination. Adenosine-to-inosine (A-to-I) RNA editing is known to change the sequence of specific pre-mRNAs in metazoans from fly to human. In mammals, both ADAR1 and ADAR2 have been shown to be able to deaminate adenosines selectively within double-stranded RNAs interrupted by bulges, mismatches, or loops (Bass, 2002). ADAR editing with low selectivity can also occur at multiple sites on completely or almost completely double-stranded RNA of extensive length. This type of hyperediting has mainly been found within introns and untranslated regions (UTRs) of mRNAs, preferentially in repetitive Alu sequences (Athanasiadis *et al.*, 2004; Blow *et al.*, 2004; Levanon *et al.*, 2004; Morse and Bass, 1999; Morse *et al.*, 2002). Only a few site-selective ADAR substrates have been detected in mammals. Most of these substrates are in pre-mRNAs expressed in the central nervous system. In *Drosophila,* A-to-I editing seems to be more frequent. Reports on searches for novel targets of A-to-I editing that increase the diversity of the proteome show edited targets in the nervous system as well as other sites like components of the actin cytoskeleton (Hoopengardner *et al.*, 2003; Stapleton *et al.*, 2006). However, it is hard to say if A-to-I editing is more frequent in fly than in human or if many of the mammalian targets simply have not been detected yet.

Most mammalian selective ADAR substrates have been found serendipitously, or have been identified by homology to edited substrates in other species. Nevertheless, several attempts have been made to find novel sites of A-to-I editing. One experimental method developed by Morse *et al.* involves the detection of ADAR candidates by cleaving poly(A)$^+$ RNA specifically after inosine and identifying the cleaved substrates by differential display (Morse and Bass, 1999). A number of bioinformatics approaches have also been developed to detect novel ADAR substrates. Using these approaches, a large number of edited substrates were detected. Most of these substrates were hyperedited in their 5′ or 3′ UTRs within Alu repetitive elements (Athanasiadis *et al.*, 2004; Blow *et al.*, 2004; Levanon *et al.*, 2004). Although editing within Alu repeats might be of importance in humans, no function has thus far been proposed. Using comparative genomics approaches, site-selectively edited sites were found and experimentally verified within the mammalian potassium channel KCNA1 (Hoopengardner *et al.*, 2003), the Filamin A mRNA (FLNA), cytoplasmic FMR1 interacting protein mRNA (CYFIP2), and the bladder cancer associated protein mRNA (BLCAP) (Levanon *et al.*, 2005). Editing of these transcripts results in recoding of single amino acids. Another bioinformatics approach identified several edited sites within an intron–exon stem of the pre-mRNA coding for BC10 (Clutterbuck et al., 2005). This substrate is identical to BLCAP. Even though some of these substrates are site-selectively edited, it is still rarely found in the mammalian transcriptome.

Our method of finding novel sites of selective editing is based on coimmunoprecipitations, a powerful tool that previously has been used to successfully precipitate other specific protein–nucleic acid complexes (Tenenbaum et al., 2002; Ule et al., 2003). Here we describe a method to find site-selectively edited A-to-I sites based on the immunoprecipitation of ADAR2–RNA complexes.

2. THE CHOICE OF ORGANISM AND TISSUE TO FIND NOVEL ADENOSINE-TO-INOSINE SUBSTRATES IN MAMMALS

To find novel A-to-I editing substrates in mammals, we have chosen mouse brain tissue as an example. The mouse genome contains fewer repetitive elements than the human genome and Alu repeats are not present in the mouse genome. It is therefore a good model organism to avoid a high background of A-to-I hyperediting in noncoding sequences. Since A-to-I editing has been found almost exclusively in the nervous system, total mouse brain was the choice of tissue. However, the method described below is applicable to any tissue expressing messenger RNA.

3. PROTEINS BINDING TO DOUBLE-STRANDED RNA

To better understand the complexity of ADAR proteins interacting with its RNA substrate, we will describe some typical characteristics of proteins that bind to double-stranded RNA (dsRNA). The ADAR proteins have multiple copies of double-stranded RNA-binding motifs (dsRBMs) in their N-terminal region. ADAR1 carries three copies of dsRBMs whereas ADAR2 and ADAR3 have two. The dsRBM consists of approximately 65 amino acids found in many other dsRNA-binding proteins, like the dsRNA-dependent protein kinase (PKR) and Dicer (Doyle and Jantsch, 2002). Several high-resolution structures of dsRBMs bound to RNA targets have revealed a highly conserved dsRBM-RNA structure (Blaszczyk et al., 2004; Ramos et al., 2000; Ryter and Schultz, 1998; Wu et al., 2004). A characteristic α-β-β-β-α fold has been shown for the dsRBM. The double-stranded RNA helix consists of a wide and shallow minor grove and a major grove that is narrow and deep. The dsRBM interacts with the sugar-phosphate backbone in a sequence-independent manner spanning two minor grooves and the intervening major groove (Ryter and Schultz, 1998). The binding site spans 16 base pairs of the RNA, making contact at three distinct locations. In accordance, the ADAR enzymes bind duplex RNA of a defined length in a largely sequence-independent fashion. A recent study of the structure of the two dsRBMs from ADAR2 verified the α-β-β-β-α fold (Stefl et al., 2006). The distinct locations in the dsRBMs that make contact with the RNA indicate that the protein can also recognize bulges and loops. Furthermore, binding studies between the recombinantly expressed dsRBMs in ADAR2 and a natural site-selectively edited RNA substrate indicate that the binding is selective (Stephens et al., 2004). In agreement with this, footprinting studies show that ADAR2 binds the stem surrounding the edited GluR-B R/G site at a discrete region, whereas binding to a mutant substrate lacking two of the internal loops is nonspecific (Öhman et al., 2000).

Swapping deaminase domains between ADAR1 and ADAR2 suggests that some substrate specificity also is determined by the catalytic domain (Wong et al., 2001). Moreover, ADAR2 exists in an autoinhibited conformation until both of its dsRBMs bind to the editing substrate. After the interaction, the autoinhibition is relieved and the deaminase domain is placed over the editing site prepared to catalyze the deamination reaction (Macbeth et al., 2004). In summary, this information indicates that the dsRNA binding motifs and the deamination domain together determine the substrate specificity for site-selective A-to-I editing.

4. Using Immunoprecipitation to Detect ADAR–RNA Complexes

We have developed a method to find novel ADAR substrates by extracting intrinsic ADAR2–RNA substrate complexes by coimmunoprecipitations using an anti-ADAR2 antibody (Ohlson et al., 2005) (Fig. 13.1). To extract ADAR2 protein complexes, without a large amount of false positives, it is important to use a strong, preferentially affinity-purified antibody. We have previously shown that ADAR2 distinguishes between binding to a selectively edited site and a random sequence of double-stranded RNA (Klaue et al., 2003). Scanning force microscopy was used to show that there is a preferred interaction between ADAR2 and the selectively edited GluR-B R/Gsite over a long completely base paired region of 400 nucleotides within the same molecule. Therefore, to minimize nonspecific binding to dsRNA, we excluded any form of crosslinking between RNA and protein prior to the immunoprecipitation (IP). Moreover, ADAR2 was shown to bind with a similar affinity to a substrate RNA as to the edited product (Öhman et al., 2000).

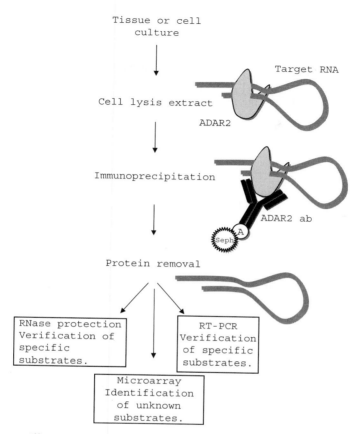

Figure 13.1 Illustration of the IP array method to find novel substrates for A-to-I editing. Cell lysis extract is prepared from mouse brain. The extract is immunoprecipitated using an ADAR2-specific polyclonal antibody. Target RNA is then extracted from the mRNP complexes upon protein removal. The RNA is amplified, labeled, and further hybridized to a mouse genomic oligo array.

   To precipitate the ADAR2–RNA complex, Protein A Sepharose 4 fast flow (Sigma-Aldrich) is coupled to affinity-purified antihuman ADAR2 antibody. The anti-ADAR2 antibody precipitates ADAR2–RNA complexes with high specificity. However, to reduce nonspecific binding, the protein A Sepharose was blocked with tRNA and bovine serum albumin (BSA) before use. It is also important to protect the RNA from degradation during this process; here we use a ribonucleoside vanadyl complex to prevent RNA degradation. Details of the immunoprecipitation are described below.

5. Using Microarray to Detect Adenosine-to-Inosine Editing Targets

A powerful method for detection of potential ADAR2 targets enriched in the specific ADAR2-IP is to use microarrays. This detection method allows us to tolerate a certain amount of background, but also to detect products of relatively low abundance since the material is amplified prior to the array. However, it should be noted that the microarray is limited in its detection of enriched transcripts, since it all depends on the quality of the array, the number of genes presented, and what species you are interested in.

To analyze the enriched RNA from the ADAR2 coimmunoprecipitation, we have used genomic microarrays from Affymetrics (Santa Clara, CA). As a background, standard preimmune serum is used in the immunoprecipitation. This immunoprecipitate is treated in the same way as the ADAR2 specific precipitate. Preparation of labeled cRNA from the immunoprecipitated RNA is done according to the Affymetrix Two-Cycle Target Labeling Assay. It is important that the RNA after the immunoprecipitation is of good quality. The RNA can be analyzed using an experion automated electrophoresis system (Bio-Rad) to determine concentration and quality of the ribosomal RNA, which will also indicate the quality of your specific RNA. It is also possible to visualize the ribosomal RNA on a 1% agarose gel. Good quality of the cRNA prior to hybridization in the array is also required, which is checked by standard affymetrix control steps. Labeled cRNA from nine mouse brains is hybridized to each Mouse Genome 430A 2.0 Array (Affymetrix). The average value from three arrays is calculated. Standard deviation for positive controls, here the GluR-B transcript, can be analyzed for consistency in the three arrays.

Further, it is good to keep in mind that the position of the oligo, targeting the gene in the array, might be important for the detection of a transcript. It is therefore important to use a microarray chip with several oligos hybridizing to each gene.

6. Detection of Known Site-Selectively Edited Sites Using Specific Immunoprecipitation

To determine if it is possible to selectively enrich A–to–I edited substrates using the ADAR2 immunoprecipitation approach, we looked at known site-selectively edited transcripts. The enrichment of these edited substrates can be identified in the genomic microarray, but also more specifically by reverse transcription and polymerase chain reaction (RT-PCR) or RNase protection. We will here show the enrichment of one known editing substrate

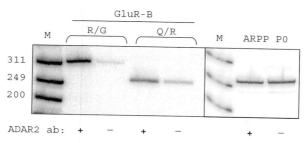

Figure 13.2 Enrichment of known substrates for A-to-I editing using ADAR2-specific immunoprecipitation (+) compared to preimmune serum (−). The GluR-B transcript is amplified by RT-PCR using primers specific for the R/G and the Q/R sites. As a control, the RT-PCR from ARPP P0 is used, a transcript that is not subjected to editing. Lane M is a marker with sizes in base pairs indicated.

by RT-PCR and microarray. The transcript that codes for the glutamate receptor subunit B (GluR-B) is selectively A-to-I edited at two sites (Q/R and R/G) within the coding sequence (Seeburg *et al.*, 1998). The specificity of the IP for GluR-B is analyzed by semiquantitative RT-PCR using radioactively labeled primers in a 25 cycles PCR. An enrichment of target substrates can be seen from the ADAR2 IP using primers in the vicinity of the edited sites (Fig. 13.2). The preimmune IP did not show an enrichment of these target RNAs. As another negative control, we used the ARPP P0 transcript coding for a ribosomal protein that is not subjected to editing. When primers specific for the ARPP P0 transcript were used in the RT-PCR, the same amount of product could be detected in the ADAR2 IP and the preimmune IP (Fig. 13.2). Further, the microarray shows that the enrichment of GluR-B in the ADAR2 IP is on average 2.6-fold (Table 13.1), whereas no enrichment of ARPP P0 could be detected. Taken together, this confirms that an RNA that is edited can be specifically enriched from a mammalian brain tissue using an anti-ADAR2 antibody in immunoprecipitations.

7. IDENTIFICATION OF SITES OF EDITING WITHIN THE SUBSTRATE CANDIDATES

Even though strong candidates for A-to-I editing are selected in the microarray, this method will reveal only the candidate substrate not the position of the edited site. The use of bioinformatic analysis to detect potential sites of editing within the candidate gene is therefore essential. We have successfully used the program EvoFold, developed to predict evolutionary conserved stem–loop structures in RNA (Pedersen *et al.*, 2006). Using this program it is possible to identify the position of regions with the potential of forming short, stable, double-stranded RNA structures

Table 13.1 Enrichment of transcripts verified by microarray[a]

Transcript	Mean (2^x-fold increase)	SD
GluR-B	1.4	0.2
ARPP P0 (control)	−0.5	0.2

[a] GluR-B, glutamate receptor subunit B; ARPP P0, acidic ribosomal phosphoprotein P0. SD equals the standard deviation.

suitable for A-to-I editing. Good substrates are stem–loop structures longer than 20 nucleotides. The sequence forming these double-stranded structures in the vicinity of an edited site is often extremely conserved between species. Within these structures adenosines suitable for editing have a 5' nearest-neighbor preference nucleotide of A = U > C > G. The edited A is often paired with a U but can also have a mismatched C as the opposing base. EvoFold will not detect edited sites within larger stem–loop structures, therefore, other bioinformatical analyses are required to detect these sites. Editing in windows of approximately 500 nucleotides can be verified by DNA sequence determination after RT-PCR. Exon–exon primers are preferred in the PCR reaction to ensure that only cDNA and not contaminating genomic DNA is amplified.

8. ADAR2 COIMMUNOPRECIPITATION USING MOUSE BRAIN

8.1. Overview

This assay describes coimmunoprecipitation using an affinity purified human anti-ADAR2 antibody. However, it is also possible to apply this method to other ADAR protein family members like ADAR1. Using this method, it would also be possible to find targets for ADAR3, which does not yet have any known targets. In the example below, we use whole brain tissue from adult mouse.

8.2. Preparation of affinity-purified antihuman ADAR2 antibody

1. Two milliliters of recombinant human ADAR2a-His$_6$ (histidine tagged) protein, expressed in *Saccharomyces cerevisiae* and purified giving a concentration of 100 µg/ml, is concentrated in a Centrion YM30 (Millipore) according to the manufacturer's instructions.

2. The protein is further purified on an 8% SDS–PAGE where the band corresponding to the ADAR2 protein is excised from the gel.

3. The gel piece containing the ADAR2 protein is immunized four times into a rabbit for antibody production (Agisera, Umeå, Sweden).

4. The serum is checked for immunoreactivity and the final bleed is used for affinity purification using Affi-Gel 10, an activated affinity support system from Bio-Rad.

5. Shake the Affi-Gel 10 Matrix until it is fully suspended and take 1.5–2 ml of the suspension for coupling of 1 mg of recombinant ADAR2a protein. Wash the resin rapidly with 6 ml of isopropanol in a scintered glass funnel under a modest vacuum. Do not let the bed resin run dry.

6. Rinse the resin in 10 ml of double-distilled water and then 10 ml of 0.1 M HEPES (pH 7.5). Remove excess liquid by vacuum.

7. Weigh the "affigel-cake" and add at least 0.5 ml of ADAR2 protein solution per milliliter of gel. The volume of the protein should not exceed 4.5 ml/ml of gel. For the purification to be efficient, 1 mg of protein is required.

8. Gently agitate on nutator for 1 h at room temperature or 4 h at 4°.

9. Block the remaining active groups by adding 0.1 ml of 1 M ethanol-amine HCl (pH 8.0) per milliliter of gel. Incubate for 1 h to complete the coupling.

10. Wash the gel with 0.1 M HEPES (pH 7.5) to remove uncoupled ligand. Wash until free of reactants as detected by OD_{280}.

11. Equilibrate the gel in 1× phosphate-buffered saline (PBS). The coupled resin can be stored at 4° in 1× PBS and 0.05% NaN_3.

12. Thaw the anti-ADAR2 serum from the final bleeds of the rabbit (use as much serum as possible). Spin at 10,000 rpm for 15 min.

13. Add NaN_3 to a final concentration of 0.05% to the serum and begin purification.

14. Cycle the entire volume of serum three times over a disposable 10-ml column (Bio-Rad) containing the coupled Affi-gel 10 resin. Use a drip rate at approximately 2 drips per 5 sec.

15. The resin is then washed with 1× PBS until the OD_{280} of the flow through is less than 0.02.

16. Label 15 tubes and add 50 μl of 1 M Tris–HCl (pH 8) to each tube.

17. Drain the column of 1× PBS and add 2 volumes (2 ml) of 150 mM NaCl. Close the column and layer 0.5 ml of 0.1 M glycine at pH 2.5 on to the resin, wait 90 sec, open the column, and collect the eluate in one of the 15 tubes prepared. Repeat 14 times.

18. Pool fractions of similar OD_{280} and dialyze against 1× PBS.

19. To test the immunoreactivity of the affinity-purified antibody, recombinant ADAR2 protein or HeLa nuclear extract can be used in a Western blot.

8.3. Isolation of RNA–protein complex from mouse brain

1. Use one mouse brain (0.4 g brain tissue) and homogenize in HBSS [1× Hanks' solution (HBSS GIBCO no. 14185–045)] including 0.01 M HEPES (pH 7.3) using a glass grinder.
2. Spin at 600×g for 10 min at 4° and wash the pellet in ice cold 1× HBSS.
3. Spin again at 600×g for 10 min at 4° and freeze the pellet in liquid nitrogen. Resuspend the pellet in PXL [1× PBS, 0.1% sodium dodecyl sulfate (SDS), 0.5% deoxycholate, and 0.5% NP-40] and 50 μl/ml ribonucleoside vanadyl complex (Sigma) on ice.
4. Sonicate the suspension and treat with 10 units DNase I (SIGMA). Spin the suspension at 10,000×g for 20 min at 4° and the supernatant is used in the immunoprecipitation.

8.4. Immunoprecipitation of RNA–ADAR2 complex

1. To reduce nonspecific binding prior to use in immunoprecipitations, 100 μg/ml tRNA and 100 μg/ml BSA in 1× PBS are added and incubated for 15 min at 4° with the Sepharose A beads. The suspension is washed once in 1× PBS and resuspended in 1 volume of 1× PBS containing 0.05% NaN_3.
2. Cell lysis extract from one mouse brain is preincubated with 50 μl of protein A Sepharose stock for 30 min at 4° under slow rotation using a vertical rotator (L29, Labinco).
3. Incubate the lysate with the affinity-purified anti-ADAR2 polyclonal antibody for 2 h at 4° under slow rotation.
4. Mix the lysate–antibody with 50 μl of prepared protein A Sepharose and incubate for 1 h at 4° under slow rotation.
5. Rinse the bead–antibody–lysate complex three times in a wash buffer containing 1× PBS, $MgCl_2$ (2 mM), EDTA (15 mM), NP-40 (1%), and Tween-20 (0.5%) including one protease Inhibitor Cocktail tablet (Roche) per 10 ml of solution. Rinse once in 1× PBS.
6. Incubate the complex in 1× PBS and 1% SDS at 65° for 10 min, spin in a table-top centrifuge for 1 min, and save the supernatant where the RNA will be eluted; repeat twice.

8.5. Preparation of RNA after immunoprecipitation

1. Add 1.8 mg of proteinase K (Roche) to the protein–RNA eluate from the immunoprecipitation and incubate at 37° for 15 min.
2. Extract with 1 volume of phenol/chloroform and precipitate using 2.5 μg glycogen (Ambion) with 2.5 volume 99.5% EtOH and 0.1 volume of 5 M NaOAc. Wash once in 70% EtOH.

3. Purify the RNA using RNeasy (Qiagen) according to the instructions from the manufacturer: guanidine isothiocyanate (GITC) containing lysis buffer and ethanol are added to the sample to create conditions that promote selective binding of the RNA to the RNeasy silica-gel membrane. Contaminants are washed away and high-quality RNA is eluted with water.

4. Measure the concentration of the RNA. The RNA is now prepared to be used in further analysis like RT-PCR, RNase protection, or microarray.

This is a powerful method that can be used to find novel site-selectively editing substrates in different tissues. It is also possible to use it for identification of editing discrepancies between different species. Another application for the method would be to use antibodies specific for other ADAR protein family members such as ADAR1 and ADAR3 in the immunoprecipitation and thereby identify specific targets for these enzymes.

REFERENCES

Athanasiadis, A., Rich, A., and Maas, S. (2004). Widespread A-to-I RNA editing of Alu-containing mRNAs in the human transcriptome. PLoS Biol. 2, e391.

Bass, B. L. (2002). RNA editing by adenosine deaminases that act on RNA. Annu. Rev. Biochem. 71, 817–846.

Blaszczyk, J., Gan, J., Tropea, J. E., Court, D. L., Waugh, D. S., and Ji, X. (2004). Noncatalytic assembly of ribonuclease III with double-stranded RNA. Structure 12, 457–466.

Blow, M., Futreal, P. A., Wooster, R., and Stratton, M. R. (2004). A survey of RNA editing in human brain. Genome Res. 14, 2379–2387.

Clutterbuck, D. R., Leroy, A., O'Connell, M. A., and Semple, C. A. (2005). A bioinformatic screen for novel A-I RNA editing sites reveals recoding editing in BC10. Bioinformatics 21, 2590–2595.

Doyle, M., and Jantsch, M. F. (2002). New and old roles of the double-stranded RNA-binding domain. J. Struct. Biol. 140, 147–153.

Hoopengardner, B., Bhalla, T., Staber, C., and Reenan, R. (2003). Nervous system targets of RNA editing identified by comparative genomics. Science 301, 832–836.

Klaue, Y., Källman, A. M., Bonin, M., Nellen, W., and Öhman, M. (2003). Biochemical analysis and scanning force microscopy reveal productive and nonproductive ADAR2 binding to RNA substrates. RNA 9, 839–846.

Levanon, E. Y., Eisenberg, E., Yelin, R., Nemzer, S., Hallegger, M., Shemesh, R., Fligelman, Z. Y., Shoshan, A., Pollock, S. R., Sztybel, D., Olshansky, M., Rechavi, G., and Jantsch, M. F. (2004). Systematic identification of abundant A-to-I editing sites in the human transcriptome. Nat. Biotechnol. 22, 1001–1005.

Levanon, E. Y., Hallegger, M., Kinar, Y., Shemesh, R., Djinovic-Carugo, K., Rechavi, G., Jantsch, M. F., and Eisenberg, E. (2005). Evolutionarily conserved human targets of adenosine to inosine RNA editing. Nucleic Acids Res. 33, 1162–1168.

Macbeth, M. R., Lingam, A. T., and Bass, B. L. (2004). Evidence for auto-inhibition by the N terminus of hADAR2 and activation by dsRNA binding. RNA 10, 1563–1571.

Morse, D. P., and Bass, B. L. (1999). Long RNA hairpins that contain inosine are present in *Caenorhabditis elegans* poly(A)+ RNA. *Proc. Natl. Acad. Sci. USA* **96,** 6048–6053.

Morse, D. P., Aruscavage, P. J., and Bass, B. L. (2002). RNA hairpins in noncoding regions of human brain and *Caenorhabditis elegans* mRNA are edited by adenosine deaminases that act on RNA. *Proc. Natl. Acad. Sci. USA* **99,** 7906–7911.

Ohlson, J., Ensterö, M., Sjöberg, B. M., and Öhman, M. (2005). A method to find tissue-specific novel sites of selective adenosine deamination. *Nucleic Acids Res.* **33,** e167.

Öhman, M., Källman, A. M., and Bass, B. L. (2000). *In vitro* analysis of the binding of ADAR2 to the pre-mRNA encoding the GluR-B R/G site. *RNA* **6,** 687–697.

Pedersen, J. S., Bejerano, G., Siepel, A., Rosenbloom, K., Lindblad-Toh, K., Lander, E. S., Kent, J., Miller, W., and Haussler, D. (2006). Identification and classification of conserved RNA secondary structures in the human genome. *PLoS Comput. Biol.* **2,** e33.

Ramos, A., Grunert, S., Adams, J., Micklem, D. R., Proctor, M. R., Freund, S., Bycroft, M., St Johnston, D., and Varani, G. (2000). RNA recognition by a Staufen double-stranded RNA-binding domain. *EMBO J.* **19,** 997–1009.

Ryter, J. M., and Schultz, S. C. (1998). Molecular basis of double-stranded RNA-protein interactions: Structure of a dsRNA-binding domain complexed with dsRNA. *EMBO J.* **17,** 7505–7513.

Seeburg, P. H., Higuchi, M., and Sprengel, R. (1998). RNA editing of brain glutamate receptor channels: Mechanism and physiology. *Brain Res. Rev.* **26,** 217–229.

Stapleton, M., Carlson, J. W., and Celniker, S. E. (2006). RNA editing in *Drosophila melanogaster*: New targets and functional consequences. *RNA* **12,** 1922–1932.

Stefl, R., Xu, M., Skrisovska, L., Emeson, R. B., and Allain, F. H. (2006). Structure and specific RNA binding of ADAR2 double-stranded RNA binding motifs. *Structure* **14,** 345–355.

Stephens, O. M., Haudenschild, B. L., and Beal, P. A. (2004). The binding selectivity of ADAR2's dsRBMs contributes to RNA-editing selectivity. *Chem. Biol.* **11,** 1239–1250.

Tenenbaum, S. A., Lager, P. J., Carson, C. C., and Keene, J. D. (2002). Ribonomics: Identifying mRNA subsets in mRNP complexes using antibodies to RNA-binding proteins and genomic arrays. *Methods* **26,** 191–198.

Ule, J., Jensen, K. B., Ruggiu, M., Mele, A., Ule, A., and Darnell, R. B. (2003). CLIP identifies Nova-regulated RNA networks in the brain. *Science* **302,** 1212–1215.

Wong, S. K., Sato, S., and Lazinski, D. W. (2001). Substrate recognition by ADAR1 and ADAR2. *RNA* **7,** 846–858.

Wu, H., Henras, A., Chanfreau, G., and Feigon, J. (2004). Structural basis for recognition of the AGNN tetraloop RNA fold by the double-stranded RNA-binding domain of Rnt1p RNase III. *Proc. Natl. Acad. Sci. USA* **101,** 8307–8312.

CHAPTER FOURTEEN

PURIFICATION AND ASSAY OF ADAR ACTIVITY

Liam P. Keegan,* Joshua J. Rosenthal,† Loretta M. Roberson,‡ *and* Mary A. O'Connell*

Contents

Abstract

ADAR editing enzymes are found in all multicellular animals and are conserved in sequence and protein organization. The number of ADAR genes differs between animals, ranging from three in mammals to one in *Drosophila*. ADAR is also alternatively spliced to generate isoforms that can differ significantly in enzymatic activity. Therefore, to study the enzyme *in vitro*, it is essential to have an easy and reliable method of expressing and purifying recombinant ADAR protein. To add to the complexity of RNA editing, the number of transcripts that are edited by ADARs differs in different organisms. In humans there is extensive editing of Alu sequences, whereas in invertebrates transcripts expressed in the central nervous

* MRC Human Genetics Unit, Western General Hospital, Edinburgh, United Kingdom
† Institute of Neurobiology, University of Puerto Rico-Medical Sciences Campus, San Juan, Puerto Rico
‡ Department of Biology, University of Puerto Rico-Rio Piedras, San Juan, Puerto Rico

Methods in Enzymology, Volume 424
ISSN 0076-6879, DOI: 10.1016/S0076-6879(07)24014-5

system are edited and this editing increases during development. It is possible to quantify site-specific RNA editing by sequencing of clones derived from RT-PCR products. However, for routine assaying of an edited position within a particular transcript, this is both expensive and time consuming. Therefore, a nonradioactive method based on poison primer extension assay is an ideal alternative.

1. INTRODUCTION

Adenosine deaminase acting on RNA (ADAR) enzymes convert adenosine to inosine in double-stranded (ds) RNA. Inosine is read as if it were guanosine by the translational machinery (Basilio *et al.*, 1962) and can result in another amino acid being inserted at the edited position if it falls within an exon (Valente and Nishikura, 2005). *In vitro* this reaction does not require energy or any additional cofactors. ADARs do not recognize a consensus sequence surrounding the edited position; instead their only requirement is that the editing site has to be within a duplex region of the transcript (Higuchi *et al.*, 1993). ADARs can deaminate many adenosines within a transcript when it forms a long perfect duplex, and this is referred to as promiscuous editing. However, they can also deaminate a few particular adenosines within a transcript, and this is called specific editing.

What is remarkable about the ADAR enzymes is that they have such base precision and that this accuracy can be recapitulated *in vitro* with purified recombinant proteins and *in vitro* transcribed RNA. For this reason, many groups have overexpressed different recombinant ADAR proteins for *in vitro* assays (Lai *et al.*, 1997; Melcher *et al.*, 1996; Xu *et al.*, 2006). As ADAR proteins are not active when expressed in *Escherichia coli*, some groups have chosen yeast as an alternative expression system (Gerber *et al.*, 1997; Macbeth *et al.*, 2005). This chapter will describe the expression in and purification of ADARs from the methylotrophic yeast *Pichia pastoris*. This yeast was chosen due to its ability to grow to high cell density and to provide a high yield of protein. This chapter will also describe a new nonradioactive assay for measuring editing activity. Chapter 15 (this volume) describes the expression of ADARs in the yeast *Saccharomyces cerevisiae,* which can also be used for protein expression.

2. OVEREXPRESSION OF RECOMBINANT PROTEINS IN *PICHIA PASTORIS*

2.1. Overview

A frequently asked question concerns the advantages of using the yeast *P. pastoris* over other eukaryotic expression systems. One major reason is that it is relatively easy to set up. *P. pastoris* is easy to manipulate, and many of the techniques developed for *S. cerevisiae* can be used with it. The expression system is available as a kit from Invitrogen Corporation and, compared to mammalian cell culture, it does not require complex or expensive media for growth. It can also grow to high density using minimal media and express recombinant proteins either intracellularly or extracellularly. As *P. pastoris* secretes very low levels of endogenous proteins, a secreted recombinant protein can be a major component in the medium. Most importantly, since it has a eukaryotic protein synthesis pathway, it performs many eukaryotic posttranslational modifications.

In the 1970s Philip's Petroleum Company first began to grow *P. pastoris* to high density, as it was hoped it could be developed as a source of single-cell protein for use in animal foods (Macauley-Patrick *et al.*, 2005). The major advantage of *Pichia* was that it was inexpensive to produce as it grew on methanol. However, this venture was not successful due to the increase in methanol prices caused by the oil crisis. Philip's Petroleum then contracted the Salk Institute Biotechnology/Industrial Associate Inc. (SIBIA) to develop *P. pastoris* as a system for heterologous protein expression. SIBIA isolated the alcohol oxidase (*AOX1*) gene and promoter and developed vectors and strains. Invitrogen Corporation subsequently obtained a license to market the *P. pastoris* expression system.

The inducible *AOX1* promoter is the most widely used for heterologous protein expression in *P. pastoris*, as transcription occurs only when the yeast is grown on methanol (Macauley-Patrick *et al.*, 2005). *AOX* catalyzes the first step in the metabolism of methanol within the cell, and although there are two genes, *AOX1* accounts for the majority of the activity.

There are three steps in the procedure for generating recombinant ADAR proteins in *P. pastoris*: (1) the cloning in *E. coli* of the ADAR sequence into the *P. pastoris* expression vector pPICZ A, (2) transformation into yeast, and (3) screening of yeast colonies for multiple integration events.

2.2. Solutions required for yeast media

Yeast extract and peptone (YE + PEP): add 5 g of yeast extract to 10 g of peptone, place in a 500-ml bottle, add 350 ml of distilled water, and autoclave.

Yeast nitrogen base (10× YNB): 134 g yeast nitrogen base with ammonium sulfate (without amino acids). Dissolve in 1 liter of distilled water and filter sterilize. It may be necessary to warm the YNB to dissolve it. Store at 4°.

1 M phosphate buffer: prepare 1 M K_2HPO_4 and 1 M KH_2PO_4, add 132 ml of 1 M K_2HPO_4 to 868 ml of 1 M KH_2PO_4, and adjust the pH with either phosphoric acid or potassium hydroxide to pH 6.0. Autoclave and store at room temperature.

10% glycerol: 10% glycerol diluted in distilled water, autoclave, and store at room temperature.

10× methanol: 5% methanol solution diluted in distilled water, filter sterilize and store at 4°.

500× biotin; 0.02% (20 mg/100 ml) biotin diluted in distilled water, filter sterilize and store at 4°.

BMGY: to a 500-ml bottle containing 350 ml of YE + PEP add the following:

    50 ml of 10× YNB

    50 ml of 1 M phosphate buffer

    50 ml of 10% glycerol

    1 ml of 500× biotin

Kanamycin at 25 mg/ml may also be added to prevent bacterial contamination.

BMMY: The same as for BMGY but with 50 ml of 10× methanol instead of 10% glycerol.

YPDS plates: 1% yeast extract, 2% peptone, 2% dextrose (glucose), 1 M sorbitol, and 2% agar. It is important not to add the dextrose (which is filtered sterilized) until after the solution has been autoclaved. When cool, add the Zeocin™ to the required concentration. After pouring the plates, store them in the dark, as Zeocin™ is light sensitive.

2.3. Cloning of ADARs in the *Pichia pastoris* expression vector pPICZ A

The original *Pichia* expression vectors from Invitrogen used gene replacement to integrate a single copy of the plasmid construct stably at the *his4* locus (Ohi *et al.*, 1998). However, the Easyselect™ pPICZ vectors are designed to give multicopy transformants. We made a derivative of pPICZ A called pPICZ A-FLIS6 (Fig. 14.1). This derivative contains a short open reading frame encoding a FLAG epitope tag at the N-terminus and a histidine hexamer (6× His) epitope tag at the carboxy-terminus. FLAG and 6× His were chosen because they encode small epitopes that do not interfere with ADAR enzymatic activity. Furthermore, an epitope at either terminus facilitates the purification of only full-length protein. This cassette was inserted at the *Eco*RI site in the pPICZ A multiple cloning site (Fig. 14.1). The ADAR sequence is amplified by polymerase chain reaction (PCR) with primers containing either the *Spe*I or *Nhe*I restriction site and cloned into the *Spe*I site between the two epitope tags so that the ADAR coding sequence is in frame with the epitope tags under the control of the methanol-inducible *AOX1* promoter. If there is an *Spe*I site present within the coding sequence of the ADAR cDNA, then *Nhe*I or *Xba*I can be used instead.

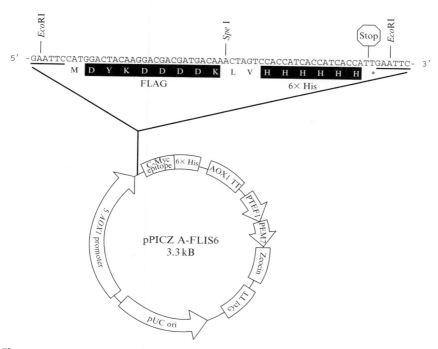

5'-GAATTCCATGGACTACAAGGACGACGATGACAAACTAGTCCACCATCACCATCACCATTGAATTC- 3'

M D Y K D D D D K L V H H H H H H *

FLAG   6× His

pPICZ A-FLIS6
3.3 kB

Figure 14.1 A schematic diagram of pPICZ A-FLIS6 derived from the Invitrogen pPICZ A vector. This vector is used for expression of recombinant ADARs. A cassette encoding both FLAG and 6× His has been inserted in the polylinker at the EcoR I restriction site. (Copyright 2006 Invitrogen Corporation. All rights reserved. Used with permission.)

The original pPICZ A vector encodes a myc and 6× His epitope tag sequences; however, they are not expressed in these constructs. For efficient 3' mRNA processing and increased mRNA stability, the native transcription termination and polyadenylation signal from the AOX1 gene are downstream of the multiple cloning site. The pPICZ A vector encodes a Zeocin^{TM} resistance gene under the control of an EM7 constitutive promoter for expression in E. coli and of a TEF1 promoter from S. cerevisiae for expression in Pichia. This permits selection in both E. coli and P. pastoris of Zeocin^{TM}-resistant transformants. Zeocin^{TM} is a DNA damaging agent related to bleomycin.

The E. coli DH5a is routinely used for transformations and is plated on low salt LB agar containing 25 µg/ml Zeocin^{TM}. Single colonies are selected and grown overnight in low salt LB media with Zeocin^{TM}. Plasmid DNA is isolated and screened for inserts by restriction enzyme digestion. Positive clones are sequenced to confirm the presence of the correct open reading frame.

2.4. Transformation of *Pichia pastoris* with pPICZ ADAR clone

1. For transformation into *P. pastoris*, 5–10 μg of pPICZ A-FLIS6-ADAR is linearized by digestion at one of three unique restriction enzymes such as *Sac*I or *Pme*I that cleave 5′ of the *AOX1* promoter region. A small aliquot of the plasmid is electrophoresed on an agarose gel to check that it is linearized. However, it is not a problem if it has not digested to completion and it is not necessary to gel purify the plasmid. The volume of the DNA is increased to 100 μl with water and an equal volume of phenol/chloroform is added. The mixture is vortexed and centrifuged for 3 min at room temperature. The DNA, which is in the upper aqueous phase, is precipitated with 10 μl 3 *M* sodium acetate and 900 μl 100% ethanol. After centrifugation for 30 min at 4°, the pellet is washed with 70% ethanol and dried. The pellet is resuspended in 10 μl dH$_2$O. Transformation occurs by integration of one or many copies of the linearized plasmid at the homologous sequence in the chromosomal *AOX1* locus in the *Pichia* genome.

2. To prepare electrocompetent *Pichia*, a single colony of *P. pastoris* KM71 is picked and 5 ml of YPD broth is inoculated in a 50-ml conical tube. The starter culture is grown overnight at 30° with shaking. The starter culture is subsequently used to inoculate 250 ml YPD medium in a 1-liter flask. The yeast is grown overnight to an OD$_{600}$ of 1.3–1.5.

3. The yeast is harvested by centrifugation at 1500×*g* for 5 min at +4°, washed twice with 250 ml ice-cold sterile dH$_2$O and once with 10 ml ice-cold 1 *M* sorbitol. Finally, the pellet is resuspended in 1 ml 1 *M* sorbitol to give a final volume of approximately 1.5 ml. Competent cells remain usable for several days.

4. Eighty microliters of competent cells is mixed with 5–10 μg of linearized DNA and transferred to a 0.2-cm electroporation cuvette on ice. A control with no DNA is included. The cells and plasmid are left on ice for 5 min and then electroporated at 1.5 kV, 25 μF, 200 Ω (Bio-Rad Gene Pulser). Immediately, 1 ml of ice-cold 1 *M* sorbitol is added to the cuvette, transferred to a 15-ml sterile tube, and incubated without shaking at 30° for 1 h. An increase in the efficiency of transformation was observed when 1 ml of YPDS was added to the cells and they were allowed to grow for a further 2 h at 30° with shaking.

5. Of the transformation mix, 200 μl is plated on separate YPDS plates containing 100 μg/ml Zeocin[TM]. Colonies form after 2–5 days incubation at 30°.

6. Approximately 50 colonies are picked from the transformation plates and restreaked for single colonies on fresh YPDS agar plates containing 100, 200, 500, and 1000 μg/ml Zeocin[TM]. They are incubated for 2–3 days at 30° and the plates are checked regularly so that the difference in growth of the transformant lines on the high concentrations of Zeocin[TM] can be

observed. Those showing the highest resistance to Zeocin™ should have the highest copy number of the expression construct. Transformant lines can be tested by induction of small-scale cultures to determine which line gives the best expression of ADAR recombinant protein.

2.4.1. Small-scale cultures

Before growing a large culture, it is good practice to check for protein expression from a number of lines. Cultures of 5 ml are grown in BMGY medium overnight at 30°. To induce expression, the cultures are then centrifuged at 3000×g, resuspended in 10 ml BMMY in a loosely capped 50-ml Falcon tube, and grown overnight. The next day fresh methanol is added to a final concentration of 0.5% and the cultures are allowed to grow overnight. Then 1 ml of the culture is centrifuged in an Eppendorf at maximum speed for 10 min to pellet the yeast. The pellet is then resuspended in 100 μl ice-cold dH$_2$O. Then 100 μl of glass beads (Sigma 425–600 μm) and 20 μl of 100% trichloroacetic acid (TCA) are added and the yeast cell wall is broken by vortexing for 5 × 1 min at 4°. One milliliter of ice-cold 5% TCA is then added and the samples are vortexed again before being centrifuged. The pellets are resuspened in sodium dodecyl sulfate (SDS) loading buffer and the pH is neutralized before electrophoresis on an SDS polyacrylamide gel. The amount of pellet loaded on the gel depends on whether the gel will be stained with Coomassie blue (10 μl) or an immunoblot analysis will be performed (1 μl).

2.4.2. Large-scale cultures

This procedure may not necessarily be the optimum for protein production of a particular ADAR. Each new construct should be tested independently by taking aliquots at different time points after methanol induction and comparing the amount of recombinant protein.

1. A single colony of *P. pastoris* is picked and grown in 10 ml of BMGY medium in a 50-ml tube overnight at 30°; the cap is left loose to allow good aeration. Then 250 ml of BMGY medium is inoculated with this starter culture in an autoclaved 2-liter flask, preferably a baffled flask. The top of the flask is loosely covered with tin foil or muslin for good aeration and it is grown overnight at 30°.

2. When the OD$_{600}$ of the culture is 2–6 (approximately 16–18 h), the culture is pelleted by centrifugation at 3000×g for 5 min at room temperature. The decanting and resuspension of the culture are performed under sterile conditions.

3. The pellet is resuspended in 500 ml of BMMY medium and incubated overnight at 30°. The next day 2.5 ml of 100% methanol is added under sterile conditions. It is presumed that all of the methanol in the BMMY

has been metabolized or has evaporated since the previous day. The cells are incubated for a an additional night.

4. The cells are harvested by centrifugation at 3000×*g* for 5 min. Sterile conditions are no longer necessary, but the cells must be kept on ice.
5. The pellet is washed twice with 250 ml of autoclaved ice-cold water, ensuring the pellet is fully resuspended to wash out the BMMY. The cells are centrifuged again and the supernatant is removed.
6. Finally, cells are resuspended in 40 ml of Buffer A [50 m*M* Tris–HCl, pH 7.9, 200 m*M* KCl, 20% glycerol, 1 m*M* dithiothreitol (DTT), 0.5 m*M* phenylmethylsulfonyl fluoride (PMSF), 0.7 mg/ml pepstatin, and 0.4 mg/ml leupeptin] and centrifuged as previously. The pellet can be frozen in liquid nitrogen and stored at −70° for use later or kept on ice for immediate use.

2.5. Purification of ADAR protein

Recombinant ADAR proteins can be purified to homogeneity in 1 day from *Pichia* pellets (Fig. 14.2).

1. The yeast pellets are weighed and 1 ml of Buffer A is added per gram of pellet. After resuspending, the thawed pellets are broken by a French press at 4°. If the French press is not in a cold room, then the cylinder should be precooled by being left in a cold room overnight. The French press cylinder has a dead volume of approximately 8 ml, so the yeast suspension has to be greater than this and maximum pressure is used to

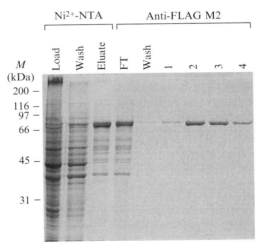

Figure 14.2 SDS–PAGE analysis of the purification of dADAR. A 1-*μ*l aliquot of the load fraction and 6 *μ*l of all the other fractions were electrophoresed on an SDS 10% polyacrylamide gel and stained with Coomassie blue.

break the yeast. It is normally necessary to pass the cells through the cylinder two or three times for complete breakage, at which point the material coming out dropwise from the cylinder will be viscous. The yeast extract is then centrifuged at $10,000 \times g$ for 20 min at $4°$. If no French press is available, then the yeast can be broken manually by grinding the yeast suspension in liquid nitrogen using a mortar and pestle.

2. While the yeast extract is being centrifuged, the Ni^{2+}-NTA resin (Qiagen) can be prepared. One to two milliliters of packed Ni^{2+}-NTA resin is sufficient to purify recombinant protein from a 500-ml culture. It is important not to use too much Ni^{2+}-NTA resin, as additional *Pichia* proteins will bind to it. To equilibrate the resin, 10 ml of Buffer A is added to it and it is spun in a Falcon tube at low speed in a table-top centrifuge for 2–3 min to pellet the resin. This washing step is repeated two times with Buffer A.

3. Before the yeast extract (the supernatant after the centrifugation step) is added to the Ni^{2+}-NTA resin, an aliquot of 0.5 ml should be taken. All subsequent steps are performed at $4°$. The yeast extract is then incubated with the Ni^{2+}-NTA resin on a rotating wheel for at least 1 h. The extract and resin are poured into a 10-ml column and the flow through from the column is reapplied two times. An aliquot is kept of the flow through and the remainder is frozen in liquid nitrogen. The resin is then washed with 15 mM imidizole in Buffer A. An aliquot of the wash is kept while the remainder is frozen. The protein is then eluted from the Ni^{2+}-NTA resin by the addition of 5 ml of Buffer A containing 250 mM imidazole, pH 7.9, and an aliquot is taken of the eluate.

4. The eluate from the Ni^{2+}-NTA resin is further purified by applying it directly to an anti-FLAG M2 matrix (Sigma), as the imidazole does not interfere with binding. The anti-FLAG M2 resin must be activated according to the manufacturer's instructions before being used. Briefly, 0.4 ml of a 50% slurry is placed in a 10-ml column to give a packed volume of approximately 0.2 ml. The resin is activated by washing with approximately 20 ml of 0.1 M glycine, pH 3.5, for no longer than 20 min. The resin is subsequently washed with 20 ml Buffer A to equilibrate it. The eluate from the Ni^{2+}-NTA column is loaded three times to the resin. An aliquot of the flow through is kept and the remainder is frozen. The resin is then washed with 5 ml Buffer A, again taking an aliquot of the wash and freezing the remainder. The protein is eluted with 5×0.5 ml Buffer A containing 100 μg/ml FLAG peptide (Sigma). The final step is to wash the resin with 0.5 ml of 0.1 M glycine, pH 3.5, to remove any protein that is bound tightly to the resin. The resin can be regenerated; however, it should not remain in the glycine for longer than 20 min.

5. The aliquots are then electrophoresed on a 10% SDS polyacrylamide gel; typically, 1 μl of the total extract and the Ni^{2+}-NTA flow through

aliquot and 6 μl of all the other fractions are used. Recombinant ADAR usually elutes in fraction two from the anti–FLAG M2 resin (Fig. 14.2).

3. Quantification of the Efficiency of RNA Editing

3.1. Overview

Seldom are all mRNAs edited at a specific site, and the proportion of transcripts edited is important for a number of reasons. First, from the perspective of the proteome, only incomplete editing creates diversity; editing must be partial or switched on or off over time. Second, certain mRNAs, such as those encoding the serotonin receptor $5HT_{2C}$ (Burns *et al.*, 1997) or the K^+ channel Kv1.1 (Burns *et al.*, 1997), are edited in a brain region-specific manner. Therefore, diversity of function, particularly in different parts of the nervous system, can be generated by RNA editing. Third, to understand how editing enzymes themselves work, it is essential to quantify their activity. In this way, structure–function relationships can be ascertained. Finally, recent data suggest that a variety of pathologies may be linked to small changes in editing efficiency (Akbarian *et al.*, 1995; Gurevich *et al.*, 2002; Iwamoto and Kato, 2003; Sodhi *et al.*, 2001; Sprengel *et al.*, 1999). By closely examining the extent of editing at specific sites, diseases can be better understood and potential therapies evaluated. For all of these reasons, simple, accurate assays to measure editing efficiency are needed. Unfortunately, many of the methods used to date have proven inadequate. This section describes the basic approaches to quantification and a recently developed assay that meets these criteria (Roberson and Rosenthal, 2006).

3.2. Methods for quantification of editing

Quantification of editing percentages is usually accomplished by either DNA sequencing or assays designed to measure nucleotide alternatives at a specific site in bulk RNA samples. Sequencing multiple reverse transciptase polymerase chain reaction (RT-PCR) products is probably the most straightforward and frequently employed method. Provided that enough clones are sequenced, the results are accurate and the technology is accessible to most laboratories. An additional advantage is that multiple editing sites can be assayed at the same time if they are contained within a typical sequencing reaction ($\sim$500–1000 bp). In terms of cost and labor, however, this approach becomes problematic when comparing multiple samples. For example, comparing five different samples could require hundreds of DNA preparations and sequencing reactions. This number could easily escalate into the thousands if the difference in editing percentage was small or if multiple samples for each point were required. For these reasons, this approach can be used to measure the initial editing percentage when a new site is identified, but is generally avoided for more complex experimental designs. As an alternative to bulk sequencing of individual clones, some laboratories have turned to directly sequencing PCR products. This approach has the advantage of sampling many clones at once and consequently greatly reducing the number of sequencing reactions. However, with standard dye-terminator chemistry, peak heights vary between positions and samples. Therefore, this approach is not very accurate and can be used only to detect large, qualitative differences.

The poison–primer extension (PPE) assay is a standard method for quantifying nucleotide alternatives at a single position. This assay is based on designing a primer close to an edited position. After the primer, the first occurrence of one of the alternative bases should be at the edited position. Taking an ADAR RNA editing site as an example, this assay has been carried out in the past using reverse transcriptase and an edited RNA substrate. The primer is extended through the edited position in the presence of three deoxynucleotide triphosphates (dA, dC, and dG) and one dideoxynucleotide triphosphate (dT). If the template has an A at the edited position, then the chain will terminate. If not, it will extend to the next instance of the A.

When PPE is instead used on double-stranded DNA products using DNA polymerases, either strand can be used as the template for primer extension and the terminating nucleotide can be chosen to terminate when either the edited or the unedited base is present at the edited site. This gives four assay design choices for quantitating either adenosine deamination or cytosine deamination; it is possible to use any base as the terminator. In general, each editing site is different, and the base that yields the greatest size difference between the edited and unedited product is selected, but the range of terminator choices is useful because different bases lead to different levels of failure to terminate. Many variants of this assay have been described, and in general there are four important considerations: the template, the polymerase, the label, and the terminator.

Both the experimental design and the available equipment can greatly influence the specific elements of a PPE assay. Early assays used either mRNA or total RNA as a template. This approach has significant drawbacks. First, the amount of the specific target mRNA is very small compared to the background, necessitating the use of a radioactive label. Second, as a polymerase, the researcher is limited to reverse transcriptase, a comparatively low-fidelity enzyme (see the discussion of run-through); there is no choice in base terminator and little option on primer choice. These factors tend to lead to a poor signal-to-noise ratio that generates extraneous bands. RT-PCR products are a preferable template. Minute amounts of RNA starting material can be used for cDNA synthesis, enabling editing to be assayed in small, specific regions. Because the template is basically homogeneous, the signal-to-noise ratio is substantially improved. Another advantage to using a DNA template is that there is a greater selection of polymerases available to carry out the PPE, most with greater fidelity than reverse transcriptases.

The particular label used for the PPE is another important consideration. Many investigators have opted for radioactive labels and end labeling with $[\gamma\text{-}^{32}P]ATP$, in particular. The advantage to this approach is high sensitivity and the fact that most laboratories are able to develop autoradiograms. A disadvantage is that the labeled primers must be used before the signal decays, normally necessitating relabeling prior to each experiment. In addition, there are considerations of radioactive waste, safety, and paperwork. Fluorescent labels [e.g., hexachlorofluorescein (HEX) or other fluorescein derivatives] are an attractive alternative because they last many years frozen and can be inexpensively ordered already attached to oligonucleotides. When using cDNA templates from RT-PCR reactions, their sensitivity is more than adequate. They offer the further advantage that multiple oligos can be used within a single reaction, provided that they are tagged with fluors that emit different wavelengths. The principal disadvantage is that the researcher needs access to a fluorescent imager.

Probably the most important consideration for a PPE assay is the choice of polymerase and terminator. For the assay to be accurate, the extension product must stop at the intended position. In a percentage of templates, however, the extension "runs through" this position to the next instance of the terminator, or beyond. It is generally thought that this results from the relative affinity that a polymerase has for dNTPs versus chain terminators, a balance that can lead to the misincorporation of an inappropriate dNTP. The particular polymerase–terminator combination strongly influences the amount of run-through. Figure 14.3A gives an example of run-through produced by the enzyme Pfu used with a dideoxynucleotide triphosphate terminator. In this case 73% of the products ran through the intended stop, rendering this particular combination useless. Similarly, Taq, probably the most commonly used DNA polymerase, produces ~40% run-through whether using dideoxynucleotides or acyclonucleotides as a chain terminator (Fig. 14.3B). The polymerase Vent (exo-) produced high run-through (~67%) when coupled with dideoxynucleotides, but gives a very low percentage (6%) when coupled with acyclonucleotides. In general, the Vent (exo-)-acyclonucleotide combination produces low run-through that varies between about 0 and 8%, depending on the specific acyclonucleotide (Fig. 14.4) and the template. Other combinations may also prove useful, but they must first be tested empirically. For most templates, acyGTP gives the lowest run-through, followed by acyCTP and acyTTP, which were about equal. acyATP generates the most run-through.

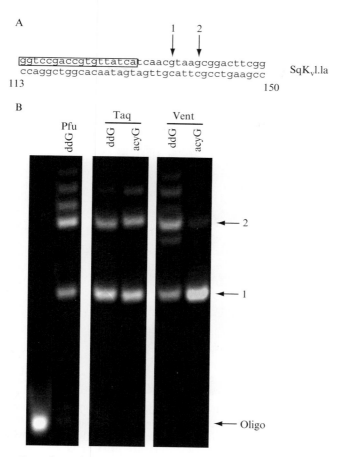

A

```
                                                    1        2
                                                    ↓        ↓
     ┌─────────────────────┐
     │ggtccgaccgtgttatca│ccaacgtaagcggacttcgg
     ccaggctggcacaatagtagttgcattcgcctgaagcc          SqK_v1.1a
     113                                          150
```

B

Figure 14.3 Run-through produced by different polymerase–terminator combinations. (A) Map of a K$^+$ channel SqKv1.1a (gbU50543) sequence showing relevant features for a PPE assay. The oligonucleotide primer is boxed. In this reaction "G" was used as a terminator. Arrow 1 shows the intended stop. Arrow 2 shows the major run-through product. (B) results from the assay using different polymerases and terminators. Pfu, cloned Pfu from Stratagene; Taq, HotstarTaq from Qiagen; Vent, Vent (exo-) from New England Biolabs; ddG, dideoxy G; acyG, acycloG. (A portion of this figure has been reproduced with permission from the journal *RNA*.)

3.3. Fluorescent poison primer extension assay

1. Generate a template using RT-PCR. In principle there should be no size limit to the template and multiple editing sites can be combined in the same amplicon. However, templates in the 250–1000 bp range are ideal, as they are easy to amplify in quantity. About 1 pmol of PCR product will be used for each assay. It is very important that the PCR product is

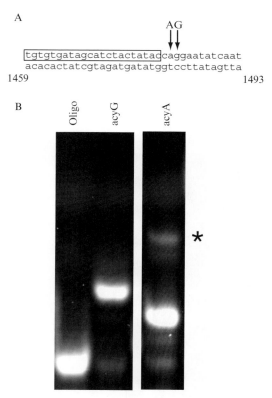

Figure 14.4 (A) Run-through produced by different bases. Map of a seagrass (*Syringodium filiforme*) *ndh*B sequence showing relevant features for a PPE assay. The oligonucleotide primer is boxed. In this reaction either acycloA or acycloG was used as a terminator. Arrows indicate the first occurrence of A or G. (B) Results from the assay. The asterisk marks the run-through band. Note that the assay can easily resolve single nucleotide differences in size. (A portion of this figure has been reproduced with permission from the journal *RNA*.)

gel purified prior to adding it to the assay. This is because even tiny amounts of contaminating dNTPs will compete with the terminator for incorporation and cause extensive run-through. An ethanol precipitation is not sufficient.

2. Combine 0.5 pmol of hexachlorofluorescein-labeled primer, 0.1–2 pmol template, 0.05 mM each of 3 dNTPs, 0.05 mM appropriate acycloNTP, 20 mM Tris–HCl, pH 8.8, 10 mM (NH$_4$)$_2$SO$_4$, 10 mM KCl, 2 mM MgSO$_4$, 0.1% Triton X-100 (this is 10× DNA POL buffer from New England Biolabs), and 1 unit Vent exo(-) DNA polymerase in a total volume of 20 μl. Reactions are cycled between 10 and 40 times using the following temperature steps: 94° for 15 sec, 55° for 20 sec, and 72° for

30 sec. The total number of cycles depends on the amount of template used. A pilot run can be performed to determine the number of cycles required to incorporate ~90% of the primer into extension products. The HEX-labeled primer should be IE-HPLC purified when ordered. Vent (exo-) and acyclonucleotides can be purchased from New England Biolabs.

3. Assay products are then electrophoresed on a large-format (>30 cm) denaturing (42% urea) acrylamide (15%) gel in 1× TBE at 25 V/cm. Depending on the product size, it may be better to leave xylene cyanol out of the loading buffer (2×: 10 mM EDTA, 95% formamide, 0.025% bromophenol blue). It is important that the gel have an aluminium backing plate to help cool it and to avoid distortions.

4. After electrophoresis, the gel is scanned directly at 532 nm using a Typhoon 9200 fluorescence/phosphorimager (GE Healthcare). Signal is collected using a 555-nm band pass-20 filter and the photomultiplier tube voltage can be varied between 475 and 500 V, depending on the signal intensity.

5. Band intensities can be analyzed using various software packages such as ImageQuant. In general, a 2–10 pixel line is drawn through the center of each lane and the integrated pixel intensity is measured for each band. It is important to assess run-through for each template/primer combination. For this, either a fully edited or unedited sample, generated from either a plasmid clone or genomic DNA, is run alongside the experimental samples. The idea is that all extension products should stop at the intended position. If there is run-through, a band will appear at the second occurrence of the terminator. The percentage of this band out of the total is the run-through. If it is small, run-through can be corrected in the experimental samples by the following equation:

$$CI_1 = BI_1 + (BI_1 \times RT)$$
$$CI_2 = BI_2 - (BI_1 \times RT) + (BI_2 \times RT)$$

where CI_1 = corrected intensity for stop 1 and CI_2 = corrected intensity for stop 2, BI_1 = measured band intensity at stop 1, BI_2 = measured band intensity at stop 2, and RT = % run-through. The corrected values were then used to calculate the editing efficiency (% edit) as the ratio between the corrected edit site intensity (either CI_1 or CI_2, depending on the edit site or terminator used) and the total intensity (sum of CI_1 and CI_2).

3.4. Evaluation of the assay

The assay presented here offers significant advantages in terms of ease of use, sensitivity, and accuracy. Figure 14.5 shows an example of the assay on templates derived from PCR amplifications of mixed (edited and unedited) plasmid clones of an *ndhB* cDNA from the seagrass *Syringodium filiforme*.

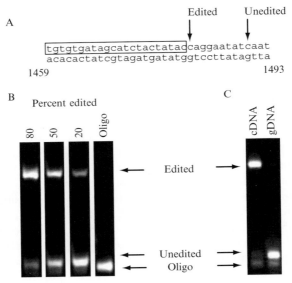

Figure 14.5 (A) Calibration of the PPE assay on mixed templates. Map of the seagrass (*Syringodium filiforme*) *ndhB* sequence showing the stop position for cDNAs that are amplified from edited or unedited mRNAs. The oligonucleotide is boxed. In this reaction, acycloC was used as a terminator. (B) Results from the assay using as templates PCR products amplified from mixed plasmid clones. These clones had either a C or a T at the arrow marked "edited," but were otherwise identical. The numbers above each lane indicate the ratio of the plasmid mix prior to amplification. (C) Results from the assay using either a RT-PCR amplification of *ndhB* or a genomic DNA amplification of *ndhB* as a template. (A portion of this figure has been reproduced with permission from the journal *RNA*.)

These plasmids either contained a C or a T at position 1481, but were otherwise the same. AcyC was used as a terminator. The primer bound immediately before position 1481 and the next C was 9 bases further down. Clearly, the band intensities accurately reflect the plasmid mixes. Furthermore, when extended to cDNA amplified from seagrass RNA (Fig. 14.5C), it was predicted that 90% of the mRNAs were edited. Sequencing 41 individual clones also showed 90% editing at this site. It was estimated that the assay can detect editing percentages as low as 2% and reliably distinguish editing differences as small as 5% (Roberson and Rosenthal, 2006). This level of accuracy is sufficient for most comparative purposes.

ACKNOWLEDGMENTS

This work was partially supported by the MRC, NSF IBN-0344070, NIH NS039405-06, and NIH NCRR RCMI Grant G12RR03051.

REFERENCES

Akbarian, S., Smith, M. A., and Jones, E. G. (1995). Editing for an AMPA receptor subunit RNA in prefrontal cortex and striatum in Alzheimer's disease, Huntington's disease and schizophrenia. *Brain Res.* **699,** 297–304.

Basilio, C., Wahba, A. J., Lengyel, P., Speyer, J. F., and Ochoa, S. (1962). Synthetic polynucleotides and the amino acid code. *Proc. Natl. Acad. Sci. USA* **48,** 613–616.

Burns, C. M., Chu, H., Rueter, S. M., Hutchinson, L. K., Canton, H., Sanders-Bush, E., and Emeson, R. B. (1997). Regulation of serotonin-2C receptor G-protein coupling by RNA editing. *Nature* **387,** 303–308.

Gerber, A., O'Connell, M. A., and Keller, W. (1997). Two forms of human double-stranded RNA-specific editase 1 (hRED1) generated by the insertion of an Alu cassette. *RNA* **3,** 453–463.

Gurevich, I., Tamir, H., Arango, V., Dwork, A. J., Mann, J. J., and Schmauss, C. (2002). Altered editing of serotonin 2C receptor pre-mRNA in the prefrontal cortex of depressed suicide victims. *Neuron* **34,** 349–356.

Higuchi, M., Single, F. N., Köhler, M., Sommer, B., Sprengel, R., and Seeburg, P. H. (1993). RNA editing of AMPA receptor subunit GluR-B: A base-paired intron-exon structure determines position and efficiency. *Cell* **75,** 1361–1370.

Iwamoto, K., and Kato, T. (2003). RNA editing of serotonin 2C receptor in human postmortem brains of major mental disorders. *Neurosci. Lett.* **346,** 169–172.

Lai, F., Chen, C.-X., Carter, K. C., and Nishikura, K. (1997). Editing of glutamate receptor B subunit ion channel RNAs by four alternatively spliced DRADA2 double-stranded RNA adenosine deaminases. *Mol. Cell. Biol.* **17,** 2413–2424.

Macauley-Patrick, S., Fazenda, M. L., McNeil, B., and Harvey, L. M. (2005). Heterologous protein production using the *Pichia pastoris* expression system. *Yeast* **22,** 249–270.

Macbeth, M. R., Schubert, H. L., Vandemark, A. P., Lingam, A. T., Hill, C. P., and Bass, B. L. (2005). Inositol hexakisphosphate is bound in the ADAR2 core and required for RNA editing. *Science* **309,** 1534–1539.

Melcher, T., Maas, S., Herb, A., Sprengel, R., Higuchi, M., and Seeburg, P. H. (1996). RED2, a brain specific member of the RNA-specific adenosine deaminase family. *J. Biol. Chem.* **271,** 31795–31798.

Ohi, H., Okazaki, N., Uno, S., Miura, M., and Hiramatsu, R. (1998). Chromosomal DNA patterns and gene stability of *Pichia pastoris*. *Yeast* **14,** 895–903.

Roberson, L. M., and Rosenthal, J. J. (2006). An accurate fluorescent assay for quantifying the extent of RNA editing. *RNA* **12,** 1907–1912.

Sodhi, M. S., Burnet, P. W., Makoff, A. J., Kerwin, R. W., and Harrison, P. J. (2001). RNA editing of the 5-HT(2C) receptor is reduced in schizophrenia. *Mol. Psychiatry* **6,** 373–379.

Sprengel, R., Higuchi, M., Monyer, H., and Seeburg, P. H. (1999). Glutamate receptor channels: A possible link between RNA editing in the brain and epilepsy. *Adv. Neurol.* **79,** 525–534.

Valente, L., and Nishikura, K. (2005). ADAR gene family and A-to-I RNA editing: Diverse roles in posttranscriptional gene regulation. *Prog. Nucleic Acid Res. Mol. Biol.* **79,** 299–338.

Xu, M., Wells, K. S., and Emeson, R. B. (2006). Substrate-dependent contribution of double-stranded RNA-binding motifs to ADAR2 function. *Mol. Biol. Cell* **17,** 3211–3220.

CHAPTER FIFTEEN

Large-Scale Overexpression and Purification of ADARs from Saccharomyces Cerevisiae for Biophysical and Biochemical Studies

Mark R. Macbeth *and* Brenda L. Bass

Contents

Abstract

Many biochemical and biophysical analyses of enzymes require quantities of protein that are difficult to obtain from expression in an endogenous system. To further complicate matters, native adenosine deaminases that act on RNA (ADARs) are expressed at very low levels, and overexpression of active protein has been unsuccessful in common bacterial systems. Here we describe the plasmid construction, expression, and purification procedures for ADARs overexpressed in the yeast *Saccharomyces cerevisiae*. ADAR expression is controlled by the Gal promoter, which allows for rapid induction of transcription when the yeast are grown in media containing galactose. The ADAR is translated with an N-terminal histidine tag that is cleaved by the tobacco etch virus protease, generating one nonnative glycine residue at the N-terminus of the ADAR protein. ADARs expressed using this system can be purified to homogeneity, are highly

Department of Biochemistry and Howard Hughes Medical Institute, University of Utah, Salt Lake City, Utah

Methods in Enzymology, Volume 424
ISSN 0076-6879, DOI: 10.1016/S0076-6879(07)24015-7

active in deaminating RNA, and are produced in quantities (from 3 to 10 mg of pure protein per liter of yeast culture) that are sufficient for most biophysical studies.

1. INTRODUCTION

Adenosine deaminases that act on RNA (ADARs) convert adenosine to inosine in regions of RNA that are largely double stranded (Bass, 2002; Keegan et al., 2004; Valente and Nishikura, 2005). Since inosine is recognized as guanosine, the result of editing by an ADAR is an A-to-G transition mutation. ADARs are found in every metazoan animal (not plants) and are essential for proper neuronal function. This is supported by evidence showing that knockouts of ADARs lead to various behavioral abnormalities (Higuchi et al., 2000; Palladino et al., 2000a; Tonkin et al., 2002) and that several known pre-mRNA substrates code for neuronal receptors and ion channels (Burns et al., 1997; Higuchi et al., 1993; Lomeli et al., 1994). Though the effects of editing coding sequences are dramatic, these editing events are rare relative to the levels of editing in noncoding regions, especially untranslated regions (UTRs) (Blow et al., 2004; Levanon et al., 2004; Morse and Bass, 1999; Morse et al., 2002). The extensive editing of noncoding regions has been proposed to affect expression of a gene posttranscriptionally, although how this might occur is unclear. Editing of double-stranded RNAs, such as those regions found in UTRs, creates I-to-U mismatches from A-to-U base pairs, and results in an RNA that is less double stranded. Although the biological consequences are not yet understood, an RNA made more single stranded by A-to-I modification can affect recognition by the RNAi machinery (Knight and Bass, 2002; Scadden and Smith, 2001).

ADARs are composed of an N-terminal RNA-binding domain and a C-terminal catalytic domain. The RNA-binding domain consists of one to three double-stranded RNA-binding motifs (dsRBM). The catalytic domain consists of a deaminase motif that resembles *Escherichia coli* cytidine deaminase and binds zinc, as well as a motif that coordinates the small molecule *myo*-inositol hexakisphosphate (IP_6, Maas et al., 2003; Macbeth et al., 2005). In addition, human and *Xenopus* ADAR1 contain two Z-DNA-binding motifs N-terminal to their three dsRBMs (Maas et al., 2003). ADARs are modular, and for hADAR2, the isolated dsRBMs will each bind RNA in the absence of the catalytic domain. Further, the catalytic domain will deaminate dsRNA in the absence of the dsRBMs, albeit poorly.

As the number of ADAR enzymes and substrates from various species rapidly increased over the past several years (Chen et al., 2000; Palladino et al., 2000b; Patton et al., 1997; Tonkin et al., 2002), it became apparent that our understanding of ADAR biochemistry was struggling to keep pace. There are several reasons for this, the most prominent being a dearth of active, pure enzyme. Purification of native enzymes was tedious and yielded low amounts of protein (Hough and Bass, 1994; O'Connell and Keller, 1994; O'Connell *et al.*, 1997). Indeed, the initial purification of *Xenopus* ADAR yielded only a few micrograms from over 4 liters of eggs (Hough and Bass, 1994).

Various expression systems have been employed to increase the yield and specific activity of ADAR proteins. Standard *E. coli* overexpression systems are useful for expressing the dsRBMs, but constructs that contain the catalytic domain are largely insoluble; any soluble protein (typically <1%) lacks deaminase activity (M. R. Macbeth and B. L. Bass, unpublished observations). The reason for this is unclear, but might be explained by the lack of IP$_6$ in *E. coli*, which is required for ADAR editing activity and perhaps proper folding. Eukaryotic expression systems such as *Pichia pastoris* and insect cell culture harboring baculovirus have been utilized with good results (Cho *et al.*, 2003; Ring *et al.*, 2004). However, each of these systems has its drawbacks. For the *Pichia* system, induction times are several days, and, in our experience, yields are low. We believe the poor yield is due to the toxic nature of ADAR overexpression, which may lead to downregulation. The baculovirus–insect cell system is tedious and requires tissue culture facilities and expensive media.

We have achieved good success using a *Saccharomyces cerevisiae* overexpression system for generating full-length and truncated ADAR proteins that contain the catalytic domain. There are many benefits to using this system: (1) There is a short (~6 h) induction time that is beneficial if the enzymes are unstable or toxic. (2) The yeasts are easy to genetically manipulate, which allows studying the effects of other factors (i.e., IP$_6$) on ADAR function. (3) There is no contamination from native editing activity, as *S. cerevisiae* does not encode an ADAR. (4) The medium is relatively inexpensive.

The ADAR proteins expressed in this system are produced in substantial quantities, are highly active, and are >99% pure. They have been used for several biophysical and biochemical assays including equilibrium sedimentation, dynamic light scattering, gel shift analyses, RNA editing, and X-ray crystallography (Macbeth *et al.*, 2004, 2005). The details of the expression and purification, as well as some of the protein analyses, are discussed in Chapter 16.

 ## 2. Methods

2.1. General considerations

Unless stated otherwise, all media and chemicals are from Sigma–Aldrich. Large-scale yeast expression cultures are grown at 30° using a 15-liter Bellco bioreactor that allows for stable regulation of oxygen levels at 22%. All media and purification buffers are either autoclaved or filter sterilized. Smaller cultures can be grown in shaker flasks with proportionate amounts of media components; however, expect disproportionately less cell mass and purified protein, an effect, we believe, that is due to unregulated air/O_2 levels. Before attempting a large-scale preparation, it is prudent to perform a small-scale growth (50–100 ml cultures) and test for expression by Western blotting using an appropriate antibody.

All purification steps are performed at 4° or on ice, and after each step the protein is analyzed by sodium dodecyl sulfate polyacrylamide gel electrophoresis (SDS–PAGE). Many of the purification steps described here use a pump-driven AKTA system (GE Healthcare); however, many columns can be eluted using gravity flow or a peristaltic pump. It is imperative that the time from cell lysis to protein storage be minimized for the highest specific activity of the enzyme.

2.2. Expression constructs and yeast strains

The vector used for overexpression of ADARs is based on the YEpTOP2P-GAL1 vector used to express DNA topoisomerase (Giaever et al., 1988). The vector has a pBR322 backbone and contains a URA3 gene, a 2 μm origin of replication, an ampicillin resistance marker, and a GAL1 promoter immediately upstream of the BamHI (5′) cloning site (Fig. 15.1A). If the ADAR gene contains a BamHI or XhoI site, several other restriction enzymes are available that will cleave different recognition sites yet leave BamHI and XhoI compatible ends. We have successfully used BsmBI (New England Biolabs) for this purpose.

The ADAR insert construct generated by polymerase chain reaction (PCR) consists of an N-terminal 10-histidine tag followed by a tobacco etch virus (TEV) protease recognition site (Fig. 15.1B). The 5′ end of the PCR product is of the sequence GGGGGG GGATCC GTAACC ATGTCA (CACCAT)$_5$ GAGAACCTCTATTTCCAGGGA (X)$_{20}$. The first six nucleotides provide space from the DNA end for the BamHI recognition site (which is the second six nucleotides). The GTAACC sequence allows for efficient translation initiation, and it is followed by the initiator methionine codon (ATG) and a +2 serine codon (TCA) that also promote efficient translation. A 10×–histidine coding sequence, (CACCAT)$_5$, is followed by a sequence that codes for TEV protease recognition, and the remaining ~20 nucleotides (X) code for the N-terminus of the ADAR gene of interest. Note that after cleavage of the His-tag with TEV, the protein will contain an N-terminal glycine, a remnant of the TEV recognition site, prior to the start of the ADAR sequence. A primer that encodes this entire sequence is exceptionally long, and we have found that splitting the primer in half and performing tandem PCR led to more efficient cloning. For example, the first round of PCR would incorporate

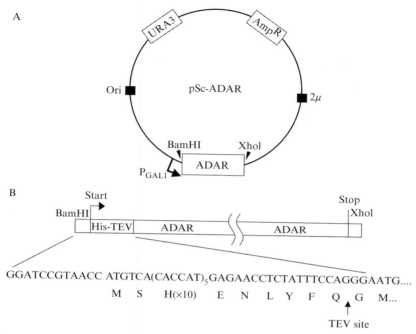

Figure 15.1 The expression vector and 5′ sequence of the construct used to tag and purify ADARs. (A) The pSc-ADAR plasmid. The empty vector (lacking an ADAR gene) is approximately 8 kb. The *Bam*HI and *Xho*I restriction sites are unique in the empty vector. Transcription of the ADAR gene is controlled by the GAL1 promoter (indicated by the arrow). (B) A schematic of the ADAR construct after PCR of an ADAR gene using the 5′ primers described in the text. The translation start and stop sites are indicated. Below the schematic are the DNA and translated protein sequences. Note that after cleavage with TEV protease, there is a glycine residue N-terminal to the first methionine of the ADAR protein of interest.

the TEV site onto the N-terminus of the ADAR gene, while the second round would incorporate the restriction site, start codon, and histidine tag. The 3′ primer is of the sequence GGGGGG CTCGAG TCA (X)$_{20}$. The six Gs provide a spacer from the end of the DNA for *Xho*I site recognition, and this is followed by a stop codon and 20 (X) of the 3′ residues of the ADAR gene of interest.

After PCR, the ends of the insert product are trimmed with *Bam*HI and *Xho*I and inserted into a vector linearized with the same enzymes. The insert in the resulting plasmid should be sequenced entirely using primers internal to the subcloned ADAR gene. The expression plasmid is then transformed, by electroporation or lithium acetate, into the haploid (type a) yeast strain BCY123, which has the genotype pep4::HIS3 prb::LEU2 bar1:HISG lys2:: GAL1/10–GAL4 can1 ade2 ura3 leu2–3,112. This strain has several defective proteases, allows selection of URA3 auxotrophs, and expresses additional Gal4 transcriptional activator during induction with galactose.

2.3. Yeast growth

After transformation, the yeasts are plated on agar plates containing $1\times$ minimal media without uracil [6.7 g/liter yeast nitrogen base containing $(NH_4)_2SO_4$ and lacking amino acids (Fisher), 10 g/liter succinic acid, 6 g/liter NaOH, 1.92 g/liter yeast synthetic dropout media without uracil, 22 mg/liter adenine hemisulfate] supplemented with 1 M sorbitol and 2% dextrose. Colonies take 2–3 days to appear. We have found it beneficial to transform the expression plasmid freshly, prior to attempting a large-scale growth, and avoid using glycerol stocks of yeast harboring the plasmid. As mentioned, prior to attempting a large-scale growth for ADAR expression, small-scale cultures should be used to test for expression of the ADAR.

The cells are grown in three stages in which the expression of the ADAR gene is repressed, derepressed, or activated. In the repressed state, the cells are grown in the presence of dextrose (glucose) in the starter culture. The cells are then switched to a media containing glycerol and lactic acid. The dextrose is depleted, and the gene is then ready for induction by the addition of galactose, which activates transcription of the ADAR gene from the GAL1 promoter.

For ADAR expression, a 10 ml starter culture consisting of $1\times$ minimal media without uracil supplemented with 2% dextrose is inoculated with one colony and grown overnight at 30° while shaking at 300 rpm. The following morning, the entire culture is used to inoculate 200 ml of the same media. After 24 h, the 200 ml culture is aseptically pumped into a 15-liter Bellco bioreactor vessel containing 13.5 liters $1\times$ minimal media without uracil and 2% glycerol/3% lactic acid as the carbon source. The oxygen levels are calibrated according to the manufacturer's instructions, set at 22%, and the culture is grown at 30° with vigorous stirring. After 24 h, the culture is induced by the addition of 1.5 liter of $2\times$ minimal media and 1.5 liter 30% galactose (final concentration of 2.7%). The final volume of the culture is 16.5 liters. Induction is for 5–8 h, but should be optimized for each construct by performing a time course and measuring protein production by Western blotting. The cells are harvested by centrifugation at 6000 rpm for 5 min in a Beckman JA-20 centrifuge equipped with a JLA 8.100 6L fixed-angle rotor. The cells are washed in a solution containing 20 mM Tris–HCl, 100 mM NaCl, 5% glycerol, 25 mM NaF, 1 mM sodium bisulfite and then centrifuged again. The wet mass of the cell pellet is determined prior to storage at −80°. The typical yield is approximately 10–12 g wet cells/liter of culture. To test for expression, we routinely lyse 0.1 g of cells by vortexing with glass beads for 30 min at 4° in 500 μl of a solution containing 20 mM Tris–HCl, 100 mM NaCl, 5% glycerol, 0.5 mM dithiothreitol (DTT), and 0.01% Triton X-100. The lysate is clarified by centrifugation at 14,000 rpm in a microfuge, and expressed proteins are detected by western blotting using the Penta-His antibody (Qiagen).

2.4. Purification

Typically, cells from 2.75 liters of the culture (~25–30 g) will yield enough pure protein for most applications, including X-ray crystallography, while the remainder of the cells can be stored at −80° indefinitely. The cells are resuspended in Buffer A (20 mM Tris–HCl, pH 8.0, 5% glycerol, 1 mM 2-mercaptoethanol) containing 750 mM NaCl, 35 mM imidazole, and 0.01% Triton X-100.[1] The cells are lysed by three passes through a Gaulin homogenizer (APV) at 20,000 psi[2] and the extract is clarified by centrifugation at 100,000×g for 1 h. The supernatant is mixed with 5 ml of Ni–NTA agarose (Qiagen) equilibrated with the same buffer used to lyse the cells. The slurry is gently rocked at 4° for 15 min and poured into a 2.5 × 20-cm column, while collecting the flowthrough by gravity. The washes are performed in four 40-ml steps: the first is with the same buffer used to lyse the cells; the three subsequent washes use Buffer A containing 35 mM imidazole and either 750 mM, 300 mM, or 100 mM NaCl, consecutively, in order to decrease the NaCl concentration. The protein is eluted in 5 × 5-ml fractions of Buffer A containing 400 mM imidazole and 100 mM NaCl. The flowthrough, wash, and elution fractions are analyzed by SDS–PAGE, and the protein is visualized by staining the gel with Coomassie blue. Figure 15.2A shows the purification of an hADAR2 protein construct after the first Ni–NTA column.

An AKTA FPLC system (GE Healthcare) is used to load the eluted protein onto a 5-ml Hi-Trap heparin column (GE Healthcare) equilibrated with Buffer A containing 100 mM NaCl. The column is washed with 50 ml Buffer A containing 100 mM NaCl and eluted with a gradient of 100 mM to 1 M NaCl in Buffer A. ADAR proteins typically elute between 300 and 500 mM NaCl, and fractions are analyzed by SDS–PAGE. At this stage, the hADAR2 protein is approximately 85–90% pure (Fig. 15.2B).

The fractions containing the ADAR protein are pooled, and TEV protease (Invitrogen) is added to cleave the His-tag and TEV recognition site from the ADAR protein. Typically, 100 units of TEV are used to cleave 5 mg of ADAR, but quantities of TEV added should be optimized empirically. The entire TEV-ADAR reaction is dialyzed against Buffer A containing 200 mM NaCl overnight at 4°. The following day, the reaction contents

---

[1] If the protein is to be analyzed by mass spectrometry, the Triton X-100 should be omitted, as detergents strongly interfere with ionization.

[2] For smaller-scale preparations, i.e., 1-liter cultures or less, a French pressure cell is used. Other methods for lysing the yeast, such as sonication and vortexing in glass beads, have been inefficient in our hands, compared to lysing the cells under high pressure.

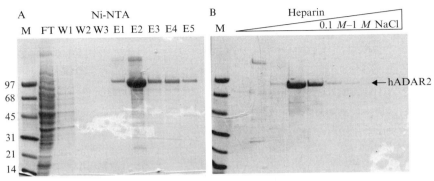

Figure 15.2 SDS–PAGE analysis of the first two purification steps of hADAR2. (A) The Coomassie blue-stained gel after the first Ni-NTA column. The lanes from left to right are molecular weight standard (M), flowthrough (FT), washes 1–3 (W1-W3), and elution fractions (E1-E5). hADAR2 migrates as a band of 80–90 kDa. (B) The stained gel containing fractions eluted from a heparin column. Every third fraction from a 0.1 to 1 M NaCl elution gradient is loaded on the gel.

are bound to 5 ml of equilibrated Ni-NTA agarose for 1 h with gentle rocking. The slurry is poured into a 2.5 × 20-cm column and washed with 80 ml of Buffer A with 200 mM NaCl. Uncleaved protein and the cleaved His-tag (as well as several other contaminants) remain bound to the resin. The flowthrough and wash (which contain the cleaved ADAR protein) are concentrated to ~1 ml using an Amicon-Ultra 15 centrifugal concentrator tube with a 30,000 Da molecular weight cut-off. If the majority of the ADAR remains bound to the resin and is not in the flowthrough or wash, it is likely the TEV reaction was inefficient. The major factor affecting cleavage by TEV is the presence of secondary structure at the N-terminus of the ADAR, which may inhibit recognition of the TEV site. If this is a problem, the N-terminus should be redesigned to include three glycine residues to act as a "spacer" between the cleavage site and the N-terminus of the ADAR.

The concentrated protein is then injected onto a Superdex 200 26/60 gel filtration column (GE Healthcare), and eluted with Buffer A containing 200 mM NaCl. One milliliter fractions are collected and analyzed by SDS–PAGE (Fig. 15.3A). Typically, our hADAR2 protein construct (calculated MW ~76.6 kDa) elutes as a symmetrical peak with a molecular weight of ~100 kDa, relative to a standard curve of proteins of a known size. The "shoulders" of the peak often contain minor contaminants and are discarded. The main peak fractions are pooled, concentrated, and dialyzed against storage buffer containing 20 mM Tris–HCl (pH 8.0), 100 mM NaCl, 20% glycerol, and 1 mM 2-mercaptoethanol. We quantify the protein using two methods. The first is by the Bradford method using a kit available from Bio-Rad; the second is by SYPRO-Red (Molecular Probes) staining of an SDS–PAGE gel

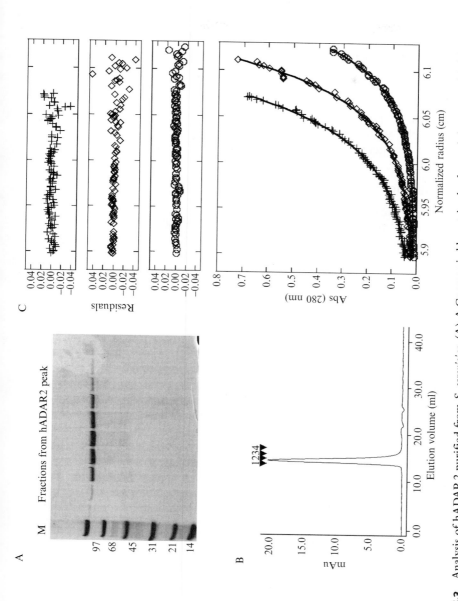

Figure 15.3 Analysis of hADAR2 purified from *S. cerevisiae*. (A) A Coomassie blue-stained gel containing every third fraction from the hADAR2 peak eluted from a Superdex 200 26/60 preparative gel filtration column. (B) The UV trace of hADAR2 eluted from a Superdex 200 10/300 analytical gel filtration column. The elution volume of hADAR2 is 14.4 ml. The elution peaks of several molecular weight standards are

and visualization using a Molecular Dynamics Storm Phosphor Imager. Yields of pure protein vary between construct, but typically range from 3 to 10 mg/liter of original yeast culture. The protein is stored as aliquots at −80°.

3. ANALYSIS OF PURIFIED hADAR2

Various ADAR proteins purified using this scheme have been used for kinetic analyses, RNA binding assays, and X-ray crystallography (Haudenschild et al., 2004; Macbeth et al., 2004, 2005). Here we present an analysis of the oligomerization state of hADAR2, as purified from S. cerevisiae, using analytical gel filtration and equilibrium sedimentation.

3.1. Analytical gel filtration

For an analytical gel filtration analysis, 50 μl of purified, concentrated (6.6 mg/ml) hADAR2 is injected onto a Superdex 200 10/300 gel filtration column using an AKTA FPLC system (GE Healthcare). The protein was eluted with 20 mM Tris–HCl (pH 8.0), 200 mM NaCl, 5% glycerol, and 1 mM 2-mercaptoethanol, and the elution profile was analyzed with the PrimeView software package (GE Healthcare, Fig. 15.3B). The elution volume of the hADAR2 peak was determined to be 14.4 ml. Relative to the elution of standard molecular weight markers, hADAR2 has an observed molecular weight of 98.1 kDa and appears to be a monomer under these conditions (Fig. 15.3B).

3.2. Equilibrium sedimentation

To observe the oligomeric state of hADAR2, equilibrium sedimentation is performed as it allows an accurate determination of the molecular weight of a macromolecule that is independent from its shape. Purified hADAR2 was dialyzed against a buffer containing 20 mM Tris–HCl (pH 8.0), 200 mM NaCl, and 1 mM 2-mercaptoethanol and diluted to 1.3, 3.3, and 6.6 μM. The protein was centrifuged in a Beckman XLA analytical ultracentrifuge, equipped with the AN-60 Ti rotor, at 16,000 and 18,000 rpm. Figure 15.3C

---

indicated and numbered 1–4. Their sizes and retention volumes are (1) aldolase 158 kDa, 13.8 ml, (2) albumin 67 kDa, 14.7 ml, (3) ovalbumin 43 kDa, 15.7 ml, and (4) chymotrypsinogen A 25 kDa, 17.5 ml. The calculated molecular weight of hADAR2, determined from a standard curve of the known molecular weight standards, is 98.1 kDa. (C) The equilibrium sedimentation data for hADAR2 at concentrations of 1.3 (○), 2.6 (△), and 6.6 (+) μM, respectively. The lower panels are the data fit to a species of MW 75,320 Da. The upper panels show the residuals corresponding to each fit.

shows data, fits, and residuals for hADAR2 spun at the three concentrations. The best overall fit of the data was to a single species molecule with an observed molecular weight of 75,320 Da. The residuals are relatively small and randomly distributed, indicating a good fit of the data. The MW_{obs}/MW_{calc} is 0.98, suggesting hADAR2 exists as a monomer under these concentrations.

4. FUTURE PERSPECTIVES

We have developed an overexpression protocol and rapid purification scheme to generate ADAR protein that is untagged and contains native sequence except for an N-terminal glycine. The protein is >99% pure, exists as a single species, and is highly active in deaminating dsRNA. The protein purified using this system can be used for many biophysical and structural studies (such as calorimetry, crystallography, and single turnover kinetics) that require large amounts of pure protein.

ACKNOWLEDGMENTS

The authors thank Herbert L. Ley III for initial development of the purification method and Debra M. Eckert for assistance with the equilibrium sedimentation analysis. This work was supported by a grant from the National Institutes of Health (GM044073). B.L.B. is a Howard Hughes Medical Institute Investigator.

REFERENCES

Bass, B. L. (2002). RNA editing by adenosine deaminases that act on RNA. *Annu. Rev. Biochem.* **71,** 817–846.

Blow, M., Futreal, P. A., Wooster, R., and Stratton, M. R. (2004). A survey of RNA editing in human brain. *Genome Res.* **14,** 2379–2387.

Burns, C. M., Chu, H., Rueter, S. M., Hutchinson, L. K., Canton, H., Sanders-Bush, E., and Emeson, R. B. (1997). Regulation of serotonin-2C receptor G-protein coupling by RNA editing. *Nature* **387,** 303–308.

Chen, C. X., Cho, D. S., Wang, Q., Lai, F., Carter, K. C., and Nishikura, K. (2000). A third member of the RNA-specific adenosine deaminase gene family, ADAR3, contains both single- and double-stranded RNA binding domains. *RNA* **6,** 755–767.

Cho, D. S., Yang, W., Lee, J. T., Shiekhattar, R., Murray, J. M., and Nishikura, K. (2003). Requirement of dimerization for RNA editing activity of adenosine deaminases acting on RNA. *J. Biol. Chem.* **278,** 17093–17102.

Giaever, G. N., Snyder, L., and Wang, J. C. (1988). DNA supercoiling *in vivo. Biophys. Chem.* **29,** 7–15.

Haudenschild, B. L., Maydanovych, O., Veliz, E. A., Macbeth, M. R., Bass, B. L., and
   Beal, P. A. (2004). A transition state analogue for an RNA-editing reaction. *J. Am. Chem.
   Soc.* **126,** 11213–11219.
Higuchi, M., Single, F. N., Kohler, M., Sommer, B., Sprengel, R., and Seeburg, P. H.
   (1993). RNA editing of AMPA receptor subunit GluR-B: A base-paired intron-exon
   structure determines position and efficiency. *Cell* **75,** 1361–1370.
Higuchi, M., Maas, S., Single, F. N., Hartner, J., Rozov, A., Burnashev, N., Feldmeyer, D.,
   Sprengel, R., and Seeburg, P. H. (2000). Point mutation in an AMPA receptor gene rescues
   lethality in mice deficient in the RNA-editing enzyme ADAR2. *Nature* **406,** 78–81.
Hough, R. F., and Bass, B. L. (1994). Purification of the *Xenopus laevis* double-stranded
   RNA adenosine deaminase. *J. Biol. Chem.* **269,** 9933–9939.
Keegan, L. P., Leroy, A., Sproul, D., and O'Connell, M. A. (2004). Adenosine deaminases
   acting on RNA (ADARs): RNA-editing enzymes. *Genome Biol.* **5,** 209.
Knight, S. W., and Bass, B. L. (2002). The role of RNA editing by ADARs in RNAi. *Mol.
   Cell* **10,** 809–817.
Levanon, E. Y., Eisenberg, E., Yelin, R., Nemzer, S., Hallegger, M., Shemesh, R.,
   Fligelman, Z. Y., Shoshan, A., Pollock, S. R., Sztybel, D., Olshansky, M.,
   Rechavi, G., *et al.* (2004). Systematic identification of abundant A-to-I editing sites in
   the human transcriptome. *Nat. Biotechnol.* **22,** 1001–1005.
Lomeli, H., Mosbacher, J., Melcher, T., Hoger, T., Geiger, J. R., Kuner, T., Monyer, H.,
   Higuchi, M., Bach, A., and Seeburg, P. H. (1994). Control of kinetic properties of
   AMPA receptor channels by nuclear RNA editing. *Science* **266,** 1709–1713.
Maas, S., Rich, A., and Nishikura, K. (2003). A-to-I RNA editing: Recent news and
   residual mysteries. *J. Biol. Chem.* **278,** 1391–1394.
Macbeth, M. R., Lingam, A. T., and Bass, B. L. (2004). Evidence for auto-inhibition by the
   N terminus of hADAR2 and activation by dsRNA binding. *RNA* **10,** 1563–1571.
Macbeth, M. R., Schubert, H. L., Vandemark, A. P., Lingam, A. T., Hill, C. P., and
   Bass, B. L. (2005). Inositol hexakisphosphate is bound in the ADAR2 core and required
   for RNA editing. *Science* **309,** 1534–1539.
Morse, D. P., and Bass, B. L. (1999). Long RNA hairpins that contain inosine are present in
   Caenorhabditis elegans poly(A)+ RNA. *Proc. Natl. Acad. Sci. USA* **96,** 6048–6053.
Morse, D. P., Aruscavage, P. J., and Bass, B. L. (2002). RNA hairpins in noncoding regions
   of human brain and *Caenorhabditis elegans* mRNA are edited by adenosine deaminases that
   act on RNA. *Proc. Natl. Acad. Sci. USA* **99,** 7906–7911.
O'Connell, M. A., and Keller, W. (1994). Purification and properties of double-stranded
   RNA-specific adenosine deaminase from calf thymus. *Proc. Natl. Acad. Sci. USA* **91,**
   10596–10600.
O'Connell, M. A., Gerber, A., and Keller, W. (1997). Purification of human double-
   stranded RNA-specific editase 1 (hRED1) involved in editing of brain glutamate receptor B pre-mRNA. *J. Biol. Chem.* **272,** 473–478.
Palladino, M. J., Keegan, L. P., O'Connell, M. A., and Reenan, R. A. (2000a). A-to-I pre-
   mRNA editing in *Drosophila* is primarily involved in adult nervous system function and
   integrity. *Cell* **102,** 437–449.
Palladino, M. J., Keegan, L. P., O'Connell, M. A., and Reenan, R. A. (2000b). dADAR, a
   Drosophila double-stranded RNA-specific adenosine deaminase is highly developmentally regulated and is itself a target for RNA editing. *RNA* **6,** 1004–1018.
Patton, D. E., Silva, T., and Bezanilla, F. (1997). RNA editing generates a diverse array of
   transcripts encoding squid Kv2 K+ channels with altered functional properties. *Neuron*
   19, 711–722.
Ring, G. M., O'Connell, M. A., and Keegan, L. P. (2004). Purification and assay of
   recombinant ADAR proteins expressed in the yeast *Pichia pastoris* or in *Escherichia
   coli. Methods Mol. Biol.* **265,** 219–238.

Scadden, A. D., and Smith, C. W. (2001). RNAi is antagonized by A → I hyper-editing. *EMBO Rep.* **2,** 1107–1111.

Tonkin, L. A., Saccomanno, L., Morse, D. P., Brodigan, T., Krause, M., and Bass, B. L. (2002). RNA editing by ADARs is important for normal behavior in *Caenorhabditis elegans. EMBO J.* **21,** 6025–6035.

Valente, L., and Nishikura, K. (2005). ADAR gene family and A-to-I RNA editing: Diverse roles in posttranscriptional gene regulation. *Prog. Nucleic Acid Res. Mol. Biol.* **79,** 299–338.

CHAPTER SIXTEEN

Mouse Models to Elucidate the Functional Roles of Adenosine-to-Inosine Editing

Elizabeth Y. Rula* *and* Ronald B. Emeson*,†,‡

Contents

* Department of Pharmacology, Vanderbilt University School of Medicine, Nashville, Tennessee
† Department of Molecular Physiology and Biophysics, Vanderbilt University School of Medicine, Nashville, Tennessee
‡ Center for Molecular Neuroscience, Vanderbilt University School of Medicine, Nashville, Tennessee

Methods in Enzymology, Volume 424
ISSN 0076-6879, DOI: 10.1016/S0076-6879(07)24016-9

Abstract

The conversion of adenosine to inosine (A-to-I) by RNA editing is a widespread RNA processing event by which genomically encoded sequences are altered through site-specific deamination of adenosine residue(s) in RNA transcripts through the actions of a family of double-stranded RNA-specific adenosine deaminases (ADARs). While significant advances have been made regarding the functional consequences of A-to-I editing using heterologous expression systems, the physiological relevance of such RNA modifications in mammals has been addressed effectively using gene-targeting strategies in mice via homologous recombination in embryonic stem (ES) cells. These gene-targeting approaches have allowed the generation of mutant mouse strains in which site-specific editing events can be fixed in the fully edited or nonedited state for individual ADAR targets, expression of ADAR proteins can be selectively ablated, or a combination of ADAR elimination and ADAR target modification can be used for a more in-depth understanding of the biological consequences of A-to-I editing dysregulation.

1. INTRODUCTION

The conversion of adenosine to inosine (A-to-I) by RNA editing is a widespread RNA processing event by which genomically encoded sequences are altered through site-specific deamination of adenosine residue(s) in precursor and mature mRNA transcripts. Since inosine preferentially base pairs with cytosine, an inosine within RNA transcripts is read as guanosine during translation, often resulting in specific change(s) in the amino acid coding potential of the mRNA that can dramatically alter the functional properties of the encoded protein product. In addition to alterations in coding potential, these editing events can alter the structure, stability, translation efficiency, and splicing patterns of the modified transcripts, thereby affecting almost all aspects of cellular RNA function (Gott and Emeson, 2000). A-to-I editing has most often been identified as an adenosine-to-guanosine (A-to-G) discrepancy between genomic and cDNA sequences due to the similar base pairing properties of inosine and guanosine during cDNA synthesis.

At least eight transcripts containing A-to-I editing events within the coding region have been identified serendipitously in mammals, based upon A-to-G discrepancies between genomic and cDNA sequences (Table 16.1); notably, transcripts encoding the 2C-subtype of the serotonin receptor ($5HT_{2C}R$), subunits of the α–amino–3–hydroxy–5–methylisoxazole–4–propionate (AMPA) subtype of ionotropic glutamate receptor (GluR–2, GluR–3, and GluR–4),

Table 16.1 Functional consequences of A-to-I editing

Transcript	Site(s)	Function	References
GluR–2	Q/R	Modulation of Ca^{2+} permeability	Sommer, 1991; Burnashev, 1992
	R/G	Alteration of recovery from receptor desensitization	Lomeli, 1994
GluR–3	R/G	Alteration of recovery from receptor desensitization	Lomeli, 1994
GluR–4	R/G	Alteration of recovery from receptor desensitization	Lomeli, 1994
GluR–5	Q/G	Modulation of Ca^{2+} permeability	Sailer, 1999
GluR–6	Q/R, I/V, Y/C	Modulation of Ca^{2+} permeability	Heinemann, 1993; Kohler, 1993
5-HT$_{2c}$R	A, B, C, D, E	Modification of constitutive activity and G-protein coupling	Niswender, 1997, 1998; Wang, 2000
K$_V$1.1	I/M	Alteration of channel inactivation	Bhalla, 2004
ADAR2	−1	Generation of alternative 3′-splice site	Rueter, 1999

subunits of the kainate subtype of glutamate-gated ion channel (GluR-5 and GluR-6), and one of the A-to-I editing enzymes (ADAR2) undergo A-to-I modifications that change the amino acid coding potential or splicing pattern of the mature mRNAs, producing protein products with altered functional properties (Emeson and Singh, 2001; Rueter and Emeson, 1998; Rueter et al., 1999). A bioinformatic strategy, based upon extensive sequence conservation surrounding editing sites, allowed the identification of 16 additional edited RNA species in Drosophila, encoding primarily ligand- and voltage-gated ion channels and components of the synaptic release machinery (Hoopengardner et al., 2003). Of these newly identified editing events, only a voltage-gated potassium channel (Kv1.1) has been validated as a target in the mammalian transcriptome (Bhalla et al., 2004; Hoopengardner et al., 2003). Additional in silico approaches, taking advantage of identified A-to-G discrepancies between genomic and cDNA sequences, predicted >12,000 editing sites in the human transcriptome that occurred largely in noncoding regions of RNA transcripts containing inverted repetitive elements of the Alu and L1 subclass (Athanasiadis et al., 2004; Blow et al., 2004; Kikuno et al., 2002; Kim et al., 2004; Levanon et al., 2004), yet the functional relevance of these editing events has not yet been explored.

2. Mammalian ADAR Enzymes

The conversion of A-to-I is catalyzed by hydrolytic deamination at the C-6 position of the purine ring (Polson et al., 1991). Studies showing that A-to-I editing in mammalian mRNAs required the presence of an extended RNA duplex (Egebjerg et al., 1994; Gerber et al., 1998) led to the purification and cloning of a family of enzymes referred to as adenosine deaminases that act on RNA (ADARs) (Hough and Bass, 1994; Kim et al., 1994a,b; Maas et al., 1996; O'Connell and Keller, 1994; O'Connell et al., 1995), which have subsequently been shown to be responsible for catalyzing A-to-I modifications in numerous RNA substrates (Bass, 2002; Emeson and Singh, 2001). In mammals, three ADAR proteins (referred to as ADAR1, ADAR2, and ADAR3) and their corresponding genes have been identified (Hough and Bass, 1997; Kim et al., 1994b; Liu et al., 1997; O'Connell et al., 1995) (Fig. 16.1). ADAR1 and ADAR2 have been shown to be expressed in almost all cell types examined (Melcher et al., 1996b; Wagner et al., 1990) and are able to covert A-to-I in extended regions of duplex RNA within pre-mRNAs and viral RNA transcripts (Emeson and Singh, 2001; Schaub and Keller, 2002). ADAR3 is expressed exclusively in the brain, but has not been shown to have any catalytic activity using synthetic dsRNA or known ADAR substrates (Chen et al., 2000; Melcher et al., 1996a). ADAR

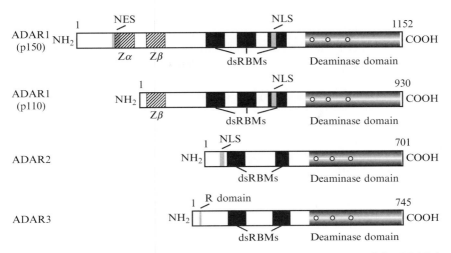

Figure 16.1 ADAR functional domains. A schematic representation of the ADAR1, ADAR2, and ADAR3 proteins is presented indicating their relative size in amino acids, as well as the location of the putative nuclear localization (NLS) and export (NES) signals, arginine-rich (R) domain, Z-DNA binding domains (Zα and Zβ), dsRNA-binding motifs (dsRBMs), and the adenosine deaminase domain. The locations of zinc-coordination residues in the deaminase domain are designated by circles.

selectivity for specific adenosine residues results from both the sequence context and the structural features of the RNA duplex. Naturally occurring hairpin loops have a variety of mismatches and bulges that can alter ADAR binding and editing (Dawson *et al.*, 2004; Lehmann and Bass, 1999). *In vitro* studies show editing of up to 50% of the adenosine residues in both strands using synthetic dsRNAs that are perfectly complementary (Cho *et al.*, 2003; Dawson *et al.*, 2004; Lehmann and Bass, 2000). The addition of mismatches reduces the overall extent of editing and increases selectivity to particular sites (Lehmann and Bass, 1999; Polson and Bass, 1994). ADAR1 and ADAR2 have overlapping yet distinct patterns of editing, with some sites edited by only one enzyme and other sites edited equally well by both (Lehmann and Bass, 2000).

The ADAR1 gene specifies two major protein isoforms, an interferon (IFN) inducible 150-kDa protein (p150) and a constitutively expressed N-terminally truncated 110-kDa protein (p110), encoded by transcripts with alternative exon 1 structures that initiate from different promoters (George and Samuel, 1999) (Fig. 16.1). The predicted protein sequence of ADAR1 reveals that it contains three copies of a dsRNA-binding motif (dsRBM), a motif shared among a number of dsRNA-binding proteins (Burd and Dreyfuss, 1994), a nuclear localization signal in the third dsRBM (Eckmann *et al.*, 2001), and a region homologous to the catalytic domain of other known adenosine and cytidine deaminases. The amino terminus of the p150 isoform also contains two Z-DNA-binding domains that have been proposed to tether ADAR1 to sites of transcription (Herbert, 1996) or mediate interactions between ADARs and other proteins (Hough and Bass, 2001). ADAR2 is an 80-kDa protein with structural features similar to those observed for ADAR1 (Melcher et al., 1996b). ADAR2 contains a nuclear localization signal and two dsRNA-binding motifs, sharing approximately 25% amino acid sequence similarity with the dsRBMs of ADAR1. ADAR2 also contains an adenosine deaminase domain sharing 70% amino acid similarity with ADAR1, as well as three zinc-chelating residues conserved in the deaminase domain of both enzymes. As with ADAR1, multiple cDNA isoforms of ADAR2 have been identified in rats, mice, and humans including alternative splicing events in mRNA regions encoding the deaminase domain and near the amino terminus (Gerber et al., 1997; Lai et al., 1997; Rueter et al., 1999).

3. Mammalian ADAR Substrates

3.1. Glutamate-gated ion channels

Ionotropic glutamate receptors (iGluRs) are involved in fast synaptic neuro-transmission and in the establishment and maintenance of synaptic plasticity critical to learning and memory. Three subtypes of iGluRs, named according to selective agonists for each receptor subtype, include N-methyl-D-aspartate (NMDA) receptors, AMPA receptors, and kainate receptors (Ozawa et al., 1998). AMPA receptors are composed of homomeric or heteromeric assemblies of four different types of subunits (GluR-1, -2, -3, and -4) (Madden, 2002). The first example of A-to-I editing in mammalian mRNAs was identified in transcripts encoding the GluR-2 subunit in which a genomically encoded glutamine codon (CAG) was altered to an arginine codon (CIG) (Melcher et al., 1995; Rueter et al., 1995; Sommer et al., 1991; Yang et al., 1995). This single amino acid alteration (Q/R site) regulates both the electrophysiological and ion-permeation properties of heteromeric AMPA receptors. AMPA channels containing the edited GluR-2(R) subunit are impermeant to calcium ions, whereas those that lack or contain a nonedited GluR-2(Q) subunit demonstrate a dramatic increase in their relative divalent cation permeability (Dingledine et al., 1992; Hollmann et al., 1991; Sommer et al., 1991; Verdoorn et al., 1991). RNA editing is also responsible for an A-to-I modification in RNAs encoding the GluR-2, -3, and -4 AMPA receptor subunits at a specific site (R/G site) that increases the rate of recovery from receptor desensitization (Lomeli et al., 1994). In addition to RNA editing events mediating the functional properties of AMPA receptors, transcripts encoding subunits of heteromeric kainate (KA) receptors (GluR-5 and GluR-6) are also modified by RNA editing to modulate the ion permeation properties of these ionotropic glutamate receptors (Egebjerg and Heinemann, 1993; Kohler et al., 1993; Sommer et al., 1991).

3.2. Serotonin 2C receptor

The 2C subtype of serotonin receptor ($5HT_{2C}R$) has been implicated in a variety of human affective disorders, including depression, obsessive–compulsive disorder, anxiety, eating disorders, and schizophrenia (Dubovsky and Thomas, 1995; Pandey et al., 1995). A member of the G–protein–coupled receptor superfamily, this receptor signals through distinct heterotrimeric G-proteins to activate downstream effector systems involving phospholipase C (PLC), phospholipase D, and mitogen–activated protein kinase (MAP kinase) (Van Oekelen et al., 2003). RNA transcripts encoding the $5HT_{2C}R$ undergo up to five A-to-I editing events that can alter the identity of three amino acids within the second intracellular loop of the receptor (Burns et al., 1997; Wang et al., 2000b), a region involved in receptor:G-protein coupling (Pin et al., 1994). Functional comparisons in heterologous expression systems, between the nonedited $5HT_{2C}R$ (encoding isoleucine, asparagines, and isoleucine at amino acids 157, 159, and 161, respectively) and the edited $5HT_{2C}R$ isoform containing valine, glycine, and valine at the analogous positions, revealed a 40-fold decrease in serotonergic potency to stimulate phosphoinositide hydrolysis due to reduced Gq/G11-protein coupling efficiency (Burns et al., 1997; Niswender et al., 1999), and decreased coupling to other signaling pathways (Burns et al., 1997; Niswender et al., 1999). In addition, cells expressing the edited $5HT_{2C}R$ demonstrate considerably reduced constitutive activation in the absence of ligand compared to cells expressing the nonedited isoform (Niswender et al., 1999).

3.3. Voltage-gated potassium channel (Kv1.1)

Kv1.1 is expressed in axons and dendrites of neurons and forms tetramers with other Kv1 subunits creating channels that regulate neuronal excitability and nerve signaling (O'Grady and Lee, 2005). A-to-I editing of transcripts encoding Kv1.1 converts a genomically encoded isoleucine (ATT) to a valine (ITT) codon in the highly conserved ion-conducting pore of the channel, a region that has been implicated in the docking of the inactivation particle (Zhou et al., 2001). Editing of Kv1.1 transcripts results in an increase in the rate of recovery from channel inactivation (Bhalla et al., 2004). Previous studies have indicated that mutant animals with alterations in the fast inactivation rate of Kv1.1 have been shown to have behavioral and neurological changes associated with hippocampal learning deficits (Giese et al., 1998) and episodic ataxia type-1 (Herson et al., 2003), demonstrating the importance of regulated inactivation for this voltage-gated channel in normal central nervous system (CNS) function.

3.4. ADAR2

Multiple cDNA isoforms of ADAR2 have been identified in rats, mice, and humans resulting from alternative splicing events in mRNA regions encoding the deaminase domain and near the amino terminus (Gerber *et al.*, 1997; Lai *et al.*, 1997; Rueter *et al.*, 1999). One such alternative splicing event introduces an additional 47 nucleotides (nt) near the 5′-end of the ADAR2 coding region, resulting in a frameshift that is predicted to produce a 9-kDa protein (82 aa) lacking the dsRBMs and catalytic deaminase domain required for protein function (Rueter *et al.*, 1999). Nucleotide sequence analysis of ADAR2 genomic DNA revealed the presence of adenosine–adenosine (AA) and adenosine–guanosine (AG) dinucleotides at proximal and distal alternative 3′-acceptor sites required for 47 nucleotide inclusion or exclusion, respectively. Use of the proximal 3′-acceptor is dependent upon the ability of ADAR2 to edit its own pre-mRNA, converting the intronic AA to an adenosine–inosine (AI) dinucleotide that effectively mimics the highly conserved AG sequence normally found at 3′-splice junctions. These observations indicate that RNA editing may serve as a novel mechanism for the regulation of alternative splicing and suggest a unique negative autoregulatory strategy by which ADAR2 can modulate its own level of expression (Feng *et al.*, 2006; Rueter *et al.*, 1999).

3.5. SINE elements

The majority of identified editing events occur within untranslated regions of RNA transcripts (Athanasiadis *et al.*, 2004; Blow *et al.*, 2004; Kikuno *et al.*, 2002; Kim *et al.*, 2004; Levanon *et al.*, 2004). Such events are prominent within short repetitive elements present at a high frequency in the genome. In mammals, the most common edited elements are categorized as short interspersed elements (SINEs), which are mobile repetitive elements of less than 500 base pairs that replicate by retrotransposition. They can be integrated into the genome in either direction, creating a propensity for RNA duplex formation when two elements of opposite orientation are present within a single transcript. Because SINEs are highly conserved, these duplexes often have few mismatches and are extensively edited. This form of editing could affect many properties of RNA beyond their amino acid coding capacity, and physiological consequences are difficult to predict since so many transcripts undergo this form of editing.

4. GENERATION OF GENETICALLY MODIFIED MICE BY HOMOLOGOUS RECOMBINATION IN ES CELLS

To study the physiological role(s) of a protein using an *in vivo* mammalian system, modification of the mouse genome represents an excellent experimental approach to introduce a specific mutation by taking advantage of homologous recombination in embryonic stem (ES) cells. Such a mutation can be introduced to completely disrupt expression of the gene of interest (null–allele or knockout) or to introduce as little as a single nucleotide alteration to affect protein coding potential or to model rare alleles that may be associated with human disease (knockin). The generation of genetically modified mice has been the subject of numerous reviews (Gerlai, 2000; Koller and Smithies, 1992; Misra and Duncan, 2002; Robertson, 1991), and the precise details for this well-established paradigm are fully described in *Manipulating the Mouse Embryo: A Laboratory Manual* (Nagy et al., 2003). The overall strategy for modification of the mouse genome begins with the generation of a targeting vector, a genomic fragment that contains the desired mutation, which is subsequently introduced into ES cells that retain the ability to contribute to all mouse cell lineages. ES cells bearing one copy of the mutant allele, having successfully undergone homologous recombination with the targeting vector, are injected into the blastocoel cavity of 3.5-day embryos (blastocysts), which are then implanted into a pseudopregnant foster mother and allowed to develop normally. The resultant offspring are chimeras, derived from both the injected mutant ES cells and wild-type ES cells present in the inner cell mass of the blastocyst. If the mutant ES cells contribute to the germinal epithelium in chimeric animals to produce eggs or sperm, breeding of these mice to wild-type animals will produce offspring that are heterozygous for the desired mutation and interbreeding of viable heterozygotes will produce mice that are homozygous for the mutation of interest.

Every targeting vector must be designed to have fundamental properties that allow it to effectively combine with a specific locus within the mouse genome including sequences that are homologous to the chromosomal locus to be targeted and a plasmid vector backbone (Fig. 16.2). The homologous sequences in the targeting vector should be derived from genomic DNA that is isogenic with that of the ES cells, because subtle differences in sequence between different mouse strains can significantly reduce the efficiency of homologous recombination (Deng and Capecchi, 1992; te Riele et al., 1992). Most available ES cell lines are derived from the 129 mouse strain, yet detailed sequence length polymorphism studies have revealed that the 129 strain is actually a diverse and complex family of substrains that have a high degree of genetic variation, resulting in differences in reproduction, behavior, antigens, and other polymorphisms (Simpson et al., 1997; Threadgill et al., 1997). For a variety of reasons, the C57BL/6 mouse strain is commonly used as a general purpose strain and a background reference strain for the generation of congenic mice carrying both spontaneous and induced mutations. As a consequence, mutant mice generated with 129-derived ES cells are routinely backcrossed to the C57BL/6 strain for at least 10 generations to ensure less than 0.1% of the genetic material from the original background strain remains. This backcrossing strategy is straightforward, yet time-consuming, as it generally requires either a commitment of 2–3 years of mouse husbandry or 1–2 years of

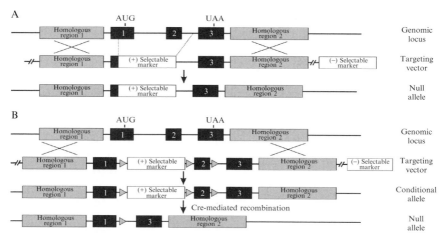

Figure 16.2   Strategies for the ablation of gene expression using a replacement-type targeting vector in ES cells. (A) A schematic diagram of a prototype gene is presented along with a targeting vector in which a region in exon 1 through exon 2 is replaced by the positive selectable marker. The positions of long arms of the homologous sequence, to promote double-reciprocal recombination events, and a negative selectable marker, outside the region of homology, are indicated. (B) A schematic diagram of a prototype gene is presented along with a targeting vector allowing the Cre-mediated ablation of gene expression by removal of loxP-flanked (▶) exon 2 and the positive selectable marker. The positions of long arms of the homologous sequence and a negative selectable marker, outside the region of homology, are indicated.

detailed genotype analysis in the case of marker-assisted selection (speed congenics) (Collins *et al.*, 2003; Visscher, 1999; Wong, 2002). The establishment of ES cell lines derived from the C57BL/6 strain has been a major step to help facilitate the direct genetic alteration of mice in a pure C57BL/6 background (Auerbach *et al.*, 2000; Baharvand and Matthaei, 2004; Cheng *et al.*, 2004); however, there are relatively few of these ES cell lines available, they require more fastidious care than their 129-derived counterparts, and they form chimeric animals less efficiently (Seong *et al.*, 2004).

4.1. Targeting vectors

A targeting vector is created by introducing a mutation into a segment of mouse genomic DNA. The mutation is flanked on both sides by sequences that are homologous to the endogenous gene, and can thus insert into the genome by a double-reciprocal crossover recombination event (Fig. 16.2A). One parameter that has clearly been shown to improve homologous recombination efficiency is the use of long regions of isogenic DNA in the targeting vector (Deng and Capecchi, 1992; Hasty *et al.*, 1991; Shulman *et al.*, 1990; te Riele *et al.*, 1992; Thomas and Capecchi, 1987).

Thomas and Capecchi (1987) first described an exponential relationship between length of homology and targeting frequency and subsequent studies have further demonstrated an increase in targeting rate with an increase in targeting arm length (Hasty et al., 1991), although this effect reached a plateau at 18.4 kb (Deng and Capecchi, 1992).

Despite the use of long regions of homology in the targeting vector, the rate of homologous recombination is relatively inefficient and requires that a positive selectable marker be included to select for ES cells that have success-fully incorporated the selection cassette into the genome; a gene cassette encoding resistance to the antibiotic neomycin (or its analogs) is generally used for this purpose (Gossler et al., 1986), although resistance cassettes for puromycin, hygromycin, and methotrexate have also been used effectively (Tucker et al., 1997). Following transfection, the ES cells are treated with a neomycin analog, G418, and only those cells that have integrated the select-able marker into the genome will survive. Because integration of the targeting vector can occur as a result of homologous recombination or random integra-tion, a negative selectable marker encoding herpes simplex virus type 1 thymidine kinase (HSV-tk) or diphtheria toxin (DT-A) is often used to enrich ES cells that have undergone homologous recombination (Borrelli et al., 1988; Wigler et al., 1977; Yanagawa et al., 1999). The negative selection cassette is placed in the targeting vector outside the region of homology and is present after random integration of the targeting vector, but is eliminated from the genome after the desired recombination event (Fig. 16.2A). When using HSV-tk as a negative selection marker, ES cells surviving positive selection can be simultaneously treated with ganciclovir to eliminate cells that have randomly integrated the targeting vector; expression of HSV-tk results in a phosphorylated ganciclovir product that can be incorporated into the DNA of replicating cells causing irreversible arrest at the G_2/M check point followed by apoptosis (Rubsam et al., 1998).

The segment of modified DNA is generally cloned into a plasmid vector that is subsequently linearized outside the region of homology, at a unique restriction site, to increase the efficiency of ES cell transfection, which is ordinarily performed by electroporation (Nagy et al., 2003). Following selection, antibiotic-resistant ES cells still require verification of proper targeting prior to blastocyst injection. Analysis of homologous recombina-tion fidelity is most often performed using a Southern blotting strategy with ES cell genomic DNA and a radiolabeled probe complementary to sequences outside the region of homology, although a polymerase chain reaction (PCR)–based strategy may also be employed for this purpose. The goals of these analyses are to distinguish between antibiotic-resistant ES cell clones that have undergone homologous recombination from those in which the targeting vector has been randomly integrated, as well as to verify that the recombination event allowed for correct integration of the desired mutation into the ES cell genome.

Common types of mutations are those that inactivate the desired gene (knockout) or that alter the gene sequence to create a more subtle mutation (knockin). Knockout mutations can be created by interrupting a functionally important exon with the *neo* gene (Fig. 16.2A), whereas knockin mice have the endogenous gene sequence replaced with an engineered sequence to study a more specific aspect of gene function, but the modified gene retains its protein coding capability. For knockin constructs in which the expression of a specific gene locus is not disrupted, the positive selectable marker must be integrated into a noncoding region of the gene, such as an intron, or within a region representing an untranslated region of the mRNA.

4.2. Conditional gene expression

A majority of the genetically modified mouse lines developed to date represent null mutants in which the positive selection cassette is introduced to disrupt the protein-coding potential of a specific gene locus. While some genes are expressed in very limited cell populations in which their role can be addressed by global deletion, other genes are expressed broadly, and their unrestricted elimination may alter the function of multiple physiological systems. For both categories of genes, there is always the problem that classic null mutations may result in embryonic lethality, interfere with development, or alter development through adaptive compensation, such that any resulting phenotypic alterations are difficult, if not impossible, to interpret accurately. To circumvent these problems, strategies to modulate the expression of specific genes in a spatiotemporal fashion have been developed that employ the use of site-specific recombinases, such as the bacteriophage P1 Cre/loxP system, in which the expression of a conditional allele containing loxP sites can be modulated by the tissue-specific expression of Cre recombinase to generate a modified (null) allele (Fig. 16.2B). Cre recombinase acts by looping out the region between two loxP sites that are present in the same orientation within DNA, leaving a single loxP site in its place (Hoess and Abremski, 1984). Cre recombinase may be transfected into correctly targeted ES cells to catalyze recombination between loxP sites prior to blastocyst injection, or mutant mice bearing a conditional allele with loxP sites may be mated with transgenic mice expressing Cre under the control of a spatiotemporally regulated promoter. The conditional regulation of Cre expression has been advanced further by the development of several inducible systems in which administration of specific inducers (e.g., tamoxifen, RU486, dexamethasone, tetracycline, Cd^{2+}) can be used to activate Cre expression during specific developmental periods (Brocard *et al.*, 1998; Danielian *et al.*, 1998; Feil *et al.*, 1996; Kellendonk *et al.*, 1996; Sauer and Henderson, 1988). The most complete database of Cre-expressing transgenic animals is maintained by the laboratory of Dr. Andras

Nagy at the Samuel Lunenfeld Research Institute as part of the CreXMice project (*http://nagy.mshri.on.ca/PubLinks/indexmain.php*).

In addition to conditional regulation of gene expression, the Cre/loxP system is also routinely used to remove the *neo* cassette following antibiotic selection of ES cells for homologous recombination. Previous studies have indicated that the *neo* gene can have long-range transcriptional effects, interfering with the expression of neighboring loci, and thus can affect the subsequent animal phenotype (Pham *et al.*, 1996). To circumvent this potential problem, a third loxP site, flanking the *neo* cassette, may be introduced such that the selectable marker can be eliminated in a Cre-dependent fashion (Fig. 16.2B). Since introduction of Cre recombinase can give rise to several recombination products when acting upon a targeted allele containing three loxP sites (recombination between loxP sites 1–2, 2–3, and 1–3), many investigators have used an alternative recombination system in which FRT recombination sites are introduced flanking the *neo* cassette so that it can subsequently be removed by the actions of FLP recombinase (Schaft *et al.*, 2001).

5. CLASSIFICATION OF TRANSGENIC APPROACHES

The field of RNA editing has advanced significantly through the use of *in vitro* and cell culture systems to explore enzyme function and the dynamics of enzyme–substrate interactions. While heterologous expression systems provide a simple model that can be manipulated easily to understand how the editing of a specific transcript may alter the function(s) of the RNA or its encoded protein, these studies provide clues regarding only the physiologic roles of editing that must be extrapolated from *in vitro* analyses. As a result, multiple groups have generated genetically modified mice through homologous recombination to examine the functional impact that editing may have in a complex organism. Mouse models that have been created to examine the role(s) of editing can be divided into three categories: substrate modulation, ADAR modulation, and concurrent substrate and ADAR modulation.

6. GENETIC MODIFICATION OF ADAR SUBSTRATES

The importance of specific editing events has been explored in genetically modified mouse models by circumventing the many levels of regulation controlling the extent of editing for a specific site. This is achieved by creating a mutant allele that solely expresses either the edited or nonedited sequence and is not subject to subsequent modification by ADAR proteins.

An edited RNA species can be generated simply by changing the adenosine at the site of interest to a guanosine using site-directed mutagenesis (Kask *et al.*, 1998; Sailer *et al.*, 1999). Guanosine cannot be modified by ADARs, and because inosine is recognized as guanosine during translation, the encoded protein sequence is identical to that encoded by the inosine-containing mRNA. An allele that expresses only a nonedited transcript must be altered such that it can no longer be bound and edited by ADAR proteins without changing the coding potential of the genomically encoded sequence. This can be achieved by disrupting the RNA duplex that forms in the pre-mRNA transcript when the region encompassing the editing site base pairs with a near complementary sequence contained within another region of the transcript. In many RNAs, this editing-site complementary sequence (ECS) is located within an intron, allowing extensive modification or deletion without altering the sequence of the mature mRNA. When the site of interest is contained within the coding region of an exon, the introduction of synonymous codon alterations that disrupt the RNA secondary structure can serve to eliminate subsequent A-to-I conversion.

6.1. Mice solely expressing edited (Q/R site) GluR-2 transcripts (GluR-2^R)

The Q/R site within the GluR-2 subunit of the AMPA receptor is critical for regulating the electrophysiological and ion permeation properties of this ligand-gated ion channel (Higuchi *et al.*, 1993; Sommer *et al.*, 1991). Editing at the Q/R site is nearly complete in the postnatal brain (Sommer *et al.*, 1991), raising questions regarding the physiological and evolutionary reasons that the critical codon for R^{607} (CIG) is conferred through editing and not through direct coding (CGG) from the genome. A functional role for this A-to-I modification may be assumed from observations that the Q/R site in GluR-2 transcripts is edited less than 100% (Higuchi *et al.*, 1993; Sommer *et al.*, 1991), suggesting that distinct neuronal populations within the brain may express only the nonedited (GluR-2^Q) isoform. Previous studies have demonstrated that Q/R site editing levels are significantly decreased at early embryonic stages, further suggesting that editing may play an important role in brain development. To determine whether the small amount of nonedited GluR-2 transcripts identified in whole mouse brain is functionally important, mice were generated that can express only the edited (GluR-2^R) form of the protein (Kask *et al.*, 1998).

The targeting vector for these mice was created from a genomic fragment containing exons 10–12 of the GluR-2 gene in which the glutamine codon (CAG) at the editing site was mutated to encode arginine (CGG) (Kask *et al.*, 1998) (Fig. 16.3, *upper*). Four silent mutations were introduced in exon 11 so that the mutant allele could be differentiated from the

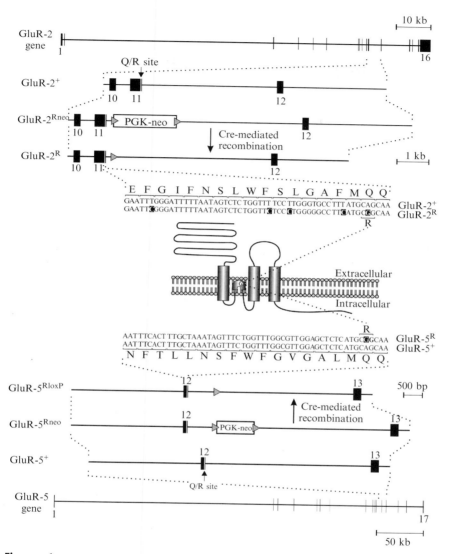

Figure 16.3 Targeting strategy for the generation of GluR-2 (Q607R) and GluR-5 (Q636R) mutant mice. A schematic diagram of the mouse GluR-2 (*upper*) and GluR-5 (*lower*) genes is shown with an expansion of the region contained in the corresponding targeting vectors. The structures of the wild-type (+) gene and the modified alleles, both before (Rneo) and after (R) Cre-mediated recombination of the loxP-flanked (►) PGK-neo cassettes, are presented. Nucleotide alterations in each mutant allele are indicated in inverse lettering, and changes in the amino acid sequence are presented using the one-letter amino acid code. The position of the introduced arginine residue in the second hydrophobic domain of each protein is mapped onto the predicted topology for a non-NMDA receptor subunit.

endogenous GluR-2 transcript. A loxP-flanked neomycin resistance cassette, under the control of a phosphoglycerate kinase (PGK) promoter, was introduced into intron 11 for positive selection of ES cells. ES cells were electroporated, and clones surviving G418 selection were checked for accurate gene targeting using PCR and Southern blotting strategies. A correctly targeted clone was injected into blastocysts that were transferred to a pseudopregnant female mouse. Highly chimeric male offspring were mated to wild-type females or females expressing Cre recombinase under transcriptional regulation of the human cytomegalovirus (CMV) minimal promoter (Schwenk et al., 1995), to eliminate the PGK-neo cassette in the germline. The mice generated from the wild-type mating were heterozygous for the GluR-2^{Rneo} allele, while the mice generated from the CMV-Cre mating were heterozygous for the GluR-2^R allele. Progeny were intercrossed to create homozygous mutant mice for each mutant allele (Kask et al., 1998).

Homozygous mice were born in the expected Mendelian ratio and appeared phenotypically normal in appearance, weight, and life span. Measurement of edited and nonedited transcripts indicated that the GluR-2^R allele was expressed to the same extent as wild-type (GluR-2^+) mRNA, although expression of GluR-2^{Rneo} RNAs was reduced by 30% (Kask et al., 1998). This difference may result from a slowed removal of intron 11 due to the presence of the PGK-neo cassette (Brusa et al., 1995) (Fig. 16.3, upper), thereby decreasing the steady-state level of mature GluR-2^{Rneo} mRNA. Electrophysiological analyses showed no differences between mice bearing the GluR-2^+ or GluR-2^R alleles, yet hippocampal pyramidal cells from mice homozygous for GluR-2^{Rneo} had a 2-fold increase in calcium permeability that may reflect the moderate reduction in expression level, yet no consequent neuronal excitotoxicity was observed. These results have suggested that Q/R site editing in GluR-2 RNAs was coopted during evolution to lower the Ca^{2+} permeability of GluRs and that this posttranscriptional modification was subsequently maintained, despite the fact that genomic alteration at this site would result in the same physiological outcome (Kask et al., 1998). Alternatively, mutant mice bearing the GluR-2^R allele may have a number of functional deficits that are not immediately obvious when using basic procedures to examine potential phenotypic alterations.

6.2. Mice solely expressing edited (Q/R site) GluR-5 transcripts (GluR-5^R)

The GluR-5 subunit of kainate receptors is edited at the Q/R site in the second hydrophobic domain, analogous to the GluR-2 Q/R site, which modulates the calcium permeability and single channel conductance of this ionotropic receptor (Bernard and Khrestchatisky, 1994; Egebjerg and

Heinemann, 1993; Kohler *et al.*, 1993; Sommer *et al.*, 1991). Adult rats show approximately 50% editing of this site in RNA isolated from whole brain; however, the levels of embryonic editing are very low, suggesting that regulation of editing at this site may be important for brain development (Bernard and Khrestchatisky, 1994). In parallel to studies of GluR-2 editing, genetically modified mice were generated that solely express the edited isoform of the subunit (GluR-5^R) to determine whether the nonedited form is important for development of the mouse nervous system and if the partial editing of GluR-5 transcripts in the adult is physiologically relevant (Sailer *et al.*, 1999).

Replacement-type targeting vectors were constructed from a genomic fragment extending from intron 11 through exon 13 in which a loxP-flanked neomycin resistance cassette was inserted in intron 12, downstream from the Q/R editing site, and the HSV-tk gene was included outside the region of homology for negative selection (Fig. 16.3, *lower*). To eliminate any interpretive complications resulting from changes in the expression or processing of the mutant gene, control mice were generated in parallel, in which no changes were made to the protein coding potential of the gene, but the loxP-flanked neomycin cassette was inserted in intron 12 (Sailer *et al.*, 1999). The edited version, R^{loxP}, was generated through PCR-mediated mutagenesis, changing the genomically encoded glutamine (CAG) to an arginine (CGG) codon. The control vector, wt^{loxP}, was subject to an equivalent process, but the genomic sequence was preserved. ES cell clones verified by Southern blotting were transfected with a plasmid encoding Cre recombinase for removal of the neomycin resistance cassette (Fig. 16.3, *lower*) (Sailer *et al.*, 1999).

Mice both heterozygous and homozygous for the R^{loxP} allele were viable and showed no visible differences from control wt^{loxP} or wild-type mice. Mice were subject to tests examining anatomy, behavior, kaanite-induced seizure susceptibility, nociception, and receptor electrophysiology. Although electrophysiological studies indicated reduced current density in wt^{loxP} mice following agonist treatment, there were no obvious physiological consequences and mice were normal in all other respects. It was concluded that the level of Q/R site editing of GluR-5 is not physiologically critical (Sailer *et al.*, 1999).

6.3. Mice deficient in GluR-2 editing (Q/R site)

The Q/R site in GluR-2 transcripts is edited to near completion in the adult mouse brain and the edited isoform is required to limit calcium entry through heteromeric AMPA receptors containing this subunit (Higuchi *et al.*, 1993; Sommer *et al.*, 1991). To determine the biological significance of this modification, mice were generated that express a modified GluR-2 allele that is not subject to deamination by ADARs (Brusa *et al.*, 1995). The targeting vector for these mice was created from a

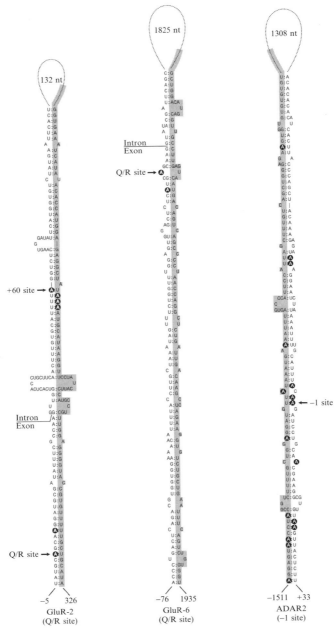

Figure 16.4   Predicted secondary structure of pre-mRNA transcripts encoding mouse ionotropic glutamate receptor subunits (GluR-2 and GluR-6) and mouse ADAR2 in the regions of major editing modifications. The positions of edited adenosine residues are

genomic fragment containing exons 9–12 of the GluR-2 locus. The ECS for the Q/R site in intron 11 (Fig. 16.4, *left*) was replaced with a loxP-flanked neo-tk selection cassette to disrupt the RNA secondary structure required for A-to-I conversion. Correctly targeted ES cells, verified by PCR, were electroporated with a Cre recombinase expression plasmid for removal of the selectable marker, creating the GluR-$2^{\Delta ECS}$ allele that was successfully transmitted to the offspring of chimeric animals (Brusa *et al.*, 1995).

Mice heterozygous for the GluR-$2^{\Delta ECS}$ allele appeared healthy until postnatal day 14 (P14), when they developed seizures that led to death by P20. This phenotype resulted from dramatically increased AMPA receptor permeability to Ca^{2+}, concomitant with neuronal degeneration. An unexpected effect of this mutation was reduced expression of the mutant allele compared to wild-type. Only 25% of transcripts in heterozygotes were nonedited, and total GluR-2 expression was only 70% of normal. This alteration in GluR-2 expression was linked to the accumulation of mutant transcripts that retained intron 11, suggesting that replacement of the ECS sequence with a loxP site interferes with splicing efficiency; splicing throughout the remainder of the transcript appeared normal. The mutation did prove successful in eliminating editing at the Q/R site without affecting editing at a different site (R/G site) within the same transcript (Brusa *et al.*, 1995).

The severe phenotype of heterozygous GluR-$2^{\Delta ECS}$ mice (Brusa *et al.*, 1995), even without equal contribution from the mutant allele, led to an investigation of the role of the expression ratio between nonedited and edited transcripts in producing the neurodegenerative phenotype (Feldmeyer *et al.*, 1999). In this case, the same targeting vector was used, but without initial removal of the selectable marker from ES cells. Mice heterozygous for this GluR-2^{neo} allele (GluR-$2^{+/neo}$) were either inter-crossed to create GluR-$2^{neo/neo}$ mice, or were bred to transgenic mice expressing CMV-Cre (Schwenk *et al.*, 1995) to remove the selectable marker *in vivo* (GluR-$2^{+/\Delta ECS^*}$). The allele generated by *in vivo* excision of the neo-tk cassette is designated GluR$^{\Delta ECS^*}$ to distinguish it from the GluR$^{\Delta ECS}$ allele that resulted from transient Cre expression in targeted ES cells (Brusa *et al.*, 1995). The allele retaining *neo* only represented about 10% of the total mature transcripts, with total GluR-2 protein expression being

indicated with closed circles, the positions of exon/intron boundaries are shown, and the number of nucleotides omitted from the figure is indicated in the loops; coordinates are relative to the Q/R editing site in GluR-2 and GluR-6 transcripts and the -1 site in ADAR2 pre-mRNA. Regions of the editing site complementary sequence (ECS), deleted to disrupt the RNA duplex structure and selectively ablate RNA editing, are indicated by a gray box.

60% and 10% of normal in heterozygous and homozygous mice, respectively. In GluR-2$^{+/\Delta ECS^*}$ mice, the mutant allele contributed 30% of total GluR-2 transcripts, and total protein was reduced to 80% of normal (Feldmeyer et al., 1999). This was comparable to GluR-2$^{+/\Delta ECS^*}$ mice previously studied (Brusa et al., 1995), and the phenotype was equivalent as well. GluR-2$^{+/neo}$ mice had a mild phenotype, showing impaired open-field behavior and slightly increased lethality. GluR-2$^{neo/neo}$ mice all died by P20 and showed lethargy, retarded growth, and dendritic deficits, but were not prone to seizures. Calcium permeability of AMPA receptors in these mice was increased 30-fold, and single-channel conductance was increased 3-fold; however, macroscopic conductance through AMPA receptors was normal, likely owing to reduced receptor density. Results of these studies indicate that both the level of editing and expression of GluR-2 are critical for normal neuronal function (Brusa et al., 1995; Feldmeyer et al., 1999).

6.4. Mice deficient in GluR-6 editing (Q/R site)

Unlike GluR-2 transcripts that are edited to near completion throughout the brain (Sommer et al., 1991), the editing of transcripts encoding KA receptor subunits is subject to spatiotemporal regulation (Bernard and Khrestchatisky, 1994; Bernard et al., 1999). Significant alterations of KA receptor editing and ADAR2 protein expression have also been observed following seizures in humans and rats, suggesting that nonedited KA receptor subunits may play a role at excitatory inputs in the embryo and that the extent of editing at synapses may modulate synaptic transmission and seizure vulnerability in the adult (Bernard and Khrestchatisky, 1994; Grigorenko et al., 1998). To explore this possibility, mice were engineered to express only a nonedited version (GluR-6^Q) of this KA receptor subunit (Vissel et al., 2001).

A replacement-type targeting vector was generated from the mouse GluR-6 gene locus in which the ECS region in the intron downstream of the Q/R editing site was replaced by a loxP-flanked *neo* cassette to disrupt the RNA secondary structure required for site-selective adenosine deamination (Fig. 16.4, *center*). An HSV-tk gene was included outside the region of homology for negative selection of randomly integrated clones. Correctly targeted ES cell clones were used for blastocyst injection and generation of chimeras that were bred to produce mice heterozygous for the ΔECS^{neo} allele. Mating with transgenic mice expressing Cre recombinase under the control of the protamine promoter (O'Gorman et al., 1997) allowed *neo* excision in the male germline and the subsequent generation of homozygous mice expressing the GluR-6$^{\Delta ECS}$ allele. The mutation was successful in abolishing editing, with homozygous mutants solely expressing the nonedited GluR-6^Q isoform of this KA receptor subunit, while the extent of editing for other sites in GluR-6 (I/V and Y/C sites) and other transcripts was not altered (Vissel *et al.*, 2001). Unlike the editing-deficient GluR-2 Q/R mice (Brusa *et al.*, 1995; Feldmeyer *et al.*, 1999), the GluR-6$^{\Delta ECS}$ allele produced RNA and protein at normal levels.

Mice homozygous for the ΔECS allele were viable and did not display obvious differences in a battery of behavioral tests. Studies of hippocampal slices indicated that the mutation conferred NMDA-independent long-term potentiation (LTP) in the medial perforant path–dentate gyrus synapse. The reduced editing also impacted the electrophysiological properties of KA receptors and resulted in an increased susceptibility to kainate-induced seizures, indicating that Q/R site editing in GluR-6 pre-mRNAs appears to play a modulatory role in synaptic plasticity and seizure vulnerability (Vissel *et al.*, 2001).

7. Enzyme Modified Mice

The generation of genetically modified mice that solely express either edited or nonedited isoforms of ADAR substrates has provided considerable insight into the importance of A-to-I conversion for specific gene products. However, ADAR1 and ADAR2 each has many substrates that have already been discovered, and additional substrates will, most likely, continue to be identified using both biochemical and bioinformatic approaches (Athanasiadis *et al.*, 2004; Blow *et al.*, 2004; Hoopengardner *et al.*, 2003; Kikuno *et al.*, 2002; Kim *et al.*, 2004; Levanon *et al.*, 2004; Morse and Bass, 1999; Morse *et al.*, 2002). While some editing sites are selectively modified by a single ADAR protein, others sites are edited by both in heterologous expression systems, so it can be difficult to differentiate the physiological roles that each enzyme may play (Emeson and Singh, 2001). To elucidate the full scope of consequences resulting from ablation of ADAR expression in a mammalian system, both ADAR1 and ADAR2 have been knocked out individually in mice (Hartner *et al.*, 2004; Higuchi *et al.*, 2000; Wang *et al.*, 2000a, 2004). In each of these mouse models, a portion of the ADAR gene that contained critical portions of the catalytic deaminase domain necessary for enzymatic function was removed. Any residual expression from the mutated genes would be predicted to render only truncated, catalytically inactive proteins.

7.1. ADAR1-null mice

ADAR1 was the first member of the ADAR gene family to be cloned (Kim *et al.*, 1994b; O'Connell *et al.*, 1995) and it contributes to, or selectively edits, many known substrates. For example, it is primarily responsible for modifying the A and B sites within pre-mRNA transcripts encoding the

5HT$_{2C}$R (Hartner *et al.*, 2004; Niswender *et al.*, 1998), editing of which has been implicated in a variety of affective disorders (Tohda *et al.*, 2006). To understand the biological importance of this enzyme, multiple lines of genetically modified mice have been developed in which the expression of ADAR1 has been selectively ablated. The first attempt to knock out ADAR1 expression generated the most extreme phenotype, characterized by embryonic lethality in chimeric animals (Wang *et al.*, 2000a). Because lethality in chimeras limited the analyses necessary to understand the full range of ADAR1 function, two groups simultaneously designed alternative targeting vectors that would allow for conditional knockout of the gene via Cre-mediated recombination (Hartner *et al.*, 2004; Wang *et al.*, 2004). Interestingly, even global ablation of ADAR1 expression using these subsequent targeting vectors allowed the generation of viable chimeric and heterozygous ADAR1-null mice, and thus more thorough physiological analyses.

The targeting vector for the first ADAR1-null allele was generated from a genomic fragment in which a portion of exon 11 through exon 13 was removed (Fig. 16.5; Δ11–13) and replaced with a PGK-neo cassette (Wang *et al.*, 2000a). The deleted gene portion corresponds to a region of the

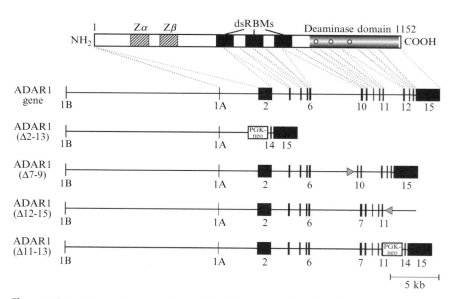

Figure 16.5 Targeted elimination of ADAR1 expression. The domain structure of mouse ADAR1 (p150) is presented, connected to the corresponding regions of the ADAR1 gene. The structure of a variety of ADAR1-null alleles is shown resulting from deletion of exons 2–13 (Δ2–13), 7–9 (Δ7–9), 12–15 (Δ12–15), and 11–13 (Δ11–13). The positions of the positive selectable marker (PGK-neo) and residual loxP sites (▶) after Cre-mediated recombination are indicated.

catalytic deaminase domain containing a critical glutamate moiety involved in proton transfer during catalysis (Lai *et al.*, 1995), and any translated protein would be prematurely terminated due to an alteration of the open reading frame. The two double-stranded RNA-binding motifs (dsRBMs; Fig. 16.5) are maintained in the targeted gene; therefore, it is possible that any residual protein product would maintain the ability to bind dsRNA targets, but not modify their sequence. Correctly targeted ES cell clones appeared normal, but no chimeric mice capable of transmitting the mutant allele were generated from blastocyst injections. Further analysis proved that the mutant allele was lethal during embryogenesis, with death occurring as early as embryonic day 9.5 (E9.5), depending on the relative contribution of wild-type and targeted ES cells to the chimeric embryo. Histologic analysis of embryos indicated defects of the hematopoietic system that prevent erythrogenesis in the liver. Prior to E12, erythrocytes are generated in the yolk sac, but this switches to the liver for the remainder of embryogenesis and into adulthood. The role that ADAR1 plays in this process was not determined, but it appeared that the functional enzyme is critical for normal development and viability (Wang *et al.*, 2000a).

Because the phenotype of the ADAR1 (Δ11–13) mice prevented analysis of additional physiological processes with ADAR1 involvement (Wang *et al.*, 2000a), additional targeting vectors were designed in which conditional null alleles were developed so that critical regions of the ADAR1 gene could be eliminated in a spatiotemporal fashion by Cre-mediated recombination (Hartner *et al.*, 2004; Wang *et al.*, 2004). If viable heterozygous or homozygous mice harboring the ADAR1-null allele could not be obtained, this would allow for tissue-specific or temporally regulated elimination of ADAR1 expression.

A second round of targeting vectors was designed in which a large portion of the ADAR1 locus, including exons 7–15 (Wang *et al.*, 2004), could be initially eliminated from the gene locus. A loxP-flanked *neo* cassette was inserted in intron 11 and an additional loxP site was engineered into intron 15. This design allowed removal of the *neo* cassette alone in ES cells, leaving one loxP site in its place, but without disruption of the coding sequence of the gene. Later, exons 11–15 could be deleted through recombination between the remaining two loxP sites (Fig. 16.5; Δ11–15). Mice homozygous for this loxP-flanked conditional allele were initially bred with a transgenic mouse strain that resulted in universal recombination through germline Cre expression under control of the adenovirus EIIa promoter (Lakso *et al.*, 1996). Because the previous knockout model could not produce viable chimeric animals, it was not expected that viable ADAR1-null heterozygotes would result from this cross. However, ADAR1$^{+/-}$ mice were born healthy in normal Mendelian ratios. Intercrossing of these to produce full knockouts was unsuccessful, generating only wild-type and heterozygous offspring due to embryonic lethality in ADAR1$^{-/-}$ mice (Wang *et al.*, 2004).

Comparable results were seen in mutant mice developed using two different targeting vectors designed by another group of investigators (Hartner *et al.*, 2004). A similar strategy was employed for the first, with a loxP-flanked *neo* cassette in intron 6 and an additional loxP site in intron 9. Full recombination of this locus would result in the deletion of exons 7 through 9 (Fig. 16.5; ADAR1 Δ7–9), comprising the end of the third dsRBM to the N-terminal portion of the deaminase domain. Mice heterozygous for the loxP-containing ADAR1 (Δ7–9) allele were crossed to CMV-Cre transgenic animals (Schwenk *et al.*, 1995) and viable offspring harboring an allele with deletion of the entire loxP-flanked region were obtained. Interbreeding of ADAR1 (Δ7–9) heterozygotes proved that this mutation results in embryonic lethality in homozygous null mice (Hartner *et al.*, 2004). Because this was a remarkably different outcome than the embryonic lethality in chimeric animals (ADAR1 Δ11–13) (Wang *et al.*, 2000a), a second targeting vector was created with the majority of the gene, exons 2–13, being replaced by PGK-neo (Fig. 16.5; ADAR1 Δ2–13). The phenotype of mice expressing this allele was indistinguishable from those with the shorter deletion (Δ7–9), indicating that the embryonic lethality in only homozygous null animals was common to multiple configurations, and not likely due to milder functional interference or an artifact resulting from residual expression of a truncated protein (Hartner *et al.*, 2004).

To summarize the phenotypes found for the three later strains of ADAR1-null animals (Hartner *et al.*, 2004; Wang *et al.*, 2004), the embryonic lethality between E11 and E12.5 was linked to defects in liver development. Live embryos recovered prior to death were small and pale in appearance, and had a significant reduction in liver size. Histological analysis showed livers had fewer hematopoietic progenitors (Hartner *et al.*, 2004) and a lower density of hepatocytes, with many possessing markers of apoptosis (Hartner *et al.*, 2004; Wang *et al.*, 2004). Other tissues, including the heart and vertebrae, also appeared to be apoptotic (Wang *et al.*, 2004). ADAR1$^{-/-}$ ES cells injected into blastocysts did not contribute to the formation of the liver or any hematopoietic tissue of resulting chimeric animals (Hartner *et al.*, 2004), demonstrating the importance of ADAR1 in normal development with regard to the liver and hematopoietic system.

ADAR1 liver function was explored more extensively through the breeding of mice homozygous for a conditional allele with loxP sites flanking exons 12–15 (Fig. 16.5; ADAR1 Δ12–15) to a transgenic mouse strain expressing Cre recombinase under the control of a liver-specific promoter (albumin-Cre). The resulting offspring demonstrated a 40–60% removal of the loxP-flanked region in the postnatal liver and were viable, but were reduced in size. The livers in these animals had disorganized architecture with abnormal hepatocytes and displayed markers of apoptosis (Wang *et al.*, 2004).

It is unclear why the chimeric lethality observed for the first ADAR1-null animals (Δ11–13) was different from those ADAR1-null animals developed later. The region of the deletion and the inclusion of the *neo* gene are not likely to be responsible, since the subsequent targeting vectors deleted overlapping regions and either removed, or included, the neo cassette in the ADAR1-null locus. Most likely, a second mutation was introduced in the ADAR1 (Δ11–13) animals that was linked to the ADAR1 locus by proximity, such that it could not effectively segregate from the null mutation. It was fortunate that the initial results were further investigated to provide more accurate data regarding the functional role(s) of ADAR1 activity *in vivo*.

7.2. ADAR2-null mice

The second member of the ADAR gene family, ADAR2, is widely expressed in mammalian tissues and has been implicated in the editing of multiple substrates (Melcher *et al.*, 1996b). Along with its contribution to the editing of sites in glutamate-gated ion channels, it is exclusively responsible for the editing of the Q/R site of GluR-2 RNAs (Melcher *et al.*, 1996b), preventing the onset of postnatal seizures and death (Brusa *et al.*, 1995; Feldmeyer *et al.*, 1999). It is also the sole enzyme that edits the C and D sites of the $5HT_{2C}R$ (Hartner *et al.*, 2004; Niswender *et al.*, 1998), the potassium channel Kv1.1 (Bhalla *et al.*, 2004), as well as a site in its own transcript involved in splicing regulation (Rueter *et al.*, 1999).

To understand the full scope of ADAR2 function in the mammalian system, ADAR2-null animals were generated using a replacement-type targeting vector in which portions of intron 6 and exon 7 were replaced by a PGK-neo cassette (Fig. 16.6). This alteration deletes a portion of the deaminase domain containing the first zinc-coordination residue, an essential glutamate moiety required for catalysis (E^{396}), and introduces a frameshift for translation (Higuchi *et al.*, 2000). ADAR2-null mice were successfully generated with a normal Mendelian ratio, indicating that the absence of ADAR2 does not lead to embryonic lethality. While mice heterozygous for the ADAR2-null allele never developed an overt phenotype, homozygous null mice died of seizures before P20. This phenotype was nearly identical to that seen in mice deficient in GluR-2 editing (Q/R site), suggesting that the ability of ADAR2 to edit the GluR-2 Q/R site supercedes its other physiological roles. The extent of editing for many sites was quantified in wild-type, heterozygous, and homozygous ADAR2-null mice, verifying the contribution of ADAR2 to *in vivo* editing patterns for numerous RNA substrates (Higuchi *et al.*, 2000).

8. ENZYME AND SUBSTRATE MODIFIED MICE

Two mouse models have been generated in which both an editing enzyme and a single ADAR substrate have been modified, allowing investigators to address two distinct issues regarding the molecular mechanism

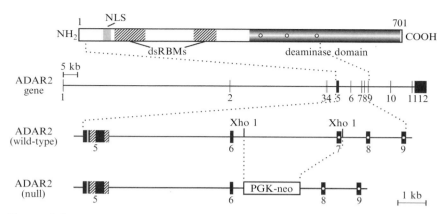

Figure 16.6 Targeting strategy for the selective ablation of ADAR2 expression in mice. The domain structure for mouse ADAR2 is presented along with the structure of the mouse ADAR2 gene. A region of the ADAR2 gene extending from exons 5 to 9, encoding a majority of the protein through the third zinc-coordination residue, was used in a targeting vector designed to eliminate ADAR2 expression by insertion of a positive selectable marker (PGK-neo) at the indicated *Xho*I restriction sites.

underlying postnatal lethality in ADAR2-null mice (Higuchi *et al.*, 2000) and whether the ability of ADAR2 to edit its own pre-mRNA represents an autoregulatory homeostatic mechanism to modulate ADAR2 activity (Feng *et al.*, 2006).

8.1. Rescue of ADAR2-null lethality with an edited GluR-2^R allele

The phenotype of ADAR2-null animals mirrored mice lacking Q/R site editing of GluR-2 transcripts so closely that it was hypothesized that Q/R site editing may be the most critical function of this enzyme (Brusa *et al.*, 1995; Higuchi *et al.*, 2000). In an attempt to rescue ADAR2-null mice from postnatal seizures and death, and to elucidate the role of ADAR2 with respect to additional RNA substrates in adult animals, an edited GluR-2^R allele was introduced into the ADAR2-null background (Higuchi *et al.*, 2000). ADAR2-null animals bearing a single copy of the GluR-2^R allele were partially rescued, as postnatal lethality was delayed until P35, whereas ADAR2-null animals containing two copies of the edited GluR-2^R allele appeared phenotypically normal in all respects including development, size, feeding, breeding, behavior, and life span. Because the previously observed decrease in editing for all other ADAR2 substrates was maintained, these results demonstrated that the postnatal lethality associated with ADAR2 ablation results from a loss of GluR-2 Q/R site editing and may represent the most crucial function of ADAR2 in mammals (Higuchi *et al.*, 2000).

8.2. Mice lacking ADAR2 autoregulation

ADAR2 has been found to edit its own transcript at an intronic site creating an adenosine-inosine (AI) dinucleotide that serves as a functional 3′-splice acceptor (Rueter *et al.*, 1999). Use of this proximal 3′-splice site leads to the inclusion of an additional 47 nucleotides in the ADAR2 coding region, resulting in premature translation termination in an altered reading frame to produce a 9-kDa protein lacking all functional domains required for A-to-I conversion. Editing at this location (−1 site) suggested an autoregulatory mechanism by which ADAR2 can reduce its own expression if it becomes too abundant in the cell (Rueter *et al.*, 1999), thereby maintaining appropriate editing levels for other important ADAR2 targets.

To test the hypothesis that by serving as its own substrate ADAR2 can modulate its own activity, a mutant mouse strain was created in which the evolutionarily conserved duplex structure in ADAR2 (Fig. 16.4, *right*) formed by base-pairing interactions between sequences surrounding the −1 site and an upstream ECS region in intron 4 was disrupted by the introduction of loxP sites flanking the ECS region and an adjacent PGK-neo cassette with a third loxP site (Fig. 16.7) (Feng *et al.*, 2006). This targeting vector design provides

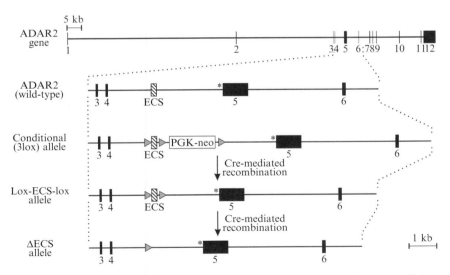

Figure 16.7 Targeting strategy for the selective ablation of ADAR2 autoediting. A schematic diagram of the mouse ADAR2 gene is presented with an expansion of the wild-type locus encompassed by the targeting vector, along with the structures of the modified alleles before and after Cre-mediated recombination. The positions of the positive selectable marker (PGK-neo), the editing-site complementary sequence (ECS), and the alternative 3′-splice site generated by ADAR2 autoediting (*) are indicated. The loxP recombination sites (▶) for selective deletion of the PGK-neo cassette and the ECS region are also shown.

the potential for generating mice with conditional ablation of ADAR2 auto-editing. Mice harboring the loxP-flanked allele (3lox) were successfully generated and bred to transgenic mice expressing EIIa-Cre (Lakso *et al.*, 1996). Offspring were analyzed for recombination to exclude either the PGK-neo cassette alone (lox-ECS-lox) or the PGK-neo cassette and the ECS region (ΔECS; Fig. 16.7). Heterozygous mice bearing the ΔECS allele were interbred, producing healthy, homozygous ΔECS offspring with a normal Mendelian frequency and lacking any overt phenotype (Feng *et al.*, 2006).

More detailed analyses of ΔECS mice confirmed that deletion of the ECS region in mice efficiently eliminated editing at the -1 site and inclusion of the 47-nt cassette in mature ADAR2 transcripts, further confirming that the ECS element is essential for -1 site editing and subsequent alternative splicing *in vivo*. Ablation of -1 site editing promoted the use of the distal $3'$-splice site immediately preceding exon 5 to allow exclusion of the 47-nt cassette and to increase the relative expression of the ADAR2 mRNA isoform that encodes the full-length, functional protein. Accordingly, increased ADAR2 protein expression in multiple tissues, and a concomitant increase in the editing of ADAR2 substrates, was observed with ΔECS mice, strongly supporting the hypothesis that ADAR2 auto-editing serves as a negative-feedback mechanism for modulating ADAR2 protein expression (Feng *et al.*, 2006). A comparison of the relationship between ADAR2 protein expression in ΔECS mice and proximal $3'$-splice site use in wild-type tissues revealed a strong correlation between predicted and observed values, confirming that ADAR2 autoediting is a key regulator for modulating ADAR2 protein expression *in vivo*. Despite the fact that these studies have identified a key mechanism for controlling ADAR2 activity in cells, the fact that ΔECS mice lack any overt phenotype is somewhat puzzling, since ADAR2 autoediting and alternative splicing are conserved in all vertebrates examined (Feng *et al.*, 2006). The absence of such overt phenotypic changes could indicate that any changes are subtle or that their absence results from adaptive alterations that could serve to circumvent the adverse phenotypic consequences resulting from increased ADAR2 activity.

ACKNOWLEDGMENTS

This work was supported by U.S. Public Health Service Grants NS33323 and NS35891 to R.B.E.

REFERENCES

Athanasiadis, A., Rich, A., and Maas, S. (2004). Widespread A-to-I RNA editing of Alu-containing mRNAs in the human transcriptome. *PLoS Biol.* **2**, e391.

Auerbach, W., Dunmore, J. H., Fairchild-Huntress, V., Fang, Q., Auerbach, A. B., Huszar, D., and Joyner, A. L. (2000). Establishment and chimera analysis of 129/SvEv- and C57BL/6-derived mouse embryonic stem cell lines. *Biotechniques* **29,** 1024–1028, 1030, 1032.

Baharvand, H., and Matthaei, K. I. (2004). Culture condition difference for establishment of new embryonic stem cell lines from the C57BL/6 and BALB/c mouse strains. *In Vitro Cell. Dev. Biol. Anim.* **40,** 76–81.

Bass, B. L. (2002). RNA editing by adenosine deaminases that act on RNA. *Annu. Rev. Biochem.* **71,** 817–846.

Bernard, A., and Khrestchatisky, M. (1994). Assessing the extent of RNA editing in the TMII regions of GluR5 and GluR6 kainate receptors during rat brain development. *J. Neurochem.* **62,** 2057–2060.

Bernard, A., Ferhat, L., Dessi, F., Charton, G., Represa, A., Ben-Ari, Y., and Khrestchatisky, M. (1999). Q/R editing of the rat GluR5 and GluR6 kainate receptors *in vivo* and *in vitro*: Evidence for independent developmental, pathological and cellular regulation. *Eur. J. Neurosci.* **11,** 604–616.

Bhalla, T., Rosenthal, J. J., Holmgren, M., and Reenan, R. (2004). Control of human potassium channel inactivation by editing of a small mRNA hairpin. *Nat. Struct. Mol. Biol.* **11,** 950–956.

Blow, M., Futreal, P. A., Wooster, R., and Stratton, M. R. (2004). A survey of RNA editing in human brain. *Genome Res.* **14,** 2379–2387.

Borrelli, E., Heyman, R., Hsi, M., and Evans, R. M. (1988). Targeting of an inducible toxic phenotype in animal cells. *Proc. Natl. Acad. Sci. USA* **85,** 7572–7576.

Brocard, J., Feil, R., Chambon, P., and Metzger, D. (1998). A chimeric Cre recombinase inducible by synthetic, but not by natural ligands of the glucocorticoid receptor. *Nucleic Acids Res.* **26,** 4086–4090.

Brusa, R., Zimmermann, F., Koh, D. S., Feldmeyer, D., Gass, P., Seeburg, P. H., and Sprengel, R. (1995). Early-onset epilepsy and postnatal lethality associated with an editing-deficient GluR-B allele in mice. *Science* **270,** 1677–1680.

Burd, C. G., and Dreyfuss, G. (1994). Conserved structures and diversity of functions of RNA-binding proteins. *Science* **265,** 615–621.

Burns, C. M., Chu, H., Rueter, S. M., Hutchinson, L. K., Canton, H., Sanders-Bush, E., and Emeson, R. B. (1997). Regulation of serotonin-2C receptor G-protein coupling by RNA editing. *Nature* **387,** 303–308.

Chen, C. X., Cho, D. S., Wang, Q., Lai, F., Carter, K. C., and Nishikura, K. (2000). A third member of the RNA-specific adenosine deaminase gene family, ADAR3, contains both single- and double-stranded RNA binding domains. *RNA* **6,** 755–767.

Cheng, J., Dutra, A., Takesono, A., Garrett-Beal, L., and Schwartzberg, P. L. (2004). Improved generation of C57BL/6J mouse embryonic stem cells in a defined serum-free media. *Genesis* **39,** 100–104.

Cho, D. S., Yang, W., Lee, J. T., Shiekhattar, R., Murray, J. M., and Nishikura, K. (2003). Requirement of dimerization for RNA editing activity of adenosine deaminases acting on RNA. *J. Biol. Chem.* **278,** 17093–17102.

Collins, S. C., Wallis, R. H., Wallace, K., Bihoreau, M. T., and Gauguier, D. (2003). Marker-assisted congenic screening (MACS): A database tool for the efficient production and characterization of congenic lines. *Mamm. Genome* **14,** 350–356.

Danielian, P. S., Muccino, D., Rowitch, D. H., Michael, S. K., and McMahon, A. P. (1998). Modification of gene activity in mouse embryos *in utero* by a tamoxifen-inducible form of Cre recombinase. *Curr. Biol.* **8,** 1323–1326.

Dawson, T. R., Sansam, C. L., and Emeson, R. B. (2004). Structure and sequence determinants required for the RNA editing of ADAR2 substrates. *J. Biol. Chem.* **279,** 4941–4951.

Deng, C., and Capecchi, M. R. (1992). Reexamination of gene targeting frequency as a function of the extent of homology between the targeting vector and the target locus. *Mol. Cell. Biol.* **12,** 3365–3371.

Dingledine, R., Hume, R. I., and Heinemann, S. F. (1992). Structural determinants of barium permeation and rectification in non-NMDA glutamate receptor channels. *J. Neurosci.* **12,** 4080–4087.

Dubovsky, S. L., and Thomas, M. (1995). Beyond specificity: Effects of serotonin and serotonergic treatments on psychobiological dysfunction. *J. Psychosom. Res.* **39,** 429–444.

Eckmann, C. R., Neunteufl, A., Pfaffstetter, L., and Jantsch, M. F. (2001). The human but not the Xenopus RNA-editing enzyme ADAR1 has an atypical nuclear localization signal and displays the characteristics of a shuttling protein. *Mol. Biol. Cell* **12,** 1911–1924.

Egebjerg, J., and Heinemann, S. F. (1993). Ca^{2+} permeability of unedited and edited versions of the kainate selective glutamate receptor GluR6. *Proc. Natl. Acad. Sci. USA* **90,** 755–759.

Egebjerg, J., Kukekov, V., and Heinemann, S. F. (1994). Intron sequence directs RNA editing of the glutamate receptor subunit GluR2 coding sequence. *Proc. Natl. Acad. Sci. USA* **91,** 10270–10274.

Emeson, R., and Singh, M. (2001). Adenosine-to-inosine RNA editing: Substrates and consequences. *In* "RNA Editing" (B. Bass, ed.), pp. 109–138. Oxford University Press, Oxford, UK.

Feil, R., Brocard, J., Mascrez, B., LeMeur, M., Metzger, D., and Chambon, P. (1996). Ligand-activated site-specific recombination in mice. *Proc. Natl. Acad. Sci. USA* **93,** 10887–10890.

Feldmeyer, D., Kask, K., Brusa, R., Kornau, H. C., Kolhekar, R., Rozov, A., Burnashev, N., Jensen, V., Hvalby, O., Sprengel, R., and Seeburg, P. H. (1999). Neurological dysfunctions in mice expressing different levels of the Q/R site-unedited AMPAR subunit GluR-B. *Nat. Neurosci.* **2,** 57–64.

Feng, Y., Sansam, C. L., Singh, M., and Emeson, R. B. (2006). Altered RNA editing in mice lacking ADAR2 autoregulation. *Mol. Cell. Biol.* **26,** 480–488.

George, C. X., and Samuel, C. E. (1999). Human RNA-specific adenosine deaminase ADAR1 transcripts possess alternative exon 1 structures that initiate from different promoters, one constitutively active and the other interferon inducible. *Proc. Natl. Acad. Sci. USA* **96,** 4621–4626.

Gerber, A., O'Connell, M. A., and Keller, W. (1997). Two forms of human double-stranded RNA-specific editase 1 (hRED1) generated by the insertion of an Alu cassette. *RNA* **3,** 453–463.

Gerber, A., Grosjean, H., Melcher, T., and Keller, W. (1998). Tad1p, a yeast tRNA-specific adenosine deaminase, is related to the mammalian pre-mRNA editing enzymes ADAR1 and ADAR2. *EMBO J.* **17,** 4780–4789.

Gerlai, R. (2000). Targeting genes and proteins in the analysis of learning and memory: Caveats and future directions. *Rev. Neurosci.* **11,** 15–26.

Giese, K. P., Storm, J. F., Reuter, D., Fedorov, N. B., Shao, L. R., Leicher, T., Pongs, O., and Silva, A. J. (1998). Reduced K^+ channel inactivation, spike broadening, and after-hyperpolarization in Kvbeta1.1-deficient mice with impaired learning. *Learn. Mem.* **5,** 257–273.

Gossler, A., Doetschman, T., Korn, R., Serfling, E., and Kemler, R. (1986). Transgenesis by means of blastocyst-derived embryonic stem cell lines. *Proc. Natl. Acad. Sci. USA* **83,** 9065–9069.

Gott, J. M., and Emeson, R. B. (2000). Functions and mechanisms of RNA editing. *Annu. Rev. Genet.* **34,** 499–531.

Grigorenko, E. V., Bell, W. L., Glazier, S., Pons, T., and Deadwyler, S. (1998). Editing status at the Q/R site of the GluR2 and GluR6 glutamate receptor subunits in the surgically excised hippocampus of patients with refractory epilepsy. *Neuroreport* **9**, 2219–2224.

Hartner, J. C., Schmittwolf, C., Kispert, A., Muller, A. M., Higuchi, M., and Seeburg, P. H. (2004). Liver disintegration in the mouse embryo caused by deficiency in the RNA-editing enzyme ADAR1. *J. Biol. Chem.* **279**, 4894–4902.

Hasty, P., Rivera-Perez, J., and Bradley, A. (1991). The length of homology required for gene targeting in embryonic stem cells. *Mol. Cell. Biol.* **11**, 5586–5591.

Herbert, A. (1996). RNA editing, introns and evolution. *Trends Genet.* **12**, 6–9.

Herson, P. S., Virk, M., Rustay, N. R., Bond, C. T., Crabbe, J. C., Adelman, J. P., and Maylie, J. (2003). A mouse model of episodic ataxia type-1. *Nat. Neurosci.* **6**, 378–383.

Higuchi, M., Single, F. N., Kohler, M., Sommer, B., Sprengel, R., and Seeburg, P. H. (1993). RNA editing of AMPA receptor subunit GluR-B: A base-paired intron-exon structure determines position and efficiency. *Cell* **75**, 1361–1370.

Higuchi, M., Maas, S., Single, F. N., Hartner, J., Rozov, A., Burnashev, N., Feldmeyer, D., Sprengel, R., and Seeburg, P. H. (2000). Point mutation in an AMPA receptor gene rescues lethality in mice deficient in the RNA-editing enzyme ADAR2. *Nature* **406**, 78–81.

Hoess, R. H., and Abremski, K. (1984). Interaction of the bacteriophage P1 recombinase Cre with the recombining site loxP. *Proc. Natl. Acad. Sci. USA* **81**, 1026–1029.

Hollmann, M., Hartley, M., and Heinemann, S. (1991). Ca^{2+} permeability of KA-AMPA–gated glutamate receptor channels depends on subunit composition. *Science* **252**, 851–853.

Hoopengardner, B., Bhalla, T., Staber, C., and Reenan, R. (2003). Nervous system targets of RNA editing identified by comparative genomics. *Science* **301**, 832–836.

Hough, R. F., and Bass, B. L. (1994). Purification of the Xenopus laevis double-stranded RNA adenosine deaminase. *J. Biol. Chem.* **269**, 9933–9939.

Hough, R. F., and Bass, B. L. (1997). Analysis of Xenopus dsRNA adenosine deaminase cDNAs reveals similarities to DNA methyltransferases. *RNA* **3**, 356–370.

Hough, R., and Bass, B. (2001). Adenosine deaminases that act on RNA. In "Frontiers in Molecular Biology" (B. Bass, ed.), pp. 77–108. Oxford University Press, Oxford, UK.

Kask, K., Zamanillo, D., Rozov, A., Burnashev, N., Sprengel, R., and Seeburg, P. H. (1998). The AMPA receptor subunit GluR-B in its Q/R site-unedited form is not essential for brain development and function. *Proc. Natl. Acad. Sci. USA* **95**, 13777–13782.

Kellendonk, C., Tronche, F., Monaghan, A. P., Angrand, P. O., Stewart, F., and Schutz, G. (1996). Regulation of Cre recombinase activity by the synthetic steroid RU 486. *Nucleic Acids Res.* **24**, 1404–1411.

Kikuno, R., Nagase, T., Waki, M., and Ohara, O. (2002). HUGE: A database for human large proteins identified in the Kazusa cDNA sequencing project. *Nucleic Acids Res.* **30**, 166–168.

Kim, D. D., Kim, T. T., Walsh, T., Kobayashi, Y., Matise, T. C., Buyske, S., and Gabriel, A. (2004). Widespread RNA editing of embedded alu elements in the human transcriptome. *Genome Res.* **14**, 1719–1725.

Kim, U., Garner, T. L., Sanford, T., Speicher, D., Murray, J. M., and Nishikura, K. (1994a). Purification and characterization of double-stranded RNA adenosine deaminase from bovine nuclear extracts. *J. Biol. Chem.* **269**, 13480–13489.

Kim, U., Wang, Y., Sanford, T., Zeng, Y., and Nishikura, K. (1994b). Molecular cloning of cDNA for double-stranded RNA adenosine deaminase, a candidate enzyme for nuclear RNA editing. *Proc. Natl. Acad. Sci. USA* **91**, 11457–11461.

Kohler, M., Burnashev, N., Sakmann, B., and Seeburg, P. H. (1993). Determinants of Ca^{2+} permeability in both TM1 and TM2 of high affinity kainate receptor channels: Diversity by RNA editing. *Neuron* **10**, 491–500.

Koller, B. H., and Smithies, O. (1992). Altering genes in animals by gene targeting. *Annu. Rev. Immunol.* **10**, 705–730.

Lai, F., Drakas, R., and Nishikura, K. (1995). Mutagenic analysis of double-stranded RNA adenosine deaminase, a candidate enzyme for RNA editing of glutamate-gated ion channel transcripts. *J. Biol. Chem.* **270**, 17098–17105.

Lai, F., Chen, C. X., Carter, K. C., and Nishikura, K. (1997). Editing of glutamate receptor B subunit ion channel RNAs by four alternatively spliced DRADA2 double-stranded RNA adenosine deaminases. *Mol. Cell. Biol.* **17**, 2413–2424.

Lakso, M., Pichel, J. G., Gorman, J. R., Sauer, B., Okamoto, Y., Lee, E., Alt, F. W., and Westphal, H. (1996). Efficient *in vivo* manipulation of mouse genomic sequences at the zygote stage. *Proc. Natl. Acad. Sci. USA* **93**, 5860–5865.

Lehmann, K. A., and Bass, B. L. (1999). The importance of internal loops within RNA substrates of ADAR1. *J. Mol. Biol.* **291**, 1–13.

Lehmann, K. A., and Bass, B. L. (2000). Double-stranded RNA adenosine deaminases ADAR1 and ADAR2 have overlapping specificities. *Biochemistry* **39**, 12875–12884.

Levanon, E. Y., Eisenberg, E., Yelin, R., Nemzer, S., Hallegger, M., Shemesh, R., Fligelman, Z. Y., Shoshan, A., Pollock, S. R., Sztybel, D., Olshansky, M., Rechavi, G., and Jantsch, M. F. (2004). Systematic identification of abundant A-to-I editing sites in the human transcriptome. *Nat. Biotechnol.* **22**, 1001–1005.

Liu, Y., George, C. X., Patterson, J. B., and Samuel, C. E. (1997). Functionally distinct double-stranded RNA-binding domains associated with alternative splice site variants of the interferon-inducible double-stranded RNA-specific adenosine deaminase. *J. Biol. Chem.* **272**, 4419–4428.

Lomeli, H., Mosbacher, J., Melcher, T., Hoger, T., Geiger, J. R., Kuner, T., Monyer, H., Higuchi, M., Bach, A., and Seeburg, P. H. (1994). Control of kinetic properties of AMPA receptor channels by nuclear RNA editing. *Science* **266**, 1709–1713.

Maas, S., Melcher, T., Herb, A., Seeburg, P. H., Keller, W., Krause, S., Higuchi, M., and O'Connell, M. A. (1996). Structural requirements for RNA editing in glutamate receptor pre-mRNAs by recombinant double-stranded RNA adenosine deaminase. *J. Biol. Chem.* **271**, 12221–12226.

Madden, D. R. (2002). The inner workings of the AMPA receptors. *Curr. Opin. Drug Discov. Dev.* **5**, 741–748.

Melcher, T., Maas, S., Higuchi, M., Keller, W., and Seeburg, P. H. (1995). Editing of alpha-amino- 3-hydroxy-5-methylisoxazole-4-propionic acid receptor GluR-B pre-mRNA *in vitro* reveals site-selective adenosine to inosine conversion. *J. Biol. Chem.* **270**, 8566–8570.

Melcher, T., Maas, S., Herb, A., Sprengel, R., Higuchi, M., and Seeburg, P. H. (1996a). RED2, a brain-specific member of the RNA-specific adenosine deaminase family. *J. Biol. Chem.* **271**, 31795–31798.

Melcher, T., Maas, S., Herb, A., Sprengel, R., Seeburg, P. H., and Higuchi, M. (1996b). A ssmammalian RNA editing enzyme. *Nature* **379**, 460–464.

Misra, R. P., and Duncan, S. A. (2002). Gene targeting in the mouse: Advances in introduction of transgenes into the genome by homologous recombination. *Endocrine* **19**, 229–238.

Morse, D. P., and Bass, B. L. (1999). Long RNA hairpins that contain inosine are present in *Caenorhabditis elegans* poly(A)+ RNA. *Proc. Natl. Acad. Sci. USA* **96**, 6048–6053.

Morse, D. P., Aruscavage, P. J., and Bass, B. L. (2002). RNA hairpins in noncoding regions of human brain and *Caenorhabditis elegans* mRNA are edited by adenosine deaminases that act on RNA. *Proc. Natl. Acad. Sci. USA* **99**, 7906–7911.

Nagy, A., Gertsenstein, M., Vintersten, K., and Behringer, R. (2003). "Manipulating the Mouse Embryo: A Laboratory Manual." Cold Spring Laboratory Press, Woodbury, NY.

Niswender, C. M., Sanders-Bush, E., and Emeson, R. B. (1998). Identification and characterization of RNA editing events within the 5-HT2C receptor. *Ann. N.Y. Acad. Sci.* **861,** 38–48.

Niswender, C. M., Copeland, S. C., Herrick-Davis, K., Emeson, R. B., and Sanders-Bush, E. (1999). RNA editing of the human serotonin 5-hydroxytryptamine 2C receptor silences constitutive activity. *J. Biol. Chem.* **274,** 9472–9478.

O'Connell, M. A., and Keller, W. (1994). Purification and properties of double-stranded RNA-specific adenosine deaminase from calf thymus. *Proc. Natl. Acad. Sci. USA* **91,** 10596–10600.

O'Connell, M. A., Krause, S., Higuchi, M., Hsuan, J. J., Totty, N. F., Jenny, A., and Keller, W. (1995). Cloning of cDNAs encoding mammalian double-stranded RNA-specific adenosine deaminase. *Mol. Cell. Biol.* **15,** 1389–1397.

O'Gorman, S., Dagenais, N. A., Qian, M., and Marchuk, Y. (1997). Protamine-Cre recombinase transgenes efficiently recombine target sequences in the male germ line of mice, but not in embryonic stem cells. *Proc. Natl. Acad. Sci. USA* **94,** 14602–14607.

O'Grady, S. M., and Lee, S. Y. (2005). Molecular diversity and function of voltage-gated (Kv) potassium channels in epithelial cells. *Int. J. Biochem. Cell Biol.* **37,** 1578–1594.

Ozawa, S., Kamiya, H., and Tsuzuki, K. (1998). Glutamate receptors in the mammalian central nervous system. *Prog. Neurobiol.* **54,** 581–618.

Pandey, S. C., Davis, J. M., and Pandey, G. N. (1995). Phosphoinositide system-linked serotonin receptor subtypes and their pharmacological properties and clinical correlates. *J. Psychiatry Neurosci.* **20,** 215–225.

Pham, C. T., MacIvor, D. M., Hug, B. A., Heusel, J. W., and Ley, T. J. (1996). Long-range disruption of gene expression by a selectable marker cassette. *Proc. Natl. Acad. Sci. USA* **93,** 13090–13095.

Pin, J. P., Joly, C., Heinemann, S. F., and Bockaert, J. (1994). Domains involved in the specificity of G protein activation in phospholipase C-coupled metabotropic glutamate receptors. *EMBO J.* **13,** 342–348.

Polson, A. G., and Bass, B. L. (1994). Preferential selection of adenosines for modification by double-stranded RNA adenosine deaminase. *EMBO J.* **13,** 5701–5711.

Polson, A. G., Crain, P. F., Pomerantz, S. C., McCloskey, J. A., and Bass, B. L. (1991). The mechanism of adenosine to inosine conversion by the double-stranded RNA unwinding/modifying activity: A high-performance liquid chromatography-mass spectrometry analysis. *Biochemistry* **30,** 11507–11514.

Robertson, E. J. (1991). Using embryonic stem cells to introduce mutations into the mouse germ line. *Biol. Reprod.* **44,** 238–245.

Rubsam, L. Z., Davidson, B. L., and Shewach, D. S. (1998). Superior cytotoxicity with ganciclovir compared with acyclovir and 1-beta-D-arabinofuranosylthymine in herpes simplex virus-thymidine kinase-expressing cells: A novel paradigm for cell killing. *Cancer Res.* **58,** 3873–3882.

Rueter, S., and Emeson, R. (1998). Adenosine-to-inosine conversion in mRNA. *In* "Modification and Editing of RNA" (H. Grosjean and R. Benne, eds.), pp. 343–361. ASM Press, Washington, DC.

Rueter, S. M., Burns, C. M., Coode, S. A., Mookherjee, P., and Emeson, R. B. (1995). Glutamate receptor RNA editing *in vitro* by enzymatic conversion of adenosine to inosine. *Science* **267,** 1491–1494.

Rueter, S. M., Dawson, T. R., and Emeson, R. B. (1999). Regulation of alternative splicing by RNA editing. *Nature* **399,** 75–80.

Sailer, A., Swanson, G. T., Perez-Otano, I., O'Leary, L., Malkmus, S. A., Dyck, R. H., Dickinson-Anson, H., Schiffer, H. H., Maron, C., Yaksh, T. L., Gage, F. H., O'Gorman, S., and Heinemann, S. F. (1999). Generation and analysis of GluR5 (Q636R) kainate receptor mutant mice. *J. Neurosci.* **19,** 8757–8764.

Sauer, B., and Henderson, N. (1988). Site-specific DNA recombination in mammalian cells by the Cre recombinase of bacteriophage P1. *Proc. Natl. Acad. Sci. USA* **85,** 5166–5170.

Schaft, J., Ashery-Padan, R., van der Hoeven, F., Gruss, P., and Stewart, A. F. (2001). Efficient FLP recombination in mouse ES cells and oocytes. *Genesis* **31,** 6–10.

Schaub, M., and Keller, W. (2002). RNA editing by adenosine deaminases generates RNA and protein diversity. *Biochimie* **84,** 791–803.

Schwenk, F., Baron, U., and Rajewsky, K. (1995). A cre-transgenic mouse strain for the ubiquitous deletion of loxP-flanked gene segments including deletion in germ cells. *Nucleic Acids Res.* **23,** 5080–5081.

Seong, E., Saunders, T. L., Stewart, C. L., and Burmeister, M. (2004). To knockout in 129 or in C57BL/6: That is the question. *Trends Genet.* **20,** 59–62.

Shulman, M. J., Nissen, L., and Collins, C. (1990). Homologous recombination in hybridoma cells: Dependence on time and fragment length. *Mol. Cell. Biol.* **10,** 4466–4472.

Simpson, E. M., Linder, C. C., Sargent, E. E., Davisson, M. T., Mobraaten, L. E., and Sharp, J. J. (1997). Genetic variation among 129 substrains and its importance for targeted mutagenesis in mice. *Nat. Genet.* **16,** 19–27.

Sommer, B., Kohler, M., Sprengel, R., and Seeburg, P. H. (1991). RNA editing in brain controls a determinant of ion flow in glutamate-gated channels. *Cell* **67,** 11–19.

te Riele, H., Maandag, E. R., and Berns, A. (1992). Highly efficient gene targeting in embryonic stem cells through homologous recombination with isogenic DNA constructs. *Proc. Natl. Acad. Sci. USA* **89,** 5128–5132.

Thomas, K. R., and Capecchi, M. R. (1987). Site-directed mutagenesis by gene targeting in mouse embryo-derived stem cells. *Cell* **51,** 503–512.

Threadgill, D. W., Yee, D., Matin, A., Nadeau, J. H., and Magnuson, T. (1997). Genealogy of the 129 inbred strains: 129/SvJ is a contaminated inbred strain. *Mamm. Genome* **8,** 390–393.

Tohda, M., Nomura, M., and Nomura, Y. (2006). Molecular pathopharmacology of 5-HT2C receptors and the RNA editing in the brain. *J. Pharmacol. Sci.* **100,** 427–432.

Tucker, K. L., Wang, Y., Dausman, J., and Jaenisch, R. (1997). A transgenic mouse strain expressing four drug-selectable marker genes. *Nucleic Acids Res.* **25,** 3745–3746.

Van Oekelen, D., Luyten, W. H., and Leysen, J. E. (2003). 5-HT2A and 5-HT2C receptors and their atypical regulation properties. *Life Sci.* **72,** 2429–2449.

Verdoorn, T. A., Burnashev, N., Monyer, H., Seeburg, P. H., and Sakmann, B. (1991). Structural determinants of ion flow through recombinant glutamate receptor channels. *Science* **252,** 1715–1718.

Visscher, P. M. (1999). Speed congenics: Accelerated genome recovery using genetic markers. *Genet. Res.* **74,** 81–85.

Vissel, B., Royle, G. A., Christie, B. R., Schiffer, H. H., Ghetti, A., Tritto, T., Perez-Otano, I., Radcliffe, R. A., Seamans, J., Sejnowski, T., Wehner, J. M., Collins, A. C., et al. (2001). The role of RNA editing of kainate receptors in synaptic plasticity and seizures. *Neuron* **29,** 217–227.

Wagner, R. W., Yoo, C., Wrabetz, L., Kamholz, J., Buchhalter, J., Hassan, N. F., Khalili, K., Kim, S. U., Perussia, B., McMorris, F. A., and Nishikura, K. (1990). Double-stranded RNA unwinding and modifying activity is detected ubiquitously in primary tissues and cell lines. *Mol. Cell. Biol.* **10,** 5586–5590.

Wang, Q., Khillan, J., Gadue, P., and Nishikura, K. (2000a). Requirement of the RNA editing deaminase ADAR1 gene for embryonic erythropoiesis. *Science* **290,** 1765–1768.

Wang, Q., O'Brien, P. J., Chen, C. X., Cho, D. S., Murray, J. M., and Nishikura, K. (2000b). Altered G protein-coupling functions of RNA editing isoform and splicing variant serotonin2C receptors. *J. Neurochem.* **74,** 1290–1300.

Wang, Q., Miyakoda, M., Yang, W., Khillan, J., Stachura, D. L., Weiss, M. J., and Nishikura, K. (2004). Stress-induced apoptosis associated with null mutation of ADAR1 RNA editing deaminase gene. *J. Biol. Chem.* **279,** 4952–4961.

Wigler, M., Silverstein, S., Lee, L. S., Pellicer, A., Cheng, Y., and Axel, R. (1977). Transfer of purified herpes virus thymidine kinase gene to cultured mouse cells. *Cell* **11,** 223–232.

Wong, G. T. (2002). Speed congenics: Applications for transgenic and knock-out mouse strains. *Neuropeptides* **36,** 230–236.

Yanagawa, Y., Kobayashi, T., Ohnishi, M., Kobayashi, T., Tamura, S., Tsuzuki, T., Sanbo, M., Yagi, T., Tashiro, F., and Miyazaki, J. (1999). Enrichment and efficient screening of ES cells containing a targeted mutation: The use of DT-A gene with the polyadenylation signal as a negative selection maker. *Transgenic Res.* **8,** 215–221.

Yang, J. H., Sklar, P., Axel, R., and Maniatis, T. (1995). Editing of glutamate receptor subunit B pre-mRNA *in vitro* by site-specific deamination of adenosine. *Nature* **374,** 77–81.

Zhou, M., Morais-Cabral, J. H., Mann, S., and MacKinnon, R. (2001). Potassium channel receptor site for the inactivation gate and quaternary amine inhibitors. *Nature* **411,** 657–661.

CHAPTER SEVENTEEN

Probing Adenosine-to-Inosine Editing Reactions Using RNA-Containing Nucleoside Analogs

Olena Maydanovych, LaHoma M. Easterwood, Tao Cui,
Eduardo A. Véliz, Subhash Pokharel, *and* Peter A. Beal

Contents

Abstract

Advances in chemical synthesis and characterization of nucleic acids allows for atom-specific modification of complex RNAs, such as present in RNA editing substrates. By preparing substrates for ADARs by chemical synthesis, it is possible to subtly alter the structure of the edited nucleotide. Evaluating the effect these changes have on the rate of enzyme-catalyzed deamination reveals features of the editing reaction and guides the design of inhibitors. We describe the synthesis of select nucleoside analog phosphoramidites and their incorporation into RNAs that mimic known editing sites by solid phase synthesis, and

Department of Chemistry, University of Utah, Salt Lake City, Utah

Methods in Enzymology, Volume 424
ISSN 0076-6879, DOI: 10.1016/S0076-6879(07)24017-0

analyze the interaction of these synthetic RNAs with ADARs using deamination kinetics and quantitative gel mobility shift assays.

 ## 1. INTRODUCTION

The RNA editing adenosine deaminase acting on RNA (ADAR) enzymes convert adenosines to inosines in mRNAs, changing the coding properties of these messages (Burns et al., 1997; Grosjean and Benne, 1998; Higuchi et al., 1993). Thus, ADARs play a pivotal role in the basic process of information transfer that takes place during protein expression. Moreover, proteins translated from edited messages have been implicated in a number of neurodegenerative, psychiatric, and behavioral disorders such as stroke, epilepsy, Alzheimer's disease, schizophrenia, and episodic ataxia (Bhalla et al., 2004; D'Adamo et al., 1998; Gurevich et al., 2002; Iwamoto and Kato, 2003; Niswender et al., 2001; Pellegrini-Giampietro et al., 1997). Indeed, deletion of genes encoding the ADARs leads to significant behavioral defects in model organisms (Higuchi et al., 2000; Palladino et al., 2000; Tonkin et al., 2002; Wang et al., 2000). Our laboratory has carried out studies directed at defining the molecular basis for the fundamental steps in this editing reaction (Easterwood et al., 2000; Haudenschild et al., 2004; Maydanovych and Beal, 2006; Pokharel and Beal, 2006; Stephens et al., 2004, 2000; Veliz et al., 2001, 2003; Yi-Brunozzi et al., 1999, 2001). Much of what we have learned about the ADAR reaction mechanism has come from studies with ADAR2 and RNA substrates bearing nucleoside analogs. In this chapter, we describe the use of nucleoside analogs in the study of the ADAR-catalyzed editing reaction.

Advances in chemical synthesis of nucleic acids allows for "atom-specific mutagenesis" of complex RNAs (Das et al., 2005). These RNAs can be prepared solely from automated synthesis methods where the practical length limit is approximately 50 nucleotides or through the use of a combination of automated synthesis and template-directed ligation strategies (see below). By preparing ADAR substrates in this way, it is possible to alter the structure of the edited nucleotide in subtle ways. Evaluating the effect these changes have on the rate of enzyme-catalyzed deamination can reveal features of the editing reaction. In addition, the insight provided from these studies can inform the design of mechanism-based inhibitors. In our laboratory, this has been accomplished using $3'$-N,N-diisopropylamino-β-cyanoethyl phosphoramidites protected with $5'$-dimethoxytrityl and $2'$-tert-butyldimethylsilyl groups and RNA editing substrates prepared from these compounds. With this approach we have shown that $2'$-O-methylation at the editing site greatly reduces the deamination rate (Yi-Brunozzi et al., 1999). This suggested a possible regulatory mechanism whereby editing could be inhibited by snoRNA-guided $2'$-O-methylation, which was subsequently shown to be likely for the $5HT_{2C}R$ pre-mRNA "C" site (Vitali *et al.*, 2005). In addition, we demonstrated that 7-deaza modification of the reactive adenosine does not inhibit the ADAR2 reaction, whereas purine C2 amination does (Easterwood *et al.*, 2000; Veliz *et al.*, 2003). This is in stark contrast to the known structure/activity relationships for the reaction catalyzed by the nucleoside-modifying enzyme adenosine deaminase (ADA), indicating these two classes of adenosine deaminase recognize substrates differently (Baer *et al.*, 1966; Frederiksen, 1966). These studies showed that the ADA reaction was an imperfect model for the ADAR reaction and suggested that inhibitors might be prepared that are selective for ADARs and do not bind ADA. Another significant observation coming from analysis of substrate analogs was that 8-aza modification of the purine facilitates interaction with the ADAR2 active site (Haudenschild *et al.*, 2004; Veliz *et al.*, 2003). Since 8-aza modification was known to facilitate hydration of the purine ring, these results emphasized the importance of a substrate's susceptibility to covalent hydration in the ADAR reaction and guided the design of high-affinity ligands for ADAR2 (8-azanebularine containing RNA, see below). The study of RNA editing substrates prepared with phosphoramidites of other adenosine analogs has revealed additional features of the ADAR reaction (Stephens *et al.*, 2000; Yi-Brunozzi *et al.*, 2001).

2. Synthesis of a Substrate Analog Phosphoramidite

Phosphoramidites of several adenosine analogs are commercially available (e.g., $2'$-O-methyladenosine, 7-deazaadenosine, and $2'$-deoxyadenosine). Short synthetic routes can be used to prepare others. An example is shown in Scheme 17.1 for the phosphoramidite of 8-azaadenosine. 8-Azaadenine is glycosylated with tetraacetyl ribose using a procedure reported by Seela *et al.* (1998). Deacylation is followed by protection of the 8-azaadenine exocyclic amine. The resulting compound is protected at the $5'$ position with a dimethoxytrityl (DMT) ether and at the $2'$ position with *tert*-butyldimethysilyl (TBDMS) ether. Phosphoramidite formation at the $3'$ position is the last step in the process. A detailed protocol for the synthesis of this compound is provided below.

3. General Synthetic Procedures

All synthetic reagents were purchased from Sigma/Aldrich (St. Louis, MO) or Fischer Scientific (Pittsburgh, PA) and were used without further purification unless otherwise noted. Glassware for all reactions is oven dried at 125° overnight and cooled to room temperature in a desiccator prior to use.

Reactions are carried out under an atmosphere of dry nitrogen when anhydrous conditions are necessary. Liquid reagents are introduced by oven-dried microsyringes. Tetrahydrofuran (THF) is distilled from sodium metal and benzophenone; acetonitrile is distilled from CaH_2. Thin-layer chromatography (TLC) is performed with Merck silica gel 60 F_{254} precoated TLC plates, eluting with the solvents indicated. Short- and long-wave visualization is performed with a Mineralight multiband ultraviolet lamp at 254 and 365 nm, respectively. Flash column chromatography is performed on Mallinckrodt Baker silica gel 150 (60–200 mesh). ^1H, ^{13}C, and ^{31}P nuclear magnetic resonance spectra of pure compounds are acquired at 300, 75, and 121 MHz, respectively. All high-resolution fast-atom bombardment mass spectra (HRFABMS) are obtained on a Finnigan MAT 95 (Department of Chemistry, University of Utah).

3.1. Protocol

2′,3′,5′-Tri-O-acetyl-8-azaadenosine (**3**): 8-Azaadenine (**1**) (1.2 g, 8.82 mmol) and 1,2,3,5-tetra-O-acetyl-β-D-ribofuranose (**2**) (2.81 g, 8.82 mmol) are dissolved in acetonitrile (30 ml). To this suspension is

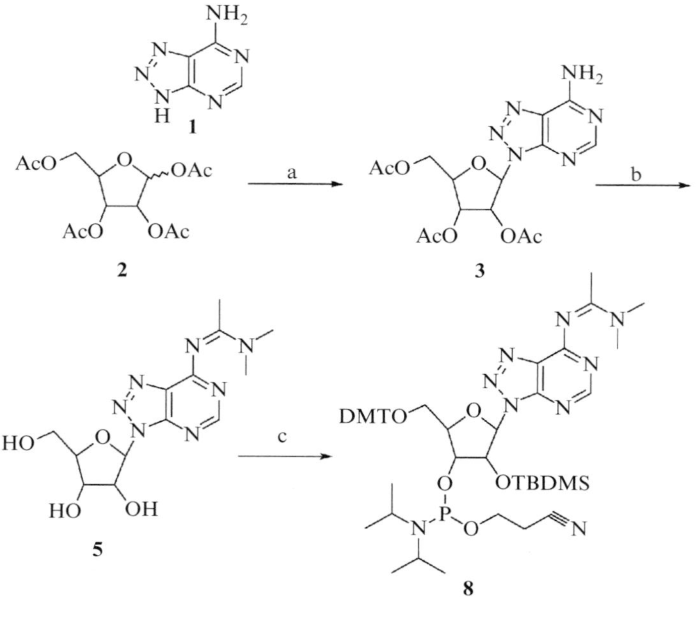

Scheme 17.1 (a) i. $SnCl_4$, CH_3CN, rt, 41%; (b) i. NH_3/CH_3OH; ii. $CH_3C(OCH_3)_2N$ $(CH_3)_2$, CH_3OH, rt, 75%; (c) i. DMTCl, pyridine, $AgNO_3$, THF; ii. TBDMSCl, $AgNO_3$, TEA, THF, 30%; iii. $ClP(OCH_2CH_2CN)[N(iPr)_2]$, DIEA, THF, 94%.

added SnCl$_4$ (3.1 ml, 26.4 mmol) within 5 min and the resulting solution is allowed to react under argon for 24 h at room temperature. The resulting mixture is washed with saturated aqueous NaHCO$_3$ (1 × 96 ml), filtered, and washed with water (2 × 30 ml). The obtained filtrate and washings are combined, extracted with CH$_2$Cl$_2$ (4 × 45 ml), dried over Na$_2$SO$_4$, and evaporated to give a slightly yellow foam. Crude material is purified by flash column chromatography that requires a 4-cm–diameter and 40-cm–height glass column packed with silica gel. The crude product is loaded on the top of the silica gel and eluted with solvent system (2–5%-CHCl$_3$/CH$_3$OH) by passing a low pressure of compressed air through the column. The eluted solvent was collected in small test tube fractions (16 × 125 mm). The resulting fractions were analyzed by TLC. The fractions that contained desired product (**3**) were combined and concentrated under reduced pressure using a rotary evaporator to give a colorless foam (1.41 g, 41%). (The majority of the remaining mass is the N^8-glycosylated product, which is removed during the chromatography step.) 1H, 13C NMR and elemental analysis data for this compound have been reported (Seela *et al.*, 1998).

8–Azaadenosine ribonucleoside (**4**): A solution of methanolic ammonia (10 ml) is added to 2′,3′,5′-tri-O-acetyl-8-azaadenosine (1.22 g, 3.09 mmol) and allowed to react for 5 h. The resulting mixture is concentrated under reduced pressure using a rotary evaporator to give the product in quantitative yield. 1H, 13C NMR data for this compound have been reported (Seela *et al.*, 1998).

N-(Dimethylacetamidine)-8-azaadenosine (**5**): N,N-Dimethylacetamide dimethyl acetal (1.36 ml, 9.28 mmol) is added to a solution of 8-azaadenosine ribonucleoside (829.8 mg, 3.09 mmol) in CH$_3$OH (15 ml). After 14 h at room temperature the mixture is concentrated to dryness and then codistilled with toluene. The resulting crude product is redissolved in CH$_3$OH and stirred for 2 h at room temperature followed by evaporation using a rotary evaporator and purification by flash column chromatography to give a colorless foam (790 mg, 75%). 1H, 13C NMR and elemental analysis data for this compound have been reported (Seela *et al.*, 1998).

N-(Dimethylacetamidine)-5′-O-(4,4′-dimethoxytrityl)-8-azaadenosine (**6**): N-(Dimethylacetamidine)-8-azaadenosine (790 mg, 2.33 mmol) is dissolved in freshly distilled THF (20 ml). Anhydrous pyridine (1.13 ml, 14 mmol), 4,4′-dimethoxytrityl chloride (867.6 mg, 2.56 mmol), and AgNO$_3$ (435.0 mg, 2.56 mmol) are added sequentially to this solution. The reaction mixture is stirred at room temperature overnight. The mixture is diluted with ethyl acetate (EtOAc) (35 ml), filtered, and washed with saturated aqueous NaHCO$_3$ (1 × 20 ml). The organic layer is dried over Na$_2$SO$_4$, filtered, and concentrated under reduced pressure using a rotary evaporator. The crude product is redissoved in a 1:1 mixture of CH$_3$CN/toluene followed by concentration twice to give a dark yellow foam. ^{1}H NMR and elemental analysis data have been reported for this compound (Seela *et al.*, 1998).

N-(Dimethylacetamidine)-5′-O-(4,4′-dimethoxytrityl)-2′-O-(t-butyldimethyl-silyl)-8-azaadenosine(7):N-(Dimethylacetamidine)-5′-O-(4,4′-dimethoxytrityl)-8-azaadenosine (1.49 g, 2.32 mmol) is dissolved in freshly distilled THF (20 ml) and to this solution is added triethylamine (TEA) (614.9 μl, 4.41 mmol), followed by the addition *of tert*-butylchlorodimethylsilane (TBDMSCl) (384.9 mg, 2.55 mmol). After 5 min, AgNO₃ (433.9 mg, 2.55 mmol) is added to the solution. After 8 h of stirring at room temperature, the reaction mixture is diluted with EtOAc (75 ml), filtered, and washed with 5% aqueous NaHCO₃ (1 × 90 ml). The organic layer is dried over Na₂SO₄, filtered, and concentrated under reduced pressure using a rotary evaporator. The crude product is purified by flash column chromatography on silica gel using EtOAc/hexanes 4:1 to produce a white foam (526.6 mg, 30%). ¹H NMR, ¹³C NMR, and HRFABMS data have been reported for this compound (Veliz *et al.*, 2003).

N-(Dimethylacetamidine)-5′-O-(4,4′-Dimethoxytrityl)-3′-O-[(2-cyanoethoxy)(N,N-diisopropylamino)phosphino]-2′-O-(t-butyldimethylsilyl)-8-azaadenosine(8):   N-(Dimethylacetamidine)-5′-O-(4,4′-dimethoxytrityl)-2′-O-(t-butyldimethyl-silyl)-8-azaadenosine (237 mg, 0.313 mmol) is dissolved in freshly distilled THF (2.0 ml). N,N-Diisopropylethylamine (327.8 μl, 1.88 mmol) followed by 2-cyanoethyl-(N,N-diisopropylamino)chlorophosphite (140 μl, 0.627 mmol) are added to the solution. The solution is allowed to stir at room temperature. After 8 h, the reaction mixture is diluted with EtOAc (50 ml), filtered, and washed with 5% (w/v) aqueous NaHCO₃ (2 × 30 ml). The organic layer is dried over Na₂SO₄, filtered, and concentrated under reduced pressure using a rotary evaporator. The crude products are purified by flash column chromatography on silica gel using EtOAc/hexanes 3:1 to afford a white foam (281.0 mg, 94%). ³¹P NMR and HRFABMS data were previously reported (Veliz *et al.*, 2003).

4. SYNTHESIS, DEPROTECTION, AND PURIFICATION OF CHEMICALLY MODIFIED ADAR SUBSTRATE RNAs

Once the phosphoramidite of an adenosine analog is available, it is used to prepare an oligoribonucleotide by automated, solid-phase RNA synthesis. The resulting strand is deprotected, purified, ³²P labeled, and hybridized to its complement to generate an RNA duplex mimicking a known editing site. In some cases, the edited nucleotide is at the 5′ end of one strand making labeling at its 5′-phosphate possible simply using T4 polynucleotide kinase. For substrates that have the editing site at an internal position, a templated ligation step is required to generate an RNA strand with a single, internal nucleotide position ³²P labeled. Detailed protocols for each step in this process are provided below.

5. General Biochemical Procedures

Distilled, deionized water is used for all aqueous reactions and dilutions. Biochemical reagents are obtained from Sigma/Aldrich unless otherwise noted. Common enzymes are purchased from Roche, Promega, or New England Biolabs. [γ-^{32}P]ATP (6000 Ci/mmol) is obtained from Perkin-Elmer Life Sciences. Storage phosphor autoradiography is performed using imaging plates from Eastman Kodak Co. and a Molecular Dynamics Typhoon 9400. Full-length human ADAR2: [hADAR2a-LV (H)$_6$], R-D: [MS(H)$_{10}$ENLYFQG-hADAR2a$_{216-701}$] are overexpressed in *Saccharomyces cerevisiae* and purified as described (Macbeth *et al.*, 2004).

5.1. Protocol

5.1.1. RNA synthesis and purification

Solid-phase synthesis of RNA oligonucleotides using 5′-DMT, 2′-O-TBDMS protected β-cyanoethyl phosphoramidites is carried out on a 1.0 μmol or 200 nmol scale with coupling times of 25 min for more efficient coupling using an ABI 394 synthesizer (DNA/Peptide Core Facility, University of Utah). For oligonucleotide deprotection, the CPG-bound oligonucleotide is treated with ~3 ml of fresh solution of ammonium hydroxide in ethanol (3:1) for approximately 2 h at room temperature and at 55° for approximately 5 h. The solution containing oligo is collected by pipetting it off into an Eppendorf tube and concentrating on a Speed-Vac concentrator. The dried pellet is treated with 1 ml of neat triethylamine trihydrofluoride and allowed to stand for 12 h at room temperature. To this mixture is added water followed by *n*-butanol for precipitation. The crude precipitate is dissolved in 0.1 M TEA-acetate, applied to an oligonucleotide purification cartridge (Glen Research), washed with water, eluted with 20% acetonitrile/water, and dried by lyophilization. Deprotected oligonucleotides are purified by polyacrylamide gel electrophoresis (PAGE) (19%), visualized by UV shadowing, and extracted from the gel via the crush and soak method with 0.5 M NH$_4$OAc, 0.1% sodium dodecyl sulfate (SDS), and 0.1 mM EDTA overnight at room temperature. Polyacrylamide particles are removed using a Centrex® filter. The RNA oligo is ethanol precipitated, concentrated to dryness, and stored at −80°. Extinction coefficients for these RNAs are calculated as the sum of the extinction coefficients of the component nucleotides using the adenosine value for the analogs studied (Sambrook *et al.*, 1989). RNA concentrations are determined by UV/VIS measurements at 260 nm using a Beckman DU 7400 spectrophotometer.

5.1.2. Preparation and purification of RNA duplexes

For the formation of labeled duplex RNA, a given oligonucleotide is labeled at the 5′ end using [γ-^{32}P]ATP (6000 Ci/mmol) and T4 polynucleotide kinase. The labeled strand is first purified on a 19% denaturing polyacrylamide gel, visualized by storage phosphor autoradiography, excised, and extracted from the gel via the crush and soak method with 0.5 M NH$_4$OAc, 0.1% SDS, and 0.1 mM EDTA. Then it is hybridized to the unlabeled complement strand in TE buffer (10 mM Tris–HCl, pH 7.5, 0.1 mM EDTA) with 50 or 200 mM NaCl. The mixture is heated at 95° for 5 min and allowed to slow cool to room temperature. The duplex is then purified on a 16% nondenaturing polyacrylamide gel. The appropriate band is visualized by storage phosphor autoradiography, excised, and extracted into TE buffer overnight at room temperature. Polyacrylamide particles are removed using a Centrex® filter. The RNA duplex is ethanol precipitated, redissolved in deionized water, and stored at −80°. The concentration of the duplex is determined using scintillation counting and the specific activity of the labeled strand.

5.1.3. Construction of internally labeled RNA substrates

The synthesis of substrates containing internally labeled editing sites is based on splint ligations as outlined by Moore and Sharp (1992). A 22-nt RNA oligonucleotide (30 pmol) (e.g., 5′-AGGUGGGUGGAAUAGUAUAA CA-3′) is 5′ end labeled with [γ-^{32}P]ATP and T4 polynucleotide kinase. The buffer is exchanged using a Microspin G-25. The labeled 22-mer is lyophilized into a thin-walled PCR tube at low heat. The lyophilized pellet is redissolved with 3 μl of a 10 μM solution of DNA splint (5′-CTATTCCACCCACCT-TAATGAGGATCCTTTAGG-3′), 2 μl of 20 μM 18-mer oligonucleotide (e.g., DNA/RNA chimera 5′-dCdCdTdAdAdAdGdGdAdTdCdCdTr-CrArUrUrA-3′, where dC, dT, dA, and dG are deoxyribonucleotides and rC, rA, and rU are ribonucleotides), 0.5 μl of RNasin from Promega (1.6 units/μl), and 1 μl of NEB T4 DNA ligase 10× buffer [1× conditions: 50 mM Tris–HCl (pH 7.5), 10 mM MgCl$_2$, 10 mM dithiothreitol (DDT), 1 mM ATP, 25 μg/ml bovine serum albumin (BSA)]. This reconstituted solution is heated to 65° for 2 min, 60° for 2 min, 55° for 5 min, 52° for 7 min, 42° for 7 min, 32° for 5 min, and 22° for 5 min using a Perkin-Elmer GeneAmp PCR System 2400. To the solution are added 0.5 μl of RNasin, 0.5 μl of 5.5 mM ATP, and 1.5 μl of T4 DNA ligase (45 Weiss units). The reaction is incubated at 30° for 6 h. RQ1 RNase-free DNase (2 units) is added. The solution is heated to 37° for 10 min, cooled to 22° for 2 min, and heat inactivated at 65° for 10 min. The sample is purified to nucleotide resolution by PAGE. The product band is visualized by storage phosphor autoradiography and excised. The labeled and ligated oligonucleotide is extracted from the gel slice as described above. To the labeled strand is added a known amount (> 70-fold excess) of unlabeled strand of the same sequence. The specific activity of the sample is then determined by liquid scintillation counting. The strands are hybridized, and the resulting duplex is gel purified and quantified as described above.

6. EVALUATION OF ADAR SUBSTRATE ANALOGS

The rate of ADAR-catalyzed deamination of the adenosine analog within a duplex RNA substrate is measured using a TLC assay described in detail elsewhere (Bass and Weintraub, 1988). Our quantitative reactions are typically carried out under single turnover conditions ([enzyme]>>[substrate]) at a concentration of enzyme well above the dissociation constant for substrate binding estimated in gel mobility shift assays (Stephens *et al.*, 2000). Aliquots of the reaction mixture are removed at different time points and the reaction is quenched by heat denaturation. For each time point, the RNA is digested with nuclease to give nucleoside 5′-monophosphates, which are resolved by TLC. Since only the editing site nucleotide is ^{32}P labeled, visualizing the TLC plate by storage phosphor autoradiography shows spots only for the starting adenosine analog and the deamination product (Fig. 17.1). The fraction of product formed at each time point is quantified and plotted versus time. The data are fit to a single exponential equation to extract the observed rate constant for deamination of that analog

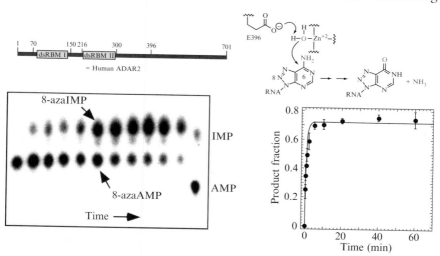

Figure 17.1 Substrate analogs (e.g., 8–azaadenosine shown) are evaluated by measuring the rate of ADAR2-catalyzed deamination under single turnover conditions. Human ADAR2 overexpressed in *S. cerevisiae* is used for these experiments. At various time points, reaction aliquots are removed and nuclease digestion to nucleoside monophosphates is carried out. The product is then separated from unreacted substrate by thin layer chromatography. Plotting the reaction yield versus time and data fitting provide the reaction rate constant. (See color insert.)

Table 17.1 Deamination rate constants for ADAR2 substrate analogs[a]

```
                          R/G site
                          ↙
                   *X
        5'-  GGUGGGUGG AUA UAUAACAAUGU-3'
        3'-U CCAUCCACC UAU AUAUUGUUGUA-5'
           C           C   G
```

X	Rate constant (k_{obs}), min^{-1}	k_{rel}
A[b]	0.066 ± 0.029	1.0
2'-OMe[b]	0.00024 ± 0.00006	0.004
2'-H[b]	0.019 ± 0.002	0.3
2'-F[b]	0.034 ± 0.017	0.5
N^6-Me[c]	0.001 ± 0.0001	0.02
N^6-Et[c]	$<10^{-4}$	<0.0015
2,6-DAP[c]	$<10^{-4}$	<0.0015
8-azaA[c]	1.2 ± 0.2	18
7-deazaA[d]	0.06 ± 0.003	0.91
6-O-MeI[d]	0.001 ± 0.0001	0.02

```
                              R/G site
                              ↙
                        *X         A   G
     5'-CCUAAAGGAUCCUCAUUA GGUGGGUGG AUA UAUAACA-3'
     3'-GUAGU CCAUCCACC UAU AUAUUGU-5'
              C           C   G
```

X	Rate constant (k_{obs}), min^{-1}	k_{rel}
A[c]	0.88 ± 0.09	1.0
8-azaA[c]	2.5 ± 0.29	2.8

```
                            R/G site
                            ↙
                      * A            A   G
  5'-CCUAAAGGAUCCUCAUUX GGUGGGUGG AUA UAUAACA-3'
  3'-GUAGU CCAUCCACC UAU AUAUUGU-5'
            C           C   G
```

X	Rate constant (k_{obs}), min^{-1}	k_{rel}
A[c]	0.0020 ± 0.0001	1.0
8-azaA[c]	0.034 ± 0.002	17

[a] *, ^{32}P; A, adenosine; 2'-O-Me, 2'-O-methyladenosine; N^6-Me, N^6-methyladenosine; N^6-Et, N^6-ethyladenosine; 2,6-DAP, 2,6-diaminopurine ribonucleoside; 8-azaA, 8-azaadenosine; 7-deazaA, 7-deazaadenosine; 6-O-MeI, 6-O-methylinosine.
[b] Yi-Brunozzi et al. (1999).
[c] Veliz et al. (2003).
[d] Easterwood et al. (2000).

(k_{obs}). Often this value is used to generate the k_{rel}, or rate constant relative to adenosine in the same RNA substrate (Table 17.1).

6.1. Protocol

6.1.1. Deamination kinetics and data analysis

Single turnover editing reactions are carried out with 250 nM ADAR2, 25 nM labeled RNA duplex, and assay buffer containing 15 mM Tris–HCl, pH 7.1, 3% glycerol, 0.5 mM DTT, 60 mM KCl, 1.5 mM EDTA, 0.003% NP-40, 160 units/ml RNasin, and 1.0 μg/ml yeast tRNA^{Phe}. Before mixing, both RNA and ADAR2 protein are incubated at 30° for 2 min. Reaction mixtures are then incubated at 30° for varying times. At each time point, an aliquot is removed and the reaction is quenched by the addition of 0.5% SDS at 95°, followed by digestion with nuclease P1 and resolution of the resulting 5'-mononucleotides by TLC as previously described (Bass and Weintraub, 1988). Storage phosphor imaging plates (Kodak) are pressed flat against TLC plates and exposed in the dark. The data are analyzed by performing volume integrations of the regions corresponding to starting material, product, and background sites using ImageQuant software. The data are fit to the equation $[P]_t = \alpha[1 - e^{(-k_{obs} \cdot t)}]$, where $[P]_t$ is the fraction of deamination product at time t, α is the fitted reaction end point, and k_{obs} is the fitted rate constant.

7. SYNTHESIS OF A TRANSITION STATE ANALOG PHOSPHORAMIDITE FOR MECHANISM-BASED TRAPPING

The observation that the 8–aza modification of adenosine significantly accelerated the ADAR2 reaction suggested that the corresponding purine derivative might be capable of mechanism-based trapping of ADAR2 (Haudenschild *et al.*, 2004; Veliz *et al.*, 2003). 8–Aza modification would facilitate covalent hydration of the purine ring, but with only a hydrogen at C6, no good leaving group would be present to allow the reaction to proceed to product. In addition, since the covalent hydrate is an excellent mimic of the proposed reaction transition state, the enzyme would bind tightly to it. This type of trapping of a nucleic acid–modifying enzyme at points along the reaction coordinate, such as at the transition state, is extremely valuable in bringing about a greater understanding of the modifying reaction. For instance, use of 2'-fluorouridine in DNA to trap uracil DNA glycosylase allowed Stivers and colleagues to study the base flipping step of that reaction in detail using fluorescence spectroscopy (Stivers *et al.*, 1999). In addition, several trapped nucleic acid/enzyme complexes have been crystallized and their structures solved using X-ray diffraction methods (Daniels *et al.*, 2004; Kilmasauskas *et al.*, 1994; Losey *et al.*, 2006).

We have prepared nucleoside analogs as phosphoramidites and tested their ability to trap ADAR2 in a variety of RNA substrates. One example is 8-azanebularine, the purine analog of 8-azaadenosine (Haudenschild *et al.*, 2004). The synthesis of this phosphoramidite is shown in Scheme 17.2 with a detailed protocol provided below.

7.1. Protocol

2′,3′,5′-Tri-O-acetyl-6-bromo-8-azanebularine (9): 2′,3′,5′-Tri-O-acetyl-8-azaadenosine (**3**) (470 mg, 1.19 mmol) is dissolved in freshly distilled acetonitrile (40 ml). To this solution *tert*-butyl nitrite (*t*-BuONO) (1.43 ml, 11.9 mmol) and trimethylsilyl bromide (TMSBr) (315 μl, 2.38 mmol) are added. The reaction mixture is allowed to stir at 0° for 1 h. The mixture is diluted with EtOAc (15 ml) and washed with water (1 × 15 ml) and brine (saturated aqueous solution of NaCl) (1 × 15 ml). The organic layer is dried (Na$_2$SO$_4$), filtered, and concentrated under reduced pressure using a rotary evaporator. The crude product is purified by flash column chromatography (CHCl$_3$ followed by CHCl$_3$/CH$_3$OH 99:1) to give a white foam (328 mg, 60%). ^{1}H NMR, ^{13}C NMR, and HRFABMS data have been reported for this compound (Haudenschild *et al.*, 2004).

2′,3′,5′-Tri-O-acetyl-8-azanebularine (10): (Method A) 2′,3′,5′-Tri-O-acetyl-6-bromo-8-azanebularine (200 mg, 0.436 mmol) is dissolved in methanol (7 ml). Then 10% palladium on carbon (Pd-C) (12 mg) and anhydrous NaOAc (72 mg, 0.873 mmol) are added to the solution. The flask is evacuated and refilled with hydrogen gas. The mixture is shaken in a

Scheme 17.2 (a) i. *t*BuONO, TMSBr, CH$_3$CN, 0°, 60%; ii. H$_2$, 10% Pd/C, NaOAc, CH$_3$OH, 80%; (b) *t*BuONO, DMF, 65°, 44%; (c) i. NH$_3$/CH$_3$OH, 4°, 81%; ii. DMTCl, pyridine, AgNO$_3$, THF, 65%; iii. TBDMSCl, AgNO$_3$, TEA, THF, 42%; iv. ClP (OCH$_2$CH$_2$CN)[N(*i*Pr)$_2$], DIEA, THF, 78%.

Parr apparatus for 8 h under 2.5 atm of pressure. The mixture is filtered through Celite, and the Celite portion is washed with methanol. Organic filtrates are then combined and concentrated under reduced pressure using a rotary evaporator. The residue is redissolved in EtOAc (4 ml) and washed with brine (1 × 3 ml). The organic layer is dried (Na$_2$SO$_4$), filtered, and concentrated under reduced pressure using a rotary evaporator. The crude product is purified by flash column chromatography (CH$_2$Cl$_2$) to give a light yellow syrup (133 mg, 80%) and taken to the deacylation step as described below.

2′,3′,5′-Tri-O-acetyl-8-azanebularine (10): (Method B) Freshly distilled t-BuONO (517.8 mg, 5.02 mmol) is added to a degassed hot solution (65°) of 2′,3′,5′-tri-O-acetyl-8-azaadenosine (**3**) (990 mg, 2.51 mmol) in anhydrous N,N-dimethylformamide (DMF) (15 ml) and stirred at this temperature for 1.5 h. The reaction mixture is diluted with EtOAc/hexanes (50 ml, 7:3), successively washed with water (3 × 20 ml) and brine (1 × 25 ml), dried (Na$_2$SO$_4$), filtered, and concentrated under reduced pressure using a rotary evaporator. The syrup obtained is purified by flash column chromatography (1% CH$_3$OH/CHCl$_3$) to afford product (420.5 mg, 44%) as a light yellow syrup and taken to the deacylation step as described below.

8-Azanebularine (11): A solution of 2′,3′,5′-Tri-O-acetyl-8-azanebularine (690 mg, 1.82 mmol) in methanolic ammonia (8 ml) is stored at 4° overnight. The mixture is concentrated under reduced pressure using a rotary evaporator and purified by flash column chromatography (pre-adsorbed on silica gel, 10% CH$_3$OH/CHCl$_3$) to afford product (370.6 mg, 81%) as a light yellow soft solid. ^{1}H NMR, ^{13}C NMR, and HRFABMS data for this compound have been reported (Nair and Chamberlain, 1984).

5′-O-(4,4′-Dimethoxytrityl)-8-azanebularine (12): 8-Azanebularine ribonucleoside (360 mg, 1.42 mmol) is dissolved in freshly distilled THF (15 ml). Anhydrous pyridine (690 μl, 8.53 mmol), 4,4′-dimethoxytrityl chloride (530 mg, 1.56 mmol), and AgNO$_3$ (265.7 mg, 1.56 mmol) are added sequentially to this solution. The reaction mixture is stirred at room temperature overnight. The mixture is diluted with EtOAc (25 ml), filtered, and washed with saturated aqueous NaHCO$_3$ (1 × 40 ml). The organic layer is dried (Na$_2$SO$_4$), filtered, and concentrated under reduced pressure using a rotary evaporator. The crude product is purified by flash column chromatography (CH$_2$Cl$_2$/CH$_3$OH/TEA 98:1:1) to give a light orange foam (511 mg, 65%). ^{1}H NMR, ^{13}C NMR, and HRFABMS data have been reported for this compound (Haudenschild et $al.$, 2004).

5′-O-(4,4′-Dimethoxytrityl)-2′-O-(t-butyldimethylsilyl)-8-azanebularine (13): 5′-O-(4,4′-Dimethoxytrityl)-8-azanebularine (460 mg, 0.828 mmol) is dissolved in freshly distilled THF (10 ml). Triethylamine (0.219 mL, 1.57 mmol) is added to the solution followed by the addition of TBDMSCl (137 mg, 0.911 mmol). After 5 min, AgNO$_3$ (155 mg, 0.911 mmol) is added to the mixture. After 8 h of stirring at room temperature, the reaction mixture is diluted with EtOAc (30 ml), filtered, and washed with saturated aqueous NaHCO$_3$ (1 × 35 ml). The organic layer is dried (Na$_2$SO$_4$), filtered, and concentrated under reduced pressure using a rotary evaporator. The crude product is purified by flash column chromatography on silica gel (EtOAc/hexanes 1:8) to give a white foam (234 mg, 42%). ^{1}H NMR, ^{13}C NMR, and HRFABMS data have been reported for this compound (Haudenschild *et al.*, 2004).

5′-O-(4,4′-Dimethoxytrityl)-3′-O-[(2-cyanoethoxy)(N,N-diiso-propylamino)phosphino]-2′-O-(t-butyldimethylsilyl)-8-azanebula-rine (14): 5′-O-(4,4′-Dimethoxytrityl)-2′-O-(t-butyldimethylsilyl)-8-azane-bularine (230 mg, 0.343 mmol) is dissolved in freshly distilled THF (2.0 ml). N,N-Diisopropylethylamine (359 μl, 2.06 mmol) followed by 2-cyanoethyl-(N,N-diisopropylamino)chlorophosphite (153 μl, 0.687 mmol) is added to the solution. The solution is allowed to stir at room temperature. After 8 h, the reaction mixture is diluted with EtOAc (45 ml), filtered, and washed with 5% (w/v) aqueous NaHCO$_3$ (2 × 25 ml). The organic layer is dried (Na$_2$SO$_4$), filtered, and concentrated under reduced pressure using a rotary evaporator. The crude product is purified by flash column chromatography on silica gel (EtOAc/hexanes/TEA 14:85:1) to give a light straw foam (232 mg, 78%). ^{31}P NMR and HRFABMS were previously reported (Haudenschild *et al.*, 2004).

8. MECHANISM-BASED TRAPPING OF ADAR–RNA COMPLEXES

Once the phosphoramidite of a trapping analog is available, a 5′-end-labeled RNA duplex is prepared mimicking a known ADAR substrate with the analog at the editing site. The protocol used to prepare these RNA duplexes is the same as previously described for substrate analogs that are 5′ end labeled. The affinity of analog-containing RNA for the editing enzyme is then measured using a quantitative gel mobility shift assay (Kerr, 1995) (Fig. 17.2). High-affinity binding dependent on the presence of the analog is taken as evidence of successful trapping. Given the high "nonspecific" affinity of ADARs for duplex RNA, we have found it useful to measure binding to a deletion mutant of ADAR2 that lacks dsRBM I, referred to as R–D, since it contains one dsRBM and the deaminase domain (Macbeth *et al.*, 2005). This mutant has lower affinity for duplex RNA than does full length ADAR2, and its binding is more sensitive to the presence of a trapping analog, like 8-azanebularine, in the RNA (Haudenschild *et al.*, 2004). Indeed, observation of a stable protein–RNA complex involving R–D requires that the RNA contain the trapping analog. Dissociation constants for the binding of the ADAR2 R–D mutant to several modified

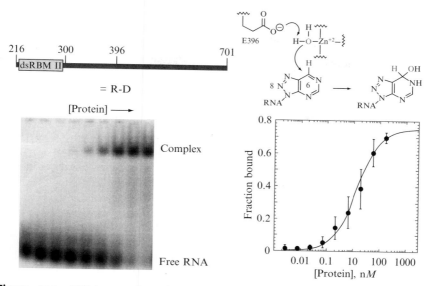

Figure 17.2 Efficiency of mechanism-based trapping (e.g., by 8-azanebularine shown) is assessed using quantitative gel mobility shift assays and a deletion mutant of ADAR2 lacking dsRBM I (R-D, human ADAR2a aa 216–701) overexpressed in *S. cerevisiae* (Macbeth *et al.*, 2005). Complexes formed are separated from free RNA using nondenaturing polyacrylamide gels. Plotting the fraction of RNA bound versus protein concentration and data fitting provide the equilibrium constant. (See color insert.)

RNAs are provided in Table 17.2. In addition, a detailed protocol for the gel mobility shift assay and data analysis is given below.

8.1. Protocol

8.1.1. Gel mobility shift assay and data analysis

Varying concentrations of the R–D deletion mutant of ADAR2 are added to ∼10 pM (8-azaN at editing site) 5′-^{32}P end-labeled RNA duplex in 15 mM Tris–HCl, pH 7.5, 3% glycerol, 0.5 mM DTT, 60 mM KCl, 1.5 mM EDTA, 0.003% NP-40, 160 units/ml RNasin, 0.1 mg/ml BSA, and 1.0 mg/ml yeast tRNA^{Phe}, and the reactions are incubated at 30° for 10 min. Variation in the incubation period from 10 to 30 min does not affect the measured dissociation constant. Samples are loaded onto a running 6% nondenaturing polyacrylamide gel (79:1 acrylamide:bisacrylamide) and electrophoresed in 0.5× TBE buffer at 4° for 45 min. Storage phosphorimaging plates (Kodak) are pressed flat against the dried electrophoresis gels and exposed in the dark. The data are analyzed by performing volume

Table 17.2   Dissociation constants for binding of R-D to analog-containing RNA[a]

GluR-B pre-mRNA (R/G site)	

R/G site
X ↙
   X   A   G
5′-CAUUA GGUGGGUGG AUA UAUAACA-3′
3′-GUAGU CCAUCCACC UAU AUAUUGU-5′
      C C G	

X	**Dissociation constant (K_d), nM**
A[b]	~1000
8-azaN[b]	2.0 ± 1.6
8-azaN(R-D E396A)[b]	308 ± 39
6-Me-8-azaN[c]	149 ± 15

GluR-B pre-mRNA (Q/R site)	

Q/R site   Q/R +4 site
  ↘   ↙   A   G
5′-UAUGCXGCAXGGAUGCG UAUUUC CCAAG-3′
3′-AUACGUCGUUUUUGUGC AUGGGG GGUUC-5′
                          C A	

X	**Dissociation constant (K_d), nM**
Q/R, 8-azaN[d]	81 ± 52
Q/R +4, 8-azaN[d]	18 ± 12

5HT$_{2C}$R pre-mRNA (D site)	

D site
  ↙   U
5′-ACGUAAUCC--UXUUGAGCAUAGCCGUU CAAUUC-3′
3′-UGUAUUAGG AUAACUCGUAUCGGCGA GUUAAG-5′
         C C   C
         U U
           U	

X	**Dissociation constant (K_d), nM**
8-azaN[d]	70 ± 37

[a] A, adenosine; 8-azaN, 8-azanebularine; 6-Me-8-azaN, 6-methyl-8-azanebularine.
[b] Haudenschild *et al.* (2004).
[c] Maydanovych and Beal (2006).
[d] This work.

integrations of the regions corresponding to free RNA, protein–RNA complex, and background sites using ImageQuant software. The data are fit to the following equation: fraction bound = A^*[protein]/([protein] + K_d), where K_d is the fitted dissociation constant and A is the fitted maximum fraction RNA bound at that dissociation constant.

REFERENCES

Baer, H.-P., Drummond, G. I., and Duncan, E. L. (1966). Formation and deamination of adenosine by cardiac muscle enzymes. *Mol. Pharmacol.* **2,** 67–76.

Bass, B. L., and Weintraub, H. (1988). An unwinding activity that covalently modifies its double-stranded RNA substrate. *Cell* **55,** 1089–1098.

Bhalla, T., Rosenthal, J. J. C., Holmgren, M., and Reenan, R. (2004). Control of human potassium channel inactivation by editing of a small mRNA hairpin. *Nat. Struct. Mol. Biol.* **11,** 950–956.

Burns, C. M., Chu, H., Rueter, S. M., Hutchinson, L. K., Canton, H., Sanders-Bush, E., and Emeson, R. B. (1997). Regulation of serotonin-2C receptor G-protein coupling by RNA editing. *Nature* **387,** 303–308.

D'Adamo, M. C., Liu, Z., Adelman, J. P., Maylie, J., and Pessia, M. (1998). Episodic ataxia type-1 mutations in the hKv1.1 cytoplasmic pore region alter the gating properties of the channel. *EMBO J.* **17,** 1200–1207.

Daniels, D. S., Woo, T. T., Luu, K. X., Noll, D. M., Clarke, N. D., Pegg, A. E., and Tainer, J. A. (2004). DNA binding and nucleotide flipping by the human DNA repair protein AGT. *Nat. Struc. Mol. Biol.* **11,** 714–720.

Das, S. R., Fong, R., and Piccirilli, J. A. (2005). Nucleotide analogues to investigate RNA structure and function. *Curr. Opin. Chem. Biol.* **9,** 585–593.

Easterwood, L. M., Veliz, E. A., and Beal, P. A. (2000). Demethylation of 6-O-methylinosine by an RNA-editing adenosine deaminase. *J. Am. Chem. Soc.* **122,** 11537–11538.

Frederiksen, S. (1966). Specificity of adenosine deaminase toward adenosine and 2'-deoxyadenosine analogs. *Arch. Biochem. Biophys.* **113,** 383–388.

Grosjean, H., and Benne, R. (1998). "Modification and Editing of RNA." Washington, DC: ASM Press, Washington, DC.

Gurevich, I., Tamir, H., Arango, V., Dwork, A. J., Mann, J. J., and Schmauss, C. (2002). Altered editing of serotonin 2C receptor pre-mRNA in the prefontal cortex of depressed suicide victims. *Neuron* **34,** 349–356.

Haudenschild, B. L., Maydanovych, O., Veliz, E. A., Macbeth, M. R., Bass, B. L., and Beal, P. A. (2004). A transition state analogue for an RNA-editing reaction. *J. Am. Chem. Soc.* **126,** 11213–11219.

Higuchi, M., Single, F. N., Kohler, M., Sommer, B., Sprengel, R., and Seeburg, P. H. (1993). RNA editing of AMPA receptor subunit GluR-B: A base-paired intron-exon structure determines position and efficiency. *Cell* **75,** 1361–1370.

Higuchi, M., Maas, S., Single, F. N., Hartner, J., Rozov, A., Burnashev, N., Feldmeyer, D., Sprengel, R., and Seeburg, P. H. (2000). Point mutation in an AMPA receptor gene rescues lethality in mice deficient in the RNA-editing enzyme ADAR2. *Nature* **406,** 78–81.

Iwamoto, K., and Kato, T. (2003). RNA editing of serotonin 2C in human postmortem brains of major mental disorders. *Neurosci. Lett.* **346,** 169–172.

Kerr, L. D. (1995). Electrophoretic mobility shift assay. *Methods Enzymol.* **254,** 619–632.

Kilmasauskas, S., Kumar, S., Roberts, R. J., and Cheng, X. (1994). HhaI methyltransferase flips its target base out of the DNA helix. *Cell* **76,** 357–369.

Losey, H. C., Ruthenburg, A. J., and Verdine, G. L. (2006). Crystal structure of Staphylococcus aureus tRNA adenosine deaminase TadA in complex with RNA. *Nat. Struc. Mol. Biol.* **13,** 153–159.

Macbeth, M. R., Lingam, A. T., and Bass, B. L. (2004). Evidence for auto-inhibition by the N-terminus of hADAR2 and activation by dsRNA binding. *RNA* **10,** 1563–1571.

Macbeth, M. R., Schubert, H. L., VanDemark, A. P., Lingam, A. T., Hill, C. P., and Bass, B. L. (2005). Inositol hexakisphosphate is bound in the ADAR2 core and required for RNA editing. *Science* **309,** 1534–1539.

Maydanovych, O., and Beal, P. A. (2006). C6-substituted analogues of 8-azanebularine: Probes of an RNA-editing enzyme active site. *Org. Lett.* **8,** 3753–3756.

Moore, M. J., and Sharp, P. A. (1992). Site-specific modification of pre-mRNA: The 2′-hydroxyl groups at the splice sites. *Science* **256,** 992–997.

Nair, V., and Chamberlain, S. D. (1984). Reductive deamination of aminopurine nucleosides. *Synthesis* **5,** 401–403.

Niswender, C. M., Herrick-Davis, K., Dilley, G. E., Meltzer, H. Y., Overholser, J. C., Stockmeier, C. A., Emeson, R. B., and Sanders-Bush, E. (2001). RNA editing of the human serotonin 5-HT2C receptor alterations in suicide and implications for serotonergic pharmacotherapy. *Neuropsychopharmacology* **24,** 478–491.

Palladino, M. J., Keegan, L. P., O'Connell, M. A., and Reenan, R. A. (2000). A-to-I pre-mRNA editing in *Drosophila* is primarily involved in adult nervous system function and integrity. *Cell* **102,** 437–449.

Pellegrini-Giampietro, D. E., Gorter, J. A., Bennett, M. V. L., and Zukin, R. S. (1997). The GluR2 (GluR-B) hypothesis: Ca2+-permeable AMPA receptors in neurological disorders. *Trends Neurosci.* **20,** 464–470.

Pokharel, S., and Beal, P. A. (2006). High-throughput screening for functional adenosine to inosine RNA editing systems. *ACS Chem. Biol.* **1,** 761–765.

Sambrook, J., Fritsch, E. F., and Maniatis, T. (1989). "Molecular Cloning: A Laboratory Manual." Cold Spring Harbor Laboratory Press, Plainview, NY.

Seela, F., Munster, I., Lochner, U., and Rosemeyer, H. (1998). 8-Azaadenosine and its 2′-deoxyribonucleoside: Synthesis and oligonucleotide base-pair stability. *Helv. Chim. Acta* **81,** 1139–1155.

Stephens, O. M., Yi-Brunozzi, H.-Y., and Beal, P. A. (2000). Analysis of the RNA-editing reaction of ADAR2 with structural and fluorescent analogues of the GluR-B R/G editing site. *Biochemistry* **39,** 12243–12251.

Stephens, O. M., Haudenschild, B. L., and Beal, P. A. (2004). The binding selectivity of ADAR2's dsRBMs contributes to RNA-editing selectivity. *Chem. Biol.* **11,** 1239–1250.

Stivers, J. T., Pankiewicz, K. W., and Watanabe, K. A. (1999). Kinetic mechanism of damage site recognition and uracil flipping by Escherichia coli uracil DNA glycosylase. *Biochemistry* **38,** 952–963.

Tonkin, L. A., Saccomanno, L., Morse, D. P., Brodigan, T., Krause, M., and Bass, B. L. (2002). RNA editing by ADARs is important for normal behavior in Caenorhabditis elegans. *EMBO J.* **21,** 6025–6035.

Veliz, E. A., Stephens, O. M., and Beal, P. A. (2001). Synthesis and analysis of RNA containing 6-trifluoromethylpurine ribonucleoside. *Org. Lett.* **3,** 2969–2972.

Veliz, E. A., Easterwood, L. M., and Beal, P. A. (2003). Substrate analogues for an RNA-editing adenosine deaminase: Mechanistic investigation and inhibitor design. *J. Am. Chem. Soc.* **125,** 10867–10876.

Vitali, P., Basyuk, E., LeMeur, E., Bertrand, E., Muscatelli, F., Cavaille, J., and Huttenhofer, A. (2005). ADAR2-mediated editing of RNA substrates in the nucleolus is inhibited by C/D small nucleolar RNAs. *J. Cell. Biol.* **169,** 745–753.

Wang, Q., Khiltan, J., Gadue, P., and Nishikura, K. (2000). Requirement of the RNA editing deaminase ADAR1 gene for embryonic erythropoiesis. *Science* **290,** 1765–1768.

Yi-Brunozzi, H.-Y., Easterwood, L. M., Kamilar, G. M., and Beal, P. A. (1999). Synthetic substrate analogs for the RNA-editing adenosine deaminase ADAR-2. *Nucleic Acids Res.* **27,** 2912–2917.

Yi-Brunozzi, H.-Y., Stephens, O. M., and Beal, P. A. (2001). Conformational changes that occur during an RNA-editing adenosine deamination reaction. *J. Biol. Chem.* **276,** 37827–37833.

NUCLEAR C-TO-U EDITING

MEASURING EDITING ACTIVITY AND IDENTIFYING CYTIDINE-TO-URIDINE MRNA EDITING FACTORS IN CELLS AND BIOCHEMICAL ISOLATES

Harold C. Smith

Contents

Abstract

Cytidine deaminases with the capacity to act on nucleic acids play a critical role in regulating the proteome through diversification of expressed sequence beyond that encoded in the genome. A family of these enzymes, known as the APOBEC family of cytidine deaminases, has been identified in mammalian cells. APOBEC-1 edits messenger RNA, whereas other family members affect

Departments of Biochemistry and Biophysics, Pathology and Toxicology, University of Rochester School of Medicine and Dentistry, Rochester, New York

Methods in Enzymology, Volume 424
ISSN 0076-6879, DOI: 10.1016/S0076-6879(07)24018-2

mRNA coding capacity by editing single-stranded DNA in expressed regions of the genomes. Biochemical isolation and analysis of APOBEC proteins and their interacting factors have led to an understanding of the diverse cellular processes including lipoprotein metabolism, antibody production, viral infectivity, and cancer. Practical approaches will be described for the measurement of editing activity and the analysis of proteins involved in C-to-U and dC-to-dU editing.

1. INTRODUCTION TO MAMMALIAN CYTIDINE-TO-URIDINE mRNA EDITING

Mammalian C-to-U mRNA editing involves a hydrolytic deamination of cytidine at the C4 position resulting in a C-to-U transition (Johnson et al., 1993). Apolipoprotein B (apoB) mRNA was the first example of mammalian C-to-U mRNA editing (Chen et al., 1987; Powell et al., 1987). Subsequently, the mRNAs encoding the NF1 tumor suppressor (Mukhopadhyay et al., 2002; Skuse et al., 1996), translation factor eIF4G (Yamanaka et al., 1997), and the HIV RNA genome (Bishop et al., 2004) were shown to be C-to-U edited. The minimal requirements for mammalian C-to-U mRNA editing are the demarcation of the cytidine to be edited by a 3′ proximal RNA recognition motif known as the mooring sequence [in apoB mRNA the mooring sequence is UGAUCAGUAUA (Backus and Smith, 1991, 1992; Driscoll et al., 1993; Shah et al., 1991; Sowden et al., 1996a)], a cytidine deaminase, and an RNA-binding protein capable of binding to both the editing site and the cytidine deaminase (Smith, 1993, 1998; Smith et al., 1997) (Fig. 18.1). Through these three requirements the catalytic site of the deaminase is restricted in its interactions with the RNA substrate for site-specific editing. Therefore, in characterizing C-to-U editing systems, it is necessary to consider the sum of protein–protein and protein–RNA interactions governing substrate- and site-specific editing activity (Smith, 2005, 2006).

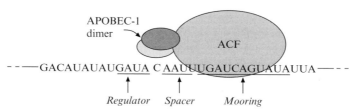

Figure 18.1 Minimal editosome model. A segment of apoB mRNA containing the C-to-U editing site at nucleotide 6666 shows the mooring sequence, a spacer element, and a regulator element. APOBEC-1 dimer is moored in the vicinity of C6666 for editing by its association with ACF. ACF binds to the editing site through its interaction with the mooring sequence.

The mechanism of site-specific C-to-U *apoB* mRNA editing involves a macromolecular assembly of proteins (an editosome) minimally consisting of a catalytic subunit, the cytidine deaminase APOBEC-1 (27 kDa) (Teng *et al.*, 1993), and an essential complementation factor, ACF (64 kDa) (Lellek *et al.*, 2000; Mehta *et al.*, 2000) (Fig. 18.1). APOBEC-1 is a highly conserved zinc-dependent enzyme (Hadjiagapiou *et al.*, 1994; Lau *et al.*, 1994; Nakamuta *et al.*, 1995; Teng *et al.*, 1994; Yamanaka *et al.*, 1994). The catalytic domain shares similarity with a variety of nucleoside/nucleotide deaminases (Anant *et al.*, 1998; Mian *et al.*, 1998; Navaratnam *et al.*, 1993; Wedekind *et al.*, 2003). ApoB editing activity in cells is absolutely dependent on APOBEC-1, as *apobec-1$^{-/-}$* knockout mice cannot edit apoB mRNA (Hirano, 1996; Nakamuta *et al.*, 1996; Xie *et al.*, 2003) even though they express other APOBEC family members (Anant *et al.*, 2001b; Jarmuz *et al.*, 2002; Wedekind *et al.*, 2003).

The specificity of APOBEC-1's activity is determined by a balance of its expression relative to other factors in the cell. Hyperphysiological levels of APOBEC-1 expression can induce promiscuous editing of additional cytidines in *apoB* mRNA (Siddiqui *et al.*, 1999; Sowden *et al.*, 1996a, 1998; Yang *et al.*, 2000) and hyperediting of other mRNAs (Yamanaka *et al.*, 1996, 1997) and may mutate DNA by deaminating dC-to-dU within ssDNA regions of the genome (Harris *et al.*, 2002; Petersen-Mahrt and Neuberger, 2003).

The editosome contains a head-to-head dimer of APOBEC-1 (Lau *et al.*, 1994; Xie *et al.*, 2004) bound to ACF cosediments from nuclear extracts of liver and intestinal cells as 27S complexes (~300–500 kDa) (Backus and Smith, 1991; Harris *et al.*, 1993; Smith, 1998; Smith *et al.*, 1991; Sowden *et al.*, 2002). ACF is a serine/threonine phosphoprotein whose serine hyperphosphorylated leads to nuclear retention, enhanced APOBEC-1 interactions, and the formation of 27S complexes active in editing (Lehmann *et al.*, 2006a,b). Hypophosphorylated ACF can be recovered from cytoplasmic extracts as 60S high-molecular-mass complexes that are catalytically inactive (Lehmann *et al.*, 2006a,b). Although these complexes contain APOBEC-1, its interaction with ACF is not as strong or may be indirect as cytoplasmic ACF cannot coimmunoprecipitate APOBEC-1 (Lehmann *et al.*, 2006a).

From this brief overview, it is apparent that editing factors participate in multiple interactions, many of which are dynamic, and their function varies depending on where in the cell they reside. Biochemical methods have been integral to the characterization of apoB mRNA editing factors. A few of the routine methods that have been used in studies of APOBEC-1 and its associated auxiliary protein ACF are described.

2. Poisoned Primer Extension

2.1. Poisoned primer extension assay

The "poisoned" primer extension assay is the method of choice for identifying and quantifying C-to-U transitions in mRNAs isolated from cells or
tissues (Driscoll *et al.*, 1989). This method enables the analysis of mRNA
editing using reverse transcriptase, polymerase chain reaction (RT-PCR)
products from cellular RNA or RNA reporters used for *in vitro* editing. The
method has a low detection limit (0.3% editing), has good accuracy and
precision ($\pm < 5\%$), can be used as a high-throughput method, and is
internally controlled and quantitative (Driscoll *et al.*, 1989; Smith, 1998).
In this method, a radio or fluorescent end-labeled primer, complementary
to sequences within 2–6 nucleotides 3′ (downstream) of the candidate
edited cytidine in the target mRNA, is annealed to the RT-PCR product
(or RNA reporter) and extended by reverse transcriptase in the presence of
ddGTP (Fig. 18.2). The sequence of the target between the 3′ end of the

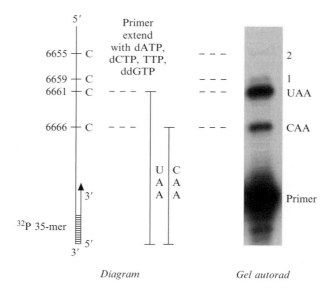

Figure 18.2   Poisoned primer extension assay on rat apoB cDNA. A diagram of apoB
cDNA with a ^{32}P end-labeled primer annealed to its 3′ end is shown with the positions of relevant cytidines indicated. Primer extension leads to a long (UAA) and short (CAA) radiolabeled primer extension product (center). Electrophoretic separation of primer and CAA
and UAA extension products followed by autoradiography (right) resolves the products.
Primer extension products at positions 1 and 2 correspond to the position of additional cytidines that could be due to promiscuous editing, or potentially the presence of dGTP impu-
rities in the dNTPs, resulting in minor read through products. This possibility can be tested
by performing primer extension on control unedited or fully edited substrates.

primer and the candidate cytidine should be devoid of cytidines to avoid chain termination of the primer extension product prior to the edited site. Primer extension continues beyond the predicted editing site when it has been edited to U and chain termination occurs at the next unedited cytidine (Fig. 18.2). Primer extension products resolved by 10% denaturing acrylamide gel electrophoresis can be imaged and quantified by a variety of means (typically by Phosphorimager analysis) and the percent of editing calculated as the percent of the total signal in the unedited and edited primer extension products represented in the edited primer extension product.

The method is ideal for comparing editing efficiency of a given target expressed in different cell types or assayed under different conditions because quantification of editing is not affected by differences in sample cell number, RNA yields, the efficiency of RT or PCR, or nonlinear amplification. The reason for this is that editing is quantified as a percent of total primer extension products for each sample, and therefore the percent editing in each sample is already normalized. Moreover, the primer extension method is typically performed on 25 ng of RT-PCR product, but the assay yields comparable quantification over a 5-fold range of input RT-PCR product.

2.2. Example of a poisoned primer extension protocol

Typically RNA is isolated from cellular or tissue sources using TriReagent (MRC) or a comparable method. The cytidine suspected of being edited should be identified and PCR primers selected to amplify a 200–400 nucleotide segment of the mRNA with the candidate edited site centrally located. Amplification of genomic DNA would bias the quantification toward more unedited cytidine. Consequently, the RNA should be digested with both RNase-free DNase I (Promega) and a restriction enzyme that will cut contaminating genomic DNA between the PCR primer annealing sites. Sufficient reverse transcript can be produced from 0.5–1 μg of total cellular RNA and 1 μg of oligo(dT) (15 nt) or a gene-specific primer. At the conclusion of the RT reaction, one-fifth to one-tenth of the reaction volume can be added directly to a PCR reaction (100 μl final volume). The PCR should be optimized according to the T_m of the primers, magnesium optimum concentration, and cycle number necessary to visualize product by ethidium bromide staining in an agarose gel. The PCR product can be purified by resin binding or gel electrophoresis and quantified by electrophoresis relative to a mass ladder (Invitrogen, CA). Contamination by the primers prohibits accurate quantification by spectroscopy.

Primer extension can be performed using either a fluorescence or ^{32}p end-labeled primer. A one nucleotide extension is difficult to quantify, so if there is no cytidine free sequence between the 3′ end of the primer and the edited cytidine, design a primer that will anneal to the opposite strand of the

PCR product such that its $3'$ end is 2–6 nucleotides $3'$ of quanosine-free sequence downstream of the candidate editing site (now G-to-A transition). If this alternate strategy is taken, primer extension must be carried out with ddCTP TTP, dATP, and dGTP rather than ddGTP, dATP, dCTP, and TTP.

- Heat 25 ng template with 2.5 p mols of end-labeled primer to 90° for 5 min in a final volume of $5\mu l$.
- Add 20 μl of the primer extension reaction mix to each tube of annealed primer/DNA following ramp cooling and incubate the reaction at 37° for 90 min.
- Carry out reverse transcription for 60-90 min at 37°.
- At the end of the reaction, add 75 μl of H_2O, 1 μl of 20 mg/ml glycogen (Roche), and 300 μl of 99% ethanol.
- Precipitate at $-80°$ for 30 min or at $-20°$ overnight.
- Recover the primer extension products by centrifugation ($14,000 \times g$, 10 min, 7°), dry the pellet (do not overdry), and resuspend the pellet in 5–15 μl of denaturing sample loading dye. [Sample loading dye consists of 90% formamide, 10 mM Tris, pH 8, and trace quantities of tracking dyes (0.025% w/v each bromophenol blue, phenol red, and xylene cyanol).]
- Heat the sample at 100° for 3 min, place quickly into an ice slurry for 3 min, microfuge pulse spin, and load all of the sample onto a 10% acrylamide (acrylamide:bisacrylamide 30:1 w/w), 8 M urea denaturing gel for electrophoresis.
- Imaging and quantification can be done without drying the gel by autoradiography followed by scanning densitometry, phosphorimager analysis, or fluoroimager analysis.

3. BIOCHEMICAL PURIFICATION OF EDITING FACTORS

3.1. Subcellular compartments of editing factors

The individual proteins or complexes of these proteins involved in C-to-U mRNA editing have been obtained from subcellular fractions or as recombinant proteins overexpressed in appropriate cell systems. Both approaches require optimization, as each of the C-to-U editing factors has unique properties that have to be understood and accommodated in the protocol. For example, nuclear extracts contain the most active form of apoB mRNA editing activity per microgram of total extract protein (Sowden et al., 2002; Yang and Smith, 1997; Yang et al., 2000), as this is the site of editing in vivo. Moreover, hyperphosphorylation of nuclear ACF enhances the assembly of active 27S editosomes (Lehmann et al., 2006a,b), and therefore phosphatase inhibitors such as sodium fluoride (10–50 mM) can be added during extract preparation. On the other hand, APOBEC-1 homologs such as APOBEC3G (Bennett *et al.*, 2006) do not traffic to the nucleus and cytoplasmic extracts contain all of the editing activity for these enzymes.

3.2. Whole cell extract preparation

The following protocol has been successfully applied to diverse cell types including McArdle 7777 rat hepatoma, HepG2 human hepatoma, HeLa human cervical carcinoma, Chinese hamster ovary (CHO), COS7 monkey kidney, 293T human embryonic kidney, A172 human astrocytoma and NGP human neurofibroma, H9 and MT2 human T cell lines, and peripheral blood mononuclear cells (PBMC). The extracts obtained by this method are whole cell S100 preparations and therefore must be considered to include both nuclear and cytoplasmic proteins. The three essential aspects of this protocol are (1) the harvesting of cells without trypsin and in the presence of proteinase inhibitors, (2) the maintenance of high cell concentrations during shearing, and (3) extraction with at least 250 mM monovalent salt. Extracts should be maintained on ice until they are frozen. The extracts will be stable for approximately 2 months but will lose approximately 10–15% of their activity with each thaw and refreezing at $-20°$.

A preparative scale extract (cultures at 80% confluency on 30 150-cm-diameter dishes) is described below.

3.2.1. Harvesting cells

Penetration of the proteinase inhibitors is ensured by scraping the cells from the plates or disrupting tissues in the presence of 5× extraction buffer [50 mM HEPES (pH 8.0), 150 mM NaCl, 1.5 mM MgCl$_2$, 2 mM EGTA, 1 mM dithiothreitol (DTT), 1 mM phenylmethylsulfonyl fluoride (PMSF) (Sigma), 15 mg/ml leupeptin (Sigma), and 15 mg/ml aprotinin (Sigma) or, alternatively, Complete Protease Inhibitor (Roche)]. This technique is referred to as "scrape-loading." A variety of cell scrapers can be used to remove adherent cells from the plates including the rubber end of a syringe plunger.

- Once all plates have been harvested, the cells can be recovered by centrifugation (500×g, 5 min, 7°). The supernatants can be discarded and the cell pellets should be resuspended into 10 packed cell volumes of 5× extraction buffer by pooling them into a single tube. Following gentle mixing, the cells should be recovered again and the cell pellets drained of supernatant and placed on ice.
- Estimate the volume of the packed cells (typically 3 ml) and resuspend the cells in 10 packed cell volumes of 1× extraction buffer and transfer them to a centrifuge tube that can withstand a high g force. The cells should be allowed to swell for 30 min on ice and then recovered by centrifugation

($6000 \times g$, 10 min at $7°$, a swinging bucket rotor works best). Drain the cell pellets thoroughly and resuspend them in two original packed cell volumes with $1 \times$ extraction buffer.

3.2.2. Preparing the extract

Disruption of the cells can be accomplished by graded gauge syringing starting with six passes through 18 gauge, followed by six passes through 20 gauge, and finally six passes through 22 gauge. It is important that the shearing be performed in a lur-lok™ (Becton Dickinson) syringe, as the success of this procedure depends on forcible expulsion of the cells through the needle against the centrifuge tube wall and, if done right, would dislodge a 22-gauge needle from a conventional slip-on syringe.

- An efficient extraction of protein from within organelles requires detergent and elevated monovalent salt concentrations and reducing conditions. To accomplish this, accurately determine the volume of sheared cells and add sufficient Triton X-100 (Bio-Rad) to a final concentration of 0.4%, mix gently, and incubate on ice for 10 min. Gradually add 4 M NaCl to bring the extract to 250 mM and incubate for 20 min on ice with intermittent mixing. Following this extraction, the insoluble material can be cleared by centrifugation ($6000 \times g$ for 20 min at $7°$) and an S100 supernatant prepared by additional centrifugation at $100,000 \times g$ (40 min, $7°$).
- Following centrifugation, draw off the supernatants, pool them, and store them as 50- to 100-μl aliquots. Extracts with 2–6 mg protein/ml can be obtained by this protocol.

3.2.3. Sample storage

Sample storage is a very important issue for preserving editing activity in extracts and biochemical fractions and avoiding precipitation of proteins. Unlike nuclear extracts used for *in vitro* premRNA splicing that are quick frozen in liquid nitrogen and stored at $-80°$, C-to-U editing factors are cryosensitive. Extracts should be stored at $-20°$ or used fresh, as storage at $-80°$ can result in 50–70% loss of activity overnight. Do not thaw and refreeze the aliquots more than once. Moreover, purified or recombinant editing factors tend to aggregate and fall out of solution when they are purified at $4–7°$ or after a freeze and thaw from $-20°$. Purification of recombinant factors at room temperature and use as fresh material are recommended.

3.3. Nuclear and cytoplasmic extracts

There may be circumstances in which nuclear or cytoplasmic extracts are required, e.g., obtaining maximal editing activity or evaluating editing factor–substrate interaction or trafficking of editing components. In studies where the distribution of editing factors between the nucleus and cytoplasm is at issue, it is important to validate that the subcellular factions are free of cross-contamination. Microscopic analysis will reveal the morphology of the nuclei and presence of unbroken cells, but marker enzyme or marker protein analysis should be used to demonstrate the purity of each fraction. Typically this is done by Western blotting a representative proportion of nuclear and cytoplasmic extract with antibodies reactive with histones (as a marker for nuclear proteins) and GAPDH or actin (as a marker for cytoplasmic proteins). The protocol described above can be used with the following modifications:

- Omit the reducing agent and prepare 5× extraction buffer at pH 6.8–7.0, as this reduces leaching of nuclear proteins during the extraction of the cytoplasm.
- Evaluate different $MgCl_2$ concentrations from 1 to 5 mM, as this divalent maintains chromatin condensation during nuclear isolation and high salt extract, but the optimal concentration for this varies with cell type.
- Do not shear the cells beyond passage through a 22-gauge needle.
- After adding Triton X-100 and before adding high salt, shear the cells by passage through an 18-22-gauge needle three times. Recover the nuclei by centrifugation (6000×g, 10 min, 7°).
- Draw off the cytoplasm-containing supernatant.
- The nuclear pellet is not completely free of cytoplasmic contamination. Sedimentation of the nuclei through dense sucrose is recommended to reduce cytoplasmic contamination. Resuspend the nuclear pellet in 1× extraction buffer by shearing through an 18-gauge needle and mix thoroughly with 9 volumes of 1 M sucrose in the isolation buffer. Sediment the nuclei by centrifugation at 100,000×g, 30 min, 7°. Wash the nuclear pellet once in isolation buffer to rehydrate them and continue with the extraction.
- Resuspend the nuclear pellet in one-half original packed cell volume of 1× extraction buffer by passage through an 18-gauge needle or vortexing and bring the solution to a final monovalent salt concentration of 250 mM (with or without reducing agents). Incubate at 7° for 30 min with occasional mixing.
- Clear residual nuclear material by centrifugation (6000×g, 10 min, 7°) and draw off the supernatant containing nuclear extract.
- Reextract the nuclear pellet in one-half original packed cell volume of 1× extraction buffer containing 250 mM monovalent salt as above, pool the nuclear extract supernatants, and prepare an S100 extract. Store the nuclear and cytoplasmic extracts as described above.

A suitable alternative for some adherent or suspension tissue culture cells is the protocol described for use with the NE-PER® kit (Pierce) for nuclear and cytoplasmic extract preparation. Solid tissues, however, require greater physical force for disruption such as glass-to-glass or Teflon-to-glass homogenization. For this purpose, replace 5× extraction buffer with 0.25 M STM (0.25 M sucrose, 50 mM Tris, pH 7.0, and 5 mM $MgCl_2$ with proteinase inhibitors) and homogenize the tissue by 8–10 passes with a tight-fitting homogenizer of choice, typically a Type A glass-to-glass homogenizer (Kontes) or a Teflon-to-glass homogenizer (Thomas) with a clearance time of 3.5 sec at room temperature (Harris *et al.*, 1993; Smith *et al.*, 1984).

Fibrous material from the tissue can be removed by filtering the homogenate through three layers of gauze. A crude nuclear pellet is obtained by centrifugation (5000×g, 10 min, 7°). An S100 supernatant should be prepared from the cytoplasmic extract as described above. To purify nuclei, the crude nuclear pellet should be thoroughly resuspended in one pellet volume of 0.25 M STM and eight pellet volumes of 2.2 M sucrose STM and the mixture sedimented at 100,000×g (1 h, 7°). The purified nuclei should be washed and rehydrated in 10 pellet volumes of 0.25 M STM followed by centrifugation (6000×g, 10 min, 7°) and then subjected to high salt extraction as described above.

3.4. Ammonium sulfate enrichment of editing factors

Although nuclear extracts are most active in apoB mRNA editing, the bulk of total cellular APOBEC-1 and ACF is cytoplasmic (Sowden *et al.*, 2002). This fraction can be activated to edit apoB mRNA *in vitro* by high salt dissociation of the 60S cytoplasmic aggregates recovered from glycerol gradient sedimentation and reassembling editosomes on RNA reporters *in vitro* (Harris *et al.*, 1993). Ammonium sulfate fractionation of cytoplasmic extracts serves as a rapid means of dissociating cytoplasmic 60S aggregates while enriching for editing activity (Fig. 18.3).

A saturated ammonium sulfate solution should be prepared by adding excess enzyme grade ammonium sulfate (USB) to a stirred buffer (for the apoB mRNA editing system, this is 50 mM Tris, pH 8) at room temperature overnight. The saturated solution should then be recovered from the insoluble ammonium sulfate and stored at 4–7° for at least 48 h prior to use to achieve an equilibrium of saturation at the lower temperature. The saturated solution of ammonium sulfate is then added to the extract at a progressively higher percentage (v/v ratios). After addition of ammonium sulfate, samples should be precipitated for 15 min on ice before the insoluble protein is recovered by centrifugation (14,000×g, 15 min, 7°).

The supernatant should be processed to the next percentage ammonium sulfate, while the pellet should be resuspended by gently pipetting in buffer (50 mM Tris, pH 8, 50 mM KCl, 0.1 mM DTT) to avoid

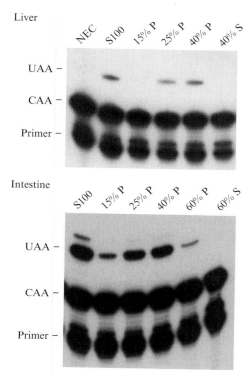

Liver

Figure 18.3 Ammonium sulfate fractionation of liver and intestine cytoplasmic extracts. Rat liver and small intestinal cell (enterocyte) cytoplasmic extracts were sequentially treated with increasing amounts of saturated ammonium sulfate to the indicated concentrations and at each percentage, the insoluble proteins were removed by centrifugation and the supernatant was processed further. The upper panel and lower panels show autoradiographs of poisoned primer extension assays performed on liver and enterocyte cytoplasmic extract ammonium sulfate "cuts," respectively. P, pellet resulting from salting out; S, residual supernatant; NEC, no edit control where the apoB reporter was assayed without reacting it with extract; S100, starting extract. Percent ammonium sulfate is indicated above each lane.

bubbles and dialyzed against 2 liters of the same buffer containing 10% glycerol with one buffer change over a 6- to 12-h period at 7°. The dialysate should be cleared of insoluble material (14,000×g, 15 min, 7°) and stored at an appropriate temperature (see below). Dialysis is essential, as high concentrations of ammonium sulfate can activate ribonuclease activity in liver cytoplasmic extracts that will cleave RNA reporters and could be confused as stops due to ddGTP in the primer extension assay. Divalent ions should be omitted from the dialysis buffer to avoid the formation of aggregates; for this reason, we frequently add 30 mM EDTA to the dialysis buffer.

3.5. Purification of recombinant C-to-U editing factors

Recombinant editing factors have proven to be a difficult class of proteins to express and purify in sufficient quantities for structural analysis, although small quantities have been expressed in *Escherichia coli*, yeast, or through baculovirus transduction of insect cells for analytical work (Anant and Davidson, 2000; Anant *et al.*, 1995, 2001a; Blanc *et al.*, 2001b, 2003; Chester *et al.*, 2003, 2004; Dance *et al.*, 2000, 2001, 2002; Galloway *et al.*, 2003; Lau *et al.*, 2001a,b; Lehmann *et al.*, 2006a,b; Lellek *et al.*, 2000; Navaratnam *et al.*, 1995, 1998; Sowden *et al.*, 2004; Xie *et al.*, 2004; Yang and Smith, 1996; Yang *et al.*, 2000, 2001). A major contributor to this is the ability of APOBEC-1 and family members to be promutagenic for single-stranded DNA in transcribed regions of the genome (Coker *et al.*, 2006; Harris *et al.*, 2002; Ichikawa *et al.*, 2006; Petersen-Mahrt and Neuberger, 2003; Petersen-Mahrt *et al.*, 2002; and see below); therefore, they are toxic to *E. coli* and/or are shunted into inclusion bodies or insoluble aggregates.

The second problem encountered with APOBEC family members and ACF is the formation of insoluble aggregates during and/or following purification. Even when monomeric proteins have been purified by size exclusion chromatography or electrophoretic methods, they will, within hours, form insoluble aggregates that cannot be dissociated with 6 M urea or 1% sodium dodecyl sulfate (SDS). This tendency is exacerbated by low temperatures or freezing.

The third difficulty encountered with these proteins is their sensitivity to the placement of affinity purification and epitope tags either on the N-terminus or C-terminus (see references above). This is particularly true for the APBOEC-1 homolog known as activation-induced deaminase (AID) (Muramatsu *et al.*, 1999, 2000) but is infrequently seen with APO-BEC-1 itself. In fact, APOBEC-1 chimeras containing twice the native mass of APOBEC still retain most of their enzymatic activity (Siddiqui *et al.*, 1999; Yang *et al.*, 2000). Guidelines and alterative approaches that have been useful in expressing recombinant proteins are listed below:

- First and foremost, carry out all cell lysing, extractions, and affinity purifications steps at room temperature.
- The yield of soluble protein may be increased by expressing the protein in insect cells or yeast.
- When expressing proteins in *E. coli*, grow the culture to late log phase (1.6–1.8 OD_{600}) prior to induction of recombinant protein expression and reduce the temperature of the culture to 24–30° during the induction.
- Induce protein expression using low induction conditions [0.1 mM isopropyl thiogalactopyranoside (IPTG)] for varying durations (30 min to 2 h) and then place the culture on ice.
- Scale up the yield by inducing multiple cultures (6 × 700 ml) rather than trying to maximize the yield from a single culture.

- Raise antibodies to peptides or to the whole protein as a detection system rather than using epitope tags.
- Use a minimal affinity tag for purification. We have found that four consecutive histidines placed at the C-terminus are sufficient for nickel NTA (Qiagen) purification.
- If tags are used, evaluate multiple tags at either the N- or C-terminus of the recombinant protein.
- Avoid protein expression conditions that give rise to truncated proteins or internal translation starts, as these partial proteins may retain their capacity to multimerize and may cause dominant negative effects on the full length protein's activity.
- Use physical methods to evaluate the maintenance of folded protein structure and monodispersity of the protein such as circular dichroism or dynamic light scattering, respectively (not just protein yield), when determining conditions for extracting the recombinant protein.
- DNase and/or RNase digestion of the cell lysate may yield more protein, but the recombinant protein may fall out more readily upon purification (such is the case with APOBEC-3G).
- The use of 1% Triton X-100, 1 M NaCl, or 1 M urea may increase the yield of recombinant protein and may not denature the protein.

4. ACTIVITY ASSAYS

The *in vitro* assay for editing activity has been essential to the rapid progress in identifying and characterizing the mechanism and factors involved in apoB mRNA editing. The ability to determine the amount of substrate edited per unit factor over time is clearly the most important assay requiring development. It is also important to establish assays for factor–substrate and factor–factor interactions (protein–nucleic acid or nucleic acid–nucleic acid and protein–protein interactions). Success in establishing an *in vitro* system depends on the development of three key components: (1) identification of an editing site and its optimal expression as a reporter substrate, (2) identification of conditions for fractionating or purifying editing competent factors, and (3) establishing an assay for quantifying editing activity.

4.1. RNA editing

In vitro apoB mRNA editing activity is typically assayed on an RNA reporter transcribed from a linearized plasmid using RNA polymerase (T7, T3, or SP6). ApoB cDNA inserts of various lengths containing the editing site at C6666 were evaluated for flanking sequence requirements (required *cis*-acting elements and ideal overall length) leading to the routine use of reporters 55–500 nt long (Backus and Smith, 1991, 1992, 1994; Driscoll *et al.*, 1989, 1993; Harris *et al.*, 1993; Hersberger and Innerarity, 1998; Hersberger *et al.*, 1999; Shah *et al.*, 1991; Smith, 1993, 1998; Smith *et al.*, 1991; Sowden *et al.*, 1996b; Yamanaka *et al.*, 1996). The reporter should be gel purified and quantified prior to use (Smith, 1998).

An appropriate starting point for an editing reaction is to evaluate the activity in 20–80 μg of protein as cell or organelle extract or 0.5–20 μg of purified recombinant protein on 10–20 fmol of reporter RNA. Important considerations in optimizing the *in vitro* assay include the following.

- There is no divalent requirement for apoB mRNA editing [editing reaction buffer: 10 mM HEPES, pH 7.9, 10% glycerol (v/v), 50 mM KCl, 50 mM EDTA, 0.25 mM DTT, and 40 units of RNasin®(Promega)], but this should be evaluated for other editing activities. The pH and monovalent salt optima should be determined.
- The temperature optima should be determined. The ideal temperature for apoB mRNA *in vitro* editing is 30°.
- A kinetic study should be performed to determine parameters such as K_m and V_{max}. The kinetics of *in vitro* apoB mRNA editing using extracts under the conditions described above reach a plateau by 1–3 h. This is highly dependent on input protein, the purity of the editing factors, and optimization of the substrate. The conditions described previously are nonsaturating in terms of substrate and higher concentrations of substrate (5- to 10-fold molar excess relative to the editing complex) should be used to determine the specific activity of editing factors.

Editing reactions can be stopped with 2× STOP buffer (2× STOP Buffer: 100 mM Tris, pH 7.5, 100 mM EDTA, 0.4% SDS, 0.2 μg/ml proteinase K) and the substrate recovered by phenol–chloroform extraction, chloroform–isoamyl alcohol extraction, and ethanol precipitation with 20 μg of glycogen/100 μl of editing reaction. The entire substrate recovered from each *in vitro* editing reaction should be quantified by the poisoned primer extension assay described above.

4.2. Mutagenic assay

It was not until the APOBEC family of related cytidine deaminases had been discovered that the field appreciated that these enzymes had significant deoxycytidine activity on single-stranded DNA (Harris *et al.*, 2003; Petersen-Mahrt *et al.*, 2002). In fact, at this time, APOBEC-1 is the only family member with RNA editing activity. Other members of the APOBEC-1 family that have been characterized are RNA-binding proteins, whose activity is directed to single-stranded DNA, largely within regions of the genome that are actively transcribed (Ichikawa et al., 2006; Muramatsu et al., 1999, 2000; Petersen-Mahrt et al., 2002) or on reverse transcripts of viral genomes (Harris et al., 2003; Petersen-Mahrt and Neuberger, 2003; Petersen-Mahrt et al., 2002; Sheehy et al., 2002; Yu et al., 2004; Zhang et al., 2003; Zheng et al., 2004). Despite biological activity as an RNA editing enzyme, APOBEC-1 is >10-fold more active as a DNA mutator than any other member of the family, a characteristic that perhaps explains why it is so difficult to overexpress APOBEC-1 in *E. coli* and why high levels of APOBEC-1 overexpression can cause cancer (Yamanaka et al., 1995).

The DNA mutator assay has been reviewed (Coker et al., 2006) and was originally used to determine DNA deaminase activity of AID as well as other members of the APOBEC family of proteins (Harris et al., 2002; Petersen-Mahrt et al., 2002). It has subsequently gained widespread use, but it is not without limitations. The assay measures the induction of rifampicin resistance in *E. coli* that is mediated by the accumulation of mutations in the rpoB gene encoding the β-subunit of RNA polymerase (Coker et al., 2006). These mutations tend to cluster to specific regions of the gene, designated clusters I, II, and III (Jin and Gross, 1988). This property also allows for sequencing of specific regions of the rpoB gene to characterize the exact nature of the deaminase-dependent mutations and the preference for nearest-neighbor nucleotides. We have found that the following methods yield the most reproducible results:

- The deaminase of interest should be inserted into an appropriate *E. coli* expression vector, e.g., pTrc99A (Amann et al., 1988). Unlike many more recently derived expression vectors such as the pET series (Novagen), pTrc99 carries no purification or epitope tags. It has been the author's experience that both 5′ and 3′ tag sequences can impair the deaminase activity of AID; therefore, whether such tags affect the biological activity of other deaminase proteins should be carefully evaluated.
- The deaminase construct should be transformed into a mutation repair deficient, uracil-DNA glycosylase-deficient (ung⁻) *E. coli* strain such as BW310 (Hfr[PO-45] relA1 spoT1 thi-1 ung-1). Vector alone should be transformed into the host strain, as these clones will serve as negative controls for mutator activity. The parental strain KL16, which expresses uracil-DNA glycosylase and can repair deaminase-induced dC-to-dU mutations, can also be used as a control that will demonstrate how many of the colonies were the result of deaminase DNA mutation activity.
- L-Broth (1.5 ml) containing 100 μg/ml carbenicillin (Gene Therapy Systems) plus 1 mM IPTG (Invitrogen) should be inoculated with a single colony and cultured at 37°, with shaking, overnight. To obtain statistically significant values for mutation rates, multiple cultures of each clone

should be inoculated (Fig.18.4). Carbenicillin is a semisynthetic analog of the naturally occurring penicillin. It is a more stable derivative than ampicillin and is not degraded as readily by secreted β-lactamase; it is therefore the antibiotic of choice for this assay in which retention of deaminase is critical. Other appropriate antibiotics should be added if alternative vectors to pTrc99 are used.

- From the overnight cultures, 10-fold serial dilutions of each culture should be prepared in L-broth (typically to 10^{-7}). Aliquots (100 μl) of the 10^{-5}, 10^{-6}, and 10^{-7} dilutions should be plated onto LB plates with the antibiotic to determine the number of viable cells in each culture. Also plate 100-μl aliquots of the 10^0, 10^{-1}, 10^{-2}, and 10^{-3} dilutions onto LB plates containing both carbenicillin 100 μg/ml and rifampicin (MP Biomedicals) 100 μg/ml to determine the number of rifampicin-resistant colonies in the culture. The plates should be incubated at 37° overnight and the number of colonies counted the next day.

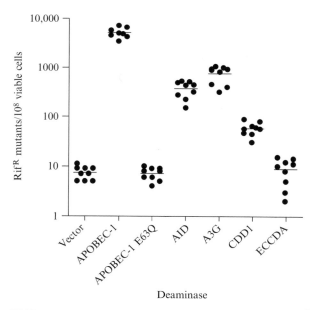

Deaminase

Figure 18.4 DNA mutator assay. Mutations in the rifampicin gene (Rif^R) due to the expression of cytidine deaminases in *E. coli* result in antibiotic resistance and colony formation. Colonies formed in bacteria transformed with each deaminase from replicate experiments are shown as scatter diagrams with the median value drawn through. Bacteria transformants tested were Vector (vector alone); APOBEC-1; APOBEC-1 E63Q (a catalytically inactivated mutant of APOBEC-1); AID; A3G, APOBEC-3G; CDD1 (yeast cytidine deaminase with RNA editing activity on test substrates); ECCDA (*E. coli* cytidine deaminase).

- The number of rifampicin–resistant colonies per 10^8 viable cells can be calculated from the colonies of each deaminase transformant plated under each condition after making corrections for the number of colonies on control plates.

Generally this assay works well to determine if a particular protein expresses DNA deaminase activity. However, caution must be exercised when comparing DNA mutator activity between deaminases, as colony counts are not normalized for deaminase expression levels or factors such as tags or truncations that might inhibit editing activity. The expression of full-length proteins and the amount of each deaminase expressed should be evaluated by Western blotting. Given the tendency of APOBEC family members to enter inclusion bodies where they are typically inactivated, there is great uncertainty as to the amount of each deaminase that exists in an active form, hence this assay is largely qualitative.

4.3. UV crosslinking

Ultraviolet (UV) light crosslinking of protein to RNA and its use in identifying ACF have been reviewed (Smith, 1998) and will be described only briefly here. The technique relies on combining a radiolabeled RNA reporter and an extract or recombinant protein of interest under appropriate buffer and temperature conditions (usually the RNA editing assay conditions are appropriate). After complexes have formed, RNA–protein interactions are fixed by inducing covalent crosslinks with 260-nm UV light (Smith, 1998). The bulk of uncrosslinked RNA should be removed by RNase digestion, leaving a radiolabeled nucleotide tag attached to the RNA binding protein. Precipitate the reaction with five volumes of high-performance liquid chromatography (HPLC) grade acetone (Baker) and resolve the proteins by sodium dodecyl sulfate polyacrylamide gel electrophoresis (SDS–PAGE).

The RNA-binding proteins will migrate homogeneously and nearly true to their native molecular weights and therefore can be readily identified (unlike the smears on gels seen in gel shift assays). Without RNase digestion, the crosslinked protein will be shifted to the top of the running gel and stacking gel–running gel interface and will be absent from its native Rf region of the gel. The UV crosslinking method is not only appropriate for detecting RNA binding proteins in crude extracts, but has also been used to validate that purified recombinant proteins have retained their activity (Galloway et al., 2003).

Typically, the reporter RNA is transcribed and uniformly radiolabeled in vitro with [α-^{32}P]ATP (Smith, 1998). Other radiolabel nucleotides can be used for labeling the reporter as shown in Fig. 18.5; however, the recovery of crosslinked protein–RNA complexes may be influenced by the choice of label through the inherent efficiency of UV crosslinking that follows a hierarchy of adenosine > uridine > guanidine > cytidine (Hockensmith *et al.*, 1993). In the case of ACF binding to the mooring sequence (an AU-rich sequence, Fig. 18.1), the yield of crosslinked ACF was greatest when the apoB reporter was radiolabeled with either ATP or UTP (Fig. 18.5). Much lower yields of crosslinked ACF were observed when GTP or CTP were used for radiolabeling. It is also apparent from these data that combined digestion with RNase A (cleaves the 3′ end of unpaired C and U residues) and T1 (cleaves the 3′ end of unpaired G residues) resulted in reduced background compared to RNase A or T1 digestion alone, although the relative signal intensities may be comparable (Fig. 18.5).

Due to the inherent differences in UV crosslinking efficiencies for each base, this approach should be used as a semiquantitative method for identifying proteins that may bind to RNA, but should not be used alone to infer site-specific interactions based on the efficiency of crosslinking to RNAs radiolabeled with different nucleotides. More appropriate methods for mapping the protein–RNA contacts should be used, such as site-specific radiolabeling of the RNA reporter (Ma *et al.*, 2005; Moore and Sharp, 1992) or evaluating reporter RNAs with site-specific mutations (Anant and Davidson, 2000; Anant *et al.*, 1995; Backus and Smith, 1992; Harris *et al.*, 1993; Navaratnam *et al.*, 1998; Shah *et al.*, 1991). The specificity of the RNA binding activity for the reporter sequence can be determined through the addition of nonradiolabeled RNAs as competitor with the radiolabeled reporter (Anant and Davidson, 2000; Harris *et al.*, 1993; Mehta and Driscoll, 1998; Mehta *et al.*, 2000). When the cold competitor is the same sequence as the reporter, near complete ablation of radiolabeled crosslinking signal at

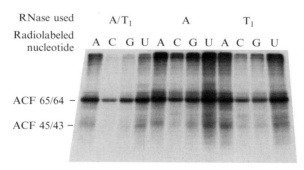

Figure 18.5 UV crosslinking efficiency on RNA substrates labeled with different nucleotides. Rat liver nuclear extract was reacted with apoB reporter uniformly labeled with [α-^{32}P]ATP, CTP, GTP, or UTP (indicated across the top of each lane as A, C, G, or U, respectively). Following UV crosslinking, the reactions were digested with RNase A and RNase T1 or each RNase separately (indicated at the top of the figure as A/T1, A, or T1). The electrophoretic migration of ACF65/64 and ACF45/43 crosslinked proteins is indicated to the left.

a 1:5 ratio can be anticipated. Nonspecific competitor RNAs or RNAs in which critical sites have been mutated may not be able to compete and reduce signal even at 100- to 1000-fold excess.

The binding constants for RNA-binding proteins can be approximated by this method or by a filter-binding assay (Mehta and Driscoll, 2002) when using purified RNA-binding proteins under defined *in vitro* conditions. However, if sufficient soluble purified protein can be obtained, kinetic parameters should be determined by Surface Plasmon Resonance (Biacore) analyses (Park-Lee *et al.*, 2003) or a related physical approach. For additional information on methods used for RNA–protein interactions, the reader is referred to the following website (www.bmb.uga.edu/bcmb8120/Technical.pdf).

4.4. Editosome assembly

Ultimately, a common goal in the editing field is to understand the structure and function of the editosome and its regulation. It is therefore necessary to characterize the individual components of the editosome and then determine the interactions that lead to substrate binding, editing site specificity, and editing activity. The analysis of RNA-binding proteins described in the preceding section is an essential step toward these objectives. Protein–protein interactions of editing factors have been studied using the yeast two hybrid system (Greeve *et al.*, 1998; Lau and Chan, 2003; Lau *et al.*, 2001a,b; Sowden *et al.*, 2004). Kits are available and commercial services for two hybrid analysis are available with a variety of expression libraries that can be probed with editing factors (www.biocompare.com.matrix3517/Two-Hybrid-System-Construction-Kits.html). The strength of the system is that selection for protein–protein interactions occurs in the context of a cellular environment, and consequently proteins that are capable of interactions are likely to be expressed in a soluble and folded condition. However, there are risks of false positives and negatives when the system is used as a screening tool to identify new factors that may interact with a "bait" editing factor (Bartel *et al.*, 1993). In fact, there are lists of frequently encountered false-positive interactions that it is necessary to be aware of at the onset of the analysis (www.uni-konstanz.de/FuF/Bio/Bioinformatik/bivi109.htm or cmbi.bjmu.edu.cn/cm bidata/proteome/method/html/2hybrid02.htm). Though not eliminated, these risks may be reduced when two known editing factors are tested in this system for their ability to interact (Fig. 18.6). Even so, it may be necessary to test the interactions with the two recombinant factors as either bait or "prey" and to test the effect of adding DNA binding or transactivating domains at either the N- or C-terminus of each factor.

Protein–protein (and protein–RNA) interactions have been evaluated using factor- or tag-specific antibodies for immunopurification of proteins involved in editing. While proteins or RNAs that coimmunopurify (or coimmunoprecipitate) with the epitope-bearing factor warrant further

A   Interactions of APOBEC-1

Histidine prototrophy β-galactosidase activity

APOBEC-1, Lamin-C   APOBEC-1, Lamin-C

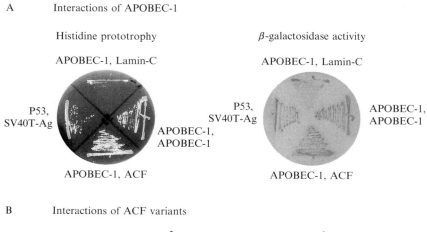

P53, SV40T-Ag   APOBEC-1, APOBEC-1   P53, SV40T-Ag   APOBEC-1, APOBEC-1

APOBEC-1, ACF   APOBEC-1, ACF

B   Interactions of ACF variants

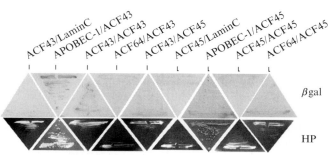

βgal

HP

Figure 18.6 Yeast two–hybrid analyses. Recombinant APOBEC-1 and ACF variants were evaluated for their ability to interact using selection in the yeast two-hybrid system. Successful interactions result in yeast growth (A, left histidine prototrophy; B, bottom HP) and the induction of β-galactosidase activity (A, right, and B, top βgal). The interaction of p53 and SV40 large T antigen is a positive control. Lamin C-APOBEC-1 serves as a negative control. (A) Strong two-hybrid interaction of APOBEC-1 with itself and with ACF64. (B) A hierarchy of interactions wherein the APOBEC-1 interaction with ACF variants is not equivalent. Moreover, the ability of ACF variants to interact with themselves (homodimers) and each other is not equivalent. For reasons that may be due to variations in plating density, some two hybrid pairs do not yield equivalent results for histidine prototrophy and the induction of β-galactosidase activity (e.g., ACF43/ACF43 and APOBEC-1/ACF45).

analysis as described above, they should be viewed with circumspection. We have found that editing factors have a tendency to aggregate following RNase digestion and during purification. In addition, there is considerable nonspecific association of proteins and bulk RNA with agarose beads or metal chelating affinity resin, evident by Commassie blue-stained SDS–PAGE or mass spectroscopy and in the sequences of copurifying RNAs determined by microarray analysis or direct sequencing. Therefore, a finding of multiple RNA-binding proteins associated with a given editing factor by the coimmunoprecipitation approach may or may not reflect biologically significant interactions. Moreover, binding of multiple proteins to RNA or DNA may result in coimmunoprecipitation of proteins where protein–protein interactions are bridged by nucleic acid and therefore are indirect. Regardless of the proteins that are identified by this approach, additional steps are usually necessary to validate the specificity of the interactions and their functional consequences (Anant et al., 2001a,b; Blanc et al., 2001a,b, 2003; Chester et al., 2004; Greeve et al., 1998; Lau and Chan, 2003; Lau et al., 2001a,b, 1997; Lellek et al., 2000; Mehta, 2002; Mehta et al., 1996; Schock et al., 1996; Yang et al., 1997).

Copurification analyses yield the most definitive results when they are carried out under defined in vitro conditions using recombinant factors (Mehta et al., 2000) or when initial steps are taken to enrich for the editosome. Glycerol gradient sedimentation of extracts has been used to enrich for complexes containing editing factors (Backus and Smith, 1991; Harris et al., 1993; Smith, 1998; Smith et al., 1991; Sowden et al., 2002) and has the following advantages: (1) discrete aggregate sizes containing the editing factors of interest can be identified by Western blotting aliquots of gradient fractions and (2) the approach is amenable to be scaled up to the preparative level. Complexes of editing factors also can be identified and analyzed by native gel electrophoresis. The in vitro editosome assembly assay is the same as a standard in vitro editing assay with the following three exceptions: (1) assembly assays are performed with ^{32}P-labeled RNA substrate, (2) assembly assays are carried out for less time than editing assays, typically less than 20–60 min [because editosome assembly precedes editing activity (Backus and Smith, 1991; Smith, 1998; Smith et al., 1991)], and (3) the assay is terminated by the addition of sample loading buffer and flash frozen in liquid nitrogen. The native gel method has been reviewed (Smith, 1998) and is summarized below.

Editosomes are high-molecular-weight complexes and as such resolve poorly in standard electrophoretic gel systems. The method of choice is a native gel [referred to as ribonucleoprotein particle or "RNP" gel system (Grabowski et al., 1985; Zillmann et al., 1988)], consisting of a very low percentage of polyacrylamide mixed with agarose as a supportive component (Backus and Smith, 1991; Harris et al., 1993; Smith, 1998; Smith et al., 1991). Gels should be run vertically, but the polymerized gel matrix does not have sufficient strength to adhere to the glass plates and will slip out of the bottom. Gels therefore should be cast with a bottom spacer consisting of two layers of Whatman^{TM} 3MM paper instead of a plastic spacer. This paper will stay in place during the electrophoresis. It is also important that the agarose not polymerize before the polyacrylamide does. For this reason, the gel needs to be poured warm and the gel plates need to be prewarmed to prevent polymerization of the agarose during pouring.

5. SUMMARY

In the foregoing sections, methods have been described that frequently have been used in the discovery of the apoB mRNA editing mechanism, the factors involved, and their regulation. The reader should consider these methods as a starting point and not necessarily inclusive of all approaches that can or should be taken. An important consideration when implementing these methods is that they are carefully evaluated for optimal conditions and their limitations are considered relative to the intended objectives of the research. The proteins involved in mammalian C-to-U editing can present unique problems during purification and analysis and therefore optimization of protein solubility, native structure, and activity should be reevaluated at each step. An effort has been made to represent the numerous contributions of investigators that have made the discovery process possible in this field.

ACKNOWLEDGMENTS

The method development and data described in this review stem from the author's research from 1991 to the present and were supported by research grants from the NIH (DK43739, AI54369, and AI58789), the Air Force (F49620), the Office of Naval Research Grant (J1915), the Alcoholic Beverage Medical Research Foundation, and the Council for Tobacco Research. The author is grateful for the many contributions of Mark P. Sowden and former postdocs and students to the research and Jenny M. L. Smith for the preparation of the figures.

REFERENCES

Amann, E., Ochs, B., and Abel, K. J. (1988). Tightly regulated tac promoter vectors useful for the expression of unfused and fused proteins in *Escherichia coli. Gene* **69,** 301–315.

Anant, S., and Davidson, N. O. (2000). An AU-rich sequence element (UUUN[A/U]U) downstream of the edited C in apolipoprotein B mRNA is a high-affinity binding site for Apobec-1: Binding of Apobec-1 to this motif in the 3' untranslated region of c-myc increases mRNA stability. *Mol. Cell. Biol.* **20,** 1982–1992.

Anant, S., MacGinnitie, A. J., and Davidson, N. O. (1995). Apobec-1, the catalytic subunit of the mammalian apolipoprotein B mRNA editing enzyme, is a novel RNA-binding protein. *J. Biol. Chem.* **270,** 14762–14767.

Anant, S., Yu, H., and Davidson, N. O. (1998). Evolutionary origins of the mammalian apolipoproteinB RNA editing enzyme, apobec-1: Structural homology inferred from analysis of a cloned chicken small intestinal cytidine deaminase. *Biol. Chem.* **379,** 1075–1081.

Anant, S., Henderson, J. O., Mukhopadhyay, D., Navaratnam, N., Kennedy, S., Min, J., and Davidson, N. O. (2001a). Novel role for RNA-binding protein CUGBP2 in mammalian RNA editing. *J. Biol. Chem.* **276,** 47338–47351.

Anant, S., Mukhopadhyay, D., Sankaranand, V., Kennedy, S., Henerson, J. O., and Davidson, N. O. (2001b). ARCD-1, an apobec-1-related cytidine deaminase, exerts a dominant negative effect on C to U mRNA editing. *Am. J. Physiol. Cell Physiol.* **281,** 1904–1916.

Backus, J. W., and Smith, H. C. (1991). Apolipoprotein B mRNA sequences 3′ of the editing site are necessary and sufficient for editing and editosome assembly. *Nucleic Acids Res.* **19,** 6781–6786.

Backus, J. W., and Smith, H. C. (1992). Three distinct RNA sequence elements are required for efficient apolipoprotein B (apoB) RNA editing *in vitro. Nucleic Acids Res.* **20,** 6007–6014.

Backus, J. W., and Smith, H. C. (1994). Specific 3′ sequences flanking a minimal apolipoprotein B (apoB) mRNA editing 'cassette' are critical for efficient editing *in vitro. Biochim. Biophys. Acta* **1217,** 65–73.

Bartel, P., Chien, C. T., Sternglanz, R., and Fields, S. (1993). Elimination of false positives that arise in using the two-hybrid system. *Biotechniques* **14,** 920–924.

Bennett, R. P., Diner, E., Sowden, M. P., Lees, J. A., Wedekind, J. E., and Smith, H. C. (2006). APOBEC-1 and AID are nucleo-cytoplasmic trafficking proteins but APOBEC3G cannot traffic. *Biochem. Biophys. Res. Commun.* **350,** 214–219.

Bishop, K. N., Holmes, R. K., Sheehy, A. M., and Malim, M. H. (2004). APOBEC-mediated editing of viral RNA. *Science* **305,** 645.

Blanc, V., Henderson, J. O., Kennedy, S., and Davidson, N. O. (2001a). Mutagenesis of apobec-1 complementation factor reveals distinct domains that modulate RNA binding, protein-protein interaction with apobec-1, and complementation of C to U RNA-editing activity. *J. Biol. Chem.* **276,** 46386–46393.

Blanc, V., Navaratnam, N., Henderson, J. O., Anant, S., Kennedy, S., Jarmuz, A., Scott, J., and Davidson, N. O. (2001b). Identification of GRY-RBP as an apolipoprotein B RNA-binding protein that interacts with both apobec-1 and apobec-1 complementation factor to modulate C to U editing. *J. Biol. Chem.* **276,** 10272–10283.

Blanc, V., Kennedy, S. M., and Davidson, N. O. (2003). A novel nuclear localization signal in the auxiliary domain of apobec-1 complementation factor (ACF) regulates nucleo-cytoplasmic import and shuttling. *J. Biol. Chem.* **278,** 21148–21204.

Chen, S. H., Habib, G., Yang, C. Y., Gu, Z. W., Lee, B. R., Weng, S. A., Silberman, S. R., Cai, S. J., Deslypere, J. P., Rosseneu, M., Gotto, A. M., Li, W.-H., and Chan, L. (1987). Apolipoprotein B-48 is the product of a messenger RNA with an organ-specific in-frame stop codon. *Science* **238,** 363–366.

Chester, A., Somasekaram, A., Tzimina, M., Jarmuz, A., Gisbourne, J., O'Keefe, R., Scott, J., and Navaratnam, N. (2003). The apolipoprotein B mRNA editing complex performs a multifunctional cycle and suppresses nonsense-mediated decay. *EMBO J.* **22,** 3971–3982.

Chester, A., Weinreb, V., Carter, C. W., Jr., and Navaratnam, N. (2004). Optimization of apolipoprotein B mRNA editing by APOBEC1 apoenzyme and the role of its auxiliary factor, ACF. *RNA* **10,** 1399–1411.

Coker, H. A., Morgan, H. D., and Petersen-Mahrt, S. K. (2006). Genetic and *in vitro* assays of DNA deamination. *Methods Enzymol.* **408,** 156–170.

Dance, G. S., Sowden, M. P., Yang, Y., and Smith, H. C. (2000). APOBEC-1 dependent cytidine to uridine editing of apolipoprotein B RNA in yeast. *Nucleic Acids Res.* **28,** 424–429.

Dance, G. S., Beemiller, P., Yang, Y., Mater, D. V., Mian, I. S., and Smith, H. C. (2001). Identification of the yeast cytidine deaminase CDD1 as an orphan C→U RNA editase. *Nucleic Acids Res.* **29,** 1772–1780.

Dance, G. S. C., Sowden, M. P., Cartegni, L., Cooper, E., Krainer, A. R., and Smith, H. C. (2002). Two proteins essential for apolipoprotein B mRNA editing are expressed from a single gene through alternative splicing. *J. Biol. Chem.* **277,** 12703–12709.

Driscoll, D. M., Wynne, J. K., Wallis, S. C., and Scott, J. (1989). An *in vitro* system for the editing of apolipoprotein B mRNA. *Cell* **58**, 519–525.

Driscoll, D. M., Lakhe-Reddy, S., Oleksa, L. M., and Martinez, D. (1993). Induction of RNA editing at heterologous sites by sequences in apolipoprotein B mRNA. *Mol. Cell. Biol.* **13**, 7288–7294.

Galloway, C. A., Sowden, M. P., and Smith, H. C. (2003). Increasing the yield of soluble recombinant protein expressed in *E. coli* by induction during late log phase. *Biotechniques* **34**, 524–526, 528, 530.

Grabowski, P. J., Seiler, S. R., and Sharp, P. A. (1985). A multicomponent complex is involved in the splicing of messenger RNA precursors. *Cell* **42**, 345–353.

Greeve, J., Lellek, H., Rautenberg, P., and Greten, H. (1998). Inhibition of the apolipoprotein B mRNA editing enzyme-complex by hnRNP C1 protein and 40S hnRNP complexes. *Biol. Chem.* **379**, 1063–1073.

Hadjiagapiou, C., Giannoni, F., Funahashi, T., Skarosi, S. F., and Davidson, N. O. (1994). Molecular cloning of a human small intestinal apolipoprotein B mRNA editing protein. *Nucleic Acids Res.* **22**, 1874–1879.

Harris, R. S., Petersen-Mahrt, S. K., and Neuberger, M. S. (2002). RNA editing enzyme APOBEC1 and some of its homologs can act as DNA mutators. *Mol. Cell* **10**, 1247–1253.

Harris, R. S., Bishop, K. N., Sheehy, A. M., Craig, H. M., Petersen-Mahrt, S. K., Watt, I. N., Neuberger, M. S., and Malim, M. H. (2003). DNA deamination mediates innate immunity to retroviral infection. *Cell* **113**, 803–809.

Harris, S. G., Sabio, I., Mayer, E., Steinberg, M. F., Backus, J. W., Sparks, J. D., Sparks, C. E., and Smith, H. C. (1993). Extract-specific heterogeneity in high-order complexes containing apolipoprotein B mRNA editing activity and RNA-binding proteins. *J. Biol. Chem.* **268**, 7382–7392.

Hersberger, M., and Innerarity, T. L. (1998). Two efficiency elements flanking the editing site of cytidine 6666 in the apolipoprotein B mRNA support mooring-dependent editing. *J. Biol. Chem.* **273**, 9435–9442.

Hersberger, M., Patarroyo-White, S., Arnold, K. S., and Innerarity, T. L. (1999). Phylogenetic analysis of the apolipoprotein B mRNA-editing region. Evidence for a secondary structure between the mooring sequence and the 3' efficiency element. *J. Biol. Chem.* **274**, 34590–34597.

Hirano, K. I., Young, S. G., Farese, R. V., Ng, J., Sande, E., Warburton, C., Powell-Braxton, L. M., and Davidson, N. O. (1996). Targeted disruption of the mouse apobec-1 gene abolishes apolipoprotein B mRNA editing and eliminates apolipoprotein B48. *J. Biol. Chem.* **271**, 9887–9890.

Hockensmith, J. W., Kubasek, W. L., Vorachek, W. R., and Von Hippel, P. H. (1993). Laser cross-linking of proteins to nucleic acids. I. Examining physical parameters of protein-nucleic acid complexes. *J. Biol. Chem.* **268**, 15712–15720.

Ichikawa, H. T., Sowden, M. P., Torelli, A. T., Bachl, J., Huang, P., Dance, G. S., Marr, S. H., Robert, J., Wedekind, J. E., Smith, H. C., and Bottaro, A. (2006). Structural phylogenetic analysis of activation-induced deaminase function. *J. Immunol.* **177**, 355–361.

Jarmuz, A., Chester, A., Bayliss, J., Gisbourne, J., Dunham, I., Scott, J., and Navaratnam, N. (2002). An anthropoid-specific locus of orphan C to U RNA-editing enzymes on chromosome 22. *Genomics* **79**, 285–296.

Jin, D. J., and Gross, C. A. (1988). Mapping and sequencing of mutations in the *Escherichia coli* rpoB gene that lead to rifampicin resistance. *J. Mol. Biol.* **202**, 45–58.

Johnson, D. F., Poksay, K. S., and Innerarity, T. L. (1993). The mechanism for apo-B mRNA editing is deamination. *Biochem. Biophys. Res. Commun.* **195**, 1204–1210.

Lau, P. P., and Chan, L. (2003). Involvement of a chaperone regulator, Bcl2-associated athanogene-4 (BAG-4), in apolipoprotein B mRNA editing. *J. Biol. Chem.* **278**, 52988–52996.

Lau, P. P., Zhu, H.-J., Baldini, H. A., Charnsangavej, C., and Chan, L. (1994). Dimeric structure of a human apo B mRNA editing protein and cloning and chromosomal localization of its gene. *Proc. Natl. Acad. Sci. USA* **91**, 8522–8526.

Lau, P. P., Zhu, H. J., Nakamuta, M., and Chan, L. (1997). Cloning of an apobec-1-binding protein that also interacts with apolipoprotein B mRNA and evidence for its involvement in RNA editing. *J. Biol. Chem.* **272**, 1452–1455.

Lau, P. P., Chang, B. H., and Chan, L. (2001a). Two-hybrid cloning identifies an RNA-binding protein, GRY-RBP, as a component of apobec-1 editosome. *Biochem. Biophys. Res. Commun.* **282**, 977–983.

Lau, P. P., Villanueva, H., Kobayashi, K., Nakamuta, M., Chang, H. J., and Chan, L. (2001b). A DnaJ protein, apobec-1-binding protein-2, modulates apolipoprotein B mRNA editing. *J. Biol. Chem.* **276**, 46445–46452.

Lehmann, D. M., Galloway, C. A., MacElrevey, C., Sowden, M. P., Wedekind, J. E., and Smith, H. C. (2006a). Functional characterization of APOBEC-1 complementation factor phosphorylation sites. *Biochim. Biophys. Acta Mol. Cell Res.* **1773**, 408–418.

Lehmann, D. M., Galloway, C. A., Sowden, M. P., and Smith, H. C. (2006b). Metabolic regulation of apoB mRNA editing is associated with phosphorylation of APOBEC-1 complementation factor. *Nucleic Acids Res.* **34**, 3299–3308.

Lellek, H., Kirsten, R., Diehl, I., Apostel, F., Buck, F., and Greeve, J. (2000). Purification and molecular cloning of a novel essential component of the apolipoprotein B mRNA editing enzyme-complex. *J. Biol. Chem.* **275**, 19848–19856.

Ma, X., Yang, C., Alexandrov, A., Grayhack, E. J., Behm-Ansmant, I., and Yu, Y. T. (2005). Pseudouridylation of yeast U2 snRNA is catalyzed by either an RNA-guided or RNA-independent mechanism. *EMBO J.* **24**, 2403–2413.

Mehta, A., and Driscoll, D. M. (1998). A sequence-specific RNA-binding protein complements apobec-1 to edit apolipoprotein B mRNA. *Mol. Cell. Biol.* **18**, 4426–4432.

Mehta, A., and Driscoll, D. M. (2002). Identification of domains in APOBEC-1 complementation factor required for RNA binding and apolipoprotein B mRNA editing. *RNA* **8**, 69–82.

Mehta, A., Banerjee, S., and Driscoll, D. M. (1996). Apobec-1 interacts with a 65-kDa complementing protein to edit apolipoprotein-B mRNA *in vitro*. *J. Biol. Chem.* **271**, 28294–28299.

Mehta, A., Kinter, M. T., Sherman, N. E., and Driscoll, D. M. (2000). Molecular cloning of apobec-1 complementation factor, a novel RNA-binding protein involved in the editing of apolipoprotein B mRNA. *Mol. Cell. Biol.* **20**, 1846–1854.

Mian, I. S., Moser, M. J., Holley, W. R., and Chatterjee, A. (1998). Statistical modelling and phylogenetic analysis of a deaminase domain. *J. Comput. Biol.* **5**, 57–72.

Moore, M. J., and Sharp, P. A. (1992). Site-specific modification of pre-mRNA: The 2′-hydroxyl groups at the splice sites. *Science* **256**, 992–997.

Mukhopadhyay, D., Anant, S., Lee, R. M., Kennedy, S., Viskochil, D., and Davidson, N. O. (2002). C→U editing of neurofibromatosis 1 mRNA occurs in tumors that express both the type II transcript and apobec-1, the catalytic subunit of the apolipoprotein B mRNA-editing enzyme. *Am. J. Hum. Genet.* **70**, 38–50.

Muramatsu, M., Sankaranand, V. S., Anant, S., Sugai, M., Kinoshita, K., Davidson, N. O., and Honjo, T. (1999). Specific expression of activation-induced cytidine deaminase (AID), a novel member of the RNA-editing deaminase family in germinal center B cells. *J. Biol. Chem.* **274**, 18470–18476.

Muramatsu, M., Kinoshita, K., Fagarasan, S., Yamada, S., Shinkai, Y., and Honjo, T. (2000). Class switch recombination and hypermutation require activation-induced cytidine deaminase (AID), a potential RNA editing enzyme. *Cell* **102**, 553–563.

Nakamuta, M., Oka, K., Krushkal, J., Kobayashi, K., Yamamoto, M., Li, W. H., and Chan, L. (1995). Alternative mRNA splicing and differential promoter utilization determine tissue-specific expression of the apolipoprotein B mRNA-editing protein (Apobec1) gene in mice. Structure and evolution of Apobec1 and related nucleoside/nucleotide deaminases. *J. Biol. Chem.* **270**, 13042–13056.

Nakamuta, M., Chang, B. H. J., Zsigmond, E., Kobayashi, K., Lei, H., Ishida, B. Y., Oka, K., Li, E., and Chan, L. (1996). Complete phenotypic characterization of the apobec-1 knockout mice with a wild-type genetic background and a human apolipoprotein B transgenic background, and restoration of apolipoprotein B mRNA editing by somatic gene transfer of Apobec-1. *J. Biol. Chem.* **271**, 25981–25988.

Navaratnam, N., Patel, D., Shah, R. R., Greeve, J. C., Powell, L. M., Knott, T. J., and Scott, J. (1991). An additional editing site is present in apolipoprotein B mRNA. *Nucleic Acids Res.* **19**, 1741–1744.

Navaratnam, N., Shah, R., Patel, D., Fay, V., and Scott, J. (1993). Apolipoprotein B mRNA editing is associated with UV crosslinking of proteins to the editing site. *Proc. Natl. Acad. Sci. USA* **90**, 222–226.

Navaratnam, N., Bhattacharya, S., Fujino, T., Patel, D., Jarmuz, A. L., and Scott, J. (1995). Evolutionary origins of apoB mRNA editing: Catalysis by a cytidine deaminase that has acquired a novel RNA-binding motif at its active site. *Cell* **81**, 187–195.

Navaratnam, N., Fujino, T., Bayliss, J., Jarmuz, A., How, A., Richardson, N., Somasekaram, A., Bhattacharya, S., Carter, C., and Scott, J. (1998). *Escherichia coli* cytidine deaminase provides a molecular model for ApoB RNA editing and a mechanism for RNA substrate recognition. *J. Mol. Biol.* **275**, 695–714.

Park-Lee, S., Kim, S., and Laird-Offringa, I. A. (2003). Characterization of the interaction between neuronal RNA-binding protein HuD and AU-rich RNA. *J. Biol. Chem.* **278**, 39801–39808.

Petersen-Mahrt, S. K., and Neuberger, M. S. (2003). *In vitro* deamination of cytosine to uracil in single-stranded DNA by apolipoprotein B editing complex catalytic subunit 1 (APOBEC1). *J. Biol. Chem.* **278**, 19583–19586.

Petersen-Mahrt, S. K., Harris, R. S., and Neuberger, M. S. (2002). AID mutates *E. coli* suggesting a DNA deamination mechanism for antibody diversification. *Nature* **418**, 99–104.

Powell, L. M., Wallis, S. C., Pease, R. J., Edwards, Y. H., Knott, T. J., and Scott, J. (1987). A novel form of tissue-specific RNA processing produces apolipoprotein-B48 in intestine. *Cell* **50**, 831–840.

Schock, D., Kuo, S. R., Steinburg, M. F., Bolognino, M., Sparks, J. D., Sparks, C. E., and Smith, H. C. (1996). An auxiliary factor containing a 240-kDa protein complex is involved in apolipoprotein B RNA editing. *Proc. Natl. Acad. Sci. USA* **93**, 1097–1102.

Shah, R. R., Knott, T. J., Legros, J. E., Navaratnam, N., Greeve, J. C., and Scott, J. (1991). Sequence requirements for the editing of apolipoprotein B mRNA. *J. Biol. Chem.* **266**, 16301–16304.

Sheehy, A. M., Gaddis, N. C., Choi, J. D., and Malim, M. H. (2002). Isolation of a human gene that inhibits HIV-1 infection and is suppressed by the viral Vif protein. *Nature* **418**, 646–650.

Siddiqui, J. F., Van Mater, D., Sowden, M. P., and Smith, H. C. (1999). Disproportionate relationship between APOBEC-1 expression and apolipoprotein B mRNA editing activity. *Exp. Cell Res.* **252**, 154–164.

Skuse, G. R., Cappione, A. J., Sowden, M., Metheny, L. J., and Smith, H. C. (1996). The neurofibromatosis type I messenger RNA undergoes base-modification RNA editing. *Nucleic Acids Res.* **24**, 478–485.

Smith, H. C. (1993). Apolipoprotein B mRNA editing: The sequence to the event. *Semin. Cell Biol.* **4**, 267–278.

Smith, H. C. (1998). Analysis of protein complexes assembled on apolipoprotein B mRNA for mooring sequence-dependent RNA editing. *Methods* **15**, 27–39.

Smith, H. C. (2005). Mammalian C to U editing. *In* "Current Topics in Genetics" (H. Grosjean, ed.), Vol. 12, pp. 365–400. Springer, Göteborg, Sweden.

Smith, H. C. (2006). "Editing Informational Content of Expressed DNA Sequences and Their Transcripts." New York: Oxford University Press, New York.

Smith, H. C., Puvion, E., Buchholtz, L. A., and Berezney, R. (1984). Spatial distribution of DNA loop attachment and replicational sites in the nuclear matrix. *J. Cell Biol.* **99**, 1794–1802.

Smith, H. C., Kuo, S. R., Backus, J. W., Harris, S. G., Sparks, C. E., and Sparks, J. D. (1991). *In vitro* apolipoprotein B mRNA editing: Identification of a 27S editing complex. *Proc. Natl. Acad. Sci. USA* **88**, 1489–1493.

Smith, H. C., Gott, J. M., and Hanson, M. R. (1997). A guide to RNA editing. *RNA* **3**, 1105–1123.

Sowden, M., Hamm, J. K., and Smith, H. C. (1996a). Overexpression of APOBEC-1 results in mooring sequence-dependent promiscuous RNA editing. *J. Biol. Chem.* **271**, 3011–3017.

Sowden, M., Hamm, J. K., Spinelli, S., and Smith, H. C. (1996b). Determinants involved in regulating the proportion of edited apolipoprotein B RNAs. *RNA* **2**, 274–288.

Sowden, M. P., Eagleton, M. J., and Smith, H. C. (1998). Apolipoprotein B RNA sequence 3' of the mooring sequence and cellular sources of auxiliary factors determine the location and extent of promiscuous editing. *Nucleic Acids Res.* **26**, 1644–1652.

Sowden, M. P., Ballatori, N., de Mesy Jensen, K. L., Hamilton Reed, L., and Smith, H. C. (2002). The editosome for cytidine to uridine mRNA editing has a native complexity of 27S: Identification of intracellular domains containing active and inactive editing factors. *J. Cell Sci.* **115**, 1027–1039.

Sowden, M. P., Lehmann, D. M., Lin, X., Smith, C. O., and Smith, H. C. (2004). Identification of novel alternative splice variants of APOBEC-1 complementation factor with different capacities to support ApoB mRNA editing. *J. Biol. Chem.* **278**, 197–206.

Teng, B., Burant, C. F., and Davidson, N. O. (1993). Molecular cloning of an apolipoprotein B messenger RNA editing protein. *Science* **260**, 1816–1819.

Teng, B., Blumenthal, S., Forte, T., Navaratnam, N., Scott, J., Gotto, A. M., Jr., and Chan, L. (1994). Adenovirus-mediated gene transfer of rat apolipoprotein B mRNA editing protein in mice virtually eliminates apolipoprotein B-100 and normal low density lipoprotein production. *J. Biol. Chem.* **269**, 29395–29404.

Wedekind, J. E., Dance, G. S., Sowden, M. P., and Smith, H. C. (2003). Messenger RNA editing in mammals: New members of the APOBEC family seeking roles in the family business. *Trends Genet.* **19**, 207–216.

Xie, K., Sowden, M. P., Dance, G. S., Torelli, A. T., Smith, H. C., and Wedekind, J. E. (2004). The structure of a yeast RNA-editing deaminase provides insight into the fold and function of activation-induced deaminase and APOBEC-1. *Proc. Natl. Acad. Sci. USA* **101**, 8114–8119.

Xie, Y., Nassir, F., Luo, J., Buhman, K., and Davidson, N. O. (2003). Intestinal lipoprotein assembly in apobec-1-/- mice reveals subtle alterations in triglyceride secretion coupled with a shift to larger lipoproteins. *Am. J. Physiol. Gastrointest. Liver Physiol.* **285**, G735–G746.

Yamanaka, S., Poksay, K. S., Balestra, M. E., Zeng, G. Q., and Innerarity, T. L. (1994). Cloning and mutagenesis of the rabbit ApoB mRNA editing protein. A zinc motif is essential for catalytic activity, and noncatalytic auxiliary factor(s) of the editing complex are widely distributed. *J. Biol. Chem.* **269**, 21725–21734.

Yamanaka, S., Balestra, M., Ferrell, L., Fan, J., Arnold, K. S., Taylor, S., Taylor, J. M., and Innerarity, T. L. (1995). Apolipoprotein B mRNA editing protein induces hepatocellular carcinoma and dysplasia in transgenic animals. *Proc. Natl. Acad. Sci. USA* **92**, 8483–8487.

Yamanaka, S., Poksay, K. S., Driscoll, D. M., and Innerarity, T. L. (1996). Hyperediting of multiple cytidines of apolipoprotein B mRNA by APOBEC-1 requires auxiliary protein(s) but not a mooring sequence motif. *J. Biol. Chem.* **271,** 11506–11510.

Yamanaka, S., Poksay, K. S., Arnold, K. S., and Innerarity, T. L. (1997). A novel translational repressor mRNA is edited extensively in livers containing tumors caused by the transgene expression of the apoB mRNA-editing enzyme. *Genes Dev.* **11,** 321–333.

Yang, Y., and Smith, H. C. (1996). *In vitro* reconstitution of apolipoprotein B RNA editing activity from recombinant APOBEC-1 and McArdle cell extracts. *Biochem. Biophys. Res. Commun.* **218,** 797–801.

Yang, Y., and Smith, H. C. (1997). Multiple protein domains determine the cell type-specific nuclear distribution of the catalytic subunit required for apolipoprotein B mRNA editing. *Proc. Natl. Acad. Sci. USA* **94,** 13075–13080.

Yang, Y., Kovalski, K., and Smith, H. C. (1997). Partial characterization of the auxiliary factors involved in apolipoprotein B mRNA editing through APOBEC-1 affinity chromatography. *J. Biol. Chem.* **272,** 27700–27706.

Yang, Y., Sowden, M. P., and Smith, H. C. (2000). Induction of cytidine to uridine editing on cytoplasmic apolipoprotein B mRNA by overexpressing APOBEC-1. *J. Biol. Chem.* **275,** 22663–22669.

Yang, Y., Sowden, M. P., Yang, Y., and Smith, H. C. (2001). Intracellular trafficking determinants in APOBEC-1, the catalytic subunit for cytidine to uridine editing of apolipoprotein B mRNA. *Exp. Cell Res.* **267,** 153–164.

Yu, Q., Konig, R., Pillai, S., Chiles, K., Kearney, M., Palmer, S., Richman, D., Coffin, J. M., and Landau, N. R. (2004). Single-strand specificity of APOBEC3G accounts for minus-strand deamination of the HIV genome. *Nat. Struct. Mol. Biol.* **11,** 435–442.

Zhang, H., Yang, B., Pomerantz, R. J., Zhang, C., Arunachalam, S. C., and Gao, L. (2003). The cytidine deaminase CEM15 induces hypermutation in newly synthesized HIV-1 DNA. *Nature* **424,** 94–98.

Zheng, Y. H., Irwin, D., Kurosu, T., Tokunaga, K., Sata, T., and Peterlin, B. M. (2004). Human APOBEC3F is another host factor that blocks human immunodeficiency virus type 1 replication. *J. Virol.* **78,** 6073–6076.

Zillmann, M., Zapp, M. L., and Berget, S. M. (1988). Gel electrophoretic isolation of splicing complexes containing U1 small nuclear ribonucleoprotein particles. *Mol. Cell Biol.* **8,** 814–821.

CHAPTER NINETEEN

MOUSE MODELS AS TOOLS TO EXPLORE CYTIDINE-TO-URIDINE RNA EDITING

Soo-Jin Cho, Valerie Blanc, *and* Nicholas O. Davidson

Contents

Abstract

RNA editing is a process through which the nucleotide sequence specified in the genomic template is modified to produce a different nucleotide sequence in the transcript. RNA editing is an important mechanism of genetic regulation that

Division of Gastroenterology, Department of Medicine, Washington University School of Medicine, St. Louis, Missouri

Methods in Enzymology, Volume 424
ISSN 0076-6879, DOI: 10.1016/S0076-6879(07)24019-4

amplifies genetic plasticity by allowing the production of alternative protein products from a single gene. There are two generic classes of RNA editing in nuclei, involving enzymatic deamination of either C-to-U or A-to-I nucleotides. The best characterized example of C-to-U RNA editing is that of apolipoprotein B (apoB), which is mediated by a holoenzyme that contains a minimal core composed of an RNA-specific cytidine deaminase apobec-1, and its cofactor apobec-1 complementation factor (ACF). C-to-U editing of apoB RNA generates two different isoforms—apoB100 and apoB48—from a single transcript. Both are important regulators of lipid transport and metabolism, and are functionally distinct. C-to-U apoB RNA editing is regulated by a range of factors including developmental, nutritional, environmental, and metabolic stimuli. Rodent models have provided a tractable system in which to study the effects of such stimuli on lipid metabolism. In addition, both transgenic and gene knockout experiments have provided important insights into gain and loss of function approaches for studying C-to-U RNA editing in a murine background.

1. INTRODUCTION

Mammalian cytidine-to-uridine (C-to-U) RNA editing was first described as a site-specific cotranslational or posttranscriptional modification to a single nucleotide residue in the intestinal isoform encoding apolipoprotein B (apoB) messenger RNA (Chen *et al.*, 1987; Powell *et al.*, 1987). Although other examples of mammalian C-to-U RNA editing have been described (Cappione *et al.*, 1997; Mukhopadhyay *et al.*, 2002; Yamanaka *et al.*, 1997), most substantive advances in understanding the molecular machinery for C-to-U RNA editing have emerged from a detailed study of apoB RNA editing. C-to-U RNA editing of apoB mRNA changes a CAA codon in the genomically templated transcript (encoding a glutamine residue in the unedited mRNA) to a UAA codon and results in translational termination of the edited transcript. The net result is that the edited apoB mRNA encodes an amino terminal, translationally truncated protein referred to on a centile scale as apoB48. By contrast, the unedited apoB mRNA encodes a full-length protein, apoB100. In considering the context and approaches that have evolved for the use of mouse models to explore the biology of C-to-U RNA editing, several key features of this process bear emphasis. These include the functional significance of apoB in relation to its lipid transport properties (Davidson and Shelness, 2000), the species- and organ-specific partitioning of mammalian C-to-U apoB mRNA editing, its physiological and evolutionary adaptive significance, and the relationship of C-to-U RNA editing to the machinery that mediates the enzymatic deamination of the targeted cytidine in apoB mRNA (Blanc and Davidson, 2003). In regard to the use of mouse models, it should be recognized that essentially all known examples of altered C-to-U RNA editing of apoB *in vivo* accompany and reflect primary alterations in cellular lipid balance (cholesterol and/or triglyceride) and thus reflect an integrated program of adaptation to developmental, hormonal, nutritional, or other cues in which tissue lipogenesis and/or lipid transport functions are changed. This chapter will consider each of these issues and illustrate with selective examples how murine models can be used to explore relevant hypotheses.

2. APOB C-TO-U RNA EDITING

2.1. Species-dependent, organ-specific partitioning and functional implications

As alluded to above, C-to-U RNA editing of apoB is regulated in both a tissue- and species-specific manner. In humans, rodents, and all placental mammals, C-to-U RNA editing of apoB takes place in the small intestine, which synthesizes and secretes apoB48. In some species such as rats and mice, apoB mRNA editing also takes place in the liver (Greeve *et al.*, 1993), which in these species synthesizes and secretes both apoB100 and apoB48 in a regulated manner. Hepatic apoB mRNA is not edited in many other mammals, including humans, hamsters, and rabbits, and in these species the liver synthesizes and secretes exclusively apoB100. This species-dependent, organ-specific partitioning of C-to-U apoB RNA editing offers investigators the opportunity to examine the molecular, genetic, and biochemical mechanisms that underlie the regulation of this process.

From a functional perspective, apoB is absolutely required for lipid export from both small intestinal enterocytes and hepatocytes (Kim and Young, 1998). The lipid-binding properties of the amino terminal half of apoB, however, are sufficient to facilitate neutral lipid binding and direct lipoprotein formation for export into the circulation, although the size and functional characteristics of apoB48-containing lipoproteins differ in important regards to those assembled with apoB100 (Veniant *et al.*, 1997; Xie *et al.*, 2003). Specifically, there are subtle functional consequences for lipid export from murine enterocytes based on the apoB genotype, with apoB48-associated lipoprotein formation and triglyceride export being more efficient than that noted in the background of apoB100 (discussed further below) (Xie *et al.*, 2003). Earlier studies in hepatoma cells and in primary rodent hepatocytes also demonstrated the superiority of apoB48 over apoB100 in directing lipid export (Davis *et al.*, 1985; Yao *et al.*, 1992). In addition, C-to-U RNA editing eliminates from apoB48 at least two key carboxyl terminus domains encoded in apoB100, including the domains that direct binding to the low-density lipoprotein receptor (LDLR) and the extreme carboxyl terminus unpaired cysteine residue that mediates formation of the atherogenic lipoprotein Lp(a) (Liu et al., 2004; McCormick et al., 1994, 1997; Sharp et al., 2004). As a result, lipoproteins with only apoB100 are directed exclusively to the LDLR while apoB48-containing lipoproteins are cleared by other pathways, including the LDLR-related protein, LRP-1. While it is beyond the scope of this chapter to detail the distinctions in lipoprotein catabolism associated with the different apoB isoforms, the important feature is that apoB48-containing lipoproteins are cleared from the circulation much faster, and by a distinct set of receptors, than lipoproteins containing apoB100, which are cleared predominantly via the LDLR. Alterations in cellular lipid metabolism will be discussed in relation to the different murine models used to examine C-to-U RNA editing.

2.2. Molecular machinery

Current evidence points to a requisite role for two gene products, specifically apobec-1, an RNA-specific cytidine deaminase (Teng et al., 1993), and apobec-1 complementation factor (ACF), the RNA-binding and targeting subunit (Lellek et al., 2000; Mehta et al., 2000). Apobec-1 expression in humans is restricted to the luminal gastrointestinal tract, whereas in rodents such as mice and rats, apobec-1 is expressed very broadly including in the liver (Greeve et al., 1993). The tissue-specific expression pattern of apobec-1 accounts for the observation that human liver (which does not express apobec-1) does not edit apoB mRNA (Giannoni et al., 1994). ACF distribution by contrast appears to be more widespread in both humans and in mice (Blanc et al., 2005; Mehta et al., 2000), with the predominant tissue expression being liver, small intestine, and kidney. Both apobec-1 and ACF are regulated in a tissue-specific and developmental pattern, which coincides with the onset of lipoprotein formation and the developmental changes in lipid metabolism (Henderson et al., 2001; Teng et al., 1990a,b; Wu et al., 1990). While the focus of this chapter is on mouse models, most of the early observations of apoB C-to-U RNA editing were made in rats. Therefore, many of the procedures outlined pertain to experimentation in rats. However, most if not all of these methods may be extended for experimentation in mice as well. This chapter will focus on murine C-to-U RNA editing, specifically its developmental, hormonal, nutritional, and other modes of physiological modulation.

3. DEVELOPMENTAL REGULATION OF C-TO-U RNA EDITING

During the course of embryogenesis as the small intestine and liver develop into lipogenic organs and an enterohepatic circulation of biliary and other luminal lipids evolves, there is a transcriptional program that directs increased expression of apoB, as well as other genes involved in lipid transport, and also an increase in the expression of both apobec-1 and ACF (Henderson et al., 2001; Teng et al., 1990a,b, 1994). Some features of the developmental program of apobec-1 expression have recently been modeled using fetal intestinal isografts to study the properties of the intrinsic program of development in the absence of luminal lipid (Patterson et al., 2003). These studies demonstrated that the expression of apobec-1 and ACF coincides with a programmed increase in the expression of the homeobox gene Cdx-1, studies confirmed in cultures of the undifferentiated rat enterocyte cell line IEC-6 (Patterson et al., 2003). These studies suggest that the developmental regulation of C-to-U apoB RNA editing is an autonomous process, distinct from cellular lipid flux, and that homeobox genes may play an important role in mediating the transcriptional increase in apobec-1 and ACF expression. The system employed by these workers offers an important opportunity to study tissue development in a dietary and biliary lipid-free setting. For this purpose, fetal intestinal isografts were obtained from BALB/c mice at embryonic day 15–16 and surgically implanted into subcutaneous capsules of congenic mice (Gutierrez et al., 1995). The isografts are harvested at timed intervals thereafter for protein and RNA extraction.

A different series of studies examined the changes in C-to-U apoB RNA editing during the aging process, using a strain of mice that demonstrates a premature aging phenotype (Higuchi et al., 1992) associated with a progressive increase in cholesterol levels. The accelerated senescence-resistant mice (SAM-R/1 strain) have a normal aging process while the accelerated senescence-prone mice (SAM-P/1 strain) have a shorter life span with earlier onset and advancement of senescence. The authors of this study measured the apoB100/apoB48 ratio as well as apoB mRNA editing beginning at day 17 of gestation (4 days before birth) to 14 (SAM-P/1) or 26 (SAM-R/1) months of age. They observed that up to 2 months after birth, before the mouse reaches maturity, C-to-U apoB RNA editing increases in both strains, but beyond 5 months, however, the pattern reverses, and as the mice age, there is a decrease in apoB C-to-U RNA editing. This decrease in editing was accelerated in the senescence-prone strain.

4. HORMONAL REGULATION OF C-TO-U APOB RNA EDITING

4.1. Effects of thyroid hormone on C-to-U apoB RNA editing

Thyroid hormone produces profound effects on lipid metabolism in all mammals. From the perspective of hepatic lipid metabolism, hypothyroidism is associated with increased levels of plasma cholesterol and decreased cholesterol catabolism, mediated through a combination of suppression of bile acid synthesis as well as through suppression of LDLR expression (Davidson *et al.*, 1988; Ness and Zhao, 1994). In addition, there is altered hepatic lipogenesis and increased accumulation of triglyceride within the liver. Experimentally-induced hypothyroidism in rats and mice results in decreased hepatic C-to-U apoB RNA editing (Davidson *et al.*, 1988; Mukhopadhyay *et al.*, 2003). By contrast, hypothyroid rats and mice that are administered thyroid hormone *in vivo* (T3 or T4) exhibit an increase in hepatic apoB mRNA editing (Davidson *et al.*, 1988; Mukhopadhyay *et al.*, 2003). In addition, gene-targeted Pax8$^{-/-}$ mice manifest spontaneous hypothyroidism with developmental delay (Mukhopadhyay *et al.*, 2003). When Pax8$^{-/-}$ mice are treated with T3 and T4 (see below), the livers of these animals exhibit an increase in C-to-U apoB RNA editing. Although apobec-1 mRNA expression has been demonstrated to be unchanged in rat liver following thyroid hormone administration (Inui *et al.*, 1994), a redistribution of ACF to the nucleus occurs in Pax8$^{-/-}$ mice treated with thyroid hormone (Mukhopadhyay *et al.*, 2003). This latter finding is mechanistically relevant since this is the compartment where C-to-U apoB RNA editing has been previously shown to occur (Lau *et al.*, 1991).

To study the effects of thyroid hormone supplementation on C-to-U apoB RNA editing, the animals must first be made hypothyroid. There are different methods that may be employed to achieve hypothyroidism in rodents, including surgical thyroidectomy in which the thyroid and parathyroid glands are removed or, as an alternative, chemically induced hypothyroidism. The latter is more readily utilized and involves feeding animals (rats or mice) a chow diet containing 0.1% (w/w) propylthiouracil (2-thio-4-hydroxy-6-*n*-propylpyrimidine, Sigma) for 21–28 days. Once the animals are hypothyroid (as assessed by plasma levels of thyroid hormones measured commercially), they are then injected intraperitoneally with 3,5,3′-triiodo-L-thyronine (T3) at 0.5 μg/100 g body weight (physiological dose; euthyroid) or 50 μg/100 g body weight (pharmacological dose; hyperthyroid) for 7 days. For this purpose, a stock solution of T3 is prepared and frozen in aliquots at −20° and dilutions are prepared fresh daily for injection. Alternatively, animals may be given a single dose of T3 intravenously at 100 μg/100 g body weight. A further approach to studying the effects of thyroid hormone supplementation is to use genetically modified mice, specifically Pax8$^{-/-}$ mice, which lack thyroid follicular cells and are hypothyroid from birth (Flamant *et al.*, 2002). These mice are injected daily for 2 or 4 days with a mixture of T4 and T3 [250 μg/100 g body weight of T4 and 25 μg/100 g of T3 in 100 μl of phosphate-buffered saline (PBS)], which has been previously shown to produce hyperthyroidism in mice (Plateroti *et al.*, 1999).

The aforementioned effects of thyroid hormone have been validated using *in vivo* animal models. It is worth noting that some of the reported effects of thyroid hormones on C-to-U apoB RNA editing have been from isolated rat hepatocytes prepared from animals following surgical hypophy-sectomy (Sjoberg et al., 1992) in which modulation of RNA editing activity requires simultaneous addition of growth hormone and cortisol in addition to thyroid hormone supplementation. These findings emphasize the com-plexity of the regulatory pathways of hepatic lipogenesis and the potential interactions of hormones and their receptors.

4.2. Effects of insulin on C-to-U apoB RNA editing

Insulin has important effects on hepatic triglyceride metabolism and the synthesis and secretion of apoB, specifically to inhibit the secretion of triglyceride-rich apoB-containing lipoprotein particles. These findings have been demonstrated in vivo in rats and mice as well as in isolated human hepatocytes. In regard to the importance of insulin in regulating C-to-U apoB RNA editing, most of the studies have been performed in rats.

Early observations indicated that chronic insulin treatment of isolated primary rat hepatocytes results in increased apoB48 secretion (Thorngate et al., 1994). For these experiments, rat primary hepatocytes are isolated from adult male Sprague–Dawley rats (Harlan Laboratories) (Elam et al., 1988) and cultured with feeding medium [William's medium E (formula-tion 79–5204); Gibco/BRL] supplemented with 0.64 mM L-ornithine, 38 mM sodium bicarbonate, 10 mM HEPES, 10 mM dextrose, 100 μg/ml streptomycin, 100 IU/ml penicillin G, 50 μg/ml gentamycin, 2.5 μg/ml amphotericin B, 100 nM dexamethasone, 50 nM triiodothyronine, 1 mg/ml bovine serum albumin, 5 μg/ml linoleic acid, and 0.1 μM CuSO$_4$, 3 nM NaSeO$_3$, 50 pM ZnSO$_4$. Bovine insulin is added to the feeding medium to a final concentration of 0–400 ng/ml (0–67 nM), and this feeding medium needs to be replaced daily. Previous results have shown that upon culturing the rat primary hepatocytes with insulin for 5 days, apoB48 secretion was increased approximately 4-fold above nontreated cells with 4 ng/ml insulin, and increased approximately 10-fold above nontreated with 400 ng/ml insulin. The increased apoB48 secretion correlated with an increased expres-sion of apoB48 protein as a result of enhanced C-to-U apoB mRNA editing. Reverse transcriptase polymerase chain reaction (RT-PCR) analysis of total RNA extracted from the hepatocytes showed an increased expression of apobec-1 mRNA (von Wronski et al., 1998). Better quantitative methods are now available, and in our laboratory, we use real-time qPCR for analysis of apobec-1 RNA expression (see below). Genetically obese, hyperinsuli-nemic Zucker (fa/fa) rats also exhibit a higher (~2-fold) level of hepatic apoB editing due to a corresponding increase in apobec-1 mRNA (Phung et al., 1996). Genetically diabetic GK (Goto-Kakizaki) rats show a similar increase in apoB editing (Yamane et al., 1995).

More recent studies in rat primary hepatocytes have focused on the ACF, which is a requisite cofactor for apoB RNA editing (Sowden et al., 2002).

ACF shuttles between the nucleus and cytoplasm (Blanc *et al.*, 2003) and has been proposed to regulate editing activity by cotransporting apobec-1 from the cytoplasm to the nucleus where editing occurs. In experiments by Sowden *et al.* (2002), treatment of the hepatocytes with insulin (10 n*M*) for 6 h stimulated apoB RNA editing activity ~1.5-fold. They observed no change in ACF protein expression, but instead observed a corresponding increase of ACF in the nuclear compartment. In addition, the phosphorylation of ACF is increased, which may further stimulate its import into the nucleus (Lehmann *et al.*, 2006). Therefore, insulin appears to have a dual effect in which apobec-1 expression is increased and its localization to the nucleus is facilitated by an increase in nuclear ACF.

4.3. Effects of estrogen on C-to-U apoB RNA editing

Estrogen treatment (pharmacological doses) of rats and mice results in upregulation of hepatic LDL receptor expression and results in decreased plasma cholesterol and cholesterol loading of the liver through augmented lipoprotein uptake and also decreased canalicular secretion of cholesterol. The net result is an increased accumulation of hepatic free and esterified cholesterol. Estrogen treatment of male Sprague–Dawley rats with pharmacological doses of estrogen does not affect total hepatic apoB synthesis, but the apoB100/apoB48 ratio increases, presumably due to a decrease in apoB editing (Seishima *et al.*, 1991). Subsequent studies in mice showed that while the plasma apoB level is altered in a strain-specific manner by estrogen treatment, in all strains studied, the proportion of apoB100 relative to apoB48 was increased, similar to rats, due to a concomitant decrease in apoB RNA editing (Srivastava, 1995). This decrease in editing correlates with a decrease in apobec-1 mRNA expression (Srivastava, 1995).

For these studies, male mice may be administered 17β-estradiol or 17α-estrinyl-estradiol (Sigma) at a dose of 3 μg/g body weight/day for 5 consecutive days. The animals should be fed *ad libitum* with a standard rodent chow diet (Ralston–Purina, St. Louis, MO). They should be sacrificed after an overnight fast to reduce fluctuations in plasma cholesterol levels.

5. NUTRITIONAL MODULATION OF C-TO-U APOB RNA EDITING

Previous studies in both rats and mice demonstrated that the nutritional state of the animal (fasting, feeding) can affect the ratio of apoB48/apoB100. This effect is likely the result of altered hepatic lipogenesis, since fasting decreases triglyceride intake and results in mobilization of adipose tissue stores for use in hepatic ketogenesis. By contrast, feeding a high carbohydrate diet induces *de novo* hepatic lipogenesis and increases triglyceride-rich lipoprotein formation and augmented secretion of hepatic lipid. Fasting has been shown to decrease C-to-U apoB RNA editing and as a result suppresses the ratio of newly synthesized apoB48/apoB100. By contrast, prolonged fasting of rats, followed by refeeding with a carbohydrate-enriched (high sucrose) diet, results in an increase in C-to-U apoB RNA editing and a corresponding increase in newly synthesized apoB48 relative to apoB100. This cascade of changes facilitates hepatic lipoprotein secretion (Baum *et al.*, 1990; Boogaerts *et al.*, 1984; Davis and Boogaerts, 1982; Davis *et al.*, 1985; Leighton *et al.*, 1990). For these experiments, male Sprague–Dawley rats (150–250 g) should be fed *ad libitum* on a standard rat chow for at least 10 days prior to the fasting protocol. All animals should be adapted to a 12-h light schedule in a temperature- and humidity-controlled environment and studied at a standardized time to minimize fluctuations in their nutritional state. All animals should be fasted for 48 h with free access to water and then refed with either the standard chow or a high-sucrose, fat-free diet (ICN Nutritional Biochemicals, catalog no. 901603, Cleveland, OH) for 24 or 48 h. Animals should be sacrificed at the mid-dark phase of their cycle (i.e., soon after consuming food). Subsequent analyses are described below.

5.1. Ethanol

Male Wistar rats fed an ethanol-liquid diet have been shown to develop hypertriglyceridemia, which is associated with increases in plasma very-low-density lipoprotein concentrations and hepatic apoB48 synthesis (Lau *et al.*, 1995). This increase in apoB48 synthesis is mediated by an increase in C-to-U apoB RNA editing ($\sim$2-fold), although the abundance of apobec-1 mRNA remains unchanged. For these experiments, the animals were fed with standard rat chow for at least 10 days prior to starting the protocol, then acclimated to a liquid-only diet. Three groups were tested: one group was fed the standard chow diet, another was fed an isocaloric liquid-only diet, and the last group was fed an ethanol-liquid diet with ethanol making up 35.5% of the total calories (Bio-Serv, Inc., Frenchtown, NJ). The animals were fed these diets for up to 40 days.

Further studies showed that similar to insulin treatment of rat primary hepatocytes, ethanol (0.9%) treatment stimulates C-to-U apoB RNA editing. This increase is independent of *de novo* protein and RNA synthesis. Rather, it is mediated by an increase in ACF phosphorylation, which enhances the ACF–apobec-1 interaction and their translocation into the nucleus (Lehmann *et al.*, 2006).

6. GENETIC MANIPULATIONS OF GENES INVOLVED IN APOB RNA EDITING

6.1. Targeted deletion and transgenic overexpression of genes involved in C-to-U apoB RNA editing

Three different laboratories simultaneously reported the development of lines of mice with targeted deletion of the catalytic cytidine deaminase, apobec-1 (Hirano *et al.*, 1996; Morrison *et al.*, 1996; Nakamuta *et al.*, 1996). All three lines utilized a germ line knockout strategy that eliminates the expression of apobec-1 mRNA and protein from all cells and tissues. The phenotype of these mice was consistent in all three reports and included the observation that the offspring were found in Mendelian proportions and were healthy and fertile with no obvious developmental or other abnormalities. Since these mice lack apobec-1 mRNA and protein in all tissues, they were demonstrated to be incapable of C-to-U apoB RNA editing and the liver and small intestine synthesize and secrete only apoB100. These results demonstrated that there is no redundancy for the catalytic cytidine deaminase in regard to the C-to-U apoB RNA editing holoenzyme. Other apobec-related genes have been identified in mice, including activation-induced deaminase (AID) (Muramatsu *et al.*, 1999), apobec-2 (Anant *et al.*, 2001; Liao *et al.*, 1999), and apobec-3 (Mariani *et al.*, 2003). Targeted deletion of AID has a major phenotype in terms of class switch recombination and somatic hypermutation, the result of altered C-to-T deamination of DNA (rather than RNA), and which results in the accumulation of IgM and failure to generate IgG or IgA (Muramatsu *et al.*, 2000). Apobec-2 is expressed almost exclusively in murine skeletal and heart muscle (Anant *et al.*, 2001). Apobec-2 deletion was not accompanied by any detectable phenotype in the heart or skeletal muscle of knockout mice (Mikl *et al.*, 2005). Apobec-3 mRNA is expressed in spleen and bone marrow–derived cells and low levels of the transcript are detectable in many murine tissues (Mikl *et al.*, 2005). Targeted deletion of apobec-3, however, was not accompanied by any overt phenotype (Mikl *et al.*, 2005). Further work will be necessary to resolve a role, if any, for this murine gene in regard to restriction of murine retroviruses.

In addition to apobec-1, apobec-2, and apobec-3 knockout mice, ACF$^{-/-}$ mice have also been generated and characterized. ACF is the requisite cofactor and presumed RNA binding subunit of the C-to-U apoB RNA editing holoenzyme (Lellek *et al.*, 2000; Mehta *et al.*, 2000). Unexpectedly, targeted deletion of the murine ACF gene results in early embryonic lethality at E3.5 (Blanc *et al.*, 2005). Blastocysts derived from heterozygous pair matings revealed homozygous deletion detectable at the preimplantation stage only. These findings suggest that ACF is required at a very early stage in embryonic development and raises the question of the nature of the RNA target(s) involved. $ACF^{+/-}$ mice are viable and fertile, and paradoxically exhibit an increase in hepatic C-to-U apoB RNA editing. Further advances in understanding the role of ACF will require development of conditional deletion strategies.

In addition to the loss-of-function phenotypes associated with targeted deletion of the genes involved in C-to-U RNA editing, there is a distinctive pattern of phenotypes associated with the gain-of-function experiments conducted using transgenic overexpression of apobec-1. Innerarity and colleagues generated transgenic mice and rabbits in which high copy number, constitutive, liver-specific expression was achieved using the human apoE promoter to drive the rabbit apobec-1 cDNA. Four independent transgenic lines were established in founder mice, with copy numbers ranging from 3 to 10 (Yamanaka et al., 1995). A transgenic rabbit was also generated, bearing 17 copies of the transgene. The findings of this study demonstrated that induction of apobec-1 expression in hepatocytes resulted in C-to-U editing of apoB mRNA and the appearance of apoB48. In addition to high levels of C-to-U RNA editing of the canonical site (6666) in apoB mRNA, transgenic overexpression of apobec-1 also resulted in aberrant editing of cytidine bases at other locations in the apoB transcript (a phenomenon referred to as hyperediting) as well as C-to-U RNA editing in other transcripts, including NAT1 (Yamanaka et al., 1995). Several features of C-to-U RNA editing associated with transgenic overexpression of apobec-1 merit emphasis. The first is that the physiological constraints on C-to-U RNA editing with respect to the cis-acting elements in apoB RNA are lost in the setting of transgenic overexpression of apobec-1. Specifically, physiological C-to-U apoB RNA editing exhibits a requirement for an AU-rich context, a downstream mooring sequence (UGAUCAGUAUA), and an optimal spacing of 5 nt from the 5' end of the mooring sequence to the targeted upstream cytidine. C-to-U apoB RNA hyperediting requires none of these cis-acting factors (Yamanaka et al., 1996). By contrast, both physiological and hyperediting of C-to-U bases in apoB RNA require the auxiliary protein ACF (Yamanaka et al., 1996). The second important observation concerning C-to-U RNA editing of apoB in apobec-1 transgenic mice was that the efficiency of hyperediting was much lower (range <10% to ~50%) than at the canonical site (>90%). The third intriguing observation was that hyperediting revealed a nearest-neighbor preference with 70% of the upstream nucleotides being thymidine and 30% adenosine, while the nearest downstream nucleotides were adenosine (58%) and thymidine (27%). Thus hyperediting of C-to-U in apoB RNA in the setting of transgenic apobec-1 overexpression demonstrates a preference for T/A bases immediately flanking the targeted cytidine (Yamanaka et al., 1996).

A further unanticipated finding in the apobec-1 transgenic animals was that high levels of apobec-1 expression induced dysplasia with nodule formation and hepatocellular carcinoma (Yamanaka *et al.*, 1995). The original I-20 line (containing seven copies) demonstrated ~50% hepatic tumors at 1 year. By contrast, the I-28 line, with 10 copies of the transgene, demonstrated ~50% developed tumors at 15 weeks and ~80% by 6 months. This copy number-dependent increase in tumorigenesis raised the possibility that alternative RNA targets of the C-to-U RNA editing machinery exist, including potential tumor suppressors or promoters. Subsequent studies on mice expressing lower levels of apobec-1 have been accomplished through transgenic expression of a 52-kb rat genomic clone [RE-4 (Qian *et al.*, 1998)] in which hepatic RNA editing activity was found to be only 3-fold above normal. Among the interesting observations from this study was that even a modest increase in hepatic C-to-U RNA editing activity resulted in almost quantitative C-to-U deamination of the canonical cytidine base in apoB mRNA but resulted in no or minimal hyperediting (Qian *et al.*, 1998). In another murine model of transgenic apobec-1 expression, conditional hepatic expression of apobec-1 was accomplished using a tetracycline-dependent promoter (tetO) fused to the rabbit apobec-1 cDNA and the founders crossed into *LAP-tTA* transgenic mice in order to generate a double transgenic line in which apobec-1 expression was suppressed by feeding doxycycline (10 μg/ml) in drinking water. This line demonstrated increased C-to-U RNA editing of both murine and human apoB RNA, but did not develop hepatic tumors. These double transgenic mice were also crossed into the apobec-1$^{-/-}$ line, permitting conditional expression of apobec-1 in an apobec-1 null background (Hersberger *et al.*, 2003). These latter animals showed only 30% C-to-U RNA editing at the canonical site of apoB RNA, but showed aberrant hyperediting of apoB RNA as well as other RNA targets (Hersberger *et al.*, 2003). The results with respect to C-to-U RNA editing at the canonical base are unexpected and suggest that either the stoichiometry of apobec-1 and ACF is disrupted as a result of forced transgenic expression or that apobec-1 is mistargeted to the cytoplasm. The results from these gain-of-function studies suggest that forced transgenic expression of apobec-1, particularly at high levels, may result in dysregulated and promiscuous C-to-U deamination of targets beyond apoB RNA and results in a tumor phenotype.

Constitutive, high-level expression of AID in transgenic mice also resulted in a tumor phenotype (Okazaki *et al.*, 2003). The murine AID cDNA for these studies was driven by a chicken β-actin promoter and five lines of mice expressing the transgene developed enlarged lymphoid organs and malignant T cell lymphomas as well as pulmonary microadenomas and adenocarcinomas. The molecular basis for this phenotype is still unclear, but multiple point mutations were demonstrated in the T cell receptor and c-myc genes, raising the possibility that forced transgenic overexpression of AID can mutate known oncogenes. Taken together, the transgenic lines expressing AID and apobec-1 may provide useful insight into relevant pathways of human malignancy.

7. ANALYSIS OF apoB RNA EDITING

The following techniques may be used for the evaluation of any of the mouse models described above.

7.1. Protein and RNA preparation

At sacrifice, liver and intestine from the animals should be harvested and snap frozen in liquid nitrogen for extraction of RNA and protein. For protein extraction, tissue samples may be homogenized using a glass dounce or tissue homogenizer in buffer containing 25 mM HEPES, pH 8.0, 150 mM NaCl, 1% Triton X-100, 0.5 mM EDTA, 0.1% sodium dodecyl sulfate (SDS), and protease inhibitors (Roche). RNA may be extracted using TRIzol, following the manufacturer's protocol (Invitrogen).

7.2. apoB, apobec-1, and ACF mRNA and protein expression

ApoB, apobec-1, and ACF mRNA levels may be evaluated by real-time qPCR using SYBR Green PCR master mix (Applied Biosystems) and the ABI7000 (Applied Biosystems). Primers are listed in Table 19.1. Following TRIzol extraction, RNA should be treated with DNase (DNA-*free* kit, Ambion) and cDNA amplification performed using random hexamers and reverse transcriptase (Superscript II, Invitrogen) prior to PCR amplification. Apobec-1 and ACF protein levels may be evaluated by Western blotting using rabbit polyclonal antibodies (see below, "Serum Lipid Content and Plasma Apolipoprotein Analysis").

Table 19.1 Oligonucleotide primers for real-time qpcr analysis of murine genes

Gene	Forward primer (5′ → 3′)	Reverse primer (5′ → 3′)
apoB	CACTGCCGTGGCCAAAA	GCTAGAGAGTTGGTCT-GAAAAATCCT
apobec-1	ACCACACGGATCAGCGAAA	TCATGATCTGGATAGT-CACACCG
ACF	GATGAAAAAAGTCACAGAA-GGAGTTG	CAAATCCCCGGTTTTT-GGT

7.3. apoB RNA editing by primer extension

Hepatic and intestinal apoB RNA editing may be analyzed by a primer extension assay (Blanc *et al.*, 2005). After RNA isolation and DNase treatment, the RNA is subjected to reverse transcription (described above) and the cDNA is PCR amplified using primers flanking a 250-bp region of murine apoB RNA surrounding the edited site (Forward: 5′-ATCT-GACTGGGAGAGACAAGTAGC-3′; Reverse: 5′-ACGGATATGATA-CTGTTCATCAAGAA-3′). Poisoned primer extension is performed on 20 ng of this PCR product using 100 pg of a 5′ end-labeled primer 39 bp downstream of the edited site (5′-CCTGTGCATCATAATTATCTC-TAATATACTGATCA-3′) using 1.5 units T7 DNA polymerase at 42° for 3 min. The products are ethanol precipitated and separated on an 8% polyacrylamide–urea gel and subjected to autoradiography. The ratio of apoB48 to apoB100 can be determined using PhosphoImager/ImageQuant (Pharmacia Biotech).

7.4. Subcellular localization of apobec-1 and ACF by immunohistochemistry

To examine the subcellular localization of apobec-1 and/or ACF, small pieces of tissue should be fixed in 10% formalin and paraffin-embedded. Apobec-1 and ACF immunohistochemistry can be performed on 5-μm sections by first rehydrating the slides then microwaving for 15 min in citrate buffer (0.01 M, pH 6) before incubation with a rabbit polyclonal anti–apobec-1 antibody (Funahashi *et al.*, 1995) or a rabbit polyclonal anti-ACF anti-body (Blanc *et al.*, 2001), followed by incubation with a secondary goat anti-rabbit biotinylated antibody and streptavidin–peroxidase detection (Histomouse, Zymed Laboratories). Hepatic lipid accumulation may be examined by Oil Red-O (Sigma) staining after immersing the section in propylene glycol and lightly counterstaining in hematoxylin. For quantitative measurements, hepatic lipids can be extracted from cell homogenates using chloroform/methanol (2:1) and enzymatic assay of triglyceride mass using a commercial kit (Wako Pure Chemical Industries, Inc., Osaka, Japan).

7.5. Serum lipid content and plasma apolipoprotein analysis

Animals should be exsanguinated via direct cardiac or aortic cannulation and the serum recovered by centrifugation at 4000×g for 20 min at 4° for analysis of serum triglyceride (L-Type TG H kit, Wako Diagnostics) and cholesterol (Cholesterol E kit, Wako Diagnostics). Apolipoprotein distribution and cholesterol and triglyceride concentrations may be determined using 200 μl plasma separated by fast pressure liquid chromatography with two tandem 25-cm Superose 6 columns (Pharmacia Biotech) connected in series and collecting 500-μl fractions in 1 mM EDTA, 150 mM NaCl, 10 mM sodium phosphate, pH 7.4 (Horie *et al.*, 1992). Samples may be stored at 4° for up to 2 weeks or frozen at −20° for later analysis. The lipid content in individual fractions can be determined with enzymatic detection kits from Sigma. ApoB100 versus apoB48 relative abundance in plasma can also be determined by 4–15% gradient gel separation of 1 ml of plasma (containing 2% SDS and 0.5% NP-40) and transfer to nitrocellulose followed by immunodetection with a rabbit polyclonal anti–apoB antibody (Bonen *et al.*, 1998).

REFERENCES

Anant, S., Mukhopadhyay, D., Sankaranand, V., Kennedy, S., Henderson, J. O., and Davidson, N. O. (2001). ARCD-1, an apobec-1-related cytidine deaminase, exerts a dominant negative effect on C to U RNA editing. *Am. J. Physiol. Cell Physiol.* **281,** C1904–C1916.

Baum, C. L., Teng, B. B., and Davidson, N. O. (1990). Apolipoprotein B messenger RNA editing in the rat liver. Modulation by fasting and refeeding a high carbohydrate diet. *J. Biol. Chem.* **265,** 19263–19270.

Blanc, V., and Davidson, N. O. (2003). C-to-U RNA editing: mechanisms leading to genetic diversity. *J. Biol. Chem.* **278,** 1395–1398.

Blanc, V., Henderson, J. O., Kennedy, S., and Davidson, N. O. (2001). Mutagenesis of apobec-1 complementation factor reveals distinct domains that modulate RNA binding, protein-protein interaction with apobec-1, and complementation of C to U RNA-editing activity. *J. Biol. Chem.* **276,** 46386–46393.

Blanc, V., Henderson, J. O., Newberry, E. P., Kennedy, S., Luo, J., and Davidson, N. O. (2005). Targeted deletion of the murine apobec-1 complementation factor (acf) gene results in embryonic lethality. *Mol. Cell. Biol.* **25,** 7260–7269.

Blanc, V., Kennedy, S., and Davidson, N. O. (2003). A novel nuclear localization signal in the auxiliary domain of apobec-1 complementation factor regulates nucleocytoplasmic import and shuttling. *J. Biol. Chem.* **278,** 41198–41204.

Bonen, D. K., Nassir, F., Hausman, A. M., and Davidson, N. O. (1998). Inhibition of N-linked glycosylation results in retention of intracellular apo[a] in hepatoma cells, although nonglycosylated and immature forms of apolipoprotein[a] are competent to associate with apolipoprotein B-100 *in vitro*. *J. Lipid Res.* **39,** 1629–1640.

Boogaerts, J. R., Malone-McNeal, M., Archambault-Schexnayder, J., and Davis, R. A. (1984). Dietary carbohydrate induces lipogenesis and very-low-density lipoprotein synthesis. *Am. J. Physiol.* **246,** E77–E83.

Cappione, A. J., French, B. L., and Skuse, G. R. (1997). A potential role for NF1 mRNA editing in the pathogenesis of NF1 tumors. *Am. J. Hum. Genet.* **60,** 305–312.

Chen, S. H., Habib, G., Yang, C. Y., Gu, Z. W., Lee, B. R., Weng, S. A., Silberman, S. R., Cai, S. J., Deslypere, J. P., Rosseneu, M., Gotto, M., Li, W.-H., and Chan, L. (1987). Apolipoprotein B-48 is the product of a messenger RNA with an organ-specific in-frame stop codon. *Science* **238,** 363–366.

Davidson, N. O., Powell, L. M., Wallis, S. C., and Scott, J. (1988). Thyroid hormone modulates the introduction of a stop codon in rat liver apolipoprotein B messenger RNA. *J. Biol. Chem.* **263,** 13482–13485.

Davidson, N. O., and Shelness, G. S. (2000). A polipoprotein B: mRNA editing, lipoprotein assembly, and presecretory degradation. *Annu. Rev. Nutr.* **20,** 169–193.

Davis, R. A., and Boogaerts, J. R. (1982). Intrahepatic assembly of very low density lipoproteins. Effect of fatty acids on triacylglycerol and apolipoprotein synthesis. *J. Biol. Chem.* **257,** 10908–10913.

Davis, R. A., Boogaerts, J. R., Borchardt, R. A., Malone-McNeal, M., and Archambault-Schexnayder, J. (1985). Intrahepatic assembly of very low density lipoproteins. Varied synthetic response of individual apolipoproteins to fasting. *J. Biol. Chem.* **260,** 14137–14144.

Elam, M. B., Simkevich, C. P., Solomon, S. S., Wilcox, H. G., and Heimberg, M. (1988). Stimulation of *in vitro* triglyceride synthesis in the rat hepatocyte by growth hormone treatment *in vivo. Endocrinology* **122,** 1397–1402.

Flamant, F., Poguet, A. L., Plateroti, M., Chassande, O., Gauthier, K., Streichenberger, N., Mansouri, A., and Samarut, J. (2002). Congenital hypothyroid Pax8($-/-$) mutant mice can be rescued by inactivating the TRalpha gene. *Mol. Endocrinol.* **16,** 24–32.

Funahashi, T., Giannoni, F., DePaoli, A. M., Skarosi, S. F., and Davidson, N. O. (1995). Tissue-specific, developmental and nutritional regulation of the gene encoding the catalytic subunit of the rat apolipoprotein B mRNA editing enzyme: functional role in the modulation of apoB mRNA editing. *J. Lipid Res.* **36,** 414–428.

Giannoni, F., Bonen, D. K., Funahashi, T., Hadjiagapiou, C., Burant, C. F., and Davidson, N. O. (1994). Complementation of apolipoprotein B mRNA editing by human liver accompanied by secretion of apolipoprotein B48. *J. Biol. Chem.* **269,** 5932–5936.

Greeve, J., Altkemper, I., Dieterich, J. H., Greten, H., and Windler, E. (1993). Apolipoprotein B mRNA editing in 12 different mammalian species: hepatic expression is reflected in low concentrations of apoB-containing plasma lipoproteins. *J. Lipid Res.* **34,** 1367–1383.

Gutierrez, E. D., Grapperhaus, K. J., and Rubin, D. C. (1995). Ontogenic regulation of spatial differentiation in the crypt-villus axis of normal and isografted small intestine. *Am. J. Physiol.* **269,** G500–G511.

Henderson, J. O., Blanc, V., and Davidson, N. O. (2001). Isolation, characterization and developmental regulation of the human apobec-1 complementation factor (ACF) gene. *Biochim. Biophys. Acta* **1522,** 22–30.

Hersberger, M., Patarroyo-White, S., Qian, X., Arnold, K. S., Rohrer, L., Balestra, M. E., and Innerarity, T. L. (2003). Regulatable liver expression of the rabbit apolipoprotein B mRNA-editing enzyme catalytic polypeptide 1 (APOBEC-1) in mice lacking endogenous APOBEC-1 leads to aberrant hyperediting. *Biochem. J.* **369,** 255–262.

Higuchi, K., Kitagawa, K., Kogishi, K., and Takeda, T. (1992). Developmental and age-related changes in apolipoprotein B mRNA editing in mice. *J. Lipid Res.* **33,** 1753–1764.

Hirano, K., Young, S. G., Farese, R. V., Jr., Ng, J., Sande, E., Warburton, C., Powell-Braxton, L. M., and Davidson, N. O. (1996). Targeted disruption of the mouse apobec-1 gene abolishes apolipoprotein B mRNA editing and eliminates apolipoprotein B48. *J. Biol. Chem.* **271,** 9887–9890.

Horie, Y., Fazio, S., Westerlund, J. R., Weisgraber, K. H., and Rall, S. C., Jr. (1992). The functional characteristics of a human apolipoprotein E variant (cysteine at residue 142) may explain its association with dominant expression of type III hyperlipoproteinemia. *J. Biol. Chem.* **267,** 1962–1968.

Inui, Y., Giannoni, F., Funahashi, T., and Davidson, N. O. (1994). REPR and complementation factor(s) interact to modulate rat apolipoprotein B mRNA editing in response to alterations in cellular cholesterol flux. *J. Lipid Res.* **35,** 1477–1489.

Kim, E., and Young, S. G. (1998). Genetically modified mice for the study of apolipoprotein B. *J. Lipid Res.* **39,** 703–723.

Lau, P. P., Cahill, D. J., Zhu, H. J., and Chan, L. (1995). Ethanol modulates apolipoprotein B mRNA editing in the rat. *J. Lipid Res.* **36,** 2069–2078.

Lau, P. P., Xiong, W. J., Zhu, H. J., Chen, S. H., and Chan, L. (1991). Apolipoprotein B mRNA editing is an intranuclear event that occurs posttranscriptionally coincident with splicing and polyadenylation. *J. Biol. Chem.* **266,** 20550–20554.

Lehmann, D. M., Galloway, C. A., Sowden, M. P., and Smith, H. C. (2006). Metabolic regulation of apoB mRNA editing is associated with phosphorylation of APOBEC-1 complementation factor. *Nucleic Acids Res.* **34,** 3299–3308.

Leighton, J. K., Joyner, J., Zamarripa, J., Deines, M., and Davis, R. A. (1990). Fasting decreases apolipoprotein B mRNA editing and the secretion of small molecular weight apoB by rat hepatocytes: Evidence that the total amount of apoB secreted is regulated post-transcriptionally. *J. Lipid Res.* **31,** 1663–1668.

Lellek, H., Kirsten, R., Diehl, I., Apostel, F., Buck, F., and Greeve, J. (2000). Purification and molecular cloning of a novel essential component of the apolipoprotein B mRNA editing enzyme-complex. *J. Biol. Chem.* **275,** 19848–19856.

Liao, W., Hong, S. H., Chan, B. H., Rudolph, F. B., Clark, S. C., and Chan, L. (1999). APOBEC-2, a cardiac- and skeletal muscle-specific member of the cytidine deaminase supergene family. *Biochem. Biophys. Res. Commun.* **260,** 398–404.

Liu, C. Y., Broadhurst, R., Marcovina, S. M., and McCormick, S. P. (2004). Mutation of lysine residues in apolipoprotein B-100 causes defective lipoprotein[a] formation. *J. Lipid Res.* **45,** 63–70.

Mariani, R., Chen, D., Schrofelbauer, B., Navarro, F., Konig, R., Bollman, B., Munk, C., Nymark-McMahon, H., and Landau, N. R. (2003). Species-specific exclusion of APOBEC3G from HIV-1 virions by Vif. *Cell* **114,** 21–31.

McCormick, S. P., Linton, M. F., Hobbs, H. H., Taylor, S., Curtiss, L. K., and Young, S. G. (1994). Expression of human apolipoprotein B90 in transgenic mice. Demonstration that apolipoprotein B90 lacks the structural requirements to form lipoprotein. *J. Biol. Chem.* **269,** 24284–24289.

McCormick, S. P., Ng, J. K., Cham, C. M., Taylor, S., Marcovina, S. M., Segrest, J. P., Hammer, R. E., and Young, S. G. (1997). Transgenic mice expressing human ApoB95 and ApoB97. Evidence that sequences within the carboxyl-terminal portion of human apoB100 are important for the assembly of lipoprotein. *J. Biol. Chem.* **272,** 23616–23622.

Mehta, A., Kinter, M. T., Sherman, N. E., and Driscoll, D. M. (2000). Molecular cloning of apobec-1 complementation factor, a novel RNA-binding protein involved in the editing of apolipoprotein B mRNA. *Mol. Cell Biol.* **20,** 1846–1854.

Mikl, M. C., Watt, I. N., Lu, M., Reik, W., Davies, S. L., Neuberger, M. S., and Rada, C. (2005). Mice deficient in APOBEC2 and APOBEC3. *Mol. Cell Biol.* **25,** 7270–7277.

Morrison, J. R., Paszty, C., Stevens, M. E., Hughes, S. D., Forte, T., Scott, J., and Rubin, E. M. (1996). Apolipoprotein B RNA editing enzyme-deficient mice are viable despite alterations in lipoprotein metabolism. *Proc. Natl. Acad. Sci. USA* **93,** 7154–7159.

Mukhopadhyay, D., Anant, S., Lee, R. M., Kennedy, S., Viskochil, D., and Davidson, N. O. (2002). C→U editing of neurofibromatosis 1 mRNA occurs in tumors that express both the type II transcript and apobec-1, the catalytic subunit of the apolipoprotein B mRNA-editing enzyme. *Am. J. Hum. Genet.* **70,** 38–50.

Mukhopadhyay, D., Plateroti, M., Anant, S., Nassir, F., Samarut, J., and Davidson, N. O. (2003). Thyroid hormone regulates hepatic triglyceride mobilization and apolipoprotein B messenger ribonucleic acid editing in a murine model of congenital hypothyroidism. *Endocrinology* **144,** 711–719.

Muramatsu, M., Kinoshita, K., Fagarasan, S., Yamada, S., Shinkai, Y., and Honjo, T. (2000). Class switch recombination and hypermutation require activation-induced cytidine deaminase (AID), a potential RNA editing enzyme. *Cell* **102,** 553–563.

Muramatsu, M., Sankaranand, V. S., Anant, S., Sugai, M., Kinoshita, K., Davidson, N. O., and Honjo, T. (1999). Specific expression of activation-induced cytidine deaminase (AID), a novel member of the RNA-editing deaminase family in germinal center B cells. *J. Biol. Chem.* **274**, 18470–18476.

Nakamuta, M., Chang, B. H., Zsigmond, E., Kobayashi, K., Lei, H., Ishida, B. Y., Oka, K., Li, E., and Chan, L. (1996). Complete phenotypic characterization of apobec-1 knock-out mice with a wild-type genetic background and a human apolipoprotein B transgenic background, and restoration of apolipoprotein B mRNA editing by somatic gene transfer of Apobec-1. *J. Biol. Chem.* **271**, 25981–25988.

Ness, G. C., and Zhao, Z. (1994). Thyroid hormone rapidly induces hepatic LDL receptor mRNA levels in hypophysectomized rats. *Arch. Biochem. Biophys.* **315**, 199–202.

Okazaki, I. M., Hiai, H., Kakazu, N., Yamada, S., Muramatsu, M., Kinoshita, K., and Honjo, T. (2003). Constitutive expression of AID leads to tumorigenesis. *J. Exp. Med.* **197**, 1173–1181.

Patterson, A. P., Chen, Z., Rubin, D. C., Moucadel, V., Iovanna, J. L., Brewer, H. B., Jr., and Eggerman, T. L. (2003). Developmental regulation of apolipoprotein B mRNA editing is an autonomous function of small intestine involving homeobox gene Cdx1. *J. Biol. Chem.* **278**, 7600–7666.

Phung, T. L., Sowden, M. P., Sparks, J. D., Sparks, C. E., and Smith, H. C. (1996). Regulation of hepatic apolipoprotein B RNA editing in the genetically obese Zucker rat. *Metabolism* **45**, 1056–1058.

Plateroti, M., Chassande, O., Fraichard, A., Gauthier, K., Freund, J. N., Samarut, J., and Kedinger, M. (1999). Involvement of T3R alpha- and beta-receptor subtypes in mediation of T3 functions during postnatal murine intestinal development. *Gastroenterology* **116**, 1367–1378.

Powell, L. M., Wallis, S. C., Pease, R. J., Edwards, Y. H., Knott, T. J., and Scott, J. (1987). A novel form of tissue-specific RNA processing produces apolipoprotein-B48 in intestine. *Cell* **50**, 831–840.

Qian, X., Balestra, M. E., Yamanaka, S., Boren, J., Lee, I., and Innerarity, T. L. (1998). Low expression of the apolipoprotein B mRNA-editing transgene in mice reduces LDL levels but does not cause liver dysplasia or tumors. *Arterioscler. Thromb. Vasc. Biol.* **18**, 1013–1020.

Seishima, M., Bisgaier, C. L., Davies, S. L., and Glickman, R. M. (1991). Regulation of hepatic apolipoprotein synthesis in the 17 alpha-ethinyl estradiol-treated rat. *J. Lipid Res.* **32**, 941–951.

Sharp, R. J., Perugini, M. A., Marcovina, S. M., and McCormick, S. P. (2004). Structural features of apolipoprotein B synthetic peptides that inhibit lipoprotein(a) assembly. *J. Lipid Res.* **45**, 2227–2234.

Sjoberg, A., Oscarsson, J., Bostrom, K., Innerarity, T. L., Eden, S., and Olofsson, S. O. (1992). Effects of growth hormone on apolipoprotein-B (apoB) messenger ribonucleic acid editing, and apoB 48 and apoB 100 synthesis and secretion in the rat liver. *Endocrinology* **130**, 3356–3364.

Sowden, M. P., Ballatori, N., Jensen, K. L., Reed, L. H., and Smith, H. C. (2002). The editosome for cytidine to uridine mRNA editing has a native complexity of 27S: Identification of intracellular domains containing active and inactive editing factors. *J. Cell Sci.* **115**, 1027–1039.

Srivastava, R. A. (1995). Increased apoB100 mRNA in inbred strains of mice by estrogen is caused by decreased RNA editing protein mRNA. *Biochem. Biophys. Res. Commun.* **212**, 381–387.

Teng, B., Black, D. D., and Davidson, N. O. (1990a). Apolipoprotein B messenger RNA editing is developmentally regulated in pig small intestine: nucleotide comparison of apolipoprotein B editing regions in five species. *Biochem. Biophys. Res. Commun.* **173,** 74–80.

Teng, B., Blumenthal, S., Forte, T., Navaratnam, N., Scott, J., Gotto, A. M., Jr., and Chan, L. (1994). Adenovirus-mediated gene transfer of rat apolipoprotein B mRNA-editing protein in mice virtually eliminates apolipoprotein B-100 and normal low density lipoprotein production. *J. Biol. Chem.* **269,** 29395–29404.

Teng, B., Burant, C. F., and Davidson, N. O. (1993). Molecular cloning of an apolipoprotein B messenger RNA editing protein. *Science* **260,** 1816–1819.

Teng, B., Verp, M., Salomon, J., and Davidson, N. O. (1990b). Apolipoprotein B messenger RNA editing is developmentally regulated and widely expressed in human tissues. *J. Biol. Chem.* **265,** 20616–20620.

Thorngate, F. E., Raghow, R., Wilcox, H. G., Werner, C. S., Heimberg, M., and Elam, M. B. (1994). Insulin promotes the biosynthesis and secretion of apolipoprotein B-48 by altering apolipoprotein B mRNA editing. *Proc. Natl. Acad. Sci. USA* **91,** 5392–5396.

Veniant, M. M., Pierotti, V., Newland, D., Cham, C. M., Sanan, D. A., Walzem, R. L., and Young, S. G. (1997). Susceptibility to atherosclerosis in mice expressing exclusively apolipoprotein B48 or apolipoprotein B100. *J. Clin. Invest.* **100,** 180–188.

von Wronski, M. A., Hirano, K. I., Cagen, L. M., Wilcox, H. G., Raghow, R., Thorngate, F. E., Heimberg, M., Davidson, N. O., and Elam, M. B. (1998). Insulin increases expression of apobec-1, the catalytic subunit of the apolipoprotein B mRNA editing complex in rat hepatocytes. *Metabolism* **47,** 869–873.

Wu, J. H., Semenkovich, C. F., Chen, S. H., Li, W. H., and Chan, L. (1990). Apolipoprotein B mRNA editing. Validation of a sensitive assay and developmental biology of RNA editing in the rat. *J. Biol. Chem.* **265,** 12312–12316.

Xie, Y., Nassir, F., Luo, J., Buhman, K., and Davidson, N. O. (2003). Intestinal lipoprotein assembly in apobec-1−/− mice reveals subtle alterations in triglyceride secretion coupled with a shift to larger lipoproteins. *Am. J. Physiol. Gastrointest. Liver Physiol.* **285,** G735–G746.

Yamanaka, S., Balestra, M. E., Ferrell, L. D., Fan, J., Arnold, K. S., Taylor, S., Taylor, J. M., and Innerarity, T. L. (1995). Apolipoprotein B mRNA-editing protein induces hepatocellular carcinoma and dysplasia in transgenic animals. *Proc. Natl. Acad. Sci. USA* **92,** 8483–8487.

Yamanaka, S., Poksay, K. S., Arnold, K. S., and Innerarity, T. L. (1997). A novel translational repressor mRNA is edited extensively in livers containing tumors caused by the transgene expression of the apoB mRNA-editing enzyme. *Genes Dev.* **11,** 321–333.

Yamanaka, S., Poksay, K. S., Driscoll, D. M., and Innerarity, T. L. (1996). Hyperediting of multiple cytidines of apolipoprotein B mRNA by APOBEC-1 requires auxiliary protein(s) but not a mooring sequence motif. *J. Biol. Chem.* **271,** 11506–11510.

Yamane, M., Jiao, S., Kihara, S., Shimomura, I., Yanagi, K., Tokunaga, K., Kawata, S., Odaka, H., Ikeda, H., Yamashita, S., *et al.* (1995). Increased proportion of plasma apoB-48 to apoB-100 in non-insulin-dependent diabetic rats: Contribution of enhanced apoB mRNA editing in the liver. *J. Lipid Res.* **36,** 1676–1685.

Yao, Z. M., Blackhart, B. D., Johnson, D. F., Taylor, S. M., Haubold, K. W., and McCarthy, B. J. (1992). Elimination of apolipoprotein B48 formation in rat hepatoma cell lines transfected with mutant human apolipoprotein B cDNA constructs. *J. Biol. Chem.* **267,** 1175–1182.

EDITING IN PLANT ORGANELLES

CHAPTER TWENTY

RNA EDITING IN PLANT MITOCHONDRIA: ASSAYS AND BIOCHEMICAL APPROACHES

Mizuki Takenaka *and* Axel Brennicke

Contents

Abstract

To analyze the C-to-U conversion of RNA editing in plant mitochondria, comple-mentary methods are required, which include *in vivo, in organello*, and *in vitro* approaches. The major obstacle for *in vitro* assays is the generally observed fragility of the activity in mitochondrial lysates and the corresponding low activity. If seen at all, this activity is often in the range of a few percent conversion of the added templates. We have developed a sensitive assay

Molekulare Botanik, Universität Ulm, Ulm, Germany

Methods in Enzymology, Volume 424
ISSN 0076-6879, DOI: 10.1016/S0076-6879(07)24020-0

system using mismatch analysis that allows detection of such low conversion rates. With this assay mitochondrial lysate preparations could be established from pea shoots and cauliflower inflorescences, which can be employed for the *in vitro* analysis of specificity requirements and biochemical parameters of RNA editing in plant mitochondria.

1. INTRODUCTION

RNA editing in flowering plant mitochondria generally involves several hundred cytidine-to-uridine (C-to-U) conversions, mostly in mRNAs, but also in several tRNAs (Bonnard *et al.*, 1992). Since its discovery more than a decade ago, progress into understanding how this change of nucleotide identity works has been hampered by the lack of manageable and reliable *in vitro* systems. The first *in vitro* system for plant mitochondrial RNA editing was successfully developed from wheat embryos, but since then this has not been exploited further (Araya *et al.*, 1992).

Conclusions about the determinants of specificity during RNA editing in mitochondria have initially been drawn from comparisons of transcribed sequence duplications (e.g., Lippok *et al.*, 1994). Such observations suggested that upstream (5'-) sequences are crucial in targeting the editing machinery and that downstream regions are less important to specify a given site. These conclusions are corroborated by experiments following the development of an electroporation protocol for mitochondria (Choury *et al.*, 2004; Farré *et al.*, 2001; Staudinger *et al.*, 2005; see Chapter 22, this volume). Mutational analysis of RNA editing templates showed that for several editing sites, 20–30 nucleotides upstream and 2–5 nucleotides downstream are sufficient to guide the editing activity, precisely what had initially been deduced for plant mitochondrial RNA editing sites from the *in vivo* observations (e.g., Lippok *et al.*, 1994). Recent analysis of the interaction between a PPR-protein and its cognate recognition site in plastid RNA editing similarly identified the upstream 25 and downstream 10 nucleotides to be important for binding and thus recognition of the editing site (Kotera *et al.*, 2005; Okuda *et al.*, 2006; Shikanai, 2006).

The biochemistry of the RNA editing activity in plant mitochondria is still largely unclear. A more detailed understanding of the enzymes involved and their mode of action requires accessible *in vitro* systems for the respective experimental investigations. Initial *in vitro* assays with plant mitochondrial extracts had indicated that the sugar phosphate backbone remains intact during the deamination step of C-to-U editing (Blanc *et al.*, 1995; Rajasekhar and Mulligan, 1993; Yu and Schuster, 1995). The most likely enzyme able to catalyze this reaction would be one of (or a homolog of) the cytidine deaminases identified in plants, e.g., in *Arabidopsis thaliana*, as well as in other organisms. The identification of this class of enzymes will allow the determination of their potential involvement in RNA editing directly (Faivre-Nitschke *et al.*, 1999). One of these deaminases is in mammals the crucial enzyme of apolipoprotein mRNA C-to-U editing (see Chapters 18 and 19, this volume). Problematic for this group of deaminases appears to be the reverse reaction, since none of the homologous enzymes from various organisms has been found to be able to catalyze the amination step leading to a U-to-C conversion. This latter reaction is frequently observed in plant species outside the seed plants such as mosses, hornworts, and ferns.

To facilitate direct investigations into the biochemistry and specificity of RNA editing in plant mitochondria, we have developed *in vitro* systems from pea and cauliflower mitochondria, similar to the methodology first established for *in vitro* RNA editing in plastid preparations (Hirose and Sugiura, 2001; Miyamoto *et al.*, 2002, 2004; Sasaki *et al.*, 2006). These *in vitro* assays now allow comparatively rapid and robust access to biochemical parameters (Takenaka and Brennicke, 2003; Takenaka *et al.*, 2004; Neuwirt *et al.*, 2005; van der Merwe *et al.*, 2006; Verbitskiy *et al.*, 2006). The system employs the advantages of mismatch detection and specific incision in double-stranded DNA by a thymine DNA glycosylase (TDG) enzyme (Taylor, 1999).

2. Preparation of Mitochondrial Lysates Active in RNA Editing

Cautionary note: Plant tissues are notoriously loaded with aggressive compounds of all kinds that will disturb whatever *in vitro* reaction is to be observed. *In vitro* RNA editing is no exception. To minimize the detrimental effects, it is essential that these inhibitory compounds are diluted and/or rendered harmless during or immediately after breaking the plant tissues and their cells. Therefore, a minimal ratio of 2–3 ml extraction buffer per 1 g plant tissue (wet weight) is recommended. If your mitochondrial lysate does not show any activity and/or your added template RNA dissappears, a new mitochondrial extraction with a higher buffer-to-plant tissue rate might be worth a try.

This ratio may need to be adapted to different sources of plant material: field grown cauliflower heads, for example, contain more secondary compounds than plants grown in the greenhouse. When shopping for cauliflower in your local market, which is easy and therefore recommended, you may thus require more buffer in summer and early autumn than in winter for the same amount of fresh weight.

2.1. Preparation of mitochondrial extracts from pea seedlings

Pea seedlings *(Pisum sativum* L., var) are grown at 25° in the dark for 6 days. All steps hereafter are performed at 4°. The white shoots are harvested and placed into a kitchen blender with 2–3 times (fresh weight/volume) extraction buffer [final concentrations: 0.3 M mannitol, 30 mM MOPS, 1 mM EGTA, adjust the pH to 7.8 with KOH; just before use, add the following compounds to the given final concentrations: 0.1% bovine serum albumin (BSA) (w/v) and 2 mM dithiothreitol (DTT)]. Several bursts at high speed of about 10 sec each will chop the shoots into small fragments and break most of the cells. The blades should be sharp to ensure that the cells are cut and not just squashed. The resulting slurry is filtered once (or twice) through four layers of muslin and two layers of miracloth to remove fibers and other crude debris.

Mitochondria are first enriched by several steps of differential centrifugation (Fig. 20.1). Cell debris and unbroken cells are removed by several consecutive low-speed centrifugation steps of 10 min at $600 \times g$, 10 min at $1300 \times g$, and 10 min at $2100 \times g$. Each time the supernatants are moved into a new centrifugation vessel. Mitochondria are pelleted at $7000 \times g$ for 20 min and the pellet is resuspended in about 3-5 ml of washing buffer [final concentrations: 0.25 M sucrose, 20 mM Tris–HCl, 5 mM EDTA, 15% glycerin (v/v); adjust the pH to 8.0 with KOH; just before use, add 1 mM DTT]. The suspension is layered on top of preformed Percoll gradients in steps of 13%/21%/45% Percoll in gradient buffer (0.25 M sucrose, 10 mM MOPS, 1 mM EGTA; adjust the pH to 7.2 with KOH; just before use, add 0.2% BSA and 2 mM DTT. Centrifugation at $70,000 \times g$ in a swing-out rotor for 45–50 min will move the mitochondria to the border of the 21%/ 45% Percoll steps. The creamy yellowish interface is collected with a pipette and the Percoll is diluted to a volume of about 500 ml with washing buffer. Mitochondria are again pelleted by centrifugation at $9000 \times g$ for 30 min. The mitochondria are transferred to an Eppendorf tube with washing buffer, pelleted at 15,000 rpm in a standard table top centrifuge for 10 min, and the buffer is pipetted off. At this step, the mitochondrial pellet can be quick-frozen in liquid nitrogen and stored at −80° or can be directly used for the lysate preparation.

Four-hundred milligrams of the mitochondrial pellet are resuspended and lysed in 1200 μl extraction buffer (30 mM HEPES-KOH, pH 7.7, 3 mM Mg-acetate, 2 M KCl, and 2 mM DTT) containing 0.2% Triton X-100. After a 30-min incubation on ice, the lysate is centrifuged at $50,000 \times g$ for 20 min. The supernatant is recovered and dialyzed against 5×100 ml dialysis buffer (30 mM HEPES-KOH pH 7.7, 3 mM magnesium acetate, 45 mM potassium acetate, 30 mM ammonium acetate, and 10% glycerol) for a total of 5 h. All steps are carried out at 4°. The resulting extract (10 ∼ 20 μg protein/μl) is rapidly frozen in liquid nitrogen. When stored at −80°, the lysate is stable for at least 3 months.

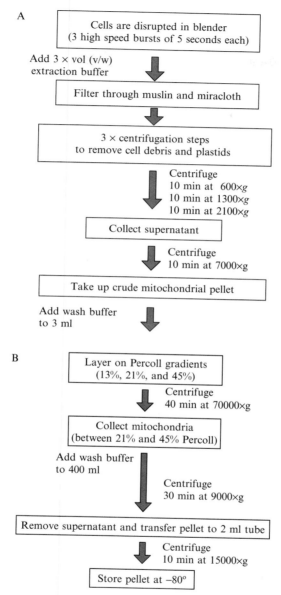

Figure 20.1 Preparation of the plant mitochondrial lysate. The procedure involves consecutive steps of differential centrifugation. Those at low speed will remove cell debris and those at higher speed will pellet the mitochondria. A subsequent Percoll gradient centrifugation clears away further contamination when mitochondria are collected at the 21%/45% interface. Any one or more of these steps including the Percoll gradient should be repeated if difficulties are encountered in the preparation of clean organelles from complex tissues such as green leaves or storage cells.

2.2. Preparation of cauliflower mitochondrial extracts

Heads of cauliflower can be purchased at local markets. About 900 g of the top tissues of the inflorescences are harvested, manually chopped into small pieces, and homogenized in a blender. Mitochondria are purified by differential centrifugation steps and a Percoll gradient basically as detailed above for the pea preparation (Takenaka *et al.*, 2004; Neuwirt *et al.*, 2005). However, there are several adaptive modifications to the procedure, especially the buffer constitution, which accommodate the specifics of the different tissue and species, such as the different secondary compounds.

The white cells are placed into a kitchen blender with at least 3 times (fresh weight/volume) extraction buffer [final concentrations: 0.3 M mannitol, 30 mM sodium pyrophosphate ($Na_4P_2O_7$), 2 mM EDTA; adjust the pH to 7.5 with HCl; just before use, add to 0.8% PVP 25 (w/v), 0.3% BSA, 3 mM cysteine, 5 mM glycine, 2 mM 2-mercaptoethanol]. Several bursts at high speed of about 10 sec each will chop the inflorescences into small fragments and break most of the cells. The blades should be sharp to ensure that the cells are cut and not just squashed. Alternatively, the use of a commercial juice extractor has turned out to be very practical for working up large amounts of tissues, a procedure that has been initiated and developed by Prof. Dr. José Gualberto, IBMP, Strasbourg. The slurry of broken and intact cells is filtered at least twice through four layers of muslin and two layers of miracloth to remove fibers and other crude debris.

Mitochondria are first enriched by several steps of differential centrifugation (Fig. 20.1). Cell debris and unbroken cells are removed by several consecutive low-speed centrifugation steps of 10 min at 600×g, 10 min at 1300×g, and 10 min at 2100×g. Each time the supernatants are moved into a new centrifugation vessel. Mitochondria are pelleted at 7000×g for 20 min and the pellet is resuspended in about 5 ml of washing buffer (final concentrations: 0.3 M mannitol, 10 mM K_2PO_4, 1 mM EDTA, 0.1% BSA, 5 mM cysteine, 15 mM glycine, pH 7.5). The suspension is layered on top of preformed Percoll gradients in steps of 13%/21%/45% Percoll in gradient buffer (0.3 M sucrose; 10 mM K_2PO_4, 1 mM EDTA, 0.1% BSA). Centrifugation at 70,000×g in a swing-out rotor for 45–50 min will move the mitochondria to the border of the 21%/45% steps. The creamy yellowish interface is collected with a pipette and the Percoll is diluted to a volume of about 500 ml with washing buffer. Mitochondria are again pelleted by centrifugation at 9000×g for 30 min. The mitochondria are transferred to an Eppendorf tube with washing buffer, pelleted at 15,000 rpm in a standard table top centrifuge for 10 min, and the buffer is pipetted off. At this step, the mitochondrial pellet can be quick-frozen in liquid nitrogen and stored at −80° or can be directly used for the lysate preparation.

Four hundred milligrams of the mitochondrial pellet are resuspended and lysed in 1200 μl extraction buffer (0.3 mM HEPES-KOH pH 7.7, 3 mM Mg-acetate, 2 M KCl, and 2 mM DTT) containing 0.2% Triton X-100. After a 30-min incubation on ice, the lysate is centrifuged at 50,000$\times g$ for 20 min. The supernatant is recovered and dialyzed against 5 $\times$ 100 ml dialysis buffer (30 mM HEPES-KOH, pH 7.7, 3 mM magnesium acetate, 45 mM potassium acetate, 30 mM ammonium acetate, and 10% glycerol) for a total of 5 h. All steps are carried out at 4°. The resulting extract (10 $\sim$ 20 μg protein/μl) is rapidly frozen in liquid nitrogen. When stored at $-80°$, the lysate is stable for at least 3 months.

The first low-speed centrifugation steps should be repeated several times if the pellet is large and loose to remove as much of the cell debris as possible. Likewise the precipitation of enriched mitochondria should be repeated until the pellet is unstructured and glassy-yellowish.

All other steps are as described above for the lysate preparation for etiolated pea seedlings.

3. PREPARATION OF TEMPLATES FOR THE RNA EDITING ASSAYS

3.1. Selection of template sites

RNA editing sites of plant mitochondria (and of other organisms) are deposited in the specialized RNA editing database REDIdb. This database can be freely queried at http://biologia.unical.it/py_script/search.html (Picardi *et al.*, 2007). Here the sites are annotated, which thus provides a good base for choosing the sites to be investigated. However, it is advisable to always confirm the presence of a given editing event in the plant species and variety that are being used. For this, genomic as well as cDNA sequences for the respective coding region have to be determined to identify *in vivo* RNA editing sites. This is a straightforward analysis by polymerase chain reaction (PCR) on genomic mitochondrial DNA and reverse transcriptase (RT)-PCR on mitochondrial RNA with primers flanking the region you want to investigate and subsequent sequencing. Having determined that this site indeed exists and is efficiently edited, a suitable fragment should be cloned so that RNA templates can be readily synthesized.

It is advisable to choose RNA editing sites in abundant RNA molecules that are efficiently edited *in vivo*. This should increase the probability of having sufficient concentrations of the specific and general *trans* factors present in the mitochondrial lysate sample for the *in vitro* reaction. Do not expect to see each site you try to be edited *in vitro*: most of them are not or barely detectably converted.

In our hands, only a few sites are edited *in vitro* sufficiently to be able to quantify reproducibly and to accommodate reductions in, e.g., competition experiments. A similar situation has been observed with the RNA editing sites in plastids, where similarly many sites are not converted *in vitro* at detectable levels (Sasaki *et al.*, 2006).

In plant mitochondria of pea and of cauliflower, we found two sites to be consistently well modified at levels of several percent: one is the first site in the atp9 open reading frame and the second is a cluster of three editing sites in the atp4 mRNA. We therefore describe the construction of the RNA editing templates for the *in vitro* assays with these two examples.

3.2. Elements of template construction

To construct the first templates for setting up the *in vitro* system, several aspects have to be taken into consideration. The mRNA fragment containing the RNA editing site should be large enough to make sure that all *cis* elements are covered. Since most *cis* elements characterized so far cover mostly upstream and only occasionally downstream sequences, usually 100–150 nucleotides upstream and 50 or so downstream sequences should cover the recognition sites adequately. The term "usually" implies that there may be exceptions and even more nucleotides may be required in either direction.

Nonplant mitochondrial primer sites will have to be accommodated within the region transcribed from the T7 or T3 promoters, respectively, to ensure that upon RT-PCR from the *in vitro* reaction only the added template RNA is used in the reverse transcription and amplification reactions and not the respective endogenous mitochondrial mRNA.

To extend the lifetime of the added template RNA, stabilizers should be added at the 3′ terminus. For this, the characterized inverted repeats of several mRNAs are recommended; one of the most efficient (and easiest to clone) is the stem-loop structure from the atp9 mRNA. This seems to operate in several species and does not have to be native to the plant to be investigated. In case antisense RNA molecules are required at some point, it is advisable to include a respective other (bacterial) promoter at the 3′ end of the construct, which can then be used to synthesize such complementary molecules without further specific cloning.

Furthermore, RNA molecules larger than 100 nt are easier to handle in all experimental steps including purification and precipitation procedures. This allows the inclusion of the above elements and the primer attachment sites, which alone will cover about 50 nucleotides. These sequences do not seem to disturb *in vitro* RNA editing, although certain sequence combinations may by chance create inhibitory motifs or may alter the efficiency at the monitored editing site.

3.3. Testing RNA editing sites for their *in vitro* activity

To investigate RNA editing sites for their potential to be edited *in vitro*, the TDG system offers the advantage that several sites can be monitored simultaneously. It is thus feasible to clone comparatively large fragments of unedited genomic DNA covering several RNA editing sites and to analyze these on appropriate sequencing gels. With such templates, the *in vitro* activity at a given site will be lower than in smaller fragments, but nevertheless detectable if a site is edited at all. By no means are all sites good *in vitro* substrates; in our experience only 1 in 20 sites analyzed (or even less) shows any *in vitro* processing. Therefore, such an initial large-scale *in vitro* screening is advisable to avoid too many futile cloning steps of editing sites that are not accessed *in vitro* at levels sufficient for identification.

Once a site is positively detected *in vitro*, subclones should be made to approach the minimal sequence requirements in *cis* and to obtain a better signal–to–noise ratio with the smaller fragments. Depending on the ultimate aim of the planned experiments, a clone size should be designed for optimal detection or, if *cis* elements are analyzed, further deletions and/or mutations will have to be introduced.

3.4. RNA substrates: atp9 mRNA as an example

Genomic DNA clones (patp9) are constructed in an adapted pBluescript SK+ vector to allow runoff transcription of the editing template RNA. The synthesized RNA molecule contains the first two pea atp9 editing sites (at nucleotide positions 19 and 49, respectively, from the AUG codon) flanked by bacterial sequences to allow specific amplification against the background of internal mRNAs. Only 154 bp of the 5′ UTR and the first 69 bp of the coding sequences of the pea mitochondrial atp9 gene are cloned into the *Pst*I site of the vector MCS between the T7 promoter followed by a KS sequence on the upstream side and the T3 promoter sequence in the downstream region. Further downstream follow 89 bp of the mRNA 3′ terminal double stem–loop sequence of the pea mitochondrial atp9 gene flanked by the 20 upstream and 22 downstream adjacent nucleotides, respectively. To facilitate easier cloning, six nucleotides are added at the 5′ flank to create an *Nde*I recognition site and at the 3′ end three additional nucleotides generate a *Pvu*II site (Fig. 20.2). The RNA substrate for the *in vitro* editing reaction is transcribed as a runoff product from *Pvu*II-digested patp9 using T7 RNA polymerase (MBI).

The resulting transcript (475 nt) contains 69 nt upstream and 183 nt downstream sequences outside the atp9 region. The upstream part corresponds to vector sequences from the T7 transcription start point to the *Pst*I restriction site. The downstream anchor consists of vector sequences from the *Pst*I site to the T3 promoter region and the downstream attached pea atp9 double stem–loop sequence.

3.5. RNA substrates: atp4 RNA editing site cluster and generation of deletion substrates

In the atp4 mRNA, three RNA editing sites are clustered within four nucleotides at positions 248, 250, and 251 downstream from the AUG codon (Verbitskiy *et al.,* 2006). Genomic DNA clones (patp4) are constructed in an adapted pBluescript SK+ to allow runoff transcription of the editing substrate RNA as described above for the atp9 templates (Neuwirt *et al.,* 2005; Takenaka *et al.,* 2004). Substrate RNAs are synthesized from the T7 RNA polymerase promoter and thus contain vector sequences at the 5′ end. Similarly, vector sequences border the 3′ end of the mitochondrial insert sequences up to the *Vsp*I site, which is used for linearization of the template DNA. These bordering bacterial sequences are employed for specific amplification of the substrate RNAs by RT-PCR after the *in vitro* assay. Mutant RNA competitors with purines instead of the native pyrimidines can be readily synthesized from PCR products with modified primer sequences.

Deletion clones with shortened native mitochondrial sequences can be constructed by successively removing original mitochondrial sequences. The 5′-deletion mutants are constructed by inverted PCR from patp4 with respective primers pairing to the −40 and −20 upstream sequences on the one side and the original primer on the other. The resulting fragments are digested with *Eco*RI to generate sticky ends in the primer-contained *Eco*RI recognition site and are self-ligated. The outside bacterial anchors for PCR amplification will move accordingly closer to the editing sites. Coincidental nucleotide similarities between these and the substituted mitochondrial sequences as well as potential secondary structures should be taken into consideration when evaluating nucleotide requirements for RNA editing.

4. SETTING UP THE *IN VITRO* TEST SYSTEM

4.1. *In vitro* RNA editing reactions

The *in vitro* RNA editing reactions are performed in a total volume of 20 μl. The reaction mixture consists of 30 m*M* HEPES-KOH pH 7.7, 3 m*M* Mg-acetate, 45 m*M* K–acetate, 30 m*M* ammonium acetate, 15 m*M* ATP, 2 m*M* DTT, 1% polyethylene glycol (PEG) 6000, 5% glycerol, 40 U RNase inhibitor (MBI), 1 × proteinase inhibitor mixture (CompleteTM, Boehringer-Mannheim), 100 amol (100 × 10⁻¹⁸ mol) mRNA substrate, and

6.0 μl mitochondrial extract (Fig. 20.2). After incubation at 28° for 4 h, the substrate mRNA (and the total RNA contained in the mitochondrial lysate) is extracted with the RNeasy kit (Nucleospin RNA II from Macherey-Nagel or the respective kit from another manufacturer such as QIAGEN or General Electric Healthcare). To accommodate variations of individual concentrations and additions of various other compounds, respective alterations will have to be made in the mixtures.

This (very low) concentration of RNA template has proven best in our hands. Variations of up to 10-fold higher concentrations are usually also tolerated, but the efficiency of the *in vitro* RNA editing reaction drops when

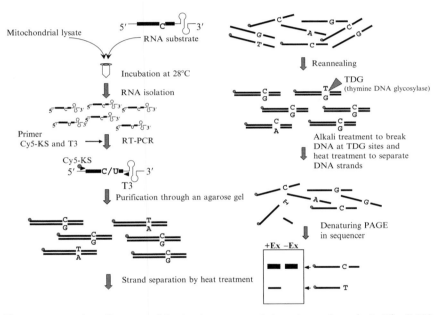

Figure 20.2   Flow diagram of the *in vitro* assay and the mismatch analysis. The RNA substrate is incubated with the lysate, which is prepared as summarized in Fig. 20.1. Subsequently, after the RNA is purified by standard methods, an unlabeled T3 primer is added. From this primer reverse transcription is initiated. Then the cDNA strand is amplified by PCR between a Cy5-labeled KS primer and the unlabeled T3 primer. The products of the RT-PCR are purified through an agarose gel and the eluted fragments are heat denatured. During reannealing at lowered temperatures, the majority of the G-containing complementary strands derived from the unedited RNA substrate will also align with the few DNA strands derived from edited RNA, forming a T•G mismatch at the editing site. This mismatch is recognized by the enzyme thymine DNA glycosylase (TDG), which removes the mismatched base. Subsequent alkaline treatment will break the nucleic acid chain at this site and, at the same time, will separate the pairing DNA strands. The ssDNA is run on a sequencing gel, where the CY5 fluorescent label is detected. The DNA strands cut at the mismatch site will appear as shorter fragments.

higher amounts of RNA are added. Probably specific *trans* factors in the mitochondrial lysate are limiting in their accessibility and thus may get titrated out by too much template RNA.

5. THE DETECTION SYSTEM

The mismatch detection system utilizes the high sensitivity of DNA repair enzymes to detect single nucleotide mismatches (Fig. 20.3). First, a double-stranded cDNA is synthesized from the population of RNA molecules after *in vitro* incubation with the mitochondrial lysate. The 5′ located KS primer initiating synthesis of the coding strand contains a fluorescent label that is later used for detection. The resulting DNA molecules are separated into single strands by heat denaturation. In the next step, the DNA strands are allowed to reanneal. This renaturation will create single nucleotide mismatches between the few molecules altered by *in vitro* editing to contain a T and the vast majority of unedited complementary strands containing a G at the editing site. These T•G mismatches are recognized by the enzyme thymine DNA glycosylase, modified, and finally broken by the subsequent alkaline treatment (Figs. 20.2, 20.3, and 20.4). The ratio of these shortened DNA strands to the long DNA molecules representing the unmodified RNA substrates reflects the efficiency of the *in vitro* editing reaction.

If multiple editing sites are present in the template RNA, the signal at the downstream site(s) will be detectable, but the signal strength there depends on the coupling to the upstream site and the efficiency of the TDG reaction. Thus the signal at downstream sites cannot be quantified directly but needs to be determined by direct sequencing of a number of clones. In cases of no coupling between the different editing sites, the signal at downstream sites will accurately reflect the editing efficiency at editing alterations of less than 10%. Higher editing will increase the probability that two editing sites are located on the same RNA molecule, in which instance only the first (upstream) editing site will be detected, provided the TDG acts with

Figure 20.3 The generation of a T•G mismatch in dsDNA is shown for the first site of the atp9 cDNA. The few strands of the cDNA derived from an edited RNA template will effectively reanneal to one of the excess complementary strands derived from the unedited template RNA and will generate a T•G mismatch. The arrowhead points to the mismatched pair of nucleotides, which are shown in bold.

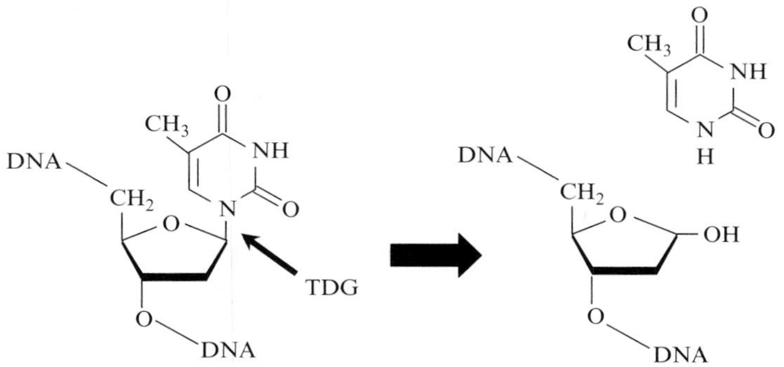

Figure 20.4 The mismatch-repair enzyme thymine DNA glycosylase (TDG) does not disrupt the sugar-phosphate backbone of the DNA strand, but removes the base from the deoxyribose moiety. The sugar-phosphate chain is disrupted by a subsequent alkaline treatment.

100% efficiency, which is of course not the case. And of course, when two editing sites are directly adjacent, TDG will not act at all, since the enzyme recognizes only single nucleotide mismatches.

5.1. Detection of the RNA editing activity by mismatch analysis

The protocol described here (Fig. 20.2) essentially follows the steps origi-nally outlined (Neuwirt *et al.*, 2005; Takenaka *et al.*, 2004). The first cDNA strand is synthesized from the T3 primer along the substrate mRNA with reverse transcriptase (e.g., StrataScript, Stratagene). The subsequent PCR reactions are performed with 0.1 U of Pwo polymerase (peqLab) and 0.5 U of Taq polymerase using a Cy5-labeled KS primer (Cy5-KS) and an unla-beled T3 primer. Cycling is performed as follows: 95° for 2 min; 5 cycles of touchdown PCR (95° for 30 sec, 65°–60° decreasing by 1° per cycle for 30 sec, and 72° for 1 min); 35–40 cycles of 95° for 30 sec, 59° for 30 sec, and 72° for 1 min; finally 72° for 5 min. The background error rate at editing versus nonediting sites is readily seen later in the sequencing gels used for analysis as the ratio of the signal at the editing site to the signals at other nucleotide positions. A major portion of the background stems from the high amplification rate of the 35–40 PCR cycles. When we investigated the false positives directly, we found in the sequence analysis of 200 cloned DNA fragments after the PCR a total of five C-to-T changes at different positions. The PCR products are purified by 1% agarose gel electrophoresis.

Denaturation and reannealing are done as follows: 95° for 10 min; 90°–70° decreasing by 5° per cycle for 5 min; 65° for 1 h. After reannealing, the resulting heteroduplexes are treated with 0.2 units of the enzyme TDG (thymine DNA glycosylase, Trevigen). The TDG-treated fragments are denatured for 5 min at 95° in alkali buffer (300 mM NaOH, 90% formamide, and 0.2% bromophenol blue), and the single strands of the DNA are separated by 6 M urea 6–10% PAGE (Fig. 20.5). The polyacrylamide concentration for the analytical sequencing gel should be adapted to the size of the analyzed DNA fragment. The gel system also needs to be adjusted to yield an optimal signal-to-noise ratio. With some templates, inadvertent background signals are seen, which may result from further TDG modifications or from other preferential break points in the DNA. These signals can by chance be near the editing site monitored and appropriate gel concentrations have to be employed to clearly separate these. To calibrate the system, it will be best to run artificial mixtures with controlled percentages of edited and unedited DNA molecules. These will give you a feeling for the real signal strength and can also show you sites of high structural background.

In this context it is essential to run a sequencing reaction alongside the TDG analysis to unambiguously identify the signal observed (Fig. 20.5B). Although most sequencing runs are quite reproducible, there are always minor differences between individual gels that alter the running time of a given fragment. To make sure that none of the above-mentioned signals outside of the editing event is erroneously followed, clear assignment of the signal to the C-nucleotide at the editing site by scaling with a sequence ladder is necessary (Fig. 20.5B).

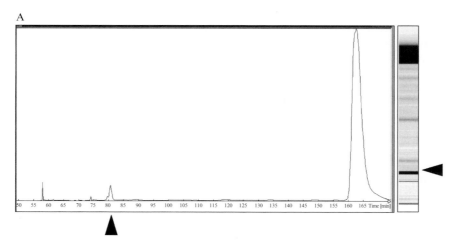

Figure 20.5 Gel picture of the product analysis after TDG treatment of an editing site incubated in the *in vitro* system. (A) The gel image generated by the automated sequencer on the right-hand side and the scan of the fluorescence intensity on the left. The signals of the editing site are marked by the arrowheads. The large peak at about 165 min in the scan (and the intense black bar in the top part of the gel image, respectively) represents the uncut cDNA strands, which are derived from the unedited

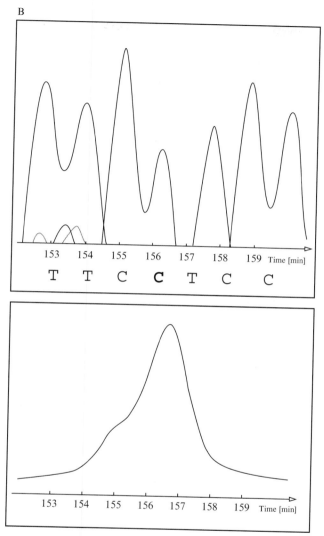

template RNA. Spurious signals appear as background; the leftover Cy5-labeled primers have run off the gel. (B) An alignment of the TDG-derived RNA editing signal with the corresponding sequence analysis from the same primer. This is necessary to identify the editing site and to unambiguously assign the signal to the editing event. The shoulder on the left-hand side of the major peak results from incomplete denaturation during the gel run; similar ghosts are sometimes seen in the sequences. These depend on the DNA sequence composition and the denaturing status. To minimize these, higher concentrations of urea, e.g., 8 M instead of 6 M in loading buffer and gel, are recommended. (See color insert.)

The Cy5 fluorescence is scanned and displayed (Fig. 20.5A) using an ALF express DNA sequencer (Amersham, now GE Healthcare). For other sequencing machines, protocol, fluorescent label, and other procedures have to be adapted to the respective requirements.

To quantify the efficiency of the *in vitro* RNA editing reaction, the area under the peaks of the cleaved and of the uncut DNA fragments, respectively, is determined. The ratio of cleaved, i.e., edited, fragment to uncut DNA is used to determine relative efficiencies of the investigated conditions in each experiment. To obtain comparable values in order to combine several independently repeated assays and to allow determination of variation bars, the ratios of cleaved to uncleaved fragments are displayed as percentages of the standard reaction conditions.

Between individual experiments, major sources of variation are the differences in RNA editing activity and RNase content of individual lysate preparations. In the gel analysis of the *in vitro* editing products, smearing of the uncut fragment signal complicates the determination of the respective signal area for comparable quantification between individual gel runs. To allow comparisons and to determine the variation between individual experiments, cotreatment and parallel resolution of the standard template under standard conditions need to be adopted for reference in each experiment. "Standard template and conditions" here means your optimal wild-type construct of the given site under optimal ion and NTP conditions for which you have control-sequenced a number of clones and know the ratio of editing that can be expected. This should be run alongside *in vitro* reactions with modified templates and/or altered buffer compositions to calibrate the investigated modified reaction.

6. CRUCIAL PARAMETERS: TROUBLESHOOTING

As usual, a number of things may go wrong. If the final sequencing run does not show a signal at the size at which the TDG cut DNA strand is expected, one or more of several common problems may have prevented a successful experiment. Here we list the problems we have encountered and sometimes had to trace back for days or even weeks to find and to rectify. This list may help you to get to the source of the problem a bit faster and to avoid some of the pitfalls we have gone through.

1. No PCR product: The template RNA may have been degraded. The RNA purification kit has gone bad.
2. Smeary PCR product: The primers are not clean.
3. Inconsistent and varying PCR product quantitites: The RNA purification kit has gone bad. Try the cheap and reliable old phenol/chloroform procedure.

4. Good PCR product, but no TDG signal: The DNase treatment after RNA template synthesis before the actual *in vitro* assay did not work properly and you are amplifying from the leftover clone DNA, which is carried along through all steps. To monitor the DNase efficiency it is advisable to include an analytical PCR/RT-PCR test of the template RNA before starting the *in vitro* assays.

5. High background noise interferes with the signal: The reasons may be either or all of the following problems. The RNA purification kit has gone bad. The primers are not clean. If you cannot get rid of some of the background signals, try running sequencing gels of different PAA percentages to move the product signal away from spurious primer signals and from chance products arising from secondary structure TDG attack points. Another source of a high background noise may be an excessive number of PCR cycles, which may amplify obscure signals. These can be suppressed by reducing the number of rounds in the PCR amplification step—more is not always better.

7. ADAPTATIONS AND ASSAYS FOR COMMON APPLICATIONS

One of the most useful tools of the *in vitro* systems that you can directly pursue are competition assays. These can be used to corroborate results obtained with mutated templates by using these as competitors. Competitors with mutated essential *cis* elements will not compete, while competing RNA molecules with mutations in nonessential regions will compete. A second application of competition assays is the titration of *trans* factors by the competing RNA molecules. Here *cis* regions addressed by different *trans* factors can potentially be differentiated—if such exist. As a third inherent advantage, competition assays allow the evaluation of the *trans* factor-binding capacity of completely edited RNA molecules. These obviously cannot be investigated directly *in vitro* or in any other way, but can be tested via their ability to compete with wild-type sequences.

7.1. Competition assays

Competitor RNAs are synthesized by T7 RNA polymerase as runoff products from the PCR products amplified with primers T7 and respective primers matching the different atp9 and/or atp4 clones. An entirely plasmid-derived control RNA is synthesized from the PCR product amplified from pBluescriptIISK+ with T7 and SK primers. Then 100 amol of substrate and the required amounts [100 times excess (10 fmol) to 1500 times excess (150 fmol)] of competitor RNA are first mixed and then incubated with the mitochondrial *in vitro* assay as described above. In the competitor RNAs the KS or T3 sequences have to be omitted/deleted to avoid contaminating the RT-PCR amplifications of the monitored substrate RNAs. This is, of course, essential and must not be forgotten during setup of the competitor RNA construction.

7.2. Biochemical alterations

Biochemical parameters can be modified in an open *in vitro* system where tests for energy requirements as well as changes in concentrations of cofactors such as pH and ions are easily done by addition or depletion of the respective compounds. It is self-evident that the volume adjustments for the final reaction volume of 20 μl have to be modified accordingly. The impact of changing, e.g., the amount of added ATP can be comparatively evaluated through quantification of the respectively resulting *in vitro* editing at the monitored site (Fig. 20.6).

Alterations of concentrations and biochemical parameters such as pH, salts, and energy requirements can be readily realized and the optimal reaction conditions for the plant species and tissue of choice can be established. Requirements of cofactors can be analyzed, from which conclusions about the enzyme(s) involved can be drawn. For example, in such assays the impact of various potential amino group acceptors has been investigated. These will be necessary if the reaction involves a transaminase, which reposits the amino group of the cytosine onto another compound. The addition of any of the various NTPs or dNTPs proved to be similarly

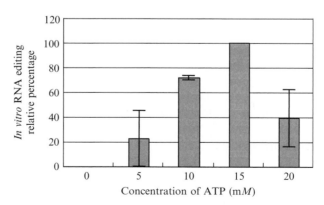

Figure 20.6 Manipulation of *in vitro* RNA editing conditions. As an example of direct manipulation of the biochemical composition of the *in vitro* assay, the results of RNA editing at several ATP concentrations in the *in vitro* reaction are depicted. For such quantifications, the ratio of TDG cut to uncut DNA strands has to be calculated. Here this is displayed as percentage relative to the highest *in vitro* editing rate, which is observed with 15 mM ATP. The error bars increase toward lower editing signals. (See color insert.)

efficient in such *in vitro* reactions, which suggests that an RNA helicase may be involved in an effective *in vitro* RNA editing reaction in plant mitochondria (Takenaka and Brennicke, 2003).

ACKNOWLEDGMENTS

We wish to thank our colleagues in the laboratory for their help and comments on all practical aspects of the real-life applications and for having worked out of and around all the potential snares in the procedure. Special thanks are due to Dré van der Merwe, Daniil Verbitskiy, Anja Zehrmann, and last but not least to Dagmar Pruchner, our specialist in the preparation of mitochondria and their lysates from all types of plants.

REFERENCES

Araya, A., Domec, C., Begu, D., and Litvak, S. (1992). An *in vitro* system for the editing of ATP synthase subunit 9 mRNA using wheat mitochondrial extracts. *Proc. Natl. Acad. Sci. USA* **89,** 1040–1044.

Blanc, V., Litvak, S., and Araya, A. (1995). RNA editing in wheat mitochondria proceeds by a deamination mechanism. *FEBS Lett.* **373,** 56–60.

Bonnard, G., Gualberto, J., Lamattina, L., and Grienenberger, J.-M. (1992). RNA editing in plant mitochondria. *Crit. Rev. Plant Sci.* **10,** 503–521.

Choury, D., Farré, J.-C., Jordana, X., and Araya, A. (2004). Different patterns in the recognition of editing sites in plant mitochondria. *Nucleic Acids Res.* **32,** 6397–6406.

Faivre-Nitschke, S. E., Grienenberger, J.-M., and Gualberto, J. M. (1999). A prokaryotic-type cytidine deaminase from Arabidopsis thaliana: Gene expression and functional characterization. *Eur. J. Biochem.* **263,** 896–903.

Farré, J.-C., Leon, G., Jordana, X., and Araya, A. (2001). Cis recognition elements in plant mitochondrion RNA editing. *Mol. Cell. Biol.* **21,** 6731–6737.

Hirose, T., and Sugiura, M. (2001). Involvement of a site-specific trans-acting factor and a common RNA-binding protein in the editing of chloroplast RNA: Development of an *in vitro* RNA editing system. *EMBO J.* **20,** 1144–1152.

Kotera, E., Tasaka, M., and Shikanai, T. (2005). A pentatricopeptide repeat protein is essential for RNA editing in chloroplasts. *Nature* **433,** 326–330.

Lippok, B., Wissinger, B., and Brennicke, A. (1994). Differential RNA editing in closely related introns in Oenothera mitochondria. *Mol. Gen. Genet.* **243,** 39–46.

Miyamoto, T., Obokata, J., and Sugiura, M. (2002). Recognition of RNA editing sites is directed by unique proteins in chloroplasts: Biochemical identification of cis-acting elements and trans-acting factors involved in RNA editing in tobacco and pea chloroplasts. *Mol. Cell. Biol.* **22,** 6726–6734.

Miyamoto, T., Obokata, J., and Sugiura, M. (2004). A site-specific factor interacts directly with its cognate RNA editing site in chloroplast transcripts. *Proc. Natl. Acad. Sci. USA* **101,** 48–52.

Neuwirt, J., Takenaka, M., van der Merwe, J. A., and Brennicke, A. (2005). An *in vitro* RNA editing system from cauliflower mitochondria: Editing site recognition parameters can vary in different plant species. *RNA* **11,** 1563–1570.

Okuda, K., Nakamura, T., Sugita, M., Shimizu, T., and Shikanai, T. (2006). A pentatricopeptide repeat protein is a site-recognition factor in chloroplast RNA editing. *J. Biol. Chem.* **281,** 37661–37667.

Picardi, E., Regina, T. M. R., Brennicke, A., and Quagliariello, C. (2007). REDIdb: The RNA editing database. *Nucleic Acids Res.* **35**(Database issue), D173–D177.

Rajasekhar, V. K., and Mulligan, R. M. (1993). RNA editing in plant mitochondria: [alpha]-phosphate is retained during C-to-U conversion in mRNAs. *Plant Cell* **5**, 1843–1852.

Sasaki, T., Yukawa, Y., Wakasugi, T., Yamada, K., and Sugiura, M. (2006). A simple *in vitro* RNA editing assay for chloroplast transcripts using fluorescent dideoxynucleotides: Distinct types of sequence elements required for editing of ndh transcripts. *Plant J.* **47**, 802–810.

Shikanai, T. (2006). RNA editing in plant organelles: Machinery, physiological function, and evolution. *Cell. Mol. Life Sci.* **63**, 698–708.

Staudinger, M., Bolle, N., and Kempken, F. (2005). Mitochondrial electroporation and in organello RNA editing of chimeric *atp6* transcripts. *Mol. Gen. Genomics* **273**, 130–136.

Takenaka, M., and Brennicke, A. (2003). *In vitro* RNA editing in pea mitochondria requires NTP or dNTP, suggesting involvement of an RNA helicase. *J. Biol. Chem.* **278**, 47526–47533.

Takenaka, M., Neuwirt, J., and Brennicke, A. (2004). Complex cis–elements determine an RNA editing site in pea mitochondria. *Nucleic Acids Res.* **32**, 4137–4144.

Taylor, G. R. (1999). Enzymatic and chemical cleavage methods. *Electrophoresis* **20**, 1125–1130.

van der Merwe, J. A., Takenaka, M., Neuwirt, J., Verbitskiy, D., and Brennicke, A. (2006). RNA editing sites in plant mitochondria can share cis–elements. *FEBS Lett.* **580**, 268–272.

Verbitskiy, D., Takenaka, M., Neuwirt, J., van der Merwe, J. A., and Brennicke, A. (2006). Partially edited RNAs are intermediates of RNA editing in plant mitochondria. *Plant J.* **47**, 408–416.

Yu, W., and Schuster, W. (1995). Evidence for a site-specific cytidine deamination reaction involved in C-to-U editing in plant mitochondria. *J. Biol. Chem.* **270**, 18227–18233.

CHAPTER TWENTY-ONE

Assay of Editing of Exogenous RNAs in Chloroplast Extracts of *Arabidopsis,* Maize, Pea, and Tobacco

Michael L. Hayes *and* Maureen R. Hanson

Contents

Abstract

Nucleotides within transcripts of chloroplasts and mitochondria are modified through C-to-U RNA editing in vascular plants. The specific protein components

Department of Molecular Biology and Genetics, Cornell University, Ithaca, New York

Methods in Enzymology, Volume 424
ISSN 0076-6879, DOI: 10.1016/S0076-6879(07)24021-2

and enzymatic machinery required for editing have not been defined. A consensus sequence is not present around all editing sites, complicating the discovery of *cis*-sequence elements critical for editing. Chloroplast extracts capable of carrying out editing *in vitro* along with precise quantification of editing extent of exogenous transcripts will facilitate identification of both *cis* and *trans* factors. We have optimized an *in vitro* assay originally developed to study editing in tobacco and pea chloroplasts and have expanded the assay to include the study of chloroplast editing in the model species *Arabidopsis* and the monocot maize. The superior genetic resources in these two species can now be utilized in conjunction with biochemical analysis to dissect the editing apparatus. We have improved the assay conditions for editing *in vitro,* achieving efficient editing (as much as 92%) with certain RNA substrates. Unlike the initial assay that relied on qualitative analysis, we are able to achieve precise quantification of editing activity within 1% through a simple poisoned primer extension (PPE) assay with radiolabeled oligonucleotides.

1. INTRODUCTION

Vascular plant organellar transcripts undergo cytidine-to-uridine (C-to-U) RNA editing. Although editing occurs in both mitochondria and chloroplasts, this chapter will focus on editing in chloroplasts. The editing of transcripts has been shown to be indispensable for the synthesis of several functional chloroplast proteins (Bock *et al.*, 1994; Sasaki *et al.*, 2001; Schmitz-Linneweber *et al.*, 2005; Zito *et al.*, 1997). The critical components of the molecular apparatus required for organellar RNA editing in plant systems are poorly understood. The *crr4* gene in *Arabidopsis* is essential for the editing of one site in *ndhD* transcripts (Kotera *et al.*, 2005), and the encoded chloroplast-targeted protein CRR4 is known to bind artificial RNAs containing *ndhD* sequences (Okuda *et al.*, 2006). Unfortunately, neither the biochemical function of this protein nor any other editing factor has been deciphered. Chloroplast extracts competent to edit exogenously synthesized RNA substrates, along with a sensitive assay of editing extent, are critical for dissecting the plant editing apparatus.

Alignment of the sequences around all the editing sites in one species has not revealed a common sequence crucial for editing. Sequence requirements for each editing site therefore must be examined on an individual site-by-site basis. Substrates have been created that are capable of being edited through the expression of exogenous transgenes in transplastomic plants (Bock *et al.*, 1994; Chaudhuri *et al.*, 1995; Reed and Hanson, 1997). The creation of transplastomic plants for extensive study of specific sequence requirements is impractical, given the number of plants required to investigate thoroughly even one chloroplast editing site in higher plants, along with the low efficiency and species' constraints of current chloroplast transformation technology. Chloroplast extracts and RNA substrates that can be edited *in vitro* and assayed with precise quantification can more rapidly identify the critical cis elements for editing at a particular site. However, a present drawback is that not all chloroplast editing sites are represented by artificial RNA substrates that can be edited *in vitro*.

The first chloroplast *in vitro* editing system was developed by Hirose and Sugiura (2001) for tobacco (*Nicotiana tabacum*) editing sites NTpsbL C2[1] and NTndhB C737. Editing analysis was determined qualitatively by inspection of autoradiograms of thin-layer chromatographs of a C nucleotide labeled at the editing site that was converted to labeled U by RNA editing. Chloroplast transcript editing *in vitro* was extended by Miyamoto *et al.* (2002) for NTpsbE C214 and NTpetB C611 and for pea (*Pisum sativum*) extracts. A major shortcoming of the method was that small changes in a substrates' editing efficiency could not be detected due to the qualitative nature of the assay. The extracts, RNA substrates, and assay were developed for tobacco and pea, species from which high-quality chloroplasts are readily isolated. However, these species do not have the extensive genetic resources that are available for certain other plants.

Quantification of editing *in vivo* has been accomplished by direct sequencing, bulk sequencing, and poisoned primer extension (PPE) assay (Bock *et al.*, 1994; Peeters and Hanson, 2002; Reed *et al.*, 2001). We developed an *in vitro* editing system based on tobacco chloroplast extracts as described by Hirose and Sugiura (2001), but we utilized RNA synthesized *in vitro* and we assayed editing extent of exogenously added substrates with the PPE method for the quantification of editing (Hegeman *et al.*, 2005b). Incubation conditions for editing *in vitro* with pea extracts and substrates made in a similar manner were optimized by Nakajima and Mulligan (2005). We further extended the number of species studied by RNA editing to include the model plant *Arabidopsis thaliana* and the monocot maize (Hayes *et al.*, 2006; Hegeman *et al.*, 2005b). These *in vitro* editing systems have proven useful for the determination of chloroplast editing site sequence requirements (Hayes and Hanson, 2007; Hayes *et al.*, 2006).

In this chapter, we describe the methods used for analysis of editing *in vitro* for four distantly related plant species. We investigated the sensitivity, accuracy, and precision of the PPE assay used for quantification. We provide detailed protocols for the isolation of chloroplasts, the creation of editing-competent chloroplast extracts, the synthesis of RNA editing substrates, the conditions for the *in vitro* assay, the quantification of editing, and the analysis of competing RNAs.

---

[1] The nomenclature used in this chapter was established by Hayes *et al.* (2006). A two-letter abbreviation denotes the species that contains the editing site, the gene that contains the editing site is listed, the identity of the nucleotide modified is indicated, and the position from the A of the initiation codon is shown.

2. GROWTH CONDITIONS FOR THE ISOLATION OF INTACT CHLOROPLASTS FROM TOBACCO, PEA, MAIZE, AND *ARABIDOPSIS*

2.1. Overview

Efficient isolation of intact plastids is critical for the *in vitro* editing system because an extract containing highly concentrated chloroplast proteins (5–20 $\mu g/\mu l$) is required. To obtain a high yield of intact chloroplasts, the quality of the leaf tissue is of utmost importance. Typically the best yields are from plants that are young and growing vigorously. The growth conditions for each plant species are listed below.

2.2. Plant growth protocol

2.2.1. Maize *(Zea mays)*

1. Imbibe maize seeds of the varieties Seneca Horizon (Seedway, Hall, NY), Jubilee (Seedway), or B73 overnight before planting for more uniform germination.
2. Sow the seeds to a depth of approximately 2.5 cm, 2.5 cm apart in 54 × 28-cm flats containing Metro Mix 360 soil (Sungro Horticulture, Bellevue, WA). Three flats are sufficient per chloroplast preparation.
3. Grow the maize seedlings at 25° under metal halide lights with a 16-h light and 8-h dark cycle for 10–14 days. The light intensity is approximately 250 W m^{-2}. Fertilize the plants daily using a continuous liquid feed of 100 ppm nitrogen, 24 ppm phosphorus, 95 ppm potassium (21-5-20, J. R. Peters, Inc., Allentown, PA), and 0.04 oz./gallon Epsom salts for 5 days. Alternate this treatment with watering plants daily for 2 days using clear water.
4. Harvest the leaf tissue when plants have three leaves and a fourth leaf is emerging. Homogenize around 80–100 g of leaf tissue per 400 ml of Buffer B. Starch is relatively less abundant in tissues grown in this manner compared to leaves of other plants that we have used to isolate chloroplasts. Therefore, we found that it is not necessary to cover plants with aluminum foil prior to harvest.

2.2.2. *Arabidopsis thaliana*

1. Imbibe the seeds of *A. thaliana*, ecotype Columbia, for 48 h at 4°.
2. Sow the seeds in Metro Mix 360 soil in 54 × 28-cm flats.

3. Thin plants to about one plant every 7.6 cm. Three flats are planted per chloroplast preparation.
4. Grow *Arabidopsis* plants for 4–6 weeks in a growth chamber with 10 h light at 21° and 14 h dark at 16° as described in Hegeman *et al.* (2005b) and Peltier *et al.* (2001). The light intensity is approximately 100 μmol photons $m^{-2} sec^{-1}$. Growth conditions are particularly important for obtaining good yields of intact chloroplasts from *A. thaliana*.
5. Harvest the leaves and float them in an ice cold water bath for 30 min as suggested by Kunst (1998) before mincing in Buffer B. Homogenize around 50 g of leaves per 400 ml of Buffer B.

2.2.3. Tobacco *(Nicotiana tabacum)*

1. Sow the tobacco seeds of the varieties Petit Havana or Samsun NN onto Metro Mix 360 soil in 54 × 28-cm flats with 4 × 9 cell dividers. Three flats are planted per chloroplast preparation. Cover the flats with flexible plastic sheeting such as Saran Wrap after planting for 1 week to promote uniform germination. Grow the tobacco plants at 25° under metal halide lights with a 16-h light and 8-h dark cycle. The light intensity is approximately 250 W m^{-2}.
2. After germination, remove the plastic and thin the plants to one seedling per cell. Grow the plants for 4–6 weeks. Fertilize and water the plants as described for maize.
3. Cover the flats with aluminum foil 48 h prior to harvest to reduce starch accumulation. Harvest around 80–100 g of leaves per 400 ml of Buffer B.

2.2.4. Pea *(Pisum sativum)*

1. Sow pea seeds of the variety Laxtons Progress (Page Seed Company, Greene, NY) about 2.5 cm deep and 7.6 cm apart in Metro Mix 360 soil in 54 × 28-cm flats. Three flats are planted per chloroplast preparation.
2. Grow the plants at 25° under metal halide lights with a 16-h light and 8-h dark cycle for 4–6 weeks. The light intensity is approximately 250 W m^{-2}. Fertilize and water the plants as described for maize.
3. Cover the plants 48 h prior to harvest with aluminum foil to reduce the accumulation of starch in the leaves. Harvest around 80–100 g of leaf tissue per 400 ml of Buffer B.

3. ISOLATION OF CHLOROPLASTS AND PRODUCTION OF EDITING COMPETENT EXTRACTS

3.1. Overview

All published methods for isolation of editing competent chloroplast extracts utilize buffers based on the ones used in the first reported *in vitro* assay (Hirose and Sugiura, 2001). For convenience, we describe these in Table 21.1 We find that the method for isolating tobacco chloroplasts as described by Hirose and Sugiura (2001) is suitable for the isolation of

Table 21.1 Buffers used for the isolation of editing competent chloroplast extracts

Solution	Composition
Buffer A	0.6 M mannitol (Fisher Scientific, Pittsburgh, PA)
	100 mM HEPES–NaOH, pH 8.0 (Fisher Scientific)
	4 mM EDTA (Sigma-Aldrich, St. Louis, MO)
Buffer B[a]	0.3 M mannitol
	50 mM HEPES–NaOH, pH 8.0
	2 mM EDTA
	15 μM BSA (EMD Chemicals Inc., Gibbstown, NJ)
	15 μM PVP-40 (Sigma-Aldrich)
	10.6 mM BME (Sigma-Aldrich)
Buffer C[a]	0.3 M mannitol
	50 mM HEPES–NaOH, pH 8.0
	2 mM EDTA
	15 μM BSA
	10.7 mM BME
Buffer D[a]	0.315M mannitol
	50 mM HEPES–NaOH, pH 8.0
	2 mM EDTA
Buffer E	30 mM HEPES-KOH, pH 7.7
	10 mM MgOAc (J.T. Baker, Phillipsburg, NJ)
	0.2% Triton-X (Sigma-Aldrich)
	2 M KCl (Sigma-Aldrich)
	2 mM DTT (Sigma-Aldrich)
Buffer F	30 mM HEPES-KOH, pH 7.7
	45 mM KOAc (Fisher Scientific)
	30 mM NH$_4$OAc (Fisher Scientific)
	10% glycerol (Fisher Scientific)
	2 mM DTT

[a] For *Arabidopsis* chloroplast isolation, the EDTA concentration in these solutions is increased to 4 mM.

chloroplasts in *Arabidopsis* and maize with only minor modifications. Higher yields of intact *Arabidopsis* chloroplasts can be obtained using a greater concentration of EDTA as suggested by Somerville (1981) in Buffer B, C, and D, compared to the original isolation conditions by Hirose and Sugiura (2001). All other plants are prepared with the same buffers as listed in Hirose and Sugiura (2001). The procedure used for all four species is similar, with differences primarily based on the growing conditions listed above. The procedure for isolation of plastids is listed below.

3.2. Chloroplast isolation and extraction protocol

Note: All solutions should be chilled on ice before use and steps should be performed as quickly as possible to alleviate premature lysis of chloroplasts and protein degradation.

1. Create the Percoll gradient solution by thoroughly mixing Percoll (Sigma–Aldrich), an equal volume of Buffer A, and a final concentration of 10.5 mM 2-mercaptoethanol (BME, Sigma–Aldrich). Add 30 ml of gradient solution to each of four clear 50-ml centrifuge tubes. The continuous gradient is formed by centrifugation of the Percoll solution in tubes at $43,140 \times g$ (Sorvall SS-34 rotor) for 30 min at 4° with the brake off.

2. Remove whole leaves using scissors and rinse them in cold water to remove any loose soil. Cover the leaves with 400 ml of ice-cold Buffer B and chop them into fine pieces using razor blades. Leaves are sufficiently chopped when pieces are smaller than 1 cm² and the Buffer B turns green in color. Add the minced leaves and Buffer B to a 500-ml bottle chilled on ice. Homogenize the leaf tissue with a Polytron (Brinkmann, Westbury, NY) at a speed between 4 and 5 until intact leaf pieces are no longer observed and the liquid is a thick soup. Strain the homogenate through four layers of cheesecloth into a 500–ml flask chilled on ice.

3. Centrifuge the strained homogenate until the rotor reaches $5858 \times g$ (Sorvall SGA rotor); then immediately stop the centrifugation and allow the rotor to come to rest with the brake on at 4°. The chloroplasts are now present in a dark green pellet at the bottom of the tube. Discard the supernatant. Gently resuspend the pelleted chloroplasts in Buffer C using a paintbrush. Layer the chloroplasts onto the prepared 50% Percoll continuous gradients and centrifuge them at $8084 \times g$ (Sorvall HB-4 rotor) for 15 min at 4° with the brake off. The intact chloroplasts segregate as a lower dark green band while chloroplasts with ruptured membranes will remain near the top of the gradient. Starch accumulates as a white pellet at the bottom of the tube. Remove the lower dark green band carefully using a Pasteur pipette without disturbing the upper band and starch pellet. Dilute the chloroplasts with four volumes of Buffer D.

Pellet the chloroplasts at $4124 \times g$ (Sorvall HB-4 rotor) for 1 min at $4°$ with the brake on. Dispose of the supernatant. Resuspend the pellet in 30 ml Buffer D and centrifuge at $4124 \times g$ (Sorvall HB-4 rotor) for 40 sec at $4°$ with the brake on. Remove the supernatant.

4. Resuspend the chloroplast pellets in a minimal volume of Buffer E and incubate them on ice for 30 min for lysis. Centrifuge the lysate at $15,600 \times g$ in an Eppendorf 5414 microfuge for 10 min at $4°$. Remove the supernatant carefully without disturbing the dark green membrane pellet to a Tube-O-Dialyzer (G-Biosciences, St. Louis, MO) 8000 MCO and dialyze in 200 ml of Buffer F for 1.5 h. Add the Tube-O-Dialyzer containing the lysate to fresh Buffer F and dialyze for an additional 3.5 h. Divide the extracts into aliquots, flash freeze them in liquid N_2, and store them at $-80°$. We have stored some extracts for 24 months without loss of activity.

4. EDITING OF EXOGENOUS RNA *IN VITRO*

4.1. Overview

The first *in vitro* editing system reported involved a series of complex steps for RNA substrate construction and thin-layer chromatography separations (Hirose and Sugiura, 2001). These made the system expensive, time-consuming, and cumbersome. Therefore, we have simplified the production of substrates for our *in vitro* editing system. RNA substrates are constructed by two rounds of polymerase chain reaction (PCR) followed by *in vitro* transcription (Fig. 21.1; Hegeman *et al.*, 2005b), without the RNA ligation of radiolabeled oligonucleotides required by the Hirose and Sugiura (2001) protocol. Bacterial sequences SK and KS on the respective 5′ and 3′ ends of each substrate allow for the specific amplification of editing substrates through reverse transcriptase (RT)-PCR with SK and KS primers without gene-specific regions. Since different substrates all share SK and KS sequences, amplification by RT-PCR following the editing reaction can be performed using the same primers. Thus, many substrates can be analyzed rapidly and easily in a cost-effective manner.

Exogenous RNA editing substrates vary considerably in their editing efficiency and the minimum amount of sequence they must contain around the C target. The sequences required in editing substrates both *in vivo* and *in vitro* are typically located 5′ near an editing site (Bock *et al.*, 1997; Chaudhuri and Maliga, 1996; Hayes and Hanson, 2007; Miyamoto *et al.*, 2002). Editing substrates expressed from transgenes *in vivo* have been reported for four sites with little endogenous 5′ and 3′ sequences surrounding the C-editing target, suggesting the sequences required for editing are

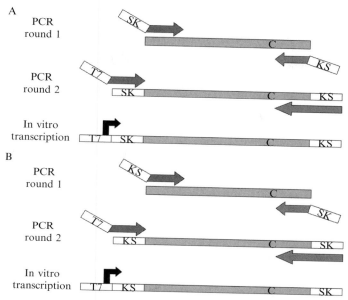

Figure 21.1 A diagram depicting the construction of (A) RNA editing substrates and (B) competitor RNAs. Gray bars represent sequences that flank an endogenous editing site. The position of the edited nucleotide is denoted by a large bolded C. Arrows and/or boxes signify primer sequences. Gray arrows indicate portions of the primer complementary to the template. Black arrows symbolize transcription start sites for T7 RNA polymerase. Oligonucleotides were obtained from IDT (Coralville, IA). Universal amplification sequences expressed in the RNA substrate are SK: 5′-CGCTCTAGAACTAGTGGATC-3′; KS: 5′-GATACCGTCGACCTCGA-3′; T7SK: 5′-TAATACGACTCACTATAGGGCGCTCTAGAACTAGTGGATC.

within 20 nt 5′ and 6 nt 3′ (Bock *et al.*, 1996; Chaudhuri and Maliga, 1996; Reed *et al.*, 2001).

In vitro RNA editing substrates have typically been constructed with at least around 100 nt of sequence 5′ of the editing site and around 15 nt of 3′ sequence in an attempt to ensure that RNAs contain the important *cis*-elements for editing (Table 21.2). Reported editing reactions *in vitro* vary in editing efficiency from 5 to 92%, depending on the substrate and chloroplast extract. We chose 5% calculated editing as the threshold to determine if RNAs are significantly edited, because we have observed that readthrough, extension past the dideoxy nucleotide, can be as high as 3% and the standard deviation for duplicate reactions is usually within 1%. Several *in vitro* substrates for NTrpoB C473 and ATpsbE C214 contain less native sequence than contained in other editing substrates but are edited as well as larger substrates for the same sites (Hayes and Hanson, 2007; Hayes *et al.*, 2006). Results with these smaller substrates indicate that a region of at

Table 21.2 Wild-type substrates that exhibit 5% or greater editing *in vitro*

Species[a]	Gene[b]	Position[c]	5' Sequence[d]	3' Sequence[e]	References
Maize	NTrpoB	473	31	22	Hayes et al., 2006
Tobacco	NTrpoB	473	31	22	Hayes et al., 2006
Tobacco	NTpsbE	214	128	10	Miyamoto et al., 2002
Tobacco	NTpsbE	214	120	20	Sasaki et al., 2006
Tobacco	NTpsbE	214	100	15	Hayes and Hanson, 2007
Tobacco	NTpsbE	214	100	10	Hayes and Hanson, 2007
Tobacco	NTndhA	341	120	21	Sasaki et al., 2006
Tobacco	NTndhB	467	120	20	Sasaki et al., 2006
Tobacco	NTndhB	586	120	20	Sasaki et al., 2006
Tobacco	NTndhB	1481	120	20	Sasaki et al., 2006
Tobacco	NTndhD	2	120	20	Sasaki et al., 2006
Tobacco	NTndhF	290	120	20	Sasaki et al., 2006
Tobacco	NTndhG	50	120	20	Sasaki et al., 2006
Tobacco	NTndhG	347	120	20	Sasaki et al., 2006
Tobacco	NTpetB	611	121	11	Miyamoto et al., 2002

Species[a]	Gene[b]	Position[c]	5′ flanking[d]	3′ flanking[e]	Reference
Tobacco	NTpsbL	2	150	15	Hirose and Sugiura, 2001
Tobacco	ATpsbE	214	150	15	Hayes and Hanson, 2007
Tobacco	ATpsbE	214	99	15	Hayes and Hanson, 2007
Tobacco	ATpsbE	214	54	15	Hayes and Hanson, 2007
Tobacco	ATpsbE	214	31	15	Hayes and Hanson, 2007
Tobacco	ATpsbE	214	22	15	Hayes and Hanson, 2007
Tobacco	ATpsbE	214	13	15	Hayes and Hanson, 2007
Arabidopsis	ATpsbE	214	150	15	Hegeman et al., 2005b
Arabidopsis	ATpsbE	214	99	15	Hegeman et al., 2005b
Arabidopsis	ATpsbE	214	54	15	Hegeman et al., 2005b
Arabidopsis	ATpsbE	214	31	15	Hegeman et al., 2005b
Pea	PSpetB	611	150	21	Nakajima and Mulligan, 2005
Pea	NTpetB	611	121	11	Miyamoto et al., 2002

[a] Species used to create the competent chloroplast extracts.

[b] Gene where the editing site and flanking sequences contained in the editable substrates originate. NT, *Nicotiana tabacum*; AT, *Arabidopsis thaliana*; ZM, *Zea mays*; PS, *Pisum sativum*.

[c] Position in nucleotides from the A of the initiation codon within the coding region containing the editing site.

[d] Number of nucleotides of the sequence flanking 5′ of the edited nucleotide contained in the *in vitro* substrate.

[e] Number of nucleotides of the sequence flanking 3′ of the edited nucleotide contained in the *in vitro* substrate.

most 30 nt 5′ and 10 nt 3′ of native sequence may be sufficient for editing of other sites *in vitro*.

Several RNAs containing more than 100 nt 5′ and ≥10 nt 3′ of native sequence around many editing sites cannot be efficiently edited *in vitro* (Table 21.3). It is not yet clear why exogenous RNAs containing certain sites cannot be edited. *In vivo* editing substrates for sites NTndhB C737 and NTndhB C746 can be edited, although exogenous RNAs containing more native sequence around the same sites cannot be edited efficiently *in vitro* (Bock *et al.*, 1996; Hirose and Sugiura, 2001; Sasaki *et al.*, 2006). It is unlikely, based on the analysis of a number of other sites *in vivo* and *in vitro*, that these RNAs do not contain sequences required for editing.

In comparing editing *in vivo* versus *in vitro*, we must consider that depending on whether RNAs are expressed from a transgene *in vivo* or are synthesized *in vitro*, different sequences flank the region of sequence containing the editing site. The substrates we use for editing assays *in vitro* contain bacterial sequences such as SK and KS, whereas *in vivo* substrates contain sequences from chloroplast promoters and terminators. Editing of a substrate for NTrpoB C473 containing a −20A → U mutation compared to wild-type differed *in vitro* depending on the flanking sequences (Hayes *et al.*, 2006). The editing efficiency of some RNAs containing required sequence elements may be affected by flanking sequences *in vitro*.

One possibility that would explain why RNAs containing some editing sites cannot be edited is that they are folded in a way that does not permit editing. Editing of NTrpoB C473 has been shown to be eliminated or reduced by expression of sequences that could create double-stranded structures within the editable substrate (Hayes *et al.*, 2006; Hegeman *et al.*, 2005a). A structural motif has not been identified in the RNAs that would explain the lack of editing. Perhaps such sites cannot be edited because critical factors are not stable or are lost during isolation of chloroplast extracts.

Another possibility is that the usual *in vitro* editing conditions are suboptimal for certain sites. For editing of ATpsbE C214 using *Arabidopsis* chloroplast extracts, elimination of MgOAc and addition of 10 mM ATP to the *in vitro* editing assay conditions increased editing (Hegeman *et al.*, 2005b). This contrasts with the 3 mM MgOAc and 3 mM ATP found to be optimal for NTpsbL C2 using tobacco extracts in Hirose and Sugiura (2001) and the 3 mM MgOAc and 1 mM ATP reported in Miyamoto *et al.* (2002) and Miyamoto *et al.* (2004) for NTpsbE C214. Possibly by altering the conditions of the *in vitro* reaction, some RNAs in Table 21.3 will be editable *in vitro*. Unfortunately, at present, the editing sites that can be analyzed *in vitro* are limited in number.

The current optimized conditions used for the *in vitro* editing assay are based on the ones first described in Hirose and Sugiura (2001). MgOAc was found to be inhibitory by Hegeman *et al.* (2005b) and is excluded from

Table 21.3 RNAs that contain wild-type sequences around an editing site that exhibit less than 5% editing *in vitro*

Species[a]	Gene[b]	Position[c]	5′ Sequence[d]	3′ Sequence[e]	Reference
Arabidopsis	ATndhB	149	150	15	This report
Arabidopsis	ATndhB	467	150	15	This report
Arabidopsis	ATndhB	586	150	15	This report
Arabidopsis	ATndhB	611	150	15	This report
Arabidopsis	ATndhB	1481	150	15	This report
Arabidopsis	ATpsbE	214	22	15	Hegeman et al., 2005b
Arabidopsis	ATpsbE	214	13	15	Hegeman et al., 2005b
Maize	ZMrpoB	545	30	30	This report
Maize	ZMrpoB	617	30	30	This report
Maize	ZMrpl2	2	30	30	This report
Maize	ZMrps8	182	30	30	This report
Maize	ZMpetB	667	30	30	This report
Maize	ZMndhB	467	30	30	This report
Tobacco	NTndhB	149	120	21	Sasaki et al., 2006
Tobacco	NTndhB	611	120	21	Sasaki et al., 2006
Tobacco	NTndhB	737	120	21	Sasaki et al., 2006
Tobacco	NTndhB	737	156	10	Hirose and Sugiura, 2001
Tobacco	NTndhB	746	120	21	Sasaki et al., 2006
Tobacco	NTndhB	830	120	21	Sasaki et al., 2006
Tobacco	NTndhB	836	120	21	Sasaki et al., 2006
Tobacco	NTndhD	383	120	21	Sasaki et al., 2006
Tobacco	NTndhD	599	120	21	Sasaki et al., 2006

(*continued*)

Table 21.3 (*continued*)

Species[a]	Gene[b]	Position[c]	5' Sequence[d]	3' Sequence[e]	Reference
Tobacco	NTndhD	674	120	21	Sasaki et al., 2006
Tobacco	NTrpoB	338	100	10	This report
Tobacco	NTndhB	467	100	10	This report
Tobacco	NTndhA	1073	120	21	Sasaki et al., 2006

[a] Species used to create the competent chloroplast extracts.

[b] Gene where the editing site and flanking sequences contained in the editable substrates originate. NT, *Nicotiana tabacum*; AT, *Arabidopsis thaliana*; ZM, *Zea mays*; PS, *Pisum sativum*.

[c] Position in nucleotides from the A of the initiation codon within the coding region containing the editing site.

[d] Number of nucleotides of the sequence flanking 5' of the edited nucleotide contained in the *in vitro* substrate.

[e] Number of nucleotides of the sequence flanking 3' of the edited nucleotide contained in the *in vitro* substrate.

our assay conditions, unlike the 3 mM reported to be optimal by Hirose and Sugiura (2001) and Miyamoto *et al.* (2002). Also concentrations of HEPES-KOH, pH 7.7, KOAc, dithiothreitol (DTT), and NH$_4$OAc have been increased and the concentration of protease inhibitors has been reduced compared to Hirose and Sugiura (2001) and Miyamoto *et al.* (2002). The concentration of RNase inhibitors is lower than Hirose and Sugiura (2001) but equal to the optimized conditions in Miyamoto *et al.* (2002). The procedure for the editing assay is listed below.

4.2. *In vitro* editing assay protocol

1. Create *in vitro* transcription templates for the production of RNA editing substrates using two rounds of PCR (Fig. 21.1). In the first round, amplify a gene fragment including an editing site and surrounding sequence using total genomic DNA as template and primers designed to flank the 5′ end of the gene fragment with the bacterial sequence SK and the 3′ end with KS (Fig. 21.1). Add the T7 sequence 5′ of the SK sequence using the amplification products from the first round of PCR as template. Purify the PCR products using the Qiaquick PCR purification kit (Qiagen Inc., Valencia, CA).

2. Produce RNAs by using the purified PCR products from the second round of PCR as template and the T7 Megashortscript kit (Ambion Inc., Austin, TX) to set up *in vitro* transcription reactions (20 μl). Incubate *in vitro* transcription reactions for 2 h at 37°. Remove the DNA templates by adding 2 U of TURBODNase (Ambion Inc.) to *in vitro* transcription reactions and incubating 15 min at 37°. After synthesis, purify RNAs using the RNA Cleanup Kit (Zymo Research, Orange, CA). Quantify RNAs using absorbance readings from a spectrophotometer at 260 nm and 280 nm. Dilute the RNAs to a working concentration of 100 pM.

3. To edit RNAs, create editing reaction mixtures containing 45 mM HEPES-KOH, pH 7.7 (Fisher Scientific), 67.5 mM KOAc (Fisher Scientific), 45 mM NH$_4$OAc (Fisher Scientific), 5% glycerol (Fisher Scientific), 1% polyethylene glycol 6000 (USB, Cleveland, OH), 1 mM ATP[2] (Sigma-Aldrich), 6 mM DTT (Sigma-Aldrich), 0.8× Complete, Mini, EDTA-free, Protease Inhibitor Cocktail (Roche Applied Science, Indianapolis, IN), 2.4 U/μl RNaseOUT Recombinant Ribonuclease Inhibitor (Invitrogen, Carlsbad, CA), 6.4 μg/μl chloroplast extract,[3] and 8 pM RNA substrate. We find 12.5 μl reactions are most economical when performed in 0.2-ml semiskirted 96-well PCR plates by Abgene (Epsom, United Kingdom). Incubate the editing reactions for 2 h at 30°,

[2] For ATpsbE C214 and NTpsbE C214, 10 mM ATP is optimal (Hayes and Hanson, 2007).
[3] For editing reactions using maize chloroplast extracts, use 1.6 μg/μl as in Hayes *et al.* (2006).

then for 5 min at 65°. Chill the reactions in an ice water bath for 5 min and centrifuge at 1509×g in a Heraeus Labofuge 400 for 1 min at room temperature to pellet precipitated proteins.

4. Use 1 μl of supernatant from each reaction as template in an RT reaction (10 μl) using the Sensiscript RT kit (Qiagen Inc.). To specifically amplify *in vitro* substrates and not endogenous nucleic acids, a primer containing the KS sequence is used as the gene specific primer in the RT reaction. Incubate the RT reactions for 1 h at 37°.

5. Amplify ssDNAs copied from edited exogenous RNAs by using 2.5 μl of the RT reaction as template, SK and KS primers, and the Taq PCR Master Mix Kit (Qiagen Inc.) in PCR reactions (25 μl). Cycle the PCR reactions 30 times at 95° for 30 sec, 55° for 30 sec, and 72° for 30 sec. Clean the PCR product from primers and nucleotides by using the Exosap-IT PCR Clean-up (USB) reaction before use as template in the PPE assay.

5. QUANTIFICATION OF EDITING OF EXOGENOUS RNAs

5.1. Overview

Direct sequencing, bulk sequencing, and radiolabeling the edited nucleotide, followed by TLC separation, thymine–DNA glycosylase (TDG) assay, or PPE using either radiolabeled or fluorescently labeled oligonucleotides have all been used for the analysis of editing in organelles both *in vivo* and *in vitro* (Bock *et al.*, 1994; Hirose and Sugiura, 2001; Peeters and Hanson, 2002; Reed *et al.*, 2001; Sasaki *et al.*, 2006; Takenaka and Brennicke, 2003). Each method has its own unique advantages and disadvantages. We have found that PPE using radiolabeled oligonucleotides is sufficiently economical, expedient, sensitive, accurate, and precise for the needs of our assay.

The PPE assay was examined to ensure that the sensitivity and accuracy of the method exist even with low editing extents encountered in some *in vitro* editing reactions (Fig. 21.2). A PPE template containing the sequence around NTrpoB C473 with a C at the editing site position was constructed through PCR as in Fig. 21.1 without *in vitro* transcription. A template with a T at the position of the editing site was created using a specific primer that overlapped the editing site containing an A nucleotide at the complementary position. The DNA concentrations of the two templates were calculated. The templates were then diluted to equal concentrations. The two templates were mixed in specific ratios and used in PPE reactions. With relative concentrations of T-template from 0 to 10%,

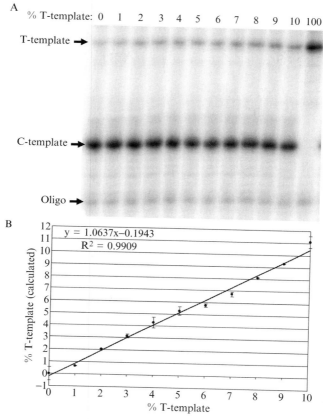

Figure 21.2 Accuracy and precision of the PPE assay. PPE templates were constructed expressing the $-31/+22$ region around the editing site NTrpoB C473 with either a C or a T nucleotide at the position of the editing site through PCR. Templates were mixed with different percentages of T-template from 1–10% in 1% intervals. (A) An electrophoretogram shows the size separated products from the PPE assay. The bands are labeled T-template and C-template according to the expected size of the extension products from priming off each template contained in the PPE reaction. The labeled oligonucleotide not extended during the reaction is labeled Oligo. The percent T-template used in each PPE reaction is indicated above the corresponding lane. (B) A graph was produced to determine the accuracy and precision of the PPE assay. The x-axis shows the percent T-template contained in each PPE reaction. The intensity of each band of the electrophoretogram was calculated and the amount of readthrough calculated in lane 0 was subtracted from experimental samples. The y-axis displays the percent T-template calculated for each PPE reaction. The equation for the regression line and the R^2 value are shown in the top left of the graph.

the calculated percent T-template is around 1% (Fig. 21.2). Each reaction was also performed in duplicate reactions and the standard deviation was within 1% for independent samples. Therefore, the linear relationship of calculated percent T-template versus percent T-template in mixed templates is clear for the range 0–10% (Fig. 21.2). In a similar reconstruction experiment, the linear relationship was observed in the range 0–100% (Peeters and Hanson, 2002).

One concern with the PPE assay is the tendency of sequencing polymerases to run through and not stop at the appropriate nucleotide (Roberson and Rosenthal, 2006). To minimize these effects, it is crucial to use ThermoSequenase (USB) as the PPE polymerase, eliminate nucleotide contamination of templates, and choose the best oligonucleotide and ddNTP for the editing site. Many polymerases were compared in a PPE fluorescent assay reported by Roberson and Rosenthal (2006). By far, the ThermoSequenase enzyme had the lowest amount of readthrough. Nucleotide contamination can also lead to significant readthrough, so it is necessary to remove dNTPs and primers that might interfere with PPE. For this purpose, we treat all PPE templates with Exosap-IT (USB), which contains alkaline phosphatase activity, before assay. The enzymes in Exosap-IT are easily deactivated by heating the reaction to 80° for 15 min. The choice of oligonucleotide and ddNTP is based on the native sequences around the editing site. Primers may be chosen so that they anneal either on the + or − strand of the PCR template. The dideoxy nucleotide to be used should be chosen so that bands can be clearly resolved on polyacrylamide gels (Fig. 21.3). C and G nucleotides are acceptable dideoxy nucleotides for terminating primer extension at most editing sites. We have experienced a low amount of readthrough, about 0–3%, which can easily be measured and accounted for by subtracting the readthrough signal from sample measurements.

Recently, two fluorescent PPE assays that rely on fluorescent labels have been reported as alternatives to radioactive methods for quantifying editing (Roberson and Rosenthal, 2006; Sasaki et al., 2006). The difference between these two fluorescent assays is that the assay of Sasaki et al. (2006) used a DNA sequencer to detect fluorescently labeled ddNTP incorporation by a one-nucleotide extension reaction, while Roberson and Rosenthal (2006) used labeled oligonucleotides in a reaction resulting in extension by several nucleotides. Extension products were separated on polyacrylamide gels, and the intensity of the bands was calculated by a Typhoon 9200 imager. Results by Roberson and Rosenthal (2006) are analogous to what we observe using radiolabeled oligonucleotides with a very high accuracy and precision. The method used in Sasaki et al. (2006) is less accurate than either our assay or Roberson and Rosenthal (2006), due most likely to the method of detection. Neither fluorescent assay has been evaluated to determine whether it is accurate within 1% in the 0–10% range.

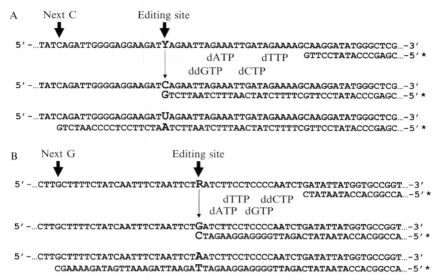

Figure 21.3 The poisoned primer extension (PPE) assay is represented using either (A) ddGTP or (B) ddCTP as the poisoned nucleotides. Characters depict nucleotides around the editing site in PPE templates, PPE products, 5′ labeled oligonucleotides, and nucleotides used in the PPE reaction. The symbol (*) signifies the 5′ radiolabeled phosphate. The editing site and the next nucleotide that can terminate the PPE reaction are indicated under large arrows. Small arrows point to the expected PPE products from the template shown above.

The procedure for our PPE assay for examining editing efficiency is described below.

5.2. Poisoned primer extension assay protocol

1. Set up PPE reactions (15 μl) using the ThermoSequenase Cycle Sequencing kit (USB) containing 50 ng of RT-PCR product, 1× ThermoSequenase buffer (USB), 4 mM each of three dNTPs and one ddNTP, 0.42 nM 5′ labeled oligo, and 4 U ThermoSequenase (USB). Incubate the PPE reactions for five cycles at 90° for 5 sec, 50° for 30 sec, and 72° for 10 sec. Add 5 μl of Stop Solution (USB) to each reaction.
2. Produce a 12% acrylamide sequencing gel 42 × 33 cm using the Sequa-gel Sequencing System (National Diagnostics, Atlanta, GA). Incubate stopped PPE reactions at 72° for 5 min to denature extension products. Load 5 μl of each PPE reaction into each lane of the polyacrylamide gel. Run the gel at 65 mA for 2 h. Remove one gel plate and wrap the gel in a flexible transparent plastic support such as Saran Wrap. Expose a

phosphorimaging cassette (Amersham Biosciences, Piscataway, NJ) to the gel either for 1 h or overnight depending on the intensity of the signal.
3. Develop electrophoretograms using a Storm 860 phosphorimager (Amersham Biosciences). Determine the intensity of the bands using ImageQuant V. 5.2 (Amersham Biosciences) software. Calculate the percent editing of RNA substrates for each reaction and one standard deviation around the mean for duplicate reactions. Establish the amount of readthrough from the intensity of bands resulting from a PPE reaction using a substrate that is not edited as template. Subtract the signal from the other measurements.

6. REDUCTION OF EDITING THROUGH THE ADDITION OF COMPETITOR RNAS

6.1. Overview

The percentage of substrates edited decreases when excessive amounts of substrates are added greater than the capacity of the reaction (Fig. 21.4). This reduction is thought to be due to competition within the reactions for editing factors. Competition experiments are useful for examining the *cis* elements within the editing substrates. The first report of competition *in vitro* was revealed with the initial *in vitro* assay (Hirose and Sugiura, 2001). In the initial assay, oligoribonucleotides were used as competitors that contained sequences around the editing site. For competition reactions, 2000× more competitor than editing substrate was required (Hirose and Sugiura, 2001). There was a reduction in editing in many reactions with only vector RNA added as competitor. This was most likely due to nonspecific interactions stemming from using such a high amount of competitor compared to substrate. The conditions for the competition assay were improved in Miyamoto *et al.* (2002) when addition of 100× and 1000× competitor RNA versus substrate substantially reduced editing compared to reactions with only vector RNA added as competitor. They did not observe a large reduction in editing reactions with vector RNA added as competitor compared to reactions without competitor RNAs. In *Arabidopsis* extracts, the same absolute amount of competitor, 1 pmol (Hegeman *et al.*, 2005b), reduced editing as in Miyamoto *et al.* (2002). This suggests a similar amount of competitor is required to reduce *in vitro* editing in *Arabidopsis* and tobacco extracts. We have further refined the competition assay to reduce nonspecific interactions by the addition of only 10× to 100× competitor for specific competition of the RNA substrate (Hayes and Hanson, 2007).

Competitor RNA substrates are constructed in a manner similar to RNA substrates (Fig. 21.1). Unlike the initial competition experiments

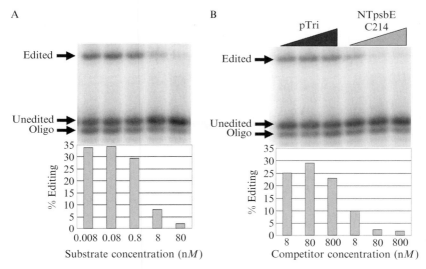

Figure 21.4 Increasing amounts of substrate (A) or self-competitor RNA (B) reduce the percentage of substrate edited *in vitro*. (A) Editing and (B) competition reactions use a substrate expressing the tobacco editing site NTpsbE C214 in tobacco chloroplast extracts. (A and B) Graphs indicating the calculated percent editing are located below the corresponding electrophoretograms. Arrows point to the expected sizes of PPE products from edited and unedited templates as well as unextended labeled oligo. Shaded triangles signify concentrations of added competitor RNAs added to editing reactions. (B) Competitor labeled NTpsbE C214 contains the same region of *psbE* as the editable substrate. Competitor pTri is an RNA of length similar to the substrate but does not share common sequences with the editing substrate. The nomenclature used was established in Hayes *et al.* (2006). A two-letter abbreviation denotes the species that contains the editing site. The gene that contains the editing site is listed, the identity of the nucleotide modified is indicated, and the position in nucleotides from the A of the initiation codon is shown. For ATpsbE C214 and NTpsbE C214, 10 m*M* ATP is optimal (Hayes and Hanson, 2007). [This figure was reproduced in modified form from Hayes and Hanson (2007).]

that utilized editing substrates that are not radiolabeled at the editing site (Hirose and Sugiura, 2001), we developed an alternative type of competitor for our *in vitro* system. The first competitors in our *in vitro* assay were identical to the RNA substrates except they did not contain the SK and KS bacterial sequences (Hegeman *et al.*, 2005b). This prevented amplification of the competitor RNAs during RT-PCR amplification of the substrate. One caveat about such competitors is that the sequences SK and KS sometimes make up a considerable portion of the editing substrate, so that competitors without these sequences could have different structural features compared to the editing substrate. Currently we construct competitor RNAs so that the SK and KS bacterial sequences are in swapped positions (Hayes and Hanson, 2007). This prevents them from being amplified by

RT–PCR along with the substrate, but allows similar sequences to be represented. The procedure for the competition experiments is given below.

6.2. *In vitro* competition assay protocol

1. Construct competitor RNAs in a manner similar to RNA substrates except in comparison the flanking sequences SK and KS are exchanged (Fig. 21.1). Create *in vitro* transcription templates for the production of RNA editing substrates through two rounds of PCR. In the first round, amplify a gene fragment including an editing site and surrounding sequence using total genomic DNA as the template. Use primers designed to flank the 5′ end of the gene fragment with the bacterial sequence KS and the 3′ end with SK. Use the amplification products from the first round as the template for the second round of PCR. In the second round, add the T7 sequence 5′ of the KS sequence and purify PCR products using the Qiaquick PCR purification kit (Qiagen Inc.).
2. Determine the maximum concentration of RNA substrate that can be edited efficiently in the *in vitro* editing reaction. Titrate substrate RNAs in editing reactions at concentrations between 8 pM and 80 nM. Use the maximum concentration of substrate that is edited as highly after 2 h at 30° as the 8 pM reaction in later competition reactions. We have observed that 0.8 nM is maximal for many different substrates in extracts from different species.
3. Add competitor RNAs at concentrations of 0.8 nM, 8 nM, or 80 nM to 0.8 nM of substrate RNA in the optimized editing reactions (12.5 μl). Conditions for competition assays are identical to optimized editing reactions except for changes to the RNA added.
4. Reverse-transcribe RNA substrates specifically in 10 μl reactions using the bacterial primer KS and 1 μl of the competition reaction as template. Perform PCR as in normal editing reactions with bacterial primers SK and KS and 2.5 μl of the RT reaction as the template. Treat RT–PCR products with Exosap-IT (USB) and use as templates in the PPE assay.

7. Quantification of Editing of Endogenous RNAs

7.1. Overview

The focus of this chapter has been to describe the *in vitro* editing assay. The amount of editing in endogenous transcripts can also be determined with the PPE assay. The procedure is based on Peeters and Hanson (2002) with the primary change being in the type of primers used for reverse transcription. Instead of using random hexamers, we find it is optimal to use gene-specific primers. The procedure for assay of editing *in vivo* is listed below.

7.2. *In vivo* editing quantification protocol

1. Isolate total RNA from 100 mg of leaf tissue using Trizol (Invitrogen).
2. Remove contaminating DNA through a DNase treatment using the TURBO DNA-free Kit (Ambion Inc.).
3. Quantify RNA using A260/280 readings from a spectrophotometer.
4. Reverse-transcribe transcripts of interest using the Sensiscript Kit (Qiagen Inc.) and a gene-specific primer. Use cDNAs as the template in a PCR reaction to amplify regions of interest. Treat PCR amplification products with Exosap-IT (USB) and use them as templates in the PPE assay.

ACKNOWLEDGMENT

This work was supported by NSF Grant MCB 0344007 to M.R.H.

REFERENCES

Bock, R., Kossel, H., and Maliga, P. (1994). Introduction of a heterologous editing site into the tobacco plastid genome: The lack of RNA editing leads to a mutant phenotype. *EMBO J.* **13,** 4623–4628.

Bock, R., Hermann, M., and Kossel, H. (1996). *In vivo* dissection of cis–acting determinants for plastid RNA editing. *EMBO J.* **15,** 5052–5059.

Bock, R., Hermann, M., and Fuchs, M. (1997). Identification of critical nucleotide positions for plastid RNA editing site recognition. *RNA* **3,** 1194–1200.

Chaudhuri, S., and Maliga, P. (1996). Sequences directing C to U editing of the plastid *psbL* mRNA are located within a 22 nucleotide segment spanning the editing site. *EMBO J.* **15,** 5958–5964.

Chaudhuri, S., Carrer, H., and Maliga, P. (1995). Site-specific factor involved in the editing of the *psbL* mRNA in tobacco plastids. *EMBO J.* **14,** 2951–2957.

Hayes, M. L., and Hanson, M. R. (2007). Identification of a sequence motif critical for editing of a tobacco chloroplast transcript. *RNA* **13,** 281–288.

Hayes, M. L., Reed, M. L., Hegeman, C. E., and Hanson, M. R. (2006). Sequence elements critical for efficient RNA editing of a tobacco chloroplast transcript *in vivo* and *in vitro*. *Nucleic Acids Res.* **34,** 3742–3754.

Hegeman, C. E., Halter, C. P., Owens, T. G., and Hanson, M. R. (2005a). Expression of complementary RNA from chloroplast transgenes affects editing efficiency of transgene and endogenous chloroplast transcripts. *Nucleic Acids Res.* **33,** 1454–1464.

Hegeman, C. E., Hayes, M. L., and Hanson, M. R. (2005b). Substrate and cofactor requirements for RNA editing of chloroplast transcripts in *Arabidopsis in vitro*. *Plant J.* **42,** 124–132.

Hirose, T., and Sugiura, M. (2001). Involvement of a site-specific trans-acting factor and a common RNA-binding protein in the editing of chloroplast mRNAs: Development of a chloroplast *in vitro* RNA editing system. *EMBO J.* **20,** 1144–1152.

Kotera, E., Tasaka, M., and Shikanai, T. (2005). A pentatricopeptide repeat protein is essential for RNA editing in chloroplasts. *Nature* **433,** 326–330.

Kunst, L. (1998). Preparation of physiologically active chloroplasts from *Arabidopsis. Methods Mol. Biol.* **82,** 43–48.

Miyamoto, T., Obokata, J., and Sugiura, M. (2002). Recognition of RNA editing sites is directed by unique proteins in chloroplasts: Biochemical identification of cis-acting elements and trans-acting factors involved in RNA editing in tobacco and pea chloroplasts. *Mol. Cell. Biol.* **22,** 6726–6734.

Miyamoto, T., Obokata, J., and Sugiura, M. (2004). A site-specific factor interacts directly with its cognate RNA editing site in chloroplast transcripts. *Proc. Natl. Acad. Sci. USA* **101,** 48–52.

Nakajima, Y., and Mulligan, R. M. (2005). Nucleotide specificity of the RNA editing reaction in pea chloroplasts. *J. Plant Physiol.* **162,** 1347–1354.

Okuda, K., Nakamura, T., Sugita, M., Shimizu, T., and Shikanai, T. (2006). A pentatrico-peptide repeat protein is a site-recognition factor in chloroplast RNA editing. *J. Biol. Chem.* **281,** 37661–37667.

Peeters, N. M., and Hanson, M. R. (2002). Transcript abundance supercedes editing efficiency as a factor in developmental variation of chloroplast gene expression. *RNA* **8,** 497–511.

Peltier, J. B., Ytterberg, J., Liberles, D. A., Roepstorff, P., and van Wijk, K. J. (2001). Identification of a 350-kDa ClpP protease complex with 10 different Clp isoforms in chloroplasts of *Arabidopsis thaliana. J. Biol. Chem.* **276,** 16318–16327.

Reed, M. L., and Hanson, M. R. (1997). A heterologous maize *rpoB* editing site is recognized by transgenic tobacco chloroplasts. *Mol. Cell. Biol.* **17,** 6948–6952.

Reed, M. L., Peeters, N. M., and Hanson, M. R. (2001). A single alteration 20 nt 5′ to an editing target inhibits chloroplast RNA editing *in vivo. Nucleic Acids Res.* **29,** 1507–1513.

Roberson, L. M., and Rosenthal, J. J. (2006). An accurate fluorescent assay for quantifying the extent of RNA editing. *RNA* **12,** 1907–1912.

Sasaki, T., Yukawa, Y., Wakasugi, T., Yamada, K., and Sugiura, M. (2006). A simple *in vitro* RNA editing assay for chloroplast transcripts using fluorescent dideoxynucleotides: Distinct types of sequence elements required for editing of *ndh* transcripts. *Plant J.* **47,** 802–810.

Sasaki, Y., Kozaki, A., Ohmori, A., Iguchi, H., and Nagano, Y. (2001). Chloroplast RNA editing required for functional acetyl-CoA carboxylase in plants. *J. Biol. Chem.* **276,** 3937–3940.

Schmitz-Linneweber, C., Kushnir, S., Babiychuk, E., Poltnigg, P., Herrmann, R. G., and Maier, R. M. (2005). Pigment deficiency in nightshade/tobacco cybrids is caused by the failure to edit the plastid ATPase alpha-subunit mRNA. *Plant Cell* **17,** 1815–1828.

Somerville, C. R., Sommerville, S. C., and Ogren, W. L. (1981). Isolation of photosynthet-ically active protoplasts and chloroplasts from *Arabidopsis thaliana. Plant Sci. Lett.* **21,** 89–96.

Takenaka, M., and Brennicke, A. (2003). *In vitro* RNA editing in pea mitochondria requires NTP or dNTP, suggesting involvement of an RNA helicase. *J. Biol. Chem.* **278,** 47526–47533.

Zito, F., Kuras, R., Choquet, Y., Kossel, H., and Wollman, F. A. (1997). Mutations of cytochrome b6 in *Chlamydomonas reinhardtii* disclose the functional significance for a proline to leucine conversion by *petB* editing in maize and tobacco. *Plant Mol. Biol.* **33,** 79–86.

CHAPTER TWENTY-TWO

IN ORGANELLO GENE EXPRESSION AND RNA EDITING STUDIES BY ELECTROPORATION-MEDIATED TRANSFORMATION OF ISOLATED PLANT MITOCHONDRIA

Jean-Claude Farré,* David Choury,[†] *and* Alejandro Araya[†]

Contents

* Section of Molecular Biology, Division of Biological Sciences, University of California, San Diego, La Jolla, California
† Laboratoire Microbiologie Cellulaire Moléculaire et Pathogenicité. UMR 5234, Centre National de la Recherche Scientifique and Université Victor Segalen, Bordeaux, France

Methods in Enzymology, Volume 424
ISSN 0076-6879, DOI: 10.1016/S0076-6879(07)24022-4

Abstract

Plant mitochondrial gene expression is a complex process involving multiple steps such as transcription, *cis-* and *trans*-splicing, RNA trimming, RNA editing, and translation. One of the main hurdles in understanding more about these processes has been the inability to incorporate engineered genes into mitochondria. We recently reported an *in organello* approach on the basis of the introduction of foreign DNA into isolated plant mitochondria by electroporation. This procedure allows the investigation of transcriptional and posttranscriptional processes, such as splicing and RNA editing, by use of site-directed mutagenesis. Foreign gene expression *in organello* is strongly dependent on the functional status of mitochondria, thus providing relevant information in conditions closer to the situation found *in vivo*. The study of mutants that affect RNA splicing and editing provides a novel and powerful method to explain the role of specific sequences involved in these processes. Here we describe a protocol to "transform" isolated plant mitochondria that has allowed us to investigate successfully some aspects of RNA editing.

1. INTRODUCTION

The concept of RNA editing was first used to describe the insertion of uridine residues within the coding region of mitochondrial mRNAs from kinetoplastid protozoa (Benne *et al.*, 1986). Subsequently, the definition of editing was extended to any cotranscriptional or posttranscriptional processes that involved changes to the primary sequence of an RNA encoded by the corresponding gene. The editing process includes changes resulting in nucleotide insertion or deletion and base conversion reactions (Bass, 1997; Blanc *et al.*, 1996; Estevez and Simpson, 1999). RNA editing has been described in the nuclear and mitochondrial compartments of eukaryotic cells, as well as in the genome of some viruses and in the mitochondria and chloroplasts of land plants.

RNA editing in plant organelles is characterized by C-to-U and a few U-to-C transitions. Evidence for RNA editing is found in all vascular plant organelles (such as seed plants, ferns, and fern allies) and in all bryophytes group except the subclass of thalloid liverworts (Marchantiidae) (Freyer *et al.*, 1997; Malek *et al.*, 1996). The complete analysis of *Arabidopsis thaliana* mtDNA and its transcripts reveals 456 C-to-U editing events (Giege and Brennicke, 1999). In wheat mitochondria, 1200 editing sites have been estimated (Bonen, 1991).

1.1. Specificity of mitochondrial RNA editing

A major concern in RNA editing studies is to understand how the editing machinery recognizes specific cytosine residues. The comparison of the sequences flanking the editing sites does not show any apparent conserved motif at the primary structure level (Giege and Brennicke, 1999), suggesting a requirement for individual *cis*-acting sequence elements determining each editing site. In plant mitochondria, chimeric genes formed by fragments of functional genes are expressed and edited at the same positions as observed in their native contexts. These observations, as well as *in vitro* studies, support the idea that only the neighboring sequences are involved in editing specificity (Kubo and Kadowaki, 1997; Kumar and Levings, 1993; Neuwirt *et al.*, 2005; Nivison *et al.*, 1994; Reed *et al.*, 2001; Takenaka and Brennicke, 2003; Takenaka *et al.*, 2004; Williams *et al.*, 1998). Similar results have been found for RNA editing in chloroplasts by use of *in vitro* systems (Hirose and Sugiura, 2001; Miyamoto *et al.*, 2002, 2004) or transplastomic lines carrying foreign genes integrated in the chloroplast genome (Bock *et al.*, 1997; Chaudhuri and Maliga, 1996).

1.2. *In vitro* and *in organello* approaches to study mitochondrial RNA editing

The mitochondrial RNA editing process is poorly understood because of the lack of appropriate experimental approaches. Most current studies on mitochondrial gene expression in plant mitochondria were either based on the analysis of intermediate molecules found *in vivo* or on laborious *in vitro* approaches. Nevertheless, many interesting insights have been reported by use of *in vitro* approaches on the promoter function, RNA processing (Binder *et al.*, 1995; Hanic-Joyce and Gray, 1991; Lupold *et al.*, 1999; Mulligan *et al.*, 1991; Rapp and Stern, 1992; Rapp *et al.*, 1993), and the RNA editing mechanism (Araya *et al.*, 1992; Blanc *et al.*, 1995; Yu and Schuster, 1995).

Unlike chloroplasts, the integration of exogenous genes into the mitochondrial genome has not been achieved probably for several reasons: the difficulty to generate site-specific integration of foreign gene, the lack of suitable selection markers, and the fact that mitochondrial function is essential for plant cell survival. In the absence of reliable methods to transform plant mitochondria, two approaches to study RNA editing have been developed: an *in vitro* RNA editing system from pea shoots and from cauliflower inflorescences (Neuwirt *et al.*, 2005; Takenaka and Brennicke, 2003), described in detail in Chapter 20 (this volume), and an *in organello* RNA editing system (electrotransformation) from wheat embryos (Farre and Araya, 2001), potato tubers (Choury *et al.*, 2005), and etiolated seedlings of maize and sorghum (Staudinger and Kempken, 2003) described in this chapter.

The electrotransformation consists of transient incorporation of DNA into purified organelles that is mediated by electroporation. This procedure has several advantages: (1) the possibility to use a site-directed mutagenesis approach to dissect out the recognition signal(s) involved in gene expression processes, (2) the possibility to analyze simultaneously a set of mutant genes with a single preparation of purified mitochondria, and (3) unlike *in vitro* editing system, *in organello* approach allows access of several molecular events of the gene expression process, such as transcriptional and posttranscriptional mechanisms.

Several assays to transform isolated organelles have been described. Collombet *et al.* (1997) reported the introduction of plasmid DNA into isolated mice mitochondria by electroporation. By use of an analogous procedure, To *et al.* (1996) studied the expression of reporter genes in isolated chloroplasts. Finally, Farre and Araya (2001) optimized this procedure by use of wheat mitochondria to obtain a detailed picture of the molecular and biochemical mechanisms occurring during transcription, splicing, and RNA editing. A major finding of this work was that a transgene could be efficiently transcribed when incorporated into mitochondria by electroporation and that the transcripts were faithfully processed and edited.

2. Purification of Mitochondria

2.1. Plant material

The plant material is often defined *a priori* to investigate a specific physiological problem. However, in case some flexibility exists in the choice of material, choose a tissue that is easy to obtain in large amounts, such tubers or hypocotyls, and choose a tissue where a good protocol is available. Useful references and protocols for purification of mitochondria are described in Leaver *et al.* (1983), Douce (1985), Day *et al.* (1985), Moller *et al.* (1996), and Kruft *et al.* (2001).

The mitochondria used for the electrotransformation protocol are isolated from wheat embryos (*Triticum aestivum* var. Fortal) and potato tubers (*Solanum tuberosum*, var. Roseval) as described in Sections 2.2 and 2.3. Isolation of mitochondria from 9-day-old etiolated seedlings of maize and sorghum has been described by Staudinger and Kempken (2003).

2.1.1. The purification of embryos from wheat seeds

The embryos are obtained from wheat seeds after the procedure of Johnston and Stern (1957) with minor modifications (Fig. 22.1). Approximately 250 g of embryos can be obtained from 20 kg of wheat seeds.

1. A Grinding Mill (Retsch GMBH SK1, Haan, Germany) is used to grind the seeds, and after two passes, a considerable proportion of seeds are broken, releasing embryos, pieces of endosperm, and aleurones.
2. The mixtures are passed across a set of three sieves (1.25, 1.0, and 0.63 mm). A rotary mixer (Retsch GMBH KS 1000, Haan, Germany) is used to sift the mixture for 15 min at maximum speed. The fraction on the 1.25-mm sieve is sifted for a second time with the same conditions.

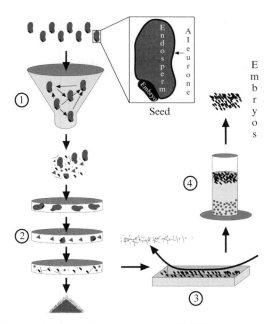

Figure 22.1 Wheat embryos purification scheme. (1) Wheat seeds are ground with an industrial grain mill. (2) The broken mixture is sifted through a set of three sieves (1.25, 1, and 0.63 mm). (3) The aleurones present in embryo-enriched fraction (0.63 mm sieve) are eliminated by blowing air softly over the mixture. (4) The contaminant pieces of endosperm are eliminated by flotation on cyclohexane/carbon tetrachloride mixture.

Finally, the 0.63-mm sieve retains the embryos enriched fraction and small pieces of endosperm and aleurones.
3. The aleurones being light are separated by blowing air softly with a hair dryer over the mixture spread out on a tray.
4. The contaminant fragments of endosperm are eliminated by flotation with a solution containing 10/25 (v/v) cyclohexane and carbon tetrachloride. The embryos remain at the surface, and the pieces of endosperm sink to the bottom. The embryos are recovered with a spoon, dried under a hood, and stored in the dark at 4°.

2.2. Purification of wheat mitochondria

The mitochondria are purified as described by Leaver *et al.* (1983) with a few modifications, as summarized in Fig. 22.2.

2.2.1. Treatment and homogenization of wheat embryos

The homogenization procedure is critical, because plant cells have rigid cell walls. Different yields of purified mitochondria can be obtained when the same protocol is used with different plant species or organs. Grinding in a Waring blender or Polytron has to be performed in a short time to break the

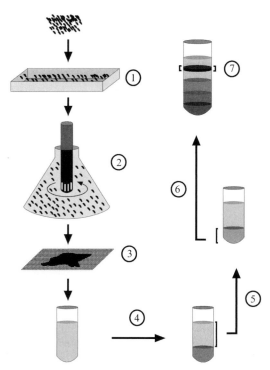

Figure 22.2 Wheat mitochondria purification scheme. (1) Embryos are imbibed in sterile water overnight in the dark. (2) The embryos are ground with a Polytron. (3) The homogenate is filtered through a nylon membrane (30 μm). (4) The homogenate is centrifuged at low speed to eliminate the nuclei, starch, and cell debris. (5) The supernatant is centrifuged at high speed to pellet the mitochondria. Steps 4 and 5 are repeated before next step. (6) The mitochondrial pellet is resuspended and layered over a sucrose gradient and centrifuged at high speed. (7) Mitochondria form a yellowish layer (bracketed) and are recovered with Pasteur pipette.

cells without disrupting mitochondria membranes. Secondary metabolites such as phenolic compounds, or fatty acids produced by released lipases, may have potential damaging effects on purified mitochondria. For a comprehensive discussion concerning these points see Douce (1985).

During the mitochondria extraction procedure (step 2 to 8), all samples are to be kept on ice, and all the steps are carried out at 4° with ice, prechilled solutions, tubes, and bottles.

1. Twenty grams of embryos is washed with a 0.6% bleach solution for 1 min to eliminate bacterial contamination and rinsed with sterile distilled water until the bleach smell disappears. The embryos are spread over four Whatman 3MM filters soaked with sterile distilled water in Petri dishes and allowed to imbibe for 18 h at room temperature in the dark.

2. The embryos are ground with a Polytron homogenizer (Kinematica CH 6010, Kriens/Luzer, Switzerland) for 15 sec at 230 rpm/min at 4°. The process is carried out in four batches, each with 5 g of embryos in 150 ml of homogenization solution (400 mM mannitol, 25 mM MOPS, 1 mM EGTA, 8 mM cysteine, 0.1 % BSA-FAF [fatty acid-free], the pH adjusted to 7.8 with KOH).

3. The homogenate is filtered through a 30-μm pore size nylon membrane (the pores are 10 times larger than the average size of a plant mitochondria).

2.2.2. Isolation of wheat mitochondria by differential centrifugation

4. The homogenate from wheat embryos is transferred to 250 ml centrifuge bottles and centrifuged in a Sorvall GSA rotor at 1000g for 10 min to remove most of the nuclei, starch, and cell debris.

5. The supernatants are decanted carefully, leaving behind the unwanted pellet, to a clean 250-ml centrifuge bottle and centrifuged at 14,000g for 15 min. The high-speed supernatants are discarded, and the pellets containing the enriched mitochondrial fraction are resuspended gently in 10 ml of wash solution (400 mM mannitol, 5 mM MOPS, 1 mM EGTA, 8 mM cysteine, 0.1% BSA-FAF, the pH adjusted to 7.8 with KOH). The resuspension is done by ejecting the solution over the pellet with a 3-ml plastic Pasteur pipet.

6. The resuspended mitochondria are transferred to four 50-ml polycarbonate centrifuge tubes, and the volumes are adjusted to 40 ml each with the wash solution. Complete resuspension of the mitochondria is achieved by inverting the tubes a couple of times. The resuspended mitochondria are centrifuged again as described in steps 4 and 5, with a Sorvall SS34 rotor. For this purpose, the mitochondria are centrifuged at 1000g for 10 min, the supernatants are transferred into clean 50-ml polycarbonate centrifuge tubes, and centrifuged at 14,000g for 15 min. After centrifugation, the supernatants are discarded, and the four mitochondrial pellets are resuspended each in 4 ml of wash solution.

2.2.3. Gradient purification of wheat mitochondria

7. The mitochondria are purified through five sucrose layers. Mitochondria purified by sucrose gradient instead of Percoll gave a higher expression of the exogenous construct after the electroporation-mediated transformation. For a starting material of 20 g of wheat embryos six centrifuge tubes containing 25 ml sucrose solution gradients are used. The centrifuge tubes are prepared by stepwise layering of sucrose solution of decreasing densities as follows: 2 ml of 1.8 M, 6 ml of 1.45 M, 6 ml of 1.2 M, 6 ml of 0.9 M, 5 ml of 0.6 M. All the sucrose solutions contain 10 mM Tricine, pH 7.2, 1 mM EGTA, and 0.1% BSA-FAF. The gradients are allowed to

equilibrate overnight at 4°. Four milliliters of resuspended mitochondria are divided and layered above the six gradients and centrifuged in a Beckman SW27 rotor at 50,000g for 1 h. After centrifugation, the intact mitochondria that accumulate as a yellowish band at a density $\sim$1.3 M sucrose (bracketed in Fig. 22.2, step 7) are recovered with a large-bore plastic Pasteur pipette.

8. The mitochondria are diluted slowly with 3 volumes of a dilution solution (200 mM mannitol, 10 mM Tricine, pH 7.2, 1 mM EGTA) under constant stirring to avoid hypotonic shock. Organelles are recovered by centrifugation at 14,000g for 15 min. The supernatant is discarded, and the purified mitochondria are resuspended in 1 ml of 0.33 M sucrose solution.

9. Mitochondrial protein concentration is measured by use of the Bio-Rad Protein Assay, and the final protein concentration is adjusted to 20 mg/ml with 0.33 M sucrose solution. By use of this procedure, we usually obtain approximately 20 mg of mitochondrial proteins, which is enough to perform 20 transformation assays. Purified mitochondria are used immediately for electrotransformation. Stored organelles rapidly lose the ability to express foreign constructs.

2.3. Potato mitochondria purification

Potato mitochondria are prepared from 2 kg of tubers in batches of 200g each with 200 ml of cold homogenization solution (same solution described in step 2). Homogenization of the potato tubers is carried out for 15 sec in a cold Waring blender at full speed. The rest of the potato mitochondria purification is done as described for wheat mitochondria (from steps 4 to 8).

3. Plasmid Construction

3.1. Overview

The electrotransformation assay requires the construction of a specific plasmid (e.g., pCOXII described in Fig. 22.3). In general, the transgene requires a promoter from the same species of the mitochondria to be electroporated. The wheat *cox II* promoter is a well-characterized promoter, able to initiate the transcription in an *in vitro* system (Hanic-Joyce and Gray, 1991). This promoter directs the transcription of different constructs used in electrotransformation experiments and might be considered as a good candidate for any future assays. Specific sequences in the transgene construct should be introduced to differentiate the endogenous from the exogenous transcript, and the presence of an intron in the transgene is an extremely useful tool to purify mature transcript in RT-PCR experiments.

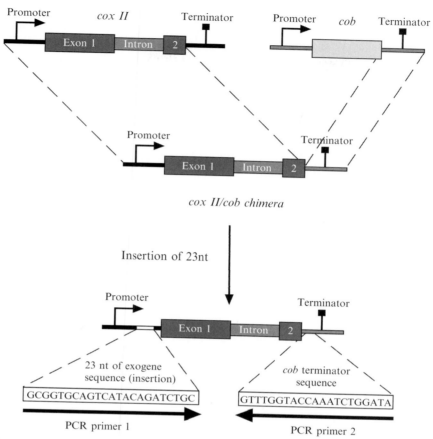

Figure 22.3 pCOXII or pCOXIISt maps. These plasmids contain the full *cox II* open-reading frame along with the *cox II* promoter region. pCOXII contains the wheat gene and pCOXIISt includes the potato gene. Both constructions have a terminal region from the wheat *cob* gene. An insertion of 23 nt is added to create a specific sequence to amplify selectively the exogenous gene. The second unique primer sequence that is absent in the endogenous *cox II* locus is located in the *cob* terminator sequence.

3.2. Plasmid

A gene from *T. timopheevi* is used to differentiate the endogenous gene from *T. aestivum*. The *cox II* gene from *T. timopheevi* contains several differences in the promoter region and in its intron sequence compared with its counter-part gene in *T. aestivum*. Moreover, the terminator region of *cox II* is replaced by *cob* sequence from *T. timopheevi*. The plasmid pCOXII (Fig. 22.3) is based on the pBluescribe vector, containing the full open-reading frame and 882 nt of the promoter region of *cox II* from *T. timopheevi* (GenBank accession numbers AF336134). The plasmid has 533 nt of the downstream (terminator) region of *T. timopheevi cob* gene (GenBank accession number AF337547). Finally a 23-nt insert is introduced into the promoter region at position −60 in plasmid pCOXII. Primers on the 23-nt insertion (PCR primer 1) and *cob* terminator sequence (PCR primer 2) provide specific sequences to amplify exclusively the transgene transcripts by RT-PCR.

A similar plasmid, based on pCOXII, is designed for potato experiments. In this plasmid, pCOXIISt, the *T. timopheevi cox II* gene, and promoter in pCOXII are replaced by the *S. tuberosum cox II* gene and promoter (GenBank accession number DQ18064), with the same 23-nt insert introduced into the promoter region.

4. ELECTROPORATION-MEDIATED TRANSFORMATION OF ISOLATED MITOCHONDRIA

The electrotransformation consists of four steps as described in Fig. 22.4: electroporation, *in organello* gene expression, RNA purification, and RT-PCR/DNA sequencing.

4.1. Electroporation

1. Electrotransfer experiments are carried out with a Bio-Rad Gene Pulser in cold 0.1-cm electrode-gap cuvettes (Bio-Rad). The electroporation settings for wheat and potato mitochondria are set at a capacitance of 25 μF, a resistance of 400 Ω ,and field strength of 13 kV/cm.
2. One microgram of plasmid was sufficient to obtain maximum expression levels. The plasmid (purified by use of Qiagen plasmid Midi kit) is added to 50 μl of purified mitochondria (1 mg of mitochondrial proteins) in 0.33 *M* sucrose solution and electroporated with the settings described previously.

Optimization notes:
   a. The electroporation field strength was optimized for potato and wheat mitochondria on the basis of a DNAse I protection assay for a foreign DNA, where the maximum protection was observed at 13 kV/cm and was reduced at lower and higher field strength (Choury *et al.*, 2005; Farre and Araya, 2001). By use of a similar electrotransformation system, Staudinger and Kempken (2003) found that a field strength of 20 kV/cm was optimum for mitochondria isolated from 9-day-old etiolated seedlings of maize and sorghum.
   b. Plasmid concentration (1 to 10 μg in a 50 μl electroporation assay) does not affect the expression level of the transgene drastically, at least in these systems. Staudinger and Kempken (2003) report optimal expression conditions with 3 μg of plasmid for 250 μg of mitochondrial proteins.

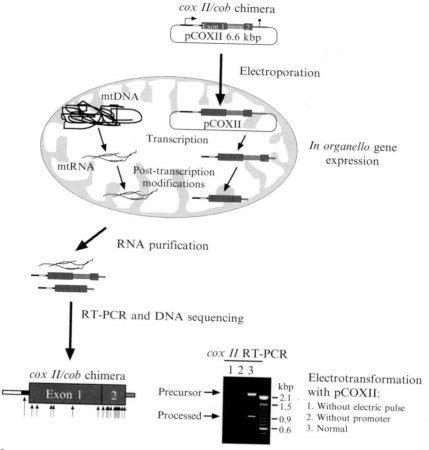

Figure 22.4 Scheme for the electrotransformation of mitochondria. Arrow in the *cox II* PCR product indicates edited sites. Agarose gel shows RT-PCR from *cox II* mRNA transcripts of mitochondria electrotransformed with either pCOXII or pCOXII without promoter sequence.

4.2. *In organello* gene expression

1. After electroporation, the mitochondrial suspension is transferred to a 1.5-ml Eppendorf tube, and the cuvette is washed with an additional 50 μl of 0.33 M sucrose, which is added to the mitochondrial suspension.
2. The mitochondria are centrifuged at 15,000g for 10 min at 4°.
3. The supernatant is discarded, and the mitochondrial pellet is resuspended in 250 μl expression solution containing 330 mM mannitol, 90 mM KCl, 10 mM MgCl$_2$, 12 mM Tricine (pH 7.2), 5 mM KH$_2$PO$_4$, 1.2 mM

EGTA, 1 mM GTP, 2 mM dithiothreitol, 2 mM ADP, 10 mM sodium succinate, and 0.15 mM (each) CTP and UTP. The expression solution should be made fresh each time from stock solutions.

4. Mitochondria are routinely incubated at 25° for 18 h with constant stirring at 150 rpm.

Optimization notes:

a. Incubation time up to 20 h was analyzed. At 3 h, the precursor was easily detected, and the mature product increased steadily up to 20 h. Because of the lengthy procedure for mitochondria purification, overnight incubation (18 h) is a convenient incubation time.

b. The expression solution was modified from a complete medium (250 mM mannitol, 90 mM KCl, 10 mM MgCl$_2$, 10 mM Tricine (pH 7.2), 5 mM KH$_2$PO$_4$, 1 mM EGTA, 1 mM GTP, 2 mM dithiothreitol, 2 mM ADP, 10 mM sodium succinate, and 25 μM amino acids pool) described for an *in organello* protein synthesis assay (Leaver *et al.*, 1983) and the requirements necessary for transgene expression were elucidated by Farre and Araya (2001). In summary, the omission of the amino acid pool had no effect on transcription and processing of the RNA, whereas lowering the potassium and magnesium ion concentration resulted in a dramatic decrease in the transcription efficiency. Furthermore, in the absence of the four rNTP precursors, transcription was abolished, indicating that the internal nucleotide pool was too low to sustain transcription of the exogenous template. A high concentration of GTP (1 mM) was necessary for expression, because transcription was found to be reduced at lower concentrations (0.15 mM) and abolished in the absence of GTP. The addition of an ATP regenerating system, ADP, and succinate was necessary for expression, but the addition of exogenous ATP did not replace the metabolically generated precursor. The latter observation indicates that during the assay the mitochondrial respiratory chain is functional, suggesting that the mitochondria are intact during the assay.

c. Staudinger and Kempken (2003) used a similar expression buffer for *in organello* gene expression of maize and sorghum except that the respiratory substrate succinate was replaced by glutamate and malate. CTP, GTP, and UTP were not required.

d. In the case of potato, mitochondria were viable only during 3 to 4 h as measured by oxygen consumption. This problem was circumvented by supplementing the expression solution with 1 mg/ml of BSA-FAF (Choury *et al.*, 2005). These results indicate that mitochondria from other plants and tissues may have different viability times and other requirements after DNA electroporation. Respiration assays can be performed as a viability control, measuring O$_2$ consumption by use of a Yellow Springs electrode system.

4.3. RNA purification

1. After incubation, the electroporated mitochondria are recovered by centrifugation at 15,000g for 10 min at 4°.
2. The RNA is isolated with 200 µl of TRIzol Reagent (Gibco BRL) according to the protocol suggested by the supplier.
3. The RNA pellet is resuspended in 20 µl DEPC-treated water. The RNA is quantified spectrophotometrically at A_{260}nm. Approximately 5 µg of mtRNA is obtained from 50 µl of purified mitochondria (equivalent to 1 mg of mitochondrial protein) used in the electrotransformation.
4. The RNA is stored at −80°.

4.4. RT-PCR and DNA sequencing

The residual DNA is eliminated from the purified RNA by treating with DNase I "Amplification grade" (Invitrogen). Two units of DNase I are used to treat 1 µg of RNA in a 10-µl solution containing 20 mM Tris-HCl, pH 8.4, 50 mM KCl, and 2 mM MgCl$_2$. The mixture is incubated at room temperature for 15 min. The reaction was terminated by addition of 1 µl of 25 µM EDTA and a further incubation at 65° for 15 min.

The cDNA synthesis is performed with a Superscript II First-strand synthesis system (Invitrogen) for RT-PCR by use of random hexamers according to the supplier's protocol.

1. Eight microliters of the DNase I–treated RNA is incubated with 1 µl of random hexamer (50 ng/µl) and 1 µl of 10 mM dNTP mix at 65° for 5 min and then on ice for 1 min.
2. The reverse transcription is carried out in a final volume of 20 µl containing 20 mM Tris·Cl, pH 8.4, 50 mM KCl, 5 mM MgCl$_2$, 10 mM dithiothreitol, 40 units of RNaseOUT Recombinant Ribonuclease Inhibitor, and the total mixture of RNA/random hexamers/dNTP mix from the previous step. The reaction mixture is incubated at room temperature for 2 min.
3. Subsequently, 200 units of Superscript II RT is added and incubated for 10 min at room temperature, followed by incubation at 42° for 50 min.
4. The reaction is stopped by heating at 75° for 15 min.
5. Finally, the sample is treated with 2 units of *E. coli* RNase H for 20 min at 37°. The cDNA can be used directly for PCR or stored at −20°.

We recommend that the PCR reactions be performed with Advantage 2 Polymerase Mix (Clontech), which provides a sensitive and robust polymerase mixture that is able to amplify rare transcripts. We observed that this enzyme gave better yields of PCR products than other polymerases in experiments that used the transgene transcript cDNA from electroporated mitochondria. The PCR reaction is carried out with 1/10 quantity of cDNA preparation following the supplier's protocol. The PCR primers

used were (1) GCGGTGCAGTCATACAGATCTGC and (2) TATCCA-
GATTTGGTACCAAAC. The PCR conditions are 95° for 1 min; 5 cycles
at 95° for 30 sec and 68° for 1 min; 30 cycles at 95° for 30 sed, 58° for 30 sec
and 68° for 30 sec; and finally 68° for 1 min.

Sequence analyses are performed either directly on the RT-PCR product or on recombinant plasmids after cloning in pGEM-T vector (Promega)
by use of either the Thermo Sequenase radiolabeled terminator cycle
sequencing kit (USB) or an automatic DNA sequencing equipment with
the BigDye Terminator Cycle Sequencing Kit (Applied Biosystem).

5. Electrotransformation Assay and Posttranscriptional Modifications of the Exogenous *cox II*

The electrotransformation protocol has been used to study different
aspects of plant mitochondria gene expression. We describe here few
examples of these studies.

5.1. Methods

a. Fifty microliters of purified mitochondria (1 mg of mitochondrial proteins from wheat or potato) was mixed with 1 μg of plasmid (pCOXII or
   pCOXIISt plasmid depending on the source of mitochondria used) or
   with a control plasmid (pCOXII or pCOXIISt lacking the promoter
   regions).
b. The mixtures were electroporated and incubated under the condition
   described earlier. A control assay with plasmid but without electric pulse
   was included in every experiment with new mitochondria preparation.
c. Total RNA was purified and cDNA was made with 1 μg of RNA and
   random primers.
d. The transcripts sequence information from *coxII/cob* chimera was obtained
   by PCR amplification of cDNA primers 1 and 2, as described earlier.

5.2. Results

a. The wheat and potato transgenic *cox II* mitochondrial genes are transcribed and correctly spliced: The exogenous *cox II* gene is formed by
   two exons linked by an intron. RT-PCR analysis of electroporated
   mitochondria with pCOXII or pCOXIISt plasmid gave two bands: a
   higher band corresponding to the precursor *cox II* transcript and a lower
   band generated by the processed *cox II* mRNA (Fig. 22.4). Control assays
   without electric field pulse or by use of a construct lacking the promoter

region showed no PCR amplification products. The sequence analysis of the processed fragment clearly showed that the exon 1–exon 2 junction corresponded to that observed *in vivo*, indicating that the foreign *cox II* gene was transcribed and it underwent a faithful splicing process (Choury and Araya, 2006; Choury *et al.*, 2004, 2005; Farré and Araya, 2001, 2002; Farré *et al.*, 2001).

b. The exogenous transcripts are edited: To analyze the editing process, the RT-PCR products from the precursor and mature *cox II* transcripts were purified with the QIAquick Gel Extraction Kit (Qiagen) and directly sequenced.

Comparing the transcript with the genomic sequences, C-to-U changes are directly visualized after DNA sequencing on autoradiography films or in the chromatogram generated by an automatic DNA sequencing equipment. The semiquantitative analysis is easily conducted by densitography of short-time exposed autoradiography films. One-day exposure may be enough to visualize [^{35}S]-labeled products.

Under the incubation conditions described, the transcripts of potato and wheat *cox II* transgene were normally edited at most of the editing sites when they were introduced in mitochondria of the respective species (Choury and Araya, 2006; Choury *et al.*, 2005; Farré and Araya, 2001; Farré *et al.*, 2001). The spliced transcripts are found strongly edited, whereas precursor molecules show partial editing.

c. *cis* Recognition elements in plant mitochondrion RNA editing: Several studies were carried out to define the recognition motif of the editing sites (Choury *et al.*, 2005; Farre and Araya, 2001, 2002). To determine the *cis* elements involved in recognition of editing sites in plant mitochondria, deletion and site-directed mutations of sequences close to the editing sites were performed in the pCOXII or pCOXIISt plasmids by use of the QuikChange II Site-Directed Mutagenesis Kit (Stratagene). These modified plasmids were introduced into purified mitochondria by electroporation. The study was achieved with the spliced transcripts only because they are the final product of the RNA maturation process and are highly edited. For this purpose, after the RT-PCR, the product of mature *cox II* transcripts was purified with the QIAquick Gel Extraction Kit (Qiagen) and sequenced directly. It was deduced that a restricted number of nucleotides in the vicinity of the target C residue were necessary for recognition by the editing machinery and that the nearest neighbor 3′ residues were crucial for the editing process (Fig. 22.5) (Farré *et al.*, 2001).

The electroporation of engineered mitochondrial genes is a suitable method to study various molecular processes involved in the mitochondrial gene expression. It is an alternative and necessary method to complement

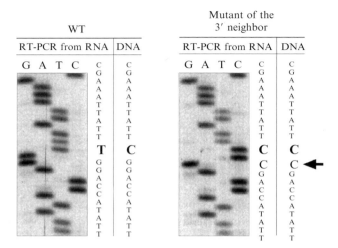

Figure 22.5 Modifications of 3′ neighbor residues of the editing site 259 located in the first exon of wheat *cox II* dramatically affect editing. Sequences of RT-PCR from electrotransformation experiments with a wild-type pCOXII (WT) or pCOXII with single mutation at the 3′ nearest neighboring residue (mutant of the 3′ neighbor) are indicated by an arrow. Editing site is shown in bold.

in vitro studies. Because the *in organello* transcription and posttranscriptional processes are strongly dependent on mitochondria integrity, the observations made by this approach closely resemble the physiological conditions found in plants.

REFERENCES

Araya, A., Domec, C., Begu, D., and Litvak, S. (1992). An *in vitro* system for the editing of ATP synthase subunit 9 mRNA using wheat mitochondrial extracts. *Proc. Natl. Acad. Sci. USA* **89**, 1040–1044.

Bass, B. L. (1997). RNA editing and hypermutation by adenosine deamination. *Trends Biochem. Sci.* **22**, 157–162.

Benne, R., Van den Burg, J., Brakenhoff, J. P., Sloof, P., Van Boom, J. H., and Tromp, M. C. (1986). Major transcript of the frameshifted coxII gene from trypanosome mitochondria contains four nucleotides that are not encoded in the DNA. *Cell* **46**, 819–826.

Binder, S., Hatzack, F., and Brennicke, A. (1995). A novel pea mitochondrial *in vitro* transcription system recognizes homologous and heterologous mRNA and tRNA promoters. *J. Biol. Chem.* **270**, 22182–22189.

Blanc, V., Jordana, X., Litvak, S., and Araya, A. (1996). Control of gene expression by base deamination: the case of RNA editing in wheat mitochondria. *Biochimie* **78**, 511–517.

Blanc, V., Litvak, S., and Araya, A. (1995). RNA editing in wheat mitochondria proceeds by a deamination mechanism. *FEBS Lett.* **373**, 56–60.

Bock, R., Hermann, M., and Fuchs, M. (1997). Identification of critical nucleotide positions for plastid RNA editing site recognition. *RNA* **3**, 1194–1200.

Bonen, L. (1991). The mitochondrial genome: So simple yet so complex. *Curr. Opin. Genet. Dev.* **1,** 515–522.

Chaudhuri, S., and Maliga, P. (1996). Sequences directing C to U editing of the plastid psbL mRNA are located within a 22 nucleotide segment spanning the editing site. *EMBO J.* **15,** 5958–5964.

Choury, D., and Araya, A. (2006). RNA editing site recognition in heterologous plant mitochondria. *Curr. Genet.* **50,** 405–416.

Choury, D., Farre, J. C., Jordana, X., and Araya, A. (2004). Different patterns in the recognition of editing sites in plant mitochondria. *Nucleic Acids Res.* **32,** 6397–6406.

Choury, D., Farre, J. C., Jordana, X., and Araya, A. (2005). Gene expression studies in isolated mitochondria: Solanum tuberosum rps10 is recognized by cognate potato but not by the transcription, splicing and editing machinery of wheat mitochondria. *Nucleic Acids Res.* **33,** 7058–7065.

Collombet, J. M., Wheeler, V. C., Vogel, F., and Coutelle, C. (1997). Introduction of plasmid DNA into isolated mitochondria by electroporation. A novel approach toward gene correction for mitochondrial disorders. *J. Biol. Chem.* **272,** 5342–5347.

Day, D. A., Neuburger, M., and Douce, R. (1985). Biochemical characterisation of chlorophyll–free mitochondria from pea leaves. *Aust. J. Plant Physiol.* **12,** 219–228.

Douce, R. (1985). "Mitochondria in Higher Plants: Structure, Function and Biogenesis." Academic Press, Orlando, FL.

Estevez, A. M., and Simpson, L. (1999). Uridine insertion/deletion RNA editing in trypanosome mitochondria—a review. *Gene* **240,** 247–260.

Farré, J. C., and Araya, A. (2001). Gene expression in isolated plant mitochondria: High fidelity of transcription, splicing and editing of a transgene product in electroporated organelles. *Nucleic Acids Res.* **29,** 2484–2491.

Farré, J. C., and Araya, A. (2002). RNA splicing in higher plant mitochondria: determination of functional elements in group II intron from a chimeric cox II gene in electroporated wheat mitochondria. *Plant J.* **29,** 203–213.

Farré, J. C., Leon, G., Jordana, X., and Araya, A. (2001). cis Recognition elements in plant mitochondrion RNA editing. *Mol. Cell Biol.* **21,** 6731–6737.

Freyer, R., Kiefer-Meyer, M. C., and Kossel, H. (1997). Occurrence of plastid RNA editing in all major lineages of land plants. *Proc. Natl. Acad. Sci. USA* **94,** 6285–6290.

Giege, P., and Brennicke, A. (1999). RNA editing in Arabidopsis mitochondria effects 441 C to U changes in ORFs. *Proc. Natl. Acad. Sci. USA* **96,** 15324–15329.

Hanic-Joyce, P. J., and Gray, M. W. (1991). Accurate transcription of a plant mitochondrial gene *in vitro. Mol. Cell Biol.* **11,** 2035–2039.

Hirose, T., and Sugiura, M. (2001). Involvement of a site-specific trans-acting factor and a common RNA-binding protein in the editing of chloroplast mRNAs: development of a chloroplast *in vitro* RNA editing system. *EMBO J.* **20,** 1144–1152.

Johnston, F. B., and Stern, H. (1957). Mass isolation of viable wheat embryos. *Nature* **179,** 160–161.

Kruft, V., Eubel, H., Jansch, L., Werhalm, W., and Braun, H.-P. (2001). Proteomic approach to identify novel mitochondrial proteins in *Arabidopsis. Plant Physiol.* **127,** 1694–1710.

Kubo, N., and Kadowaki, K. (1997). Involvement of 5' flanking sequence for specifying RNA editing sites in plant mitochondria. *FEBS Lett.* **413,** 40–44.

Kumar, R., and Levings, C. S., 3rd. (1993). RNA editing of a chimeric maize mitochondrial gene transcript is sequence specific. *Curr. Genet.* **23,** 154–159.

Leaver, C. J., Hack, E., and Forde, B. G. (1983). Protein synthesis by isolated plant mitochondria. *Methods Enzymol.* **97,** 476–484.

Lupold, D. S., Caoile, A. G., and Stern, D. B. (1999). Genomic context influences the activity of maize mitochondrial cox2 promoters. *Proc. Natl. Acad. Sci. USA* **96,** 11670–11675.

Malek, O., Lattig, K., Hiesel, R., Brennicke, A., and Knoop, V. (1996). RNA editing in bryophytes and a molecular phylogeny of land plants. *EMBO J.* **15,** 1403–1411.

Miyamoto, T., Obokata, J., and Sugiura, M. (2002). Recognition of RNA editing sites is directed by unique proteins in chloroplasts: Biochemical identification of cis-acting elements and trans-acting factors involved in RNA editing in tobacco and pea chloroplasts. *Mol. Cell Biol.* **22,** 6726–6734.

Miyamoto, T., Obokata, J., and Sugiura, M. (2004). A site-specific factor interacts directly with its cognate RNA editing site in chloroplast transcripts. *Proc. Natl. Acad. Sci. USA* **101,** 48–52.

Moller, I. M., Johansson, F., and Brodelius, P. (1996). "Plant Cell Membranes." Clarendon Press, Oxford.

Mulligan, R. M., Leon, P., and Walbot, V. (1991). Transcriptional and posttranscriptional regulation of maize mitochondrial gene expression. *Mol. Cell Biol.* **11,** 533–543.

Neuwirt, J., Takenaka, M., van der Merwe, J. A., and Brennicke, A. (2005). An *in vitro* RNA editing system from cauliflower mitochondria: Editing site recognition parameters can vary in different plant species. *RNA* **11,** 1563–1570.

Nivison, H. T., Sutton, C. A., Wilson, R. K., and Hanson, M. R. (1994). Sequencing, processing, and localization of the petunia CMS-associated mitochondrial protein. *Plant J.* **5,** 613–623.

Rapp, W. D., Lupold, D. S., Mack, S., and Stern, D. B. (1993). Architecture of the maize mitochondrial atp1 promoter as determined by linker-scanning and point mutagenesis. *Mol. Cell Biol.* **13,** 7232–7238.

Rapp, W. D., and Stern, D. B. (1992). A conserved 11 nucleotide sequence contains an essential promoter element of the maize mitochondrial atp1 gene. *EMBO J.* **11,** 1065–1073.

Reed, M. L., Peeters, N. M., and Hanson, M. R. (2001). A single alteration 20 nt 5′ to an editing target inhibits chloroplast RNA editing *in vivo*. *Nucleic Acids Res.* **29,** 1507–1513.

Staudinger, M., and Kempken, F. (2003). Electroporation of isolated higher-plant mitochondria: Transcripts of an introduced cox2 gene, but not an atp6 gene, are edited in organello. *Mol. Genet. Genomics* **269,** 553–561.

Takenaka, M., and Brennicke, A. (2003). *In vitro* RNA editing in pea mitochondria requires NTP or dNTP, suggesting involvement of an RNA helicase. *J. Biol. Chem.* **278,** 47526–47533.

Takenaka, M., Neuwirt, J., and Brennicke, A. (2004). Complex cis-elements determine an RNA editing site in pea mitochondria. *Nucleic Acids Res.* **32,** 4137–4144.

To, K. Y., Cheng, M. C., Chen, L. F., and Chen, S. C. (1996). Introduction and expression of foreign DNA in isolated spinach chloroplasts by electroporation. *Plant J.* **10,** 737–743.

Williams, M. A., Kutcher, B. M., and Mulligan, R. M. (1998). Editing site recognition in plant mitochondria: The importance of 5′-flanking sequences. *Plant Mol. Biol.* **36,** 229–237.

Yu, W., and Schuster, W. (1995). Evidence for a site-specific cytidine deamination reaction involved in C to U RNA editing of plant mitochondria. *J. Biol. Chem.* **270,** 18227–18233.

CHAPTER TWENTY-THREE

TRANSFORMATION OF THE PLASTID GENOME TO STUDY RNA EDITING

Kerry A. Lutz *and* Pal Maliga

Contents

Abstract

In this chapter we provide an overview of cytosine-to-uridine (C-to-U) RNA editing in the plastids of higher plants. Particular emphasis will be placed on the role plastid transformation played in understanding the editing process. We discuss how plastid transformation enabled identification of mRNA *cis* elements for editing and gave the first insight into the role of editing *trans* factors. The introduction will be followed by a protocol for plastid transformation, including vector design employed to identify editing *cis* elements. We also discuss how to test RNA editing *in vivo* by cDNA sequencing. At the end, we summarize the status of the field and outline future directions.

Waksman Institute, Rutgers, The State University of New Jersey, Piscataway, New Jersey

Methods in Enzymology, Volume 424
ISSN 0076-6879, DOI: 10.1016/S0076-6879(07)24023-6

1. INTRODUCTION

In land plants RNA editing has been reported in the plastids and mitochondria, two DNA-containing cytoplasmic organelles with their own prokaryotic-type transcription and translation machinery (Maier *et al.*, 1996). In lower plants, RNA editing was shown in some (Kugita *et al.*, 2003; Miyata and Sugita, 2004; Sugita *et al.*, 2006), but not all, taxonomic groups (Freyer *et al.*, 1997), indicating that editing is an evolved trait. In bryophytes both cytidine-to-uridine (C-to-U) as well as U-to-C editing have been reported; RNA editing in the plastids of higher plants involves only C-to-U conversion. RNA editing in plastids of higher plants was discovered in 1991 (Hoch *et al.*, 1991), and since then has been shown in all higher plant species tested so far (Bock, 2000; Tsudzuki *et al.*, 2001). Editing in most cases occurs in the coding region and restores a conserved amino acid at the mRNA level. Examples for the creation of translation initiation or stop codons are also known. Among the higher plant species, tobacco has the highest number (37) of identified editing sites (Hirose *et al.*, 1999; Kahlau *et al.*, 2006; Sasaki *et al.*, 2003). In bryophytes the number of editing sites can be as high as 942 (Kugita *et al.*, 2003). Comparison of editing sites indicates that some sites are highly conserved across large taxonomic groups such as monocots and dicots, while others vary even among closely related species (Calsa Junior *et al.*, 2004; Kahlau *et al.*, 2006; Sasaki *et al.*, 2003). A comprehensive listing of editing sites is available for a number of species, including the allotetrapoid *Nicotiana tabacum* (Hirose *et al.*, 1999; Kahlau *et al.*, 2006; Sasaki *et al.*, 2003) and its progenitor species *Nicotiana sylvestris* and *Nicotiana tomentosiformis* (Sasaki *et al.*, 2003) and the related solanaceous plant tomato (Kahlau *et al.*, 2006). A catalogue of editing sites is also available for the cereal crops maize (Maier *et al.*, 1995), rice (Corneille *et al.*, 2000), and sugarcane (Calsa Junior *et al.*, 2004) and the model species *Arabidopsis thaliana* (Lutz and Maliga, 2001; Tillich *et al.*, 2005) and pea (Inada *et al.*, 2004).

First we provide an overview of RNA editing in plastids. We cover how genetic approaches, such as plastid transformation, complementation by protoplast fusion, and the study of mutants, advanced our understanding of the editing process. Two protocols will be presented. In the first protocol, we discuss construction of transplastomic plants for studies on RNA editing. In the second protocol, we describe methods to detect RNA editing *in vivo*. The protocols we discuss have been developed in tobacco (*N. tabacum*), the only species in which plastid transformation is routinely obtained (Bock, 2007; Maliga, 2004).

2. IDENTIFICATION OF UNIQUE *CIS* SEQUENCES FOR RNA EDITING

Plastid transformation enabled *in vivo* testing of chimeric mRNAs containing sequences surrounding the edited C nucleotide. Testing a series of deletion derivatives defined the sequences required for editing. The most complete information is available for the *psbL* editing site. *psbL* is a plastid photosynthetic gene, in which the translation initiation codon is created by conversion of an ACG codon to an AUG codon at the mRNA level. *In vivo* dissection using plastid transformation revealed that information required for editing is contained within a 22-nucleotide fragment including 16 nucleotides upstream and 5 nucleotides downstream (−16 to +5) of the edited C (Chaudhuri and Maliga, 1996). The importance of the upstream and downstream sequences for *psbL* editing has been confirmed *in vitro* (Hirose and Sugiura, 2001). Study of *ndhB* sites IV and V indicates that essential elements are located between the −12 and −2 positions (Bock *et al.*, 1996). Scanning mutagenesis of the same sites revealed that the upstream sequences are important for editing and that spacing of the upstream recognition sequence relative to the edited C is important (Bock *et al.*, 1997; Hermann and Bock, 1999). The third site for which detailed *in vivo* transgenic data are available is the tobacco *rpoB* site II (NtrpoB C473). Sequences minimally required for editing were located in the −20 to +6 region surrounding the edited C (Reed *et al.*, 2001b), although longer sequences were edited more efficiently (Hayes *et al.*, 2006).

In the early studies, no consensus sequence was recognized by alignment of sequences adjacent to the edited C nucleotide. Thus, it was assumed that each editing site is individually recognized. Later, specificity clusters were identified, which share short (2–3 nucleotide) group-specific sequence elements (Hayes *et al.*, 2006).

3. EDITING *TRANS* FACTORS

Protein factors involved in editing have been identified in many systems. In higher plant plastids, most information for the involvement of protein factors is indirect: the existence of species-specific, organelle-specific, and site-specific factors was inferred from genetic experiments. The first evidence for a species-specific editing *trans* factor was obtained when a spinach (*psbE*) editing sequence was incorporated in tobacco plastids where it was not edited (Bock *et al.*, 1994), unless the spinach nuclear *trans* factor was provided by cell fusion (Bock and Koop, 1997). Similarly, maize *rpoB* site IV, which is absent in the tobacco *rpoB* gene, was not edited in transplastomic tobacco. However, the maize *rpoB* site I sequence, a site that is edited in tobacco, was edited when incorporated in the tobacco ptDNA (Reed and Hanson, 1997). Also, the tobacco *atpA* editing site, which is absent in *Atropa belladonna*, was not recognized in tobacco plastids when introduced by cell fusion into the *A. belladonna* nuclear background, and, as a consequence, the plants were pigment deficient. This was the first time that the molecular mechanism of nucleus–cytoplasm incompatibility could be explained by the lack of RNA editing (Schmitz-Linneweber *et al.*, 2005). Based on these findings, it is assumed that each species has the capacity to edit the sites that it carries, but lacks the capacity for editing the sites that it does not have. Exceptions to this rule have been found in *N. sylvestris,* which has the capacity to edit *ndhA* site 1, even though it does not have the editing site encoded in the DNA. This was explained by conservation of the editing capacity present in related species (Tillich *et al.*, 2006).

Evidence for organelle-specific factors was obtained when an edited *Petunia* mitochondrial *coxII* sequence was expressed in tobacco chloroplasts, where none of the seven sites was edited (Sutton *et al.*, 1995). Since editing of the cognate sequence has not been studied in tobacco, the lack of editing could also be due to species-specific differences in the editing capacity of the two closely related species.

The existence of site-specific factors was inferred from competition between transgenic and native mRNAs. In early studies with overexpressed *psbL* editing segments, competition was found only with the native *psbL* segment (Chaudhuri and Maliga, 1996; Chaudhuri *et al.*, 1995). Based on this, individual recognition of each of the plastid editing sites was proposed. More extensive studies with the *rpoB* and *ndhF* editing sites led to the conclusion that at least some of the editing sites can be grouped in specificity clusters (Hayes *et al.*, 2006; Reed *et al.*, 2001a). The best studied is the NtrpoB C473, NtpsbL C2, and Ntrps14 C80 cluster in which a group-specific response is attributed to short (2–3 nucleotide) group-specific sequence elements (Hayes *et al.*, 2006). *In vivo* competition data have been corroborated *in vitro* for the *psbL* (Hirose and Sugiura, 2001), *rpoB,* and *ndhF* genes (Hayes *et al.*, 2006; Reed *et al.*, 2001a).

Although the existence of general and site-specific factors is documented *in vitro*, none of these proteins has been cloned to date. The only exception is a pentatricopeptide repeat protein encoded in the *crr4* gene involved in editing *ndhD* site 1 of *A. thaliana*, which was identified in a mutant screen (Kotera *et al.*, 2005).

4. PROTOCOLS FOR CONSTRUCTION OF TRANSPLASTOMIC PLANTS

Transformation of the plastid genome (ptDNA) provided the first experimental tool to study RNA editing. The ptDNA of higher plants is ~120–150 kb in size, is highly polyploid, and may be present in

1000–10,000 copies per cell (Bendich, 1987; Shaver *et al.*, 2006; Wakasugi *et al.*, 2001). Plastid transformation vectors are *Escherichia coli* plasmids in which the transgenes are flanked by plastid DNA sequences. Plastid transformation involves introduction of transforming DNA into tobacco leaf chloroplasts on the surface of microscopic gold particles (0.6 μm) by the biolistic process. The transgenes integrate into the ptDNA by two homologous recombination events via flanking ptDNA. To obtain a genetically stable transplastomic plant, all ptDNA copies must be changed. Such genetically stable, homoplastomic plants are obtained through a gradual process of ptDNA replication and sorting, and preferential maintenance of transgenic ptDNA copies on a selective tissue culture medium (Fig. 23.1).

To ensure preferential maintenance of plastids carrying transgenic ptDNA, a selective marker is incorporated adjacent to the gene of interest so that the two form a heterologous block (Fig. 23.2). The selective agent in the culture medium is spectinomycin, streptomycin, or kanamycin, inhibitors of protein synthesis on the plastid's prokaryotic type 70S ribosomes. (The cells survive because protein synthesis on eukaryotic 80S cytoplasmic ribosomes is not sensitive to these drugs and the cells are supplied with a reduced carbon source, sucrose, in the culture medium.) The *aadA* gene confers resistance to spectinomycin and streptomycin (Svab and Maliga, 1993); the *neo* or *kan* (Carrer *et al.*, 1993) or *aphA-6* (Huang *et al.*, 2002) gene confers resistance to kanamycin. Cells of the bombarded leaves form scanty white callus on the selective medium due to inhibition of chlorophyll biosynthesis. Transplastomic clones are recognized by the formation of green cells and shoots (Fig. 23.1B). The cells in the regenerated shoots are chimeric with transgenic and nontransformed sectors, but leaf cells within the transgenic sectors carry only uniformly transformed ptDNA copies as the result of replication and sorting of transformed ptDNA. Homoplastomic, uniformly transformed plants are obtained during a second cycle of plant regeneration from the resistant sectors (Fig. 23.1D). The shoots are rooted and the plants are transferred to the greenhouse (Fig. 23.1E). For reviews on plastid transformation, see Bock (2007), Herz *et al.* (2005), and Maliga (2004, 2005); for protocols on tobacco plastid transformation, see Bock (1998) and Lutz *et al.* (2006b).

4.1. Plastid transformation vector design

In plastid transformation vectors, the marker gene and the gene of interest, the heterologous block to be introduced, are flanked by ptDNA "targeting sequences" to ensure integration at the target site (Fig. 23.2). Three approaches were used to test mRNA editing: minigenes, translational fusion with a reporter gene, and incorporation of an editing segment in the 3′-untranslated region (3′-UTR). Conceptually the most simple design was construction of minigenes, which involved insertion of an editing fragment

Figure 23.1 Transformation of tobacco leaves to obtain transplastomic plants. (A) Leaves prepared for bombardment with DNA-coated gold particles. (B) Resistant shoots and/or calli appear in 4–12 weeks on leaf cut into 1-cm^2 pieces on selective RMOP regeneration medium. (C) Classification of spectinomycin-resistant lines as putative transplastomic clones by resistance to streptomycin on RMOP medium containing streptomycin and spectinomycin (500 mg/liter each). The plate is shown after 6 weeks. (D) Shoot regeneration from leaf sections on RMOP spectinomycin medium (500 mg/liter) to obtain homoplastomic shoots. The plate is shown after 6 weeks. (E) Homoplastomic shoots rooted in plate or (F) in Magenta box. (See color insert.)

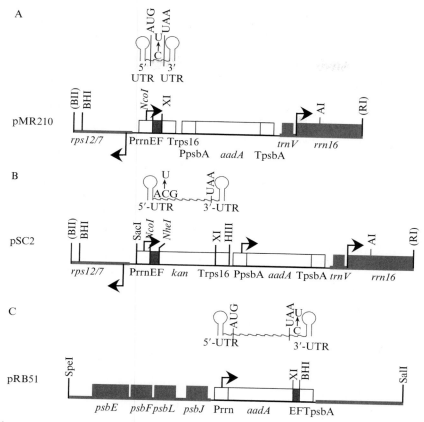

Figure 23.2   Plastid vectors to test *in vivo* RNA editing. (A) The pMR210 minigene vector with the maize *rpoB* site I editing segment (EF, *NcoI–XbaI* fragment) (Reed and Hanson, 1997). (B) The pSC2 *kan* fusion vector with the *psbL* editing segment (*NcoI–NheI* fragment) (Chaudhuri and Maliga, 1996). (C) The pRB51 3'-UTR vector to accept editing segments as *XbaI–BamHI* fragments (Bock *et al.*, 1996). The left and right plastid targeting sequences are in bold. The 5'-UTR and 3'-UTR of mRNA above the edited genes are depicted as a stem-loop structure; the position of the edited nucleotide is also marked. Shown are the spectinomycin resistance (*aadA*) and kanamycin resistance (*kan*) genes; the plastid *rrn16*, *trnV*, *psbE*, *psbF*, *psbL*, *psbJ*, and 3'-*rps12* genes; EF, editing fragment; Prrn, promoter of the plastid ribosomal RNA operon; Trps16, 3'-UTR of the plastid *rps16* gene; PpsbA, promoter of the plastid *psbA* gene; TpsbA, 3'-UTR of the plastid *psbA* gene. Arrows mark transcription initiation sites.

in a plastid expression cassette linked to a marker gene. The progenitor plasmid was vector pLAA24A, a transformation vector in which a *uidA* reporter gene is expressed in a Prrn-Trps16 promoter–terminator cassette (Zoubenko *et al.*, 1994). The editing constructs were obtained by PCR amplification of the editing segments with *NcoI* (5') and *XbaI* (3') sites at the ends that were used to replace the *uidA* reporter gene to create plasmid pMR210 (Fig. 23.2A). This approach was used to test editing, for example, of *rpoB* sites I, II, and IV (Hayes *et al.*, 2006; Hegeman *et al.*, 2005a; Reed and Hanson, 1997; Reed *et al.*, 2001a,b) and *ndhF* site 2 (Reed *et al.*, 2001a). Editing segments as short as 27 nucleotides could be studied in minigenes because fusion with the 5′-untranslated region (5′-UTR) of the promoter and the 3′-UTR of the terminator increased the transcript size to 0.2 kb and the transcript was stable (Reed *et al.*, 2001b).

The second approach, translational fusion with a reporter gene (Fig. 23.2B), was used to study the *psbL* and *ndhD* editing events that create an AUG translation initiation codon by editing of an ACG codon at the mRNA level (Chaudhuri and Maliga, 1996; Chaudhuri *et al.*, 1995). Editing of the two sites was studied in chimeric *aadA* or *kan* genes that confer spectinomycin or kanamycin resistance, respectively. Expression of the antibiotic resistance phenotype in plastids could be used to verify RNA editing because the reporter gene segment lacked an initiation codon, therefore, translation of the chimeric mRNA was dependent on RNA editing. Vector pSC2 carries an editing-dependent *kan* gene fused with a *psbL* editing fragment and allowed a series of editing fragments to be tested by replacing the *psbL* (*psbF/psbL*) editing segment (*NcoI*–*NheI* fragment) with new test segments (Chaudhuri and Maliga, 1996). Replacement of the *kan* coding segment with alternate genes, for example, a *bar* gene (Lutz *et al.*, 2006a), facilitated testing of editing fragments in a different sequence context.

The third approach to test editing was incorporation of editing segments in the 3′-UTR of the *aadA* marker gene (Fig. 23.2C) where the editing status of the segment does not affect the expression of the marker gene. In plasmid pRB51, the editing segment can be conveniently cloned in an *XbaI*–*BamHI* fragment (Bock *et al.*, 1996). Derivatives of pRB51 have been used to study editing of *ndhB* sites (Bock *et al.*, 1996, 1997; Hermann and Bock, 1999).

4.1.1. Materials

RM plant maintenance medium [MS medium (Murashige and Skoog, 1962)] (per liter add 100 ml 10× macronutrients, 10 ml 100× micronutrients, 5 ml 1% Fe-EDTA, 30 g sucrose, pH 5.6–5.8, with 1 M KOH, 7 g agar)

RMOP shoot regeneration medium [per liter add 100 ml 10× macronutrients, 10 ml 100× micronutrients, 5 ml 1% Fe-EDTA, 1 ml thiamine (1 mg/ml), 0.1 ml α–naphthaleneacetic acid (NAA 1 mg/ml in 0.1 M NaOH), 1 ml 6-benzylaminopurine (BAP, 1 mg/ml in 0.1 M HCl), 0.1 g myo-inositol, 30 g sucrose, pH 5.8, with 1 M KOH, 7 g agar]

RM medium 10× macronutrient solution [per liter add 19 g KNO_3, 3.7 g $MgSO_4 \cdot 7H_2O$, 4.4 g $CaCl_2 \cdot 2H_2O$, 1.7 g KH_2PO_4, 16.5 g $(NH_4)NO_3$]
RM medium 100× micronutrient solution [per liter add 169 mg $MnSO_4 \cdot H_2O$, 62 mg H_3BO_3, 86 mg $ZnSO_4 \cdot 7H_2O$, 8.3 mg KI, 2.5 ml $Na_2MoO_4 \cdot 2H_2O$ (1 mg/ml), 2.5 ml $CuSO_4 \cdot 5H_2O$ (1 mg/ml), 0. 25 ml $CoCl_2 \cdot 6H_2O$ (1 mg/ml)]
Nicotiana tabacum, c.v. Petit Havana
Agar, plant tissue culture tested (Sigma, A7921)
Spermidine free base 0.1 *M* (Sigma S4139)

4.2. Cleaning of gold particles

This protocol was modified from Bio-Rad Bulletin 9075.

1. Weigh out 30 mg 0.6 μm gold microcarrier (Bio-Rad Catalog No. 165-2262) in a 1.5-ml Eppendorf tube and add 1 ml ice-cold 70% ethanol.
2. Place the tube in a Vortex microtube holder and vortex vigorously for 5 min. Let the particles settle at room temperature (20–25°) for 15 min.
3. Spin in a microcentrifuge at 3000 rpm (600×g) for 1 min to compact the gold.
4. Remove ethanol with a pipette and add 1 ml ice-cold sterile distilled water.
5. Vortex the tube to suspend the particles. Allow the particles to settle at room temperature for 10 min.
6. Sediment the gold by spinning in a microcentrifuge at 3000 rpm (600×g) for 1 min.
7. Remove water with a pipette and add 1 ml ice-cold sterile distilled water.
8. Repeat washing the gold with water by repeating Steps 5, 6, and 7.
9. Resuspend the particles by vortexing and store the tube at room temperature for 10 min to let the particles settle.
10. Microcentrifuge at 5000 rpm (1700×g) for 15 sec and then remove the water completely.
11. Add 500 μl 50% glycerol and vortex for 1 min to resuspend the particles. The gold concentration will be 60 mg/ml. Clean gold can be stored for 2 weeks at room temperature.

4.3. Coating gold particles with plasmid DNA

With one DNA construct 20–30 leaf samples are bombarded. Each tube (30 mg) of gold is sufficient for 50 bombardments (two DNA constructs). Freshly prepared gold particles or stored gold may be used. If stored gold is used, vortex the tube for 5 min before coating with DNA. This protocol was modified from Bio-Rad Bulletin 9075.

1. Place the Eppendorf tube containing the gold in a Vortex microtube holder and shake at setting 3. While the tube is shaking, remove 50-μl aliquots of gold and pipette into 10 1.5-ml Eppendorf tubes in a rack.
2. Place the tubes with gold aliquots in a Vortex microtube holder and shake at setting 3. While the tubes are shaking, add 5 μl DNA (1 μg/μl), 50 μl 2.5 M CaCl$_2$, and 20 μl 0.1 M spermidine free base (Sigma S4139). Make sure to add these in the order described and make sure that the contents are thoroughly mixed before adding the next component.
3. Shake the tubes in Vortex microtube holder at setting 3 for 5 min.
4. Sediment the gold by spinning in a microcentrifuge at 3000 rpm (600$\times$$g$) for 1 min.
5. Discard the supernatant and add 140 μl 70% ethanol to each tube.
6. Tap the tube lightly until the pellet just starts to come into solution to make sure the pellet is not tightly packed. If the gold does not go into solution by gently tapping the tube, break up the pellet by pipetting up and down.
7. Sediment the gold by spinning in a microcentrifuge at 3000 rpm for 1 min.
8. Remove the supernatant and add 140 μl ice-cold 100% ethanol to each tube.
9. Lightly tap the tube until the pellet just starts to come into solution.
10. Sediment the gold by spinning in a microcentrifuge at 5000 rpm (1700$\times$$g$) for 15 sec.
11. Resuspend the coated gold pellet in 50 μl 100% ethanol by gently tapping the tube. The pellet should easily enter solution. Shake the tubes at setting 3 while waiting to use them for bombardment. If the tubes are sitting for a long period of time before bombardment, replace the ethanol in the tube with fresh 100% ethanol.

4.4. Introduction of DNA into chloroplasts by the biolistic process

The protocol we described is for transformation of tobacco leaves, collected from plants grown on RM medium in Magenta boxes under sterile conditions. Plastid transformation can also be accomplished using surface sterilized leaves of greenhouse-grown plants. The protocol we describe here is for biolistic transformation with the Bio-Rad PDS1000/He biolistic gun (catalog no. 165-2257). A suitable vacuum pump for the gun is Thermo-Savant VLP285 (ThermoSavant, Holbrook, NY). With this gun, 20–30 leaf samples are bombarded per DNA construct. We expect to obtain one transplastomic clone per plate (range 5–0.5). The Hepta–adaptor version of the gun (which is simultaneously using seven macrocarriers) is more efficient; bombardment of five plates is sufficient to obtain a similar number of transformants.

1. Place a tobacco leaf for biolistic transformation abaxial side up on two sterile Whatman No. 4 filter papers on a Petri plate (10 cm) of solid RMOP medium (20 ml) (Fig. 23.1A). Use more than one leaf if necessary to cover the central area. (If you are using greenhouse leaves, you need to cut out a segment that covers the plate.)

2. Leaf bombardment with the gun is carried out in a sterile laminar flow hood. Before bombardment, sterilize the gun main chamber, rupture disk retaining cap, microcarrier launch assembly, and target shelf by wiping off with a cloth soaked in 70% ethanol.

3. Sterilize the rupture disks (1100 psi), macrocarriers, macrocarrier holders, and stopping screens by soaking in 100% ethanol (5 min) and then air dry them in a tissue culture hood in an open container.

4. Turn on the helium tank and set the delivery pressure in a regulator (distal to tank) for 1300 psi (200–300 psi above the rupture disk value).

5. Turn on the vacuum pump and gene gun. Set the vacuum rate on the gene gun to 7 and the vent rate to 2.

6. Prepare DNA-coated gold particles as described above. Pipette 10 μl of DNA-coated gold onto one flying disk (placed in holder) and let air dry for 5 min. Five samples may be made up at one time.

7. Place the rupture disk into the retaining cap and screw in tightly.

8. Put the stopping screen and flying disk (face down) in a microcarrier launch assembly and place in a chamber just below the rupture disk. For a description, see Bio-Rad Bulletin 9075.

9. Place the leaf on a thin RMOP plate into the chamber 9 cm (fourth shelf from top) below the microcarrier launch assembly and close the door.

10. Press the vacuum button to open the valve. When the vacuum reaches 28 in. Hg, hold down the fire button until the pop from the gas breaking the rupture disk is heard. If the gun is fired at lower pressure, DNA-coated particles will lack momentum to penetrate the cells and no transplastomic lines will be obtained. If you have no experience with biolistic transformation, we recommend that you test particle coating and DNA delivery using a transient expression of the nuclear uidA gene, which encodes β-glucuronidase, an enzyme whose activity can be readily detected by histochemical staining (Gallagher, 1992; Jefferson et al., 1987).

11. Immediately release the vacuum and remove the leaf sample.

12. Repeat steps 6–11 until all leaf samples are bombarded. When finished, turn off the helium tank and release the pressure by holding down the fire button while the vacuum is on. Turn off the vacuum pump.

13. Place a plastic bag over the plates containing bombarded leaf samples and incubate in the culture room. Incubation allows time for marker gene expression before selection is started.

14. After 2 days, cut the bombarded leaves into small (1-cm-square) pieces and place them abaxial side up in deep RMOP-spectinomycin (500 mg/liter) plates (Fig. 23.1B). Place only seven pieces per plate, as the leaf pieces will grow and expand. Typically leaf pieces from one leaf bombardment fit on 3–5 thick plates. If the leaf sections are too large, there will be insufficient nutrient in the medium to support growth for up to 12 weeks, the time frame within which transplastomic clones appear. The diagnostic sign of overcrowding is the absence of spontaneous spectinomycin-resistant mutants and transplastomic clones. Overcrowding may also be caused by less than the desired 50-ml culture medium in a deep plate.

15. Individually seal each plate on the side with a strip of plastic wrap that is permeable to gas exchange (Glad ClingWrap, The Glad Products Co., Oakland, CA) and incubate the plates in a culture room for 4–12 weeks.

4.5. Identification of transplastomic plants

1. Green, spectinomycin-resistant shoots appear on the bleached leaf sections between 4 and 12 weeks after bombardment (Fig. 23.1B). Spectinomycin resistance may be due to expression of the *aadA* gene or to a spontaneous mutation in the plastid small ribosomal RNA (*rrn16*) gene (Svab and Maliga, 1993). Each shoot at a distinct location derives from an independent transformation event and, therefore, is an independently derived clone. The clones are identified by the plasmid name and a serial number.

2. Transgenic clones are resistant to both spectinomycin and streptomycin, whereas spontaneous spectinomycin-resistant mutants are resistant only to spectinomycin (Svab and Maliga, 1993). To distinguish transgenic clones from mutants, test each regenerated shoot for resistance to streptomycin by inoculating small callus or small leaf sections in deep plates on selective RMOP streptomycin–spectinomycin (500 mg/liter each) media. Putative transplastomic clones are resistant to both antibiotics and form green calli in 3–6 weeks, whereas plastid mutants will be sensitive and bleach on streptomycin-containing medium (Fig. 23.1C). Since streptomycin delays shoot regeneration, simultaneously regenerate new shoots on RMOP spectinomycin medium (500 mg/liter) to obtain homoplastomic shoots (Fig. 23.1D). The *aadA* gene rarely inserts and expresses in the nuclear genome. Transplastomic clones are positively identified by confirming incorporation of *aadA* in the plastid genome by DNA gel blot analysis.

3. Identify clones that are resistant to streptomycin and spectinomycin. Take 10 shoots of the same clone regenerating on RMOP spectinomycin plates and root them in RM deep plates, because not all shoots develop into a plant. Test plants from three or four clones because ~10% of plants regenerated in tissue culture are sterile due to somaclonal variation.

The shoots will root and form multiple leaves in ~3 weeks (Fig. 23.1E and F). Shoots regenerated from the same clone (initial shoot) are considered subclones and are distinguished with letters. Normally, we perform DNA gel blot analysis on two shoots from three to four clones. Southern analysis will show that some of the plants are homoplastomic (all ptDNA copies transformed), some are heteroplastomic (contain transformed and wild-type ptDNA copies), and some contain only wild-type ptDNA copies.

4. Although Southern analysis indicates the homoplastomic state, a small number of wild-type ptDNA copies may remain undetected. Repeating plant regeneration on RMOP spectinomycin medium (500 mg/liter) from the leaves of homoplastomic plants identified in step 3 ensures that no wild-type ptDNA is retained.

5. Transfer homoplastomic plants to soil in the greenhouse after gently breaking up the agar, and washing off the agar-solidified RM medium.

6. Cover the pots with household plastic foil (Glad ClingWrap, The Glad Products Co., Oakland, CA) to prevent desiccation. Grow the plants in shade for about a week, then remove plastic foil and expose the plants to full sunlight.

7. Collect mature seed pods from transplastomic plants and let dry on a laboratory bench at room temperature for 1 week.

8. Germinate surface-sterilized seeds (Lutz et al., 2006b) on RM-spectinomycin (500 mg/liter) medium. Transplastomic seedlings will be dark green, whereas sensitive seedlings will be white. One hundred percent green seedlings confirm the homoplasmic state of the plant; no segregation for spectinomycin resistance should be seen.

5. PROTOCOLS FOR TESTING RNA EDITING IN TOBACCO CHLOROPLASTS

Study of cDNA is the most reliable method to identify RNA editing sites in plants. Occasionally, digestion of cDNA at fortuitous restriction endonuclease sites can be used to determine if RNA editing has occurred, but the creation/absence of these sites by RNA editing is rarely feasible. The most common in vivo approach is sequencing of reverse transcriptase polymerase chain reaction (PCR)-amplified cDNA. RNA isolation can be performed using the TRIzol (Chomczynski and Sacchi, 1987) (Invitrogen, Carlsbad, CA) or lithium chloride (Stiekema et al., 1988) method or with the Qiagen kit (Qiagen, Valencia, CA). We found that good quality RNA can be obtained in a short time from 100 mg tobacco leaf tissue using Qiagen RNeasy Plant Mini Kit (catalog no. 74903). The TRIzol and lithium chloride methods yield more RNA, but contain more contaminating

DNA. We describe here a protocol for reverse transcription of RNA and PCR amplification of cDNA but not DNA sequencing, because there are no plant-template-specific protocols.

5.1. Reverse transcription of RNA and PCR amplification of cDNA

Due to the sensitivity of PCR, it is essential to remove all traces of DNA before PCR amplification of the cDNA with DNase I using the QIAGEN RNase-Free DNase Set protocol (catalog no. 79254). To verify that no contaminating DNA is present, a control PCR experiment should be performed using the RNA as template: the absence of PCR product in the control indicates the absence of contaminating DNA. The protocol to obtain cDNA is as follows.

1. Resuspend 0.5 μg RNA in RNase-free water to a total volume of 33.5 μl. Add 3 μl of random primers (Promega, Madison, WI) and incubate at 95° for 1 min and immediately place on ice to anneal the primers.
2. To perform the reverse transcription reaction, add to the tube 5 μl 10× PCR buffer (100 mM Tris–HCl, pH 8.4, 500 mM KCl, 15 mM MgCl$_2$, 0.1% w/v gelatin), 4 μl 2.5 M dNTPs, 0.5 μl RNAguard (GE Healthcare, Piscataway, NJ), and 1.5 μl AMV reverse transcriptase (USB Corporation, Cleveland, OH). Place the tube at room temperature for 10 min and then transfer to 42° for 50 min.
3. To PCR amplify the cDNA, add 1 μl each of the gene-specific oligonucleotide pair (100 pmol/μl each) and 0.5 μl of Ampiltaq polymerase to the tube and run the PCR program: 3 min at 92°, 30 cycles of 1 min at 92°, 1 min at 55°, 1 min at 72°, 1 cycle of 11 min at 72°, then keep the tube at 4°. Run 5 μl of PCR reaction on a 0.8% agarose gel to check the products.
4. To determine if all DNA has been removed from the RNA sample, perform a control PCR reaction with RNA that has not undergone reverse transcription. Resuspend 0.5 μg DNase I-treated RNA in RNase-free water to a final volume of 38.5 μl. Add 5 μl 10× PCR buffer, 4 μl 2.5 M dNTPs, 1 μl each of the oligonucleotides (100 pmol/μl), and 0.5 μl of Ampiltaq polymerase. Perform the PCR reaction as described in step 3. Run 5 μl of PCR reaction on a 0.8% agarose gel to check the products. If a gene-specific PCR fragment is obtained, the RNA sample is contaminated with DNA. The cDNA prepared with the contaminated template should be discarded and DNase I treatment of the RNA should be repeated.
5. PCR amplify a DNA sample to obtain a template for the reference sequence by mixing 1 μl (20 ng/μl) of tobacco total leaf DNA prepared with the CTAB protocol (Lutz et al., 2006b), 5 μl Ampiltaq 10×

AmpliTaq PCR Buffer (Applied Biosystems, Foster City, CA), 1 μl each of the gene-specific primers (100 pmol/μl), 4 μl 2.5 M dNTPs, 0.5 μl AmpliTaq (Applied Biosystems, Foster City, CA), and 37.5 μl H$_2$O. Use the PCR program of step 3. Run 5 μl of reaction on a 0.8% agarose gel to check the PCR products.

6. CONCLUSIONS AND FUTURE DIRECTIONS

Progress in understanding RNA editing in plastids has been driven by technical innovation. Plastid transformation reviewed here provided the first experimental tool to study *cis* sequences and *trans factors* involved in editing. The toolkit available to study plastid RNA editing has recently been expanded to include *in vitro* editing systems (Hegeman *et al.*, 2005b; Hirose and Sugiura, 2001; Sasaki *et al.*, 2006). The components of the plastid RNA editing machinery still elude identification. A breakthrough in this area may be achieved by purification and identification of the components by a proteomics approach or by genetic screens to identify mutations in nuclear genes encoding components of the editing machinery.

Future studies will focus on understanding the role of RNA editing in plastids. RNA editing is a corrective mechanism when it restores a functionally important, conserved amino acid, as reported for the *psbF* (Bock *et al.*, 1994), *accD* (Sasaki *et al.*, 2001), *atpA* (Schmitz–Linneweber *et al.*, 2005), and *ndhD* (Hirose and Sugiura, 1997; Kotera *et al.*, 2005) genes. Since both edited and nonedited mRNAs are translated, RNA editing may serve as a regulatory mechanism yielding multiple proteins from the same gene. The plastid *ndh* genes encode a significant number of editing sites; at least one of these is differentially edited in light-grown and dark-grown tissues establishing a possible link between photosynthesis, light-induced chloroplast development, and RNA editing (Karcher and Bock, 2002). Interestingly, the editing sites in the plastid RNA polymerase β subunits are clustered adjacent to the dispensable region (Corneille *et al.*, 2000). This means that multiple, catalytically active RNA polymerase β subunit polypeptides may be obtained from edited and nonedited mRNAs. Identification of the components of the RNA editing machinery will open the way to study the role of editing in plastid metabolism and in regulation of plastid gene expression.

ACKNOWLEDGMENTS

Research in P. M.'s laboratory was supported by the National Science Foundation and USDA. K. L. is the recipient of a Charles and Johanna Busch Predoctoral Fellowship.

REFERENCES

Bendich, A. J. (1987). Why do chloroplasts and mitochondria contain so many copies of their genome? *BioEssays* **6**, 279–282.

Bock, R. (1998). Analysis of RNA editing in plastids. *Methods* **15**, 75–83.

Bock, R. (2000). Sense from nonsense: How the genetic information of chloroplasts is altered by RNA editing. *Biochimie* **82**, 549–557.

Bock, R. (2007). Plastid biotechnology: Prospects for herbicide and insect resistance, metabolic engineering and molecular farming. *Curr. Opin. Biotech.* **18**, 100–106.

Bock, R., and Koop, H. U. (1997). Extraplastidic site-specific factors mediate RNA editing in chloroplasts. *EMBO J.* **16**, 3282–3288.

Bock, R., Kössel, H., and Maliga, P. (1994). Introduction of a heterologous editing site into the tobacco plastid genome: The lack of RNA editing leads to a mutant phenotype. *EMBO J.* **13**, 4623–4628.

Bock, R., Hermann, M., and Kössel, H. (1996). *In vivo* dissection of *cis*-acting determinants for plastid RNA editing. *EMBO J.* **15**, 5052–5059.

Bock, R., Hermann, M., and Fuchs, M. (1997). Identification of critical nucleotide positions for plastid RNA editing site recognition. *RNA* **3**, 1194–1200.

Calsa Junior, T., Carraro, D. M., Benatti, M. R., Barbosa, A. C., Kitajima, J. P., and Carrer, H. (2004). Structural features and transcript-editing analysis of sugarcane (*Saccharum officinarum* L.) chloroplast genome. *Curr. Genet.* **46**, 366–373.

Carrer, H., Hockenberry, T. N., Svab, Z., and Maliga, P. (1993). Kanamycin resistance as a selectable marker for plastid transformation in tobacco. *Mol. Gen. Genet.* **241**, 49–56.

Chaudhuri, S., and Maliga, P. (1996). Sequences directing C to U editing of the plastid *psbL* mRNA are located within a 22 nucleotide segment spanning the editing site. *EMBO J.* **15**, 5958–5964.

Chaudhuri, S., Carrer, H., and Maliga, P. (1995). Site-specific factor involved in the editing of the *psbL* mRNA in tobacco plastids. *EMBO J.* **14**, 2951–2957.

Chomczynski, P., and Sacchi, N. (1987). Single-step method of RNA isolation by acid guanidinium thiocyanate-phenol-chloroform extraction. *Anal. Biochem.* **162**, 156–159.

Corneille, S., Lutz, K., and Maliga, P. (2000). Conservation of RNA editing between rice and maize plastids: Are most editing events dispensable? *Mol. Gen. Genet.* **264**, 419–424.

Freyer, R., Kiefer-Meyer, M. C., and Kössel, H. (1997). Occurrence of plastid RNA editing in all major lineages of land plants. *Proc. Natl. Acad. Sci. USA* **94**, 6285–6290.

Gallagher, S. R. (1992). "GUS Protocols: Using the GUS Gene as a Reporter of Gene Expression." Academic Press, San Diego, CA.

Hayes, M. L., Reed, M. L., Hegeman, C. E., and Hanson, M. R. (2006). Sequence elements critical for efficient RNA editing of a tobacco chloroplast transcript *in vivo* and *in vitro*. *Nucleic Acids Res.* **34**, 3742–3754.

Hegeman, C. E., Halter, C. P., Owens, T. G., and Hanson, M. R. (2005a). Expression of complementary RNA from chloroplast transgenes affects editing efficiency of transgene and endogenous chloroplast transcripts. *Nucleic Acids Res.* **33**, 1454–1464.

Hegeman, C. E., Hayes, M. L., and Hanson, M. R. (2005b). Substrate and cofactor requirements for RNA editing of chloroplast transcripts in *Arabidopsis in vitro*. *Plant J.* **42**, 124–132.

Hermann, M., and Bock, R. (1999). Transfer of plastid RNA-editing activity to novel sites suggests a critical role for spacing in editing-site recognition. *Proc. Natl. Acad. Sci. USA* **96**, 4856–4861.

Herz, S., Fussl, M., Steiger, S., and Koop, H. U. (2005). Development of novel types of plastid transformation vectors and evaluation of factors controlling expression. *Transgenic Res.* **14**, 969–982.

Hirose, T., and Sugiura, M. (1997). Both RNA editing and RNA cleavage are required for translation of tobacco chloroplast *ndhD* mRNA: A possible regulatory mechanism for the expression of a chloroplast operon consisting of functionally unrelated genes. *EMBO J.* **16**, 6804–6811.

Hirose, T., and Sugiura, M. (2001). Involvement of a site-specific trans-factor and a common RNA-binding protein in the editing of chloroplast mRNAs: Development of a chloroplast *in vitro* RNA editing system. *EMBO J.* **20**, 1144–1152.

Hirose, T., Kusumegi, T., Tsudzuki, T., and Sugiura, M. (1999). RNA editing sites in tobacco chloroplast transcripts: Editing as a possible regulator of chloroplast RNA polymerase activity. *Mol. Gen. Genet.* **262**, 462–467.

Hoch, B., Maier, R. M., Appel, K., Igloi, G. L., and Kössel, H. (1991). Editing of a chloroplast mRNA by creation of an initiation codon. *Nature* **353**, 178–180.

Huang, F. C., Klaus, S. M. J., Herz, S., Zuo, Z., Koop, H. U., and Golds, T. J. (2002). Efficient plastid transformation in tobacco using the *aphA-6* gene and kanamycin selection. *Mol. Gen. Genom.* **268**, 19–27.

Inada, M., Sasaki, T., Yukawa, M., Tsudzuki, T., and Sugiura, M. (2004). A systematic search for RNA editing sites in pea chloroplasts: An editing event causes diversification from the evolutionarily conserved amino acid sequence. *Plant Cell Physiol.* **45**, 1615–1622.

Jefferson, R. A., Kavanagh, T. A., and Bevan, M. W. (1987). GUS fusions: Beta-glucuronidase as a sensitive and versatile gene fusion marker in higher plants. *EMBO J.* **6**, 3901–3907.

Kahlau, S., Aspinall, S., Gray, J. C., and Bock, R. (2006). Sequence of the tomato chloroplast DNA and evolutionary comparison of solanaceous plastid genomes. *J. Mol. Evol.* **63**, 194–207.

Karcher, D., and Bock, R. (2002). The amino acid sequence of a plastid protein is developmentally regulated by RNA editing. *J. Biol. Chem.* **277**, 5570–5574.

Kotera, E., Tasaka, M., and Shikanai, T. (2005). A pentatricopeptide repeat protein is essential for RNA editing in chloroplasts. *Nature* **433**, 326–330.

Kugita, M., Yamamoto, Y., Fujikawa, T., Matsumoto, T., and Yoshinaga, K. (2003). RNA editing in hornwort chloroplasts makes more than half the genes functional. *Nucleic Acids Res.* **31**, 2417–2423.

Lutz, K., and Maliga, P. (2001). Lack of conservation of editing sites in mRNAs that encode subunits of the NAD(P)H dehydrogenase complex in plastids and mitochondria of *Arabidopsis thaliana*. *Curr. Genet.* **40**, 214–219.

Lutz, K. A., Bosacchi, M. H., and Maliga, P. (2006a). Plastid marker gene excision by transiently expressed CRE recombinase. *Plant J.* **45**, 447–456.

Lutz, K. A., Svab, Z., and Maliga, P. (2006b). Construction of marker-free transplastomic tobacco using the Cre-*loxP* site-specific recombination system. *Nat. Protocols* **1**(2), 900–910.

Maier, R. M., Neckermann, K., Igloi, G. L., and Kössel, H. (1995). Complete sequence of the maize chloroplast genome: Gene content, hotspots of divergence and fine tuning of genetic information by transcript editing. *J. Mol. Biol.* **251**, 614–628.

Maier, R. M., Zeltz, P., Kossel, H., Bonnard, G., Gualberto, J. M., and Grienenberger, J. M. (1996). RNA editing in plant mitochondria and chloroplasts. *Plant Mol. Biol.* **32**, 343–365.

Maliga, P. (2004). Plastid transformation in higher plants. *Annu. Rev. Plant Biol.* **55**, 289–313.

Maliga, P. (2005). New vectors and marker excision systems mark progress in engineering the plastid genome of higher plants. *Photochem. Photobiol. Sci.* **4**, 971–976.

Miyata, Y., and Sugita, M. (2004). Tissue- and stage-specific RNA editing of rps 14 transcripts in moss (*Physcomitrella patens*) chloroplasts. *J. Plant Physiol.* **161**, 113–115.

Murashige, T., and Skoog, F. (1962). A revised medium for the growth and bioassay with tobacco tissue culture. *Physiol. Plant* **15,** 473–497.

Reed, M. L., and Hanson, M. R. (1997). A heterologous maize *rpoB* editing site is recognized by transgenic tobacco chloroplasts. *Mol. Cell. Biol.* **17,** 6948–6952.

Reed, M. L., Lyi, S. M., and Hanson, M. R. (2001a). Edited transcripts compete with unedited mRNAs for trans-acting editing factors in higher plant chloroplasts. *Gene* **272,** 165–171.

Reed, M. L., Peeters, N. M., and Hanson, M. R. (2001b). A single alteration 20 nt 5′ to an editing target inhibits chloroplast RNA editing *in vivo*. *Nucleic Acids Res.* **29,** 1507–1513.

Sasaki, T., Yukawa, Y., Miyamoto, T., Obokata, J., and Sugiura, M. (2003). Identification of RNA editing sites in chloroplast transcripts from the maternal and paternal progenitors of tobacco (*Nicotiana tabacum*): Comparative analysis shows the involvement of distinct trans-factors for *ndhB* editing. *Mol. Biol. Evol.* **20,** 1028–1035.

Sasaki, T., Yukawa, Y., Wakasugi, T., Yamada, K., and Sugiura, M. (2006). A simple *in vitro* RNA editing assay for chloroplast transcripts using fluorescent dideoxynucleotides: Distinct types of sequence elements required for editing of *ndh* transcripts. *Plant J.* **47,** 802–810.

Sasaki, Y., Kozaki, A., Ohmori, A., Iguchi, H., and Nagano, Y. (2001). Chloroplast RNA editing required for functional acetyl-CoA carboxylase in plants. *J. Biol. Chem.* **276,** 3937–3940.

Schmitz-Linneweber, C., Kushnir, S., Babiychuk, E., Poltnigg, P., Herrmann, R. G., and Maier, R. M. (2005). Pigment deficiency in nightshade/tobacco cybrids is caused by the failure to edit the plastid ATPase alpha-subunit mRNA. *Plant Cell* **17,** 1815–1828.

Shaver, J. M., Oldenburg, D. J., and Bendich, A. J. (2006). Changes in chloroplast DNA during development in tobacco, *Medicago truncatula*, pea, and maize. *Planta* **224,** 72–82.

Stiekema, W. J., Heidekamp, F., Dirkse, W. G., van Beckum, J., de Haan, P., ten Bosch, C., and Louwerse, J. D. (1988). Molecular cloning and analysis of four potato tuber mRNAs. *Plant Mol. Biol.* **11,** 255–269.

Sugita, M., Miyata, Y., Maruyama, K., Sugiura, C., Arikawa, T., and Higuchi, M. (2006). Extensive RNA editing in transcripts from the *PsbB* operon and *rpoA* gene of plastids from the enigmatic moss *Takakia lepidozioides*. *Biosci. Biotechnol. Biochem.* **70,** 2268–2274.

Sutton, C. A., Zoubenko, O. V., Hanson, M. R., and Maliga, P. (1995). A plant mitochondrial sequence transcribed in transgenic tobacco chloroplasts is not edited. *Mol. Cell. Biol.* **15,** 1377–1381.

Svab, Z., and Maliga, P. (1993). High-frequency plastid transformation in tobacco by selection for a chimeric *aadA* gene. *Proc. Natl. Acad. Sci. USA* **90,** 913–917.

Tillich, M., Funk, H. T., Schmitz-Linneweber, C., Poltnigg, P., Sabater, B., Martin, M., and Maier, R. M. (2005). Editing of plastid RNA in *Arabidopsis thaliana* ecotypes. *Plant J.* **43,** 708–715.

Tillich, M., Poltnigg, P., Kushnir, S., and Schmitz-Linneweber, C. (2006). Maintenance of plastid RNA editing activities independently of their target sites. *EMBO Rep.* **7,** 308–313.

Tsudzuki, T., Wakasugi, T., and Sugiura, M. (2001). Comparative analysis of RNA editing sites in higher plant chloroplasts. *J. Mol. Evol.* **53,** 327–732.

Wakasugi, T., Tsudzuki, T., and Sugiura, M. (2001). The genomics of land plant chloroplasts: Gene content and alteration of genomic information by RNA editing. *Photosynth. Res.* **70,** 107–118.

Zoubenko, O. V., Allison, L. A., Svab, Z., and Maliga, P. (1994). Efficient targeting of foreign genes into the tobacco plastid genome. *Nucleic Acids Res.* **22,** 3819–3824.

Author Index

Subject Index

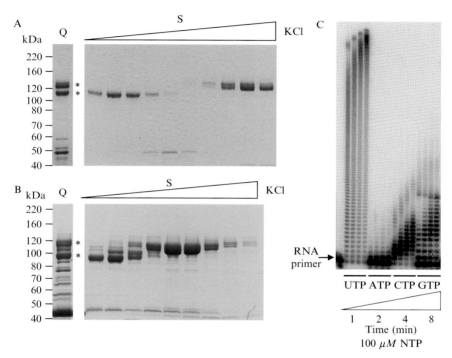

Ruslan Aphasizhev and Inna Aphasizheva, Figure 3.1 Purification of the recombinant RET1. Expression of *L. tarentolae* (A) and *T. brucei* (B) RET1 in *E. coli* produces the full-length and truncated forms (shown by asterisks), which become apparent after the anion-exchange chromatography step (Q). These forms can be separated by cation-exchange chromatography (S). The origin of the truncation, which occurs at the C-terminus, is unknown. Fractions were separated on 8–16% polyacrylamide–SDS gel and stained with Coomassie blue R250. (C) The radiolabeled 6[U] RNA was incubated with purified LtRET1 for the indicated time periods in the presence of ribonucleoside triphosphates and the products were separated on 15% acrylamide urea gel.

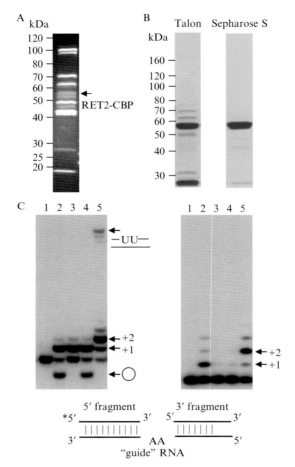

Ruslan Aphasizhev and Inna Aphasizheva, Figure 3.2 Purification of the 20S editosome–associated and recombinant RET2. (A) Protein profile of the affinity isolated RET2 complex. The fraction obtained from the calmodulin column (10 μl) was separated on a 10–20% gradient acrylamide gel and stained with Sypro Ruby (Invitrogen). The position of the RET2 fused with a calmodulin binding peptide (CBP) at the C-terminus is indicated by an arrow. (B) Purification of the recombinant RET2. Fractions after the metal affinity step (Talon) and cation–exchange chromatography (Sepharose S) were separated on 8–16% polyacrylamide–SDS gel and stained with a Coomassie blue R250. (C) Precleaved insertion activity of the affinity isolated RET2 complex (left panel) and recombinant RET2 enzyme (right panel). Schematic representation of the RNA substrate is shown underneath. Migration positions of the extended and ligated, extended by +1 and +2 uridylyl residues, and the circularized 5′ cleavage fragment are indicated by arrows. (1) Control RNA with no protein added; (2) 5′ fragment; (3) 5′ fragment plus "guide" RNA; (4) 5′ fragment plus 3′ fragment only; (5) fully assembled substrate. Note +2 additions occurring with fully assembled substrate and 5′ fragment circularization in the absence of the antisense "guide" RNA.

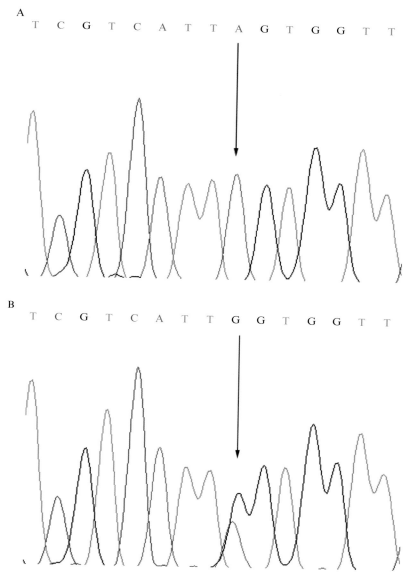

Jaimie Sixsmith and Robert Reenan, Figure 11.2 Comparison of electropherograms to display A-to-I RNA editing. (A) Electropherogram of ADAR null *Drosophlia* displaying a sequence without A-to-I RNA editing. The arrow indicates only a pure A signal (green). (B) Electropherogram of ADAR⁺ *Drosophila* of the same region of the sequence with A-to-I RNA editing. The arrow indicates a mixed A/G signal (green/black).

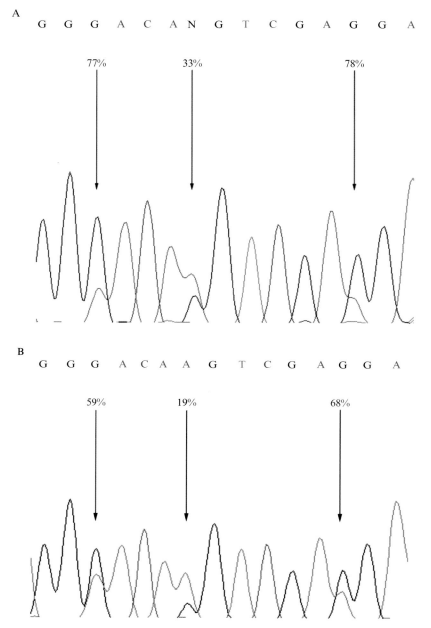

Jaimie Sixsmith and Robert Reenan, Figure 11.3 Direct sequencing method for quantification of A-to-I RNA editing sites. The presence of more than one editing site in close proximity allows comparisons of electropherograms to measure changes in levels of editing between species (A) *Drosophila hydei* and (B) *Drosophila eohydei*. The percentage of editing at each site is indicated above the relevant peak. This figure is the area of the G peak (black) as a percentage of the combined areas of both mixed peaks, including the A peak (green).

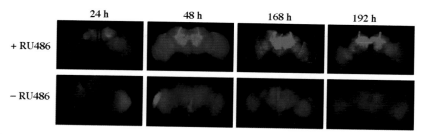

James E. C. Jepson and Robert A. Reenan, Figure 12.3   Expression of GFP in adult *Drosophila* brains using the Gene-Switch system. Females containing the UAS-GFP transgene were crossed with males carrying the RU486-activated GAL4-chimera under the control of the pan-neuronal *elav* (*embryonic lethal abnormal vision*) driver. The resulting adult offspring (UAS-GFP/*elav*-Switch) were placed on food $\pm$ 200 μM RU496. GFP expression above background levels was detected in dissected adult brains after 24–48 h, and peaked after approximately 7 days on RU486-containing food (J. E. C. Jepson and R. A. Reenan, unpublished results). No neuronal GFP expression above background was detected in flies fed food without RU486.

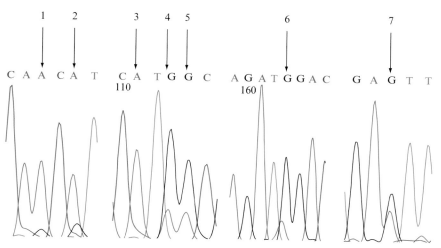

James E. C. Jepson and Robert A. Reenan, Figure 12.4   Mixed A/G peaks in the cDNA of a *Drosophila* ADAR target. Sequence chromatograms of cDNA encoding the *Drosophila* Dα6 nicotinic acetylcholine receptor amplified by RT-PCR from Canton-S flies exhibit several sites of mixed A/G peaks in exons 5 and 6 (J. E. C. Jepson and R. A. Reenan, unpublished results), corresponding to the seven known editing sites within this gene (Grauso *et al.*, 2002). The ratio of A/G peaks has been shown to agree approximately with the ratio of A- and G-containing cDNA clones for these seven sites (Grauso *et al.*, 2002). Low, medium, and high levels of editing are observed at sites 1 and 2, 7, and 4–6, respectively. Site 3 shows no detectable editing.

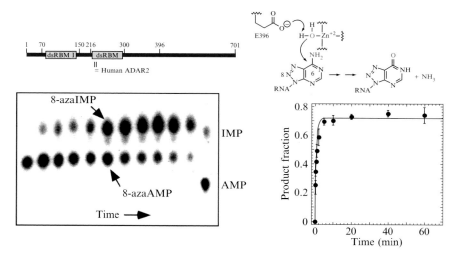

Olena Maydanovych et al., Figure 17.1   Substrate analogs (e.g., 8-azaadenosine shown) are evaluated by measuring the rate of ADAR2-catalyzed deamination under single turnover conditions. Human ADAR2 overexpressed in *S. cerevisiae* is used for these experiments. At various time points, reaction aliquots are removed and nuclease digestion to nucleoside monophosphates is carried out. The product is then separated from unreacted substrate by thin layer chromatography. Plotting the reaction yield versus time and data fitting provide the reaction rate constant.

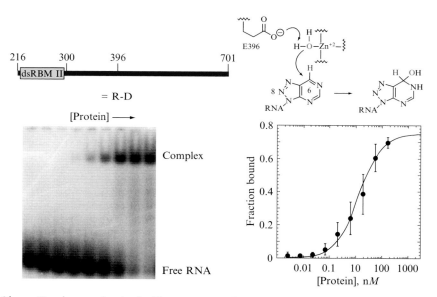

Olena Maydanovych et al., Figure 17.2   Efficiency of mechanism-based trapping (e.g., by 8-azanebularine shown) is assessed using quantitative gel mobility shift assays and a deletion mutant of ADAR2 lacking dsRBM I (R-D, human ADAR2a aa 216–701) overexpressed in *S. cerevisiae* (Macbeth *et al.*, 2005). Complexes formed are separated from free RNA using nondenaturing polyacrylamide gels. Plotting the fraction of RNA bound versus protein concentration and data fitting provide the equilibrium constant.

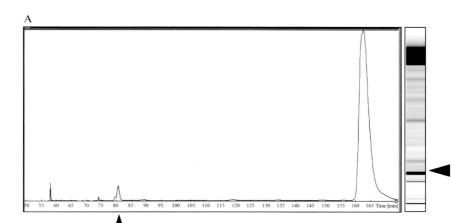

A

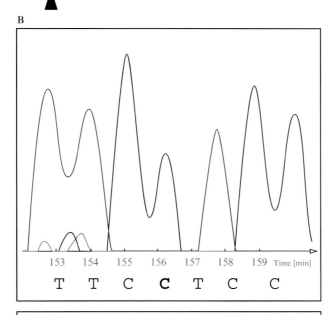

B

153 154 155 156 157 158 159 Time [min]

T T C **C** T C C

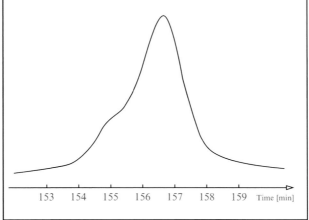

153 154 155 156 157 158 159 Time [min]

Mizuki Takenaka and Axel Brennicke, Figure 20.5 Gel picture of the product analysis after TDG treatment of an editing site incubated in the *in vitro* system. (A) The gel image generated by the automated sequencer on the right-hand side and the scan of the fluorescence intensity on the left. The signals of the editing site are marked by the arrowheads. The large peak at about 165 min in the scan (and the intense black bar in the top part of the gel image, respectively) represents the uncut cDNA strands, which are derived from the unedited template RNA. Spurious signals appear as background; the leftover Cy5-labeled primers have run off the gel. (B) An alignment of the TDG-derived RNA editing signal with the corresponding sequence analysis from the same primer. This is necessary to identify the editing site and to unambiguously assign the signal to the editing event. The shoulder on the left-hand side of the major peak results from incomplete denaturation during the gel run; similar ghosts are sometimes seen in the sequences. These depend on the DNA sequence composition and the denaturing status. To minimize these, higher concentrations of urea, e.g., 8 M instead of 6 M in loading buffer and gel, are recommended.

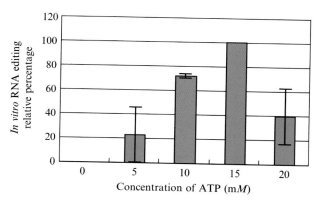

Mizuki Takenaka and Axel Brennicke, Figure 20.6 Manipulation of *in vitro* RNA editing conditions. As an example of direct manipulation of the biochemical composition of the *in vitro* assay, the results of RNA editing at several ATP concentrations in the *in vitro* reaction are depicted. For such quantifications, the ratio of TDG cut to uncut DNA strands has to be calculated. Here this is displayed as percentage relative to the highest *in vitro* editing rate, which is observed with 15 mM ATP. The error bars increase toward slower editing signals.

Kerry A. Lutz and Pal Maliga, Figure 23.1 Transformation of tobacco leaves to obtain transplastomic plants. (A) Leaves prepared for bombardment with DNA-coated gold particles. (B) Resistant shoots and/or calli appear in 4–12 weeks on leaf cut into 1-cm^2 pieces on selective RMOP regeneration medium. (C) Classification of spectinomycin-resistant lines as putative transplastomic clones by resistance to streptomycin on RMOP medium containing streptomycin and spectinomycin (500 mg/liter each). The plate is shown after 6 weeks. (D) Shoot regeneration from leaf sections on RMOP spectino-mycin medium (500 mg/liter) to obtain homoplastomic shoots. The plate is shown after 6 weeks. (E) Homoplastomic shoots rooted in plate or (F) in Magenta box.